Lecture Notes in Computer Science 8238

Commenced Publication in 1973
Founding and Former Series Editors:
Gerhard Goos, Juris Hartmanis, and Jan van Leeuwen

Editorial Board

David Hutchison
  Lancaster University, UK
Takeo Kanade
  Carnegie Mellon University, Pittsburgh, PA, USA
Josef Kittler
  University of Surrey, Guildford, UK
Jon M. Kleinberg
  Cornell University, Ithaca, NY, USA
Alfred Kobsa
  University of California, Irvine, CA, USA
Friedemann Mattern
  ETH Zurich, Switzerland
John C. Mitchell
  Stanford University, CA, USA
Moni Naor
  Weizmann Institute of Science, Rehovot, Israel
Oscar Nierstrasz
  University of Bern, Switzerland
C. Pandu Rangan
  Indian Institute of Technology, Madras, India
Bernhard Steffen
  TU Dortmund University, Germany
Madhu Sudan
  Microsoft Research, Cambridge, MA, USA
Demetri Terzopoulos
  University of California, Los Angeles, CA, USA
Doug Tygar
  University of California, Berkeley, CA, USA
Gerhard Weikum
  Max Planck Institute for Informatics, Saarbruecken, Germany

Volume Editors

Adam Jatowt, e-mail: adam@dl.kuis.kyoto-u.ac.jp
Ee-Peng Lim, e-mail: eplim@smu.edu.sg
Ying Ding, e-mail: dingying@indiana.edu
Asako Miura, e-mail: asarin@kwansei.ac.jp
Taro Tezuka, e-mail: tezuka@slis.tsukuba.ac.jp
Gaël Dias, e-mail: gael.dias@unicaen.fr
Katsumi Tanaka, e-mail: ktanaka@i.kyoto-u.ac.jp
Andrew Flanagin, e-mail: flanagin@comm.ucsb.edu
Bing Tian Dai, e-mail: btdai@smu.edu.sg

ISSN 0302-9743 e-ISSN 1611-3349
ISBN 978-3-319-03259-7 e-ISBN 978-3-319-03260-3
DOI 10.1007/978-3-319-03260-3
Springer Cham Heidelberg New York Dordrecht London

Library of Congress Control Number: 2013953425

CR Subject Classification (1998): C.2, H.5, H.4, H.3, I.2.6, J.4

LNCS Sublibrary: SL 3 – Information Systems and Application, incl. Internet/Web and HCI

© Springer International Publishing Switzerland 2013
This work is subject to copyright. All rights are reserved by the Publisher, whether the whole or part of the material is concerned, specifically the rights of translation, reprinting, reuse of illustrations, recitation, broadcasting, reproduction on microfilms or in any other physical way, and transmission or information storage and retrieval, electronic adaptation, computer software, or by similar or dissimilar methodology now known or hereafter developed. Exempted from this legal reservation are brief excerpts in connection with reviews or scholarly analysis or material supplied specifically for the purpose of being entered and executed on a computer system, for exclusive use by the purchaser of the work. Duplication of this publication or parts thereof is permitted only under the provisions of the Copyright Law of the Publisher's location, in its current version, and permission for use must always be obtained from Springer. Permissions for use may be obtained through RightsLink at the Copyright Clearance Center. Violations are liable to prosecution under the respective Copyright Law.
The use of general descriptive names, registered names, trademarks, service marks, etc. in this publication does not imply, even in the absence of a specific statement, that such names are exempt from the relevant protective laws and regulations and therefore free for general use.
While the advice and information in this book are believed to be true and accurate at the date of publication, neither the authors nor the editors nor the publisher can accept any legal responsibility for any errors or omissions that may be made. The publisher makes no warranty, express or implied, with respect to the material contained herein.

Typesetting: Camera-ready by author, data conversion by Scientific Publishing Services, Chennai, India

Printed on acid-free paper

Springer is part of Springer Science+Business Media (www.springer.com)

Adam Jatowt   Ee-Peng Lim   Ying Ding
Asako Miura   Taro Tezuka   Gaël Dias
Katsumi Tanaka   Andrew Flanagin
Bing Tian Dai (Eds.)

Social Informatics

5th International Conference, SocInfo 2013
Kyoto, Japan, November 25-27, 2013
Proceedings

Preface

This volume contains the papers presented at SocInfo 2013, the 5th International Conference on Social Informatics, held during November 25–27, 2013, in Kyoto, Japan. SocInfo is an interdisciplinary venue for researchers from computer science, informatics, social sciences, and management sciences to share ideas and opinions, and present original research work on studying the interplay between socially-centric platforms and social phenomena. After the conferences in Warsaw, Poland, in 2009, Laxenburg, Austria, in 2010, Singapore in 2011, and Lausanne, Switzerland, in 2012, the International Conference on Social Informatics returned to Asia. This year, SocInfo 2013 was co-located with WebDB Forum 2013, which is a Japanese domestic forum that covers a wide range of research areas related to database and Web studies attracting participants from industry, government, and academia.

We were delighted to present a strong technical program at the conference as a result of the hard work of the authors, reviewers, and conference organizers. We received 103 submissions of which 94 were full paper submissions and nine demonstration paper submissions. During the review process each paper was evaluated by at least three different Program Committee members. On average there were 3.63 reviews per paper. After a careful evaluation, we accepted 23 full-length papers (24% acceptance rate), 15 short papers (21% acceptance rate), and three poster papers (30% acceptance rate). We were also pleased to invite Yoshiaki Hashimoto, Paul Resnick, and Irwin King to give exciting keynote talks.

This year SocInfo 2013 included four satellite events: the First International Workshop on Social Recommendation, the First Workshop of Quality, Motivation and Coordination of Open Collaboration, the First International Workshop on Histoinformatics, as well as a tutorial entitled: "Towards a Spatial and Temporal Representation of Social Processes" given by Christophe Claramunt.

We would like to thank the authors of submitted papers and presenters as well as the participants for making this workshop a success. We express our gratitude to the Program Committee members and reviewers for their hard and dedicated work. We also thank Taro Tezuka and Gael Dias for selecting demonstration papers and Akiyo Nadamoto and Jochen Leidner for managing the workshops and tutorials. We are extremely grateful to General Co-chairs Katsumi Tanaka and Andrew Flanagin for their continuous leadership. The record number of submissions we received in this year is attributed in large part to the great effort of Publicity Co-chairs Yoshinori Hijikata, Antoine Doucet, Ricardo Campos, Goh Hoe Lian Dion, Atsuyuki Morishima, and Leonard Bolc. We want to thank Makoto Kato for managing the website, Bing Tian Dai for supervising the publication process, as well as our local organizers Takehiro Yamamoto, Toshiyuki Shimizu and our secretary Mie Ashiwa for their great work. Lastly, this conference would not be possible without the generous help of our sponsors and supporters and the effort of the financial chairs: Kazutoshi Sumiya, Hiroaki Ohshima, and Daisuke Kitayama.

September 2013

Adam Jatowt
Ee-Peng Lim
Ying Ding
Asako Miura
Keishi Tajima

Organization

Organizing Committee

General Co-chairs

Katsumi Tanaka — Kyoto University, Japan
Andrew Flanagin — University of California, Santa Barbara, USA

Program Co-chairs

Ee-Peng Lim — Singapore Management University, Singapore
Adam Jatowt — Kyoto University, Japan
Ying Ding — Indiana University, Bloomington, USA
Asako Miura — Kwansei Gakuin University, Japan
Keishi Tajima — Kyoto University, Japan

Workshop/Tutorial Co-chairs

Akiyo Nadamoto — Konan University, Japan
Jochen Leidner — Thomson Reuters, Switzerland

Demo Co-chairs

Tezuka Taro — University of Tsukuba, Japan
Gaël Dias — Normandie University, Caen, France

Publications Chair

Bing Tian Dai — Singapore Management University

Publicity Co-chairs

Yoshinori Hijikata — Osaka University, Japan
Antoine Doucet — Normandie University, Caen, France
Ricardo Campos — Polytechnic Institute of Tomar, Portugal
Goh Hoe Lian Dion — Nanyang Technological University, Singapore
Atsuyuki Morishima — University of Tsukuba, Japan
Leonard Bolc — Polish-Japanese Institute of Information Technology, Poland

Web Chair

Makoto Kato — Kyoto University, Japan

Local Arrangements Co-chairs

Toshiyuki Shimizu — Kyoto University, Japan
Takehiro Yamamoto — Kyoto University, Japan

Program Committee

Aek Palakorn Achananuparp	Singapore Management University, Singapore
Thomas Ågotnes	University of Bergen, Norway
Luca Maria Aiello	Yahoo Research, Spain
Fred Amblard	CNRS IRIT – Université des Sciences Sociales Toulouse 1, France
Stuart Anderson	University of Edinburgh, UK
Emma Angus	University of Wolverhampton, UK
Yasuhito Asano	Kyoto University, Japan
Sitaram Asur	HP Labs, USA
Ching Man Au Yeung	Noah's Ark Lab, Huawei, China
George Barnett	UC Davis, USA
Michael Baron	Baron Consulting, Australia
Abraham Bernstein	University of Zurich, Switzerland
Klemens Böhm	Karlsruhe Institute of Technology, Germany
Francesco Bolici	University of Cassino and Southern Lazio, Italy
Piotr Bródka	Wroclaw University of Technology, Poland
Michele Catasta	EPFL, Switzerland
James Caverlee	Texas A&M University, USA
Freddy Chong-Tat Chua	Singapore Management University, Singapore
Amit Chopra	Lancaster University, UK
Chris Cornelis	University of Granada, Spain
Anwitaman Datta	Nanyang Technological University, Singapore
Martine De Cock	Ghent University, Belgium
Guillaume Deffuant	Irstea, France
Jean-Yves Delort	Google Inc., Switzerland
Jana Diesner	University of Illinois at Urbana-Champaign, USA
Marios Dikaiakos	University of Cyprus, Cyprus
Pawel Dybala	Otaru University of Commerce, Japan
Andreas Ernst	University of Kassel, Germany
Alois Ferscha	University of Linz, Austria
Andreas Flache	University of Groningen, The Netherlands
Richard Forno	UMBC, USA
Timothy French	Bedfordshire University, UK
Wai-Tat Fu	University of Illinois at Urbana-Champaign, USA
Gerhard Fuchs	University of Stuttgart, Germany
Nurit Gal-Oz	Sapir Academic College, Israel
Ruth Garcia Gavilanes	Universitat Pompeu Fabra/Yahoo Research Barcelona, Spain
Armando Geller	Scensei, USA
Christophe Guéret	Data Archiving and Networked Services, The Netherlands

Ido Guy	IBM Research, Israel
Mohammed Hasanuzzaman	Normandie University, Caen, France
Takako Hashimoto	Chiba University of Commerce, Japan
Chris Hinnant	Florida State University, USA
Yu Hirate	Rakuten, Japan
Tuan-Anh Hoang	Singapore Management University, Singapore
Yuh-Jong Hu	National Chengchi University, Taiwan
Baden Hughes	Glentworth, Australia
Stephan Humer	Berlin University of the Arts, Germany
Yoshiharu Ishikawa	Nagoya University, Japan
Marco Janssen	Arizona State University, USA
Mark Jelasity	University of Szeged, Hungary
Siddhartha Jonnalagadda	Mayo Clinic, USA
James Joshi	University of Pittsburgh, USA
Radu Jurca	Google Inc., Switzerland
Kazuhiro Kazama	Wakayama University, Japan
Przemyslaw Kazienko	Wroclaw University of Technology, Poland
Kyoung-Sook Kim	NICT, Japan
Daisuke Kitayama	Kogakuin University, Japan
Andreas Koch	University of Salzburg, Austria
Dariusz Krol	Bournemouth University, UK
Walter LaMendola	University of Denver, USA
Georgios Lappas	Technological Educational Institute, T.E.I. of Western Macedonia, Greece
Yefeng Liu	ABB Corporate Research, Sweden
Nadine Lucas	Normandie University, Caen, France
Paul Lukowicz	University of Passau, Germany
Paul Martin	Normandie University, Caen, France
Christopher Mascaro	Drexel University, USA
Hisashi Miyamori	Kyoto Sangyo University, Japan
Bamshad Mobasher	DePaul University, USA
Yves-Alexandre de Montjoye	Massachusetts Institute of Technology, USA
Tony Moore	Deloitte Services, USA
José Moreno	Normandie University, Caen, France
Mikolaj Morzy	Poznan University of Technology, Poland
Tsuyoshi Murata	Tokyo Institute of Technology, Japan
Shinsuke Nakajima	Kyoto Sangyo University, Japan
Keiichi Nakata	University of Reading, UK
Wolfgang Nejdl	L3S and University of Hannover, Germany
See-Kiong Ng	Institute for Infocomm Research, Singapore
Huseyin Oktay	University of Massachusetts, Amherst, USA
Anne-Marie Oostveen	University of Oxford, UK
Mario Paolucci	Institute of Cognitive Sciences and Technologies, Italy
Nimit Pattanasri	Shinawatra University, Thailand
Gregor Petrič	University of Ljubljana, Slovenia

Michal Ptaszynski	Kitami Institute of Technology, Japan
Hemant Purohit	Wright State University, USA
Daniele Quercia	Yahoo! Labs, Spain
Alice Robbin	Indiana University, USA
Benjamin Schiller	TU Darmstadt, Germany
Axel Schulz	TU Darmstadt, Germany
Geoffery Seaver	National Defense University, USA
Xiaolin Shi	Microsoft Corporation, USA
Yukari Shirota	Gakushuin University, Japan
Carlos Nunes Silva	University of Lisbon, Portugal
Bjørnar Solhaug	SINTEF, Norway
Thanakorn Sornkaew	Ramkhamheang University, Thailand
Flaminio Squazzoni	University of Brescia, Italy
Kazutoshi Sumiya	University of Hyogo, Japan
Aixin Sun	Nanyang Technological University, Singapore
Neel Sundaresan	eBay Research Labs, USA
Jie Tang	Tsinghua University, China
Maurizio Teli	Fondazione Ahref, Italy
Klaus G. Troitzsch	University of Koblenz-Landau, Germany
Julita Vassileva	University of Saskatchewan, Canada
Miguel Vicente-Mariño	Universidad de Valladolid, Spain
Roger Whitaker	Cardiff University, UK
Yusuke Yamamoto	Kyoto University, Japan
Hayato Yamana	Waseda University, Japan
Surender Reddy Yerva	EPFL, Switzerland
Koji Zettsu	NICT, Japan
Weining Zhang	University of Texas at San Antonio, USA

Additional Reviewers

Rania Al-Sabbagh	Kerstin Bischoff
Orphée De Clercq	Ernesto Diaz-Aviles
Sarah Ebling	Golnoosh Farnadi
Rumi Ghosh	Tal Grinshpoun
Lei Jin	Nattiya Kanhabua
Jörg-Uwe Kietz	Jinseok Kim
Andres Ledesma	Xuelian Long
Nicholas Loulloudes	Anh Tuan Luu
Radosław Michalski	Geoffrey Morgan
Wichuda Pan-In	Andreas Papadopoulos
Miriam Redi	Stefanie Roos
Xiaolan Sha	Marija Slavkovik
Lubos Steskal	Shoko Wakamiya
Ghim Eng Yap	Liang Yu
Peng Yu	Yong Zheng

Platinum Sponsors

Bronze Sponsors

Other Sponsors

Supporters

Table of Contents

Modeling Analogies for Human-Centered Information Systems 1
   Christoph Lofi and Christian Nieke

Resilience of Social Networks under Different Attack Strategies 16
   Mohammad Ayub Latif, Muhammad Naveed, and Faraz Zaidi

Changing with Time: Modelling and Detecting User Lifecycle Periods
in Online Community Platforms 30
   Matthew Rowe

Metro: Exploring Participation in Public Events 40
   *Luca Chiarandini, Luca Maria Aiello, Neil O'Hare, and
Alejandro Jaimes*

Follow My Friends This Friday! An Analysis of Human-Generated
Friendship Recommendations 46
   *Ruth Garcia Gavilanes, Neil O'Hare, Luca Maria Aiello, and
Alejandro Jaimes*

A Divide-and-Conquer Approach for Crowdsourced Data
Enumeration 60
   Hideto Aoki and Atsuyuki Morishima

Social Listening for Customer Acquisition 75
   Juan Du, Biying Tan, Feida Zhu, and Ee-Peng Lim

Passive Participation in Communities of Practice:
Scope and Motivations... 81
   Azi Lev-On and Odelia Adler

An Ontology-Based Approach to Sentiment Classification of Mixed
Opinions in Online Restaurant Reviews 95
   Hea-Jin Kim and Min Song

A Novel Social Event Recommendation Method Based on Social and
Collaborative Friendships 109
   Yu-Chun Sun and Chien Chin Chen

Factors That Influence Social Networking Service Private Information
Disclosure at Diverse Openness and Scopes......................... 119
   Basilisa Mvungi and Mizuho Iwaihara

An Approach to Building High-Quality Tag Hierarchies from
Crowdsourced Taxonomic Tag Pairs 129
  Fahad Almoghim, David E. Millard, and Nigel Shadbolt

Automating Credibility Assessment of Arabic News 139
  Mohamed Hammad and Elsayed Hemayed

Polarity Detection of Foursquare Tips 153
  *Felipe Moraes, Marisa Vasconcelos, Patrick Prado, Daniel Dalip,
  Jussara M. Almeida, and Marcos Gonçalves*

The Study of Social Mechanisms of Organization, Boundary
Capabilities, and Information System 163
  Shiuann-Shuoh Chen, Pei-Yi Chen, Min Yu, and Yu-Wei Chuang

Predicting User's Political Party Using Ideological Stances 177
  Swapna Gottipati, Minghui Qiu, Liu Yang, Feida Zhu, and Jing Jiang

A Fast Method for Detecting Communities from Tripartite Networks ... 192
  Kyohei Ikematsu and Tsuyoshi Murata

Predicting Social Density in Mass Events to Prevent Crowd Disasters... 206
  *Bernhard Anzengruber, Danilo Pianini, Jussi Nieminen, and
  Alois Ferscha*

Modeling Social Capital in Bureaucratic Hierarchy for Analyzing
Promotion Decisions 216
  Jyi-Shane Liu, Zhuan-Yao Lin, and Ke-Chih Ning

Information vs Interaction: An Alternative User Ranking Model for
Social Networks.. 227
  Wei Xie, Ai Phuong Hoang, Feida Zhu, and Ee-Peng Lim

Feature Extraction and Summarization of Recipes Using Flow Graph ... 241
  *Yoko Yamakata, Shinji Imahori, Yuichi Sugiyama,
  Shinsuke Mori, and Katsumi Tanaka*

Unsupervised Opinion Targets Expansion and Modification Relation
Identification for Microblog Sentiment Analysis 255
  Jenq-Haur Wang and Ting-Wei Ye

Pilot Study toward Realizing Social Effect in O2O Commerce
Services... 268
  Tse-Ming Tsai, Ping-Che Yang, and Wen-Nan Wang

The Estimation of aNobii Users' Reading Diversity Using Book
Co-ownership Data: A Social Analytical Approach 274
  Muh-Chyun Tang, Yi-Ling Ke, and Yi-Jin Sie

An Ontology-Based Technique for Online Profile Resolution 284
   Keith Cortis, Simon Scerri, Ismael Rivera, and Siegfried Handschuh

Aspects of Rumor Spreading on a Microblog Network 299
   Sejeong Kwon, Meeyoung Cha, Kyomin Jung, Wei Chen, and Yajun Wang

Traffic Condition Is More Than Colored Lines on a Map:
Characterization of Waze Alerts 309
   Thiago H. Silva, Pedro O.S. Vaz de Melo, Aline Carneiro Viana, Jussara M. Almeida, Juliana Salles, and Antonio A.F. Loureiro

The Three Dimensions of Social Prominence 319
   Diego Pennacchioli, Giulio Rossetti, Luca Pappalardo, Dino Pedreschi, Fosca Giannotti, and Michele Coscia

Automatic Thematic Content Analysis: Finding Frames in News 333
   Daan Odijk, Björn Burscher, Rens Vliegenthart, and Maarten de Rijke

Optimal Scales in Weighted Networks 346
   Diego Garlaschelli, Sebastian E. Ahnert, Thomas M.A. Fink, and Guido Caldarelli

Why Do I Retweet It? An Information Propagation Model for
Microblogs 360
   Fabio Pezzoni, Jisun An, Andrea Passarella, Jon Crowcroft, and Marco Conti

Society as a Life Teacher – Automatic Recognition of Instincts
Underneath Human Actions by Using Blog Corpus 370
   Rafal Rzepka and Kenji Araki

Diversity-Based HITS: Web Page Ranking by Referrer and Referral
Diversity 377
   Yoshiyuki Shoji and Katsumi Tanaka

The Babel of Software Development: Linguistic Diversity in Open
Source 391
   Bogdan Vasilescu, Alexander Serebrenik, and Mark G.J. van den Brand

Using and Asking: APIs Used in the Android Market and Asked about
in StackOverflow 405
   David Kavaler, Daryl Posnett, Clint Gibler, Hao Chen, Premkumar Devanbu, and Vladimir Filkov

Temporal, Cultural and Thematic Aspects of Web Credibility 419
   Radoslaw Nielek, Aleksander Wawer, Michal Jankowski-Lorek, and Adam Wierzbicki

Social-Urban Neighborhood Search Based on Crowd Footprints
Network 429
    Shoko Wakamiya, Ryong Lee, and Kazutoshi Sumiya

A Notification-Centric Mobile Interaction Survey and Framework 443
    Jonas Elslander and Katsumi Tanaka

How Do Students Search during Class and Homework? – A Query Log
Analysis for Academic Purposes 457
    Rafael López-García, Makoto P. Kato, Yoko Yamakata, and
    Katsumi Tanaka

On Constrained Adding Friends in Social Networks.................. 467
    Bao-Thien Hoang and Abdessamad Imine

Social Sensing for Urban Crisis Management: The Case of Singapore
Haze 478
    Philips Kokoh Prasetyo, Ming Gao, Ee-Peng Lim, and Christie
    Napa Scollon

Author Index 493

Modeling Analogies for Human-Centered Information Systems

Christoph Lofi and Christian Nieke

National Institute of Informatics
2-1-2 Hitotsubashi, Chiyoda-ku, Tokyo 101-8430, Japan
{lofi,nieke}@nii.ac.jp

Abstract. This paper introduces a conceptual model for representing queries, statements, and knowledge in an analogy-enabled information system. Analogies are considered to be one of the core concepts of human cognition and communication, and are very efficient at conveying complex information in a natural fashion. Integrating analogies into modern information systems paves the way for future truly human-centered paradigms for interacting with data and information, and opens up a number of interesting scientific challenges, especially due to the ambiguous and often consensual nature of analogy statements. Our proposed conceptual analogy model therefore provides a unified model for representing analogies of varying complexity and type, while an additional layer of interpretation models adapts and adjusts the operational semantics for different data sources and approaches, avoiding the shortcomings of any single approach. Here, especially the Social Web promises to be a premier source of analogical knowledge due to its rich variety and subjective content, and therefore we outline first steps for harnessing this valuable information for future human-centered information systems.

1 Introduction

Despite the huge success of mobile and Web-based information systems, interaction with such systems still follows *system-centric* interaction paradigms such as hierarchical categorization, list browsing, or keyword searches. *Human-centered* approaches like natural language queries or question answering, which try to emulate the natural interaction of humans with each other, are still few and in the early stages of their infancy. One of the problems which hamper the development of such approaches is that human communication is often ambiguous and carries a lot of implicit information provided by context, common knowledge and interpretation or inference. In this paper, we further this cause by modeling *analogies*, one of the core principles of natural communication, and provide formal foundations for integrating this powerful concept in information systems. This challenging inter-disciplinary research brings together aspects from information systems, social systems, linguistics, and psychology.

Human cognition is largely based on processing similarities of conceptual representations. During nearly all cognitive everyday tasks like e.g., visual perception, problem solving or learning, we continuously perform *analogical inference* in order to deal with new information [1] in a flexible and cross-domain fashion. It's most striking feature is that it is performed on high-level perceptual structures and properties. Moreover, in contrast to formal reasoning or deduction, the use of

A. Jatowt et al. (Eds.): SocInfo 2013, LNCS 8238, pp. 1–15, 2013.
© Springer International Publishing Switzerland 2013

analogies and analogical inference comes easy and natural to people. As analogical reasoning plays such an important role in many human cognitive abilities, it has been suggested that this ability is the "core of cognition" [2] and the "thing that makes us smart" [3]. Due to analogy's ubiquity and importance, there is long-standing interest in researching the respective foundations in the fields of philosophy, linguistics, and in the cognitive sciences.

In general, an analogy is a cognitive process of transferring some high-level meaning from one particular subject (often called the *analogue* or the *source*) to another subject, usually called the *target*. When using analogies, one emphasizes that the *"essence"* of source and target is similar, i.e. their most discriminating and prototypical behaviors are perceived in a similar way. As a running example, consider the following analogies: "The Skytree is for Tokyo as the Eiffel Tower is for Paris" (or also "The Skytree is for Tokyo as the Statue of Liberty is for New York"). This simple analogy communicates a lot of implicit information, as for example that the Skytree is an iconic landmark of Tokyo, and a great vantage point. Also, this can lead to very intuitive analogical queries like "is there something like the Eiffel Tower in Tokyo". A more metaphorical analogy example is the famous Rutherford analogy "atoms are like the solar system", an analogy which explained the complex and newly discovered mechanics of the microcosm by pointing out similarities to the well-known celestial mechanics.

In human-centered information systems, there are several interesting and challenging application scenarios for taking advantage of analogies. The prime use case are natural language query interfaces using question answering or verbose queries (e.g. IBM Watson [4] or Apple Siri[1]), but also approaches analyzing user-generated text in Social Media as for example opinion mining, sentiment analysis, or Social Media analytics in general. Also, they can be used to explain suggestions of e-commerce or recommender systems in an easier to understand fashion. For example, in a travel booking portal, the system could use statements like "The Okinawa Islands are the Hawai'i of Japan" to explain the concept to foreign customers agnostic to Japanese holiday locations.

Capturing an abstract notion of similarity is essential for the semantics of analogies. Usually this means that source and target, while being potentially different in many respects (i.e. Tokyo Skytree does not look like the Eiffel Tower at all), *behave* similarly or show similar properties within a larger *context*. Therefore, evaluating and processing analogies relies on the concept of *relational* (or behavioral) *similarity*. Mining these similarities can be realized using different data sources like databases or ontologies, but most notably using user statements in the Social Web and in Social Media.

Our paper aims at paving the way for future research by contributing the following:

- Showcasing analogy-enabled information systems, and discussing the potential data sources with a special focus on the Social Web
- Introducing the *knowledge primitives* required for conceptually modeling analogies for information systems
- Providing a conceptual definition for general *analogy statements,* and a slightly restricted definition more suited for later implementation
- Discussing the semantics of analogy statements, and the generic *operations* required for evaluating them
- Introducing a *generic architecture* for analogy-enabled information systems

---
[1] http://www.apple.com/ios/siri/

2 Towards Analogy-Enabled Information Systems

Using analogies in natural speech allows communicating dense information easily and naturally just using few words, as most of the intended semantics will be inferred at the receiver's site. The core semantics are that the analogy source and target *behave* similarly, are *structurally* similar in their context [5], are *perceived* similar [6], or have a *shared high level abstraction* [7]. Understanding analogies is therefore a task requiring powerful cognitive abilities, and computer-based analogy processing is faced with many challenges. From a formal perspective, there have been works interpreting analogies as a special case of *induction* [7], or *hidden inductions* [8] in predicate logics. However, such strict formal modeling neglects the defining characteristic of analogies: their high-level and perceptual nature as most analogies contain a large degree of vagueness and human judgment. Also, considering our previous example query of looking for something equivalent to the Eiffel Tower in Tokyo, two candidates come to mind: the Skytree and Tokyo Tower. Tokyo Tower being similar to the Eiffel Tower can be deduced quite easily, as it is a popular landmark which is also very similar to the Eiffel Tower architecture- and construction-wise. The Skytree on the other hand has only few similarities with the Eiffel Tower, it is newly built and of a completely different design. However, it might be a better answer to the query (i.e. its analogy to the Eiffel Tower is stronger) because its *defining* relationship, i.e. being an iconic and touristically significant landmark, is stronger (and is therefore more similar to the Eiffel Tower conceptually). However, these notions of "more similar", "defining", or "stronger" depend on the common consensus of people and are therefore hard to elicit, or may even change over time or between different groups of persons. Therefore, we argue that any formal model for analogies needs to be able to incorporate these vague concepts in a suitable fashion.

We face these challenges with a two-tier approach, consisting of a *conceptual model*, and one or more respective *interpretation models*. The conceptual model, presented in this paper, allows to represent analogical knowledge, facts, or queries on an abstract level. Working towards an implementation of an analogy-enabled information system, the conceptual model needs additional *interpretation models* defining the operation semantics of the primitives and operations used in the conceptual model. Each interpretation model can therefore have slightly different semantics for operations and concepts like "similarity" or "prototypical relations", which are suited for different scenarios (e.g. defining similarity based on structural aspects in ontologies [9], or relying on the distributional hypothesis [10] for natural language texts).

Current algorithmic approaches only focus on a limited and specialized subset of the analogy semantics, and can be incorporated into our presented two-tier model as specific interpretation models. Most promising seem to be approaches using natural language processing (NLP), and they have been proven to be successful in certain areas of analogy processing. These systems rely on interpreting large (Web-based) text collections [11, 12] and are often tailored to be used with the US-based SAT challenge dataset (part of the standardized aptitude test for college admission) [13, 14]. They are particularly well-suited to capture the consensual nature of perceived similarity by relying on statistics on text written by a large number of (Social) Web users. Besides NLP approaches, there have been experiments with ontologies-based approaches [9], structure-mapping approaches [15], or approaches based on neural networks [16]. Analogical reasoning has also been leveraged for adapting and expanding database schemas [17].

The complexity of interpretation models increases with the desired expressiveness of the analogies the system is supposed to handle, which in turn relies on the semantic distance between the contexts of the analogy source and target. On the one end of this spectrum, we have simple *simile analogies* (not to be confused with the rhetoric figure of speech) which are fairly straightforward, comparing concepts which are very similar with respect to their *attributes* ("A Pony is like a small Horse") but are therefore very limited in their applicability and the amount of transferred implicit information. On the other end we have *metaphors*, which are a very powerful form of analogy that is able to cover large semantic distances between concepts, using very abstract, high-level *structural similarity* [5] ("atoms are like the solar system").

One very common form of analogy is the so-called *4-term analogy model* that consists of two pairs of terms that behave analogous. Our example "The Skytree is for Tokyo as the Eiffel Tower is for Paris" is one of those *4-term analogies*, featuring the pair [Skytree, Tokyo] that is like [Eiffel Tower, Paris]. While the *4-term analogy* is widely used in scientific research on the topic [5, 11, 12], it is actually fairly uncommon in natural speech. Rather than stating a 4-term analogy, people more frequently give statements like: "The Skytree is like the Eiffel Tower" or "Is there something like the Eiffel Tower in Tokyo?". While these statements are based on the same analogy, only parts of the analogy are expressed explicitly. This we refer to as hypocatastasis, meaning that the receiver of the message is supposed to figure out the missing parts of the statement himself, using the general context of the statement, common knowledge and his inference abilities. However, some more complex metaphorical analogies (as e.g. $[atom]::[solar\ system]$) cannot easily be represented in a 4-term form, and are significantly more difficult to process. Therefore, in section 4.3, we will focus more closely on the semantics of 4-term analogies as a trade-off between expressivity and complexity.

3 Adapting Semantics and Data Sources

In this section, we will briefly discuss the advantages and shortcoming of different data sources which could be harnessed for our model as future research challenges. For each data source, respective interpretation models are required for capturing the required information on relationships, attributes, and the perceived similarity. Also, keep in mind that analogy-enabled information systems are not limited to a single interpretation model, but can have multiple, specialized interpretation models for different types of analogies, which could even work in parallel with a subsequent combination / voting phase (as for example as in [14, 18]) to avoid the shortcomings of each individual approach.

3.1 Relational Databases

Databases (i.e. tabular data) provide precise, explicit information on attribute values of entities (the rows) and also information like their shared class (given by the table itself), e.g. a collection of famous monuments or car models with their specifications. However, they usually lack information on relationships to (abstract) concepts, apart from simple foreign key relationships to other tables. Therefore, interpretation models based on relational databases are suitable for analogies with simpler semantics, as for example similes and simpler analogies with source and target in closely related contexts. Realizing such a simple interpretation model still poses some interesting challenges:

While most similes are based on similar attributes, the partners are usually not similar in their actual values (like in regular similarity queries), but their values are similar in relation to the respective prototype of their class. For example, the Volkswagen Golf GTI is a very expensive and extremely powerful car compared to other compact cars, just as the Porsche Cayenne GTS is for SUV cars. This leads to the central challenges of discovering the correct classes for comparison (e.g. compact cars vs. SUVs), and the respective prototypical attribute values for obtaining relative statements such as "very expensive" or "extremely powerful". While this can simply be done in the given example by calculating the average for each attribute, the task can become increasingly difficult in other cases. For example, because of its popularity, most people would consider the Apple iPad as the prototypical point of reference for tablet computers. However, its specifications do not match the average product at all. Discovering which values are prototypical might requires additional steps like opinion mining in product reviews [19].

3.2 Linked Open Data and Ontologies

The main advantage of Linked Open Data (LOD) sources (e.g. DBpedia, Yago) or various available ontologies is that information on concepts and relationships is explicitly available. The sources are usually cleaned to improve data quality, which, however, may lead to many LOD sources containing only very few (but correct) relationships. For example the entry of "Tokyo Skytree" in DBpedia (one of largest available LOD sources) contains only few attributes like size, location, build date, and very few additional relationships linking for example to the owner company, the architect, or the city district. While the usefulness of current LOD sources for analogy processing is therefore still limited, this will likely improve when more information is incorporated. Furthermore, identifying relevant or prototypical concepts or relationships is more difficult with no quantitative information available, and most approaches which aim at discovering typicality on linked data fall back to text mining (e.g. [20]).

3.3 Unstructured Text and the Social Web

Most current approaches for automatically processing analogies are based on processing natural language text, usually crawled from the Web. In contrast to databases and LOD, the Web contains an astonishing amount of information, even on more obscure concepts and entities. However, extracting this information is usually an error-prone task with many challenges. Therefore, current approaches aim at dealing with simplified analogy problems as for example solving multiple-choice analogies from the SAT analogy test data [11–14, 21], basically relying on the distributional hypothesis [10] for heuristically estimating relational similarity. For example in [13], the natural language text snippets relevant to an analogy are obtained via web-search, and then concepts and relationships are identified using predefined extraction patterns. Such approaches are strictly heuristics in their nature, and apply statistics on words with no or only limited further considerations on their actual semantics or type. However, they have shown to deliver good performance for multiple-choice analogy queries, mainly due to the fact that statistically analyzing a large number of web sources allows to grasp perceived similarity and consensus of a large number of people quite well.

Besides purely textual approaches, the Web is also the premier source for establishing prototypicality and similarity measures in structured knowledge-bases, as e.g. described in [20] which discovers typical attributes and relationships for classes. Extracting this type of information from the Social Web (e.g. Blogs, Microblogs like Twitter, product reviews, or even more exotic sources like user-tagged image galleries, or recommender system feedback [22]) promises to be a valuable source for perceptual information.

4 Conceptual Model for Analogies

In this section we will first present a high-level design for an analogy-enabled information system. This design is limited by certain practical considerations, and will mostly be suitable for processing analogies which can be represented as 4-term analogies. We will then continue to develop a conceptual model for analogies and related concepts that allows representing the required information and operations.

4.1 System Design

One of the challenges of designing an analogy-enabled information system is the ability to combine a range of different data sources and diverse semantics, since we believe that no single approach will be applicable and well suited for every scenario. We therefore propose a system architecture with a layer of interpretation models to be used alternatively or even in parallel, to complement each other. Data sources can include typical knowledge bases or Linked Open Data sources such as DBpedi[2] or Yago[3], containing explicit structured information on entities, instances, their relationships, and properties. Also tabular databases containing structured data on entities and their attributes are possible (e.g. Freebase[4]), but also fully unstructured natural language sources such as Web data and document collections. Therefore, our modeling strives for containing only concepts and operators which can be applied to both structured and unstructured data sources. Basically we follow the semantics of simple structured knowledge bases, but we will sometimes require additional information not

---

[2] http://dbpedia.org/
[3] http://www.mpi-inf.mpg.de/yago-naga/yago/
[4] http://www.freebase.com/

necessarily available in an explicit form, and which has to be delivered by the interpretation model (such as how typical or similar certain relations are, a notion which is not stored in most ontologies or knowledge bases). A case study for such an interpretation model is given in section 5.

A visual overview of our general system design is presented in Figure 1. Basically, data sources relying on different data models are transformed into a common shared abstraction, the *knowledge base*. This can be done in a push or pull fashion (i.e., purposefully extracting information for each query by e.g. using on-demand web search, or preemptively extracting a large knowledge base). The primitives for modeling the knowledge base are described in section 4.2. All meta-data, intermediate results as well as final results from analogy processing can be stored in the analogy repository for later re-use. This covers information on similarity or prototypicality, but also discovered analogies and their supporting facts.

The general semantics of analogy processing are given by our conceptual model, while the specific operative semantics for different data sources, domains, or special cases are provided by respective interpretation models. Please note that in this work, we refrain from discussions about how to parse natural language queries or statements, but assume that the required language processing is provided by a suitable query parser.

4.2 Knowledge Base Primitives

In this section, we introduce the basic building blocks for modeling the semantics of analogies in information systems. Please note, our model and architecture is only intended to be used for analogical reasoning. Like in the human thinking process, this allows ignoring several constraints of logical reasoning or formal ontologies, like certain consistency requirements, and will not always give clear, precise answers, as analogies are often ambiguous by nature.

Concepts C: Concepts are our primary modeling primitives, and are identified by their unique label. The set of all concepts is denoted by $C \subseteq \mathcal{L}$, wheras $\mathcal{L}$ is the set of all unique labels. Several different types of concepts can be encountered and depending on their type, these might require special considerations while designing an interpretation model, resulting in tailored evaluation algorithms or data sources (see section 3). In the following, we distinguish between entities and abstract objects and their special cases classes and prototypes. Heuristics employed in the interpreting model may take advantage of this disambiguation.

- *Entities*: We use entities to represent potential real-world objects which can also be identified by their attribute values, similar to entities in the relational model for databases. This can encompass existing objects as for example the city "Tokyo" or the "Eiffel Tower". Furthermore, entities might also represent more abstract notions which are often used in speech to represent groups of real world instances with the same attributes, as for example the iPhone5 (all actual iPhones have closely similar attribute values, and when used in analogies, no further distinguishing between individual physical objects is usually required). Information on entities is often readily available in Deep Web or LOD data sources, especially in form of structured data or stored in relational databases (for example FreeBase, DBpedia, IMDB, etc.).

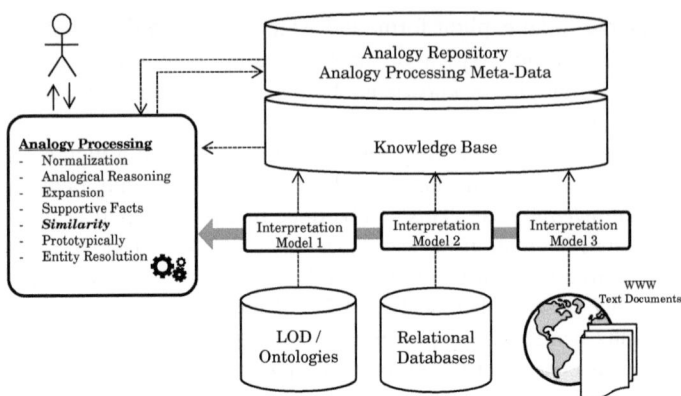

Fig. 1. General System Overview

- *Abstract Objects*: Abstract objects do not exist in a particular time or place, but rather represent ideas or high-level abstractions as for example "truth", "curiosity", but also abstractions which frequently appear as attribute dimensions like "size" or "costs". Information on abstract objects is usually harder to obtain than information on entities, and they have often only few explicit attribute values.
- *Classes* are a special case of abstract objects, and represent collections of concepts, as for example city, monument, or planet, but also more descriptive classes like "famous monuments loved by tourists which are also towers". One reason for special treatment of classes is due to their implications (especially due to class-subclass relationships) for interpreting algorithms. Class-related information is often explicitly available in web-based ontologies and taxonomies (e.g. ProBase [23]).
- *Prototypes*: Another special case of abstract objects relevant for analogies are prototypes. Prototypes are associated to certain classes, and represent a consensual abstraction of the entities grouped by the classes. For example, the prototypical tablet computer is white and made by Apple. Note that a prototype is not just the average entity, but is usually the prominent representative in peoples' mind. Also, having multiple prototypes per class is possible when perception significantly differs between groups of people. Prototypes play a crucial role when evaluating simile analogies (see section 3.1).

Attributes $\mathcal{A}$: Attributes describe the properties of an entity, and can be considered as a labeled relationship between an entity and some alpha-numeric literal values. Attributes only play a minor role for modeling analogies, and we do not further distinguish between different data types for attributes. The set of attributes is given by $\mathcal{A} \subseteq \mathcal{C} \times \mathcal{L} \times \mathcal{J}$, whereas $\mathcal{J}$ is the set of literal values.

Relationships $\mathcal{R}$: The set of relationships further describes the relations and interactions between concepts. Each relationship is labeled, and the label represents certain real-world semantics. As the knowledge base does not have to follow a strict vocabulary or schema, one challenge with respect to relationships is that there may be multiple relationships with different labels describing similar real-world semantics.

The set of relationships $\mathcal{R}$ is given by $\mathcal{R} \subseteq \mathcal{C} \times \mathcal{L} \times \mathcal{C}$. Furthermore, we define a function $r: \mathcal{C} \times \mathcal{C} \to (\mathcal{C} \times \mathcal{L} \times \mathcal{C})$ returning all relationships between two given concepts.

4.3 Analogons, General Analogies and 4-Term Analogies

An analogy is a high-level comparison between *source* and *target*. Both are represented as *analogons* and the general form of all analogies can be denoted as $A :: B$ (with A being the source and B being the target analogon). In general, analogons represent a set of concepts (usually with one dominant core concept while the others are related) and, implicitly, the relations between them. We formally define the set of all analogons as a subset of the power set of all concepts: $\mathcal{A}g_{Full} \subseteq \mathcal{P}(\mathcal{C})$. This definition is close to one of the dominant views in cognitive sciences, where analogies are often described as high-level structural mappings between two complexes mental representations [5] (concepts and relationships).

However, this complex cognitive model is difficult to realize, and therefore we will focus on the less complex special case of *4-term analogies* as the basic and *canonical form* for representing analogy statements and queries in our model. Of course, an adaption to the more general case is possible, but would strongly increase the required effort for the problem's presentation and is usually beyond the reach of most of the current and still prototypical implementations. In the case of 4-term analogies, only a restricted set of those analogons which contain exactly two concepts is used: $\mathcal{A}g_4 \subseteq \mathcal{C} \times \mathcal{C} \subseteq \mathcal{A}g_{Full}$.

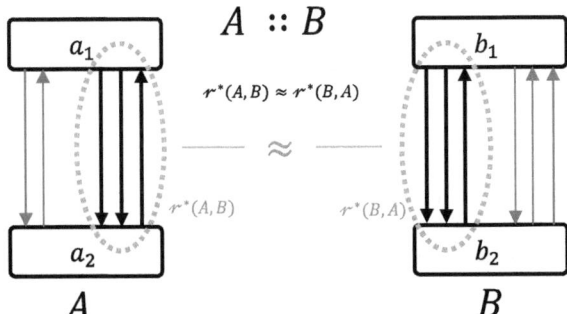

Fig. 2. Intended semantics of a 4-term analogy statement: An analogy holds true when the relevant relationships between the concepts of the analogons are sufficiently similar

Using analogons, *4-term analogy* statements (in the following also referred to as *canonical* analogy statements) can be expressed as:

$A :: B$ with $A, B \in \mathcal{A}g_4$ or $[a_1, a_2] :: [b_1, b_2]$ with $a_1, a_2, b_1, b_2 \in \mathcal{C}$

We also allow incomplete analogons with ? representing the missing concept. The shorthand notation $[a_1]$ represents $[a_1, ?]$. We refer to the set of all canonical analogy statements as $\mathcal{A}nalogy_4$, which is a subset of general analogies $\mathcal{A}nalogy_{Full}$.

4.4 Semantics of Analogy Statements

In order to define the semantics of 4-term analogy statements, several additional operators are required on the conceptual level. Analogously to logical languages and their respective interpretations, our *interpretation models* will provide the operative semantics of these operators and the involved modeling primitives. Therefore, they also decide which analogy statements actually hold true and which don't. Unfortunately, unlike logic-based systems or relational databases, these operators can only be defined heuristically and have to mimic the imprecise and vague semantics of analogies, trying to grasp the consensual opinions on concepts and their relations.

Basically, the intended semantics of a canonical analogy statement is that the statement holds true if all *relevant relationships* (in the context of the current analogy) between the concepts of each analogon are *closely similar*. Furthermore, there can be stronger analogies (i.e. good analogies which people usually consensual agree to, i.e. the Skytree-Eiffel Tower analogy) or weaker analogies (i.e. analogies people often do not agree with, like [$university, life$] :: [$strawberry, cake$]). This is reflected by applying an *interpretation function* $\mathcal{I}$, given by an interpretation model, to an analogy statement. The interpretation function will return a numerical value between 0 (weak) and 1 (strong) representing the respective strength of the analogy (analogously to interpreting logical statements which evaluate to true or false), i.e. $\mathcal{I}: \mathcal{A}nalogy_4 \to [0,1]$. An analogy *holds true* when this value is above a threshold specific to each interpretation model. Interpretation models are further discussed in section 3.

r^* **(Relevant Relationships):** The first conceptual operation required to formally define the analogy is retrieving the set of relevant relationships, which is given by:

$$r^*(A, B) = r \quad \text{with } A, B \in \mathcal{A}g_4 \text{ and } r \subseteq \mathcal{R}$$

This function returns those relationships between a_1 and a_2 (with $[a_1, a_2] = A$) which are relevant to the analogy $A :: B$. As we consider analogies being symmetric, i.e. $\mathcal{I}(A :: B) = \mathcal{I}(B :: A)$, $r^*(B, A)$ can be used to obtain all those relationships between b_1 and b_2 relevant for the analogy $A :: B$.

Unfortunately, to realize these intended semantics in an interpretation model one has to rely on heuristic assumptions for "relevant for the analogy" (please note that humans interpreting analogies also heuristically infer these relationships individually for each analogy). The simplest heuristic is to assume that all relationships between the concepts of an analogon are relevant, which will result in weaker semantics but might be sufficient in certain restricted scenarios. A practically feasible trade-off between complexity and expressivity is relying on (proto-) typicality, i.e. assuming that those relationships are relevant for the analogy which are typical for the concept pair of the analogon (i.e. for [Skytree, Tokyo], being an iconic landmark). For this task, approaches for capturing property or relationship typicality as for example [19, 20] can be adapted.

σ **(Relationship Set Similarity):** Having defined the sets of relevant relationships, the next step is to define a notion of similarity between these sets. We denote the *relationship similarity function*:

$$\sigma: R_1 \times R_2 \to [0,1] \text{ with } R_1, R_2 \subseteq \mathcal{R}$$

The function describes the similarity between two sets of relationships (with 1.0 representing maximal similarity) and can be used to describe the strength of an analogy when applied to the sets of relevant relationships:

$$\mathcal{J}(A :: B) = \sigma(r^*(A,B), r^*(B,A))$$

We further denote: $R_1 \approx R_2$ as the relationship sets R_1, R_2 being *sufficiently similar*, i.e. their similarity is above an interpretation model specific threshold:

$$R_1 \approx R_2 \equiv R_1, R_2 \subseteq \mathcal{R} \wedge \sigma(R_1, R_2) > threshold$$

Again, the operative semantics of this notion and the threshold have to be established by an interpretation model. Here, most heuristics for set similarity will have to rely on the similarity between two single relationships, which is discussed in the next section.

σ_R (**Relational Similarity**): Similarity in the context of analogies does not refer to attribute similarity, but to the more challenging concept of *relational similarity*. We will denote relational similarity between two relationships as a function returning a value ranging from 0 to 1, with 1 representing maximal similarity:

$$\sigma_R : R \times R \to [0,1]$$

One heuristic for approaching this problem is the distributional hypothesis [10] from linguistics, which claims that words frequently occurring in the same context also have similar meanings. An implementation of this heuristic can be found in [13]. Another heuristic approach relying on pattern extraction is given in [24]. Also, approaches based on crowd-sourcing can provide valuable input [18]. Still, developing effective heuristics for relational similarity needed for interpretation models remains as one of the core challenges of future research - poor implementation of similarity might capture only simple analogies, while a superior heuristic interpretation model may cover even more complex metaphors.

Please note that especially for similes, also attribute values may play a role. This is still covered by our model, as the semantics of such analogies are that attribute values are similar relative to another concept or in a certain context (e.g. compared to a prototype or reference concept), and are therefore captured by relational similarity. For example, a Porsche 911 is *expensive* for a sport car, as is the Mercedes Benz S-Class for a sedan. However, mining the correct relationships from attribute values is its own challenge (see 3.1). A graphical summary of our intended analogy semantics can be found in figure 3.

Table 1. Different Representations of Analogies

	Natural Communication	Formal Example
Implicit Analogy	"The Skytree is analog to / like the Eiffel tower"	$[a_1] :: [b_1]$ $\equiv [a_1, ?] :: [b_1, ?]$
Explicit Analogy	"The Skytree is to Tokyo what the Eiffel tower is to Paris"	$[a_1, a_2] :: [b_1, b_2]$
Expansion	Inferred by receiver: List of relevant relationships and their similarity, representing the transferred knowledge	$\{\mathcal{J}(A :: B) = s,$ $\sigma_R(r_{a,1}, r_{b,1}) = s_1, \ldots\}$
Supporting Facts	Facts supporting the decisions on relevance and similarity, e.g.: "Tokyo is a city", "Paris is a city"	$\{isCity(a_2), isCapital(b_2),$ $isA(City, Capital), \ldots\}$

4.5 Normalizing Analogies, Expansions, and Supportive Facts

As indicated before, analogies are usually given in natural speech and are therefore often not fully explicit. In our model, we propose a *normalization* of analogies by transforming an *implicit* analogy into its *explicit canonical form*, i.e. a complete 4-term analogy. This step requires to find the *implicit* information in a statement like "The Skytree is analog to the Eiffel tower" to normalize it into the explicit form "The Skytree is to Tokyo what the Eiffel tower is to Paris". Since the explicit statement contains more information than the implicit one, this transformation can obviously not be performed by applying a strict set of rules, but is in itself a non-trivial cognitive task relying on heuristic inference.

One simple heuristic for normalization, i.e. inferring the missing concepts, could be realized as follows: Given an implicit analogy statement $[a_1] :: [b_1]$ (equivalent to $[a_1, ?_1] :: [b_1, ?_2]$), the values of $?_1$ and $?_2$ can be obtained by inspecting all possible assignments of $?_1, ?_2$ (i.e. those concepts which have a relationship with a_1 or b_1), and selecting that pair for which the prototypical relationships are similar. This task is closely related to evaluating an explicit analogy statement, and the same heuristics provided by an interpretation model for relevance and similarity can be reused. The careful inclusion of crowd-sourcing could also be beneficial.

The *analogy expansion* and *supportive facts* are optional features of our model. Basically, after evaluating an analogy statement or query, they provide additional information on the evaluation process. The *expansion* represents the information that is implicitly transferred by the analogy, i.e. the relevant relations existing in both analogons ("isIconicLandmark"), with their similarity. The *supportive facts* are facts from the knowledge base used to derive the similarity and relevance of the relations. In summary, an analogy is true because of the similar and relevant relations shown in the expansion, which were derived using the data in the supportive facts. You can see an overview of the different proposed representations of analogies in Table 1.

4.6 Analogy Queries

To query an analogy based information system, we propose to state the analogy with all unknown concepts or parts replaced by a '?'. The answer returned by the system would be all analogy statements in canonical form that fulfill the request, optionally including the extension and the supporting facts, or an empty result set if no analogy could be found. This process is very similar to normalizing statements. In the following, we present some examples of possible analogy queries:

- $?[a_1, a_2] :: [b_1, b_2]$: This statement checks if the given analogy statement is true (or the result set is empty) and retrieves its extension and supporting facts.
- $?[a_1, a_2] :: [?, b_2]$: This statement can be used to find the missing concept in a 4-term analogy. ("What is to Tokyo like the Eiffel Tower is to Paris?")
- $?[a_1, a_2] :: ?$: This statement can be used to find possible matches to an analogon. This is basically the non-multiple-choice version of the SAT analogy challenge. The SAT challenges are significantly easier to solve, as they just test 5 candidate analogons and decide for the one with the highest similarity.
- $?[a_1, ?] :: [b_1, ?] \equiv [a_1] :: [b_1]$: This statement can be used to request the explicit form of an implicit analogy.

5 Case Study: Mining Analogies from the Social Web

In this section, we briefly show a case study outlining first steps towards a technical implementation of an analogy interpretation model based on mining the Social Web, and therefore demonstrate the real-world applicability of our proposed model. This study is discussed in closer detail in our later works, and is showcased here to highlight the feasibility and applicability of analogy-enabled information systems. We focus on analogies between locations as e.g. used in travel and tourism systems, e.g. "West Shinjuku (a district of Tokyo) is like Lower Manhattan". During the course of the study, we used Web search with Hearst-like patterns to obtain a set of 22.360 Web documents retrieved mostly from various discussion forums and blogs. Using automated filtering heuristics and crowd-souring-based manual filtering we extracted a Gold dataset with short text snippets containing such analogies. Then, we designed, trained, and evaluated supervised learning models with rich feature sets to *recognize analogy statements automatically*, allowing us to substitute manual crowd filtering with automated techniques. Our best performing feature set based on subsequence patterns derived from PrefixSpans [25] resulted in a precision of 0.92 with recall of 0.88, clearly demonstrating that it is possible to extract analogical statements from the Social Web reliably in an automated fashion.

From here, the next steps are to actually extract the information required to implement the interpretation model: "relevant relationships", "prototypical relationships", and "perceived similarity" between relationships. This task is particularly promising when working on Social Web data, as user's often explicitly explain the analogy they used by providing why they believe it holds true. Consider one of the text snippets from our Gold dataset: "Tokyo, like Disneyland, is sterile. It's too clean and really safe, which are admirable traits, but also unrealistic. Tokyo is like a bubble where people can live their lives in a very naive and enchanted way because real problems do not exist." Here, the speaker clearly provides that in her opinion, Tokyo´s most prototypical properties are that it is very clean and safe, and that it shares these properties with Disneyland. However, in order to use such statements for building an analogy knowledge base as described in the previous chapters, extraction models have to be designed and trained to locate the properties compared in this natural language statement. This task is actually quite closely related to trigger detection in NLP exploiting analysis of the word-dependency graph representation of the sentence. This problem, and of course aggregating the (potentially subjective and even conflicting) information obtained from multiple statements authored by different users remains as a challenge to be solved by current research in progress.

6 Summary and Outlook

During the course of this paper we outlined a generic conceptual design for an analogy-enabled information system that can be adapted to different data sources and operative semantics using one or several interpretation models. We continued with an overview on different data sources, and highlighted their challenges and prospects and provided some insights for designing respective interpretation models, including references to existing work. We then gave a formal model for analogies, including an explicit canonical form, and related problems as e.g. normalization of implicit analogies. We further defined the semantic operations needed to perform analogical reasoning and query the system. Also, we provided a brief survey outlining how an interpretation model based on text crawled from the Social Web can be realized. In our future works we plan to complete this interpretation model, and also provide several additional interpretation models based on alternative data sources for selected scenarios for a complete multi-model analogy-enabled information system, able to adapt to situations where a single model would fail.

References

1. Gentner, D., Markman, A.B.: Structure mapping in analogy and similarity. American Psychologist 52, 45–56 (1997)
2. Hofstadter, D.R.: Analogy as the Core of Cognition. The Analogical Mind, 499–538 (2001)
3. Gentner, D.: Why We're So Smart. Language in Mind: Advances in the Study of Language and Thought, pp. 195–235. MIT Press (2003)
4. Ferrucci, D., Brown, E., Chu-Carroll, J., Fan, J., Gondek, D., Kalyanpur, A.A., Lally, A., Murdock, J.W., Nyberg, E., Prager, J., Schlaefer, N., Welty, C.: Building Watson: An Overview of the DeepQA Project. AI Magazine 31, 59–79 (2010)
5. Gentner, D.: Structure-mapping: A theoretical framework for analogy. Cognitive Science 7, 155–170 (1983)
6. Chalmers, D.J., French, R.M., Hofstadter, D.R.: High-level perception, representation, and analogy: A critique of artificial intelligence methodology. Journal of Experimental & Theoretical Artificial Intelligence 4, 185–211 (1992)
7. Shelley, C.: Multiple Analogie in Science and Philosophy. John Benjamins Pub. (2003)
8. Juthe, A.: Argument by Analogy. Argumentation 19, 1–27 (2005)
9. Forbus, K.D., Mostek, T., Ferguson, R.: Analogy Ontology for Integrating Analogical Processing and First-principles Reasoning. In: Nat. Conf. on Artificial Intelligence (AAAI), Edmonton, Alberta, Canada (2002)
10. Harris, Z.: Distributional Structure. Word 10, 146–162 (1954)
11. Ishizuka, M.: Exploiting macro and micro relations toward web intelligence. In: Zhang, B.-T., Orgun, M.A. (eds.) PRICAI 2010. LNCS, vol. 6230, pp. 4–7. Springer, Heidelberg (2010)
12. Turney, P.: A Uniform Approach to Analogies, Synonyms, Antonyms, and Associations. In: 22th Int. Conf. on Computational Linguistics (COLING), Beijing, China (2008)
13. Bollegala, D.T., Matsuo, Y., Ishizuka, M.: Measuring the similarity between implicit semantic relations from the web. In: 18th Int. Conf. on World Wide Web (WWW), Madrid, Spain (2009)
14. Turney, P.D., Littman, M.L., Bigham, J., Shnayder, V.: Combining Independent Modules to Solve Multiple-choice Synonym and Analogy Problems. In: Int. Conf. on Recent Advances in Natural Language Processing, RANLP (2003)
15. Gentner, D., Gunn, V.: Structural alignment facilitates the noticing of differences. Memory & Cognition 29, 565–577 (2001)

16. Hummel, J.E., Holyoak, K.J.: Relational Reasoning in a Neurally Plausible Cognitive Architecture. An Overview of the LISA Project. Current Directions in Psychological Science 14, 153–157 (2005)
17. Breitman, K.K., Barbosa, S.D.J., Casanova, M.A., Furtado, A.L.: Conceptual modeling by analogy and metaphor. In: ACM Conf. in Information and Knowledge Management (CIKM), Lisabon, Portugal (2007)
18. Lofi, C.: Just ask a human? – Controlling Quality in Relational Similarity and Analogy Processing using the Crowd. In: CDIM Workshop at Database Systems for Business Technology and Web (BTW), Magdeburg, Germany (2013)
19. Selke, J., Homoceanu, S., Balke, W.-T.: Conceptual Views for Entity-Centric Search: Turning Data into Meaningful Concepts. Computer Science: Research and Development 27, 65–79 (2012)
20. Lee, A., Wang, Z., Wang, H., Hwang, S.: Attribute Extraction and Scoring: A Probabilistic Approach. In: Int. Conf. on Data Engineering (ICDE), Brisbane, Australia (2013)
21. Davidov, D.: Unsupervised Discovery of Generic Relationships Using Pattern Clusters and its Evaluation by Automatically Generated SAT Analogy Questions. Ass. for Computational Linguistics: Human Language Technologies (ACL:HLT), Columbus, Ohio, USA (2008)
22. Selke, J., Lofi, C., Balke, W.-T.: Pushing the Boundaries of Crowd-Enabled Databases with Query-Driven Schema Expansion. In: 38th Int. Conf. on Very Large Data Bases (VLDB), vol. 5(2), pp. 538–549, PVLDB Istanbul (2012)
23. Wu, W., Li, H., Wang, H., Zhu, K.: Probase: A Probabilistic Taxonomy for Text Understanding. In: SIGMOD Int. Conf. on Management of Data, Scottsdale, USA (2012)
24. Nakashole, N., Weikum, G., Suchanek, F.: PATTY: A Taxonomy of Relational Patterns with Semantic Types. In: Int. Conf. on Empirical Methods in Natural Language Processing (EMNLP 2012), Jeju Island, Korea (2012)
25. Pei, J., Han, J., Mortazavi-asl, B., Pinto, H., Chen, Q., Dayal, U., Hsu, M.: PrefixSpan: Mining Sequential Patterns Efficiently by Prefix-Projected Pattern Growth. IEEE Computer Society (2001)

Resilience of Social Networks under Different Attack Strategies

Mohammad Ayub Latif[1], Muhammad Naveed[2], and Faraz Zaidi[1,3]

[1] Karachi Institute of Economics and Technology, Karachi, Pakistan
{malatif,faraz}@pafkiet.edu.pk
[2] Muhammad Ali Jinnah University, Karachi, Pakistan
naveed.shaikh@jinnah.edu
[3] University of Lausanne, Lausanne, Switzerland

Abstract. Recent years have seen the world become a closely connected society with the emergence of different types of social networks. Online social networks have provided a way to bridge long distances and establish numerous communication channels which were not possible earlier. These networks exhibit interesting behavior under intentional attacks and random failures where different structural properties influence the resilience in different ways.

In this paper, we perform two sets of experiments and draw conclusions from the results pertaining to the resilience of social networks. The first experiment performs a comparative analysis of four different classes of networks namely small world networks, scale free networks, small world-scale free networks and random networks with four semantically different social networks under different attack strategies. The second experiment compares the resilience of these semantically different social networks under different attack strategies. Empirical analysis reveals interesting behavior of different classes of networks with different attack strategies.

Keywords: Resilience of Networks, Targetted Attacks, Social Networks.

1 Introduction

Online communication channels or mediums of computer mediated communication such as emails, blogs and online social networking websites represent different forms of social networks. These networks have attracted billions of users in recent years [2] adding new dimensions to socializing behavior and communication technologies. These networks provide a challenging opportunity for researchers from different domains to analyze and understand how the new age of communication is shaping the future. These networks also help us understand how information disseminates [14] in social networks and how communication plays a role in the creation of new knowledge [20].

An important aspect of these networks is that they can undergo intentional attacks or random failures which results in communication breakdown. Thus researchers have focussed on studying the stability of networks in terms of how resilient or how robust these networks are against any malicious activity or natural random failures [7,13]. Given a network with n nodes and m edges, targeted or random attacks are modeled by the removal of a series of selected nodes or edges from the network. The way these nodes or links are chosen, known as the attack strategy determines the impact and behavoir it causes on the resilience of the network.

The natural evolution of these networks has introduced several structural properties which play an important role in determining the resilience of these networks. These properties characterize the behvior of many social and other real world networks giving us two important classifications, the scale free networks [4] and small world networks [25]. Scale free networks have a degree distribution following power law[1]. Small world networks have low average path lengths (APL) scaling logarithmically with the increase in number of nodes (n) and high clustering coefficient implying the presence of large number of triads present in the network. Many social networks have both these structural properties giving us another class of networks called, small world-scale free networks.

Scale free networks have been extensively studied with respect to resilience, and Internet provides the perfect dataset for such analysis [7,8,10]. Researchers have shown that scale free networks are highly sensitive to targeted attacks and very robust against random attack strategies [7,8]. This phenomena is often termed as the 'Achilles heel of the Internet'. Resilience of networks with only small world properties, and both small world-scale free properties has not been the focus of studies even though many social networks around us exhibit both small world and scale free properties [21,18].

One example of such networks is the structure of the world wide web studied by [5]. The authors found that the web has a bow tie structure and is very robust against targeted attack on nodes. This result contradicts the findings that scale free networks are fragile to targeted attacks. The reason is that deleting nodes with high degree is not enough to cross the percolation threshold as the average edge-node ratio (also called density or average degree) of these graphs is very high. This finding is similar to our results for the case of social networks.

In this paper, we perform two sets of experiments. The first set of experiments compares the behavior of four different classes of networks, small world networks, scale free networks, small world-scale free networks and random networks with four equivalent size real social networks. These social networks are from a political blog, Epinions who-trust-whom network, Twitter social network and Co-authorship network of researchers. We study these networks under six different attack strategies which are, targeted attack on nodes and edges, random failure of nodes and edges, and almost random failure[2] of nodes and edges [10]. The idea is to see how structural organization of these different networks impact resilience when their edge-node ratio is equivalent to that of semantically different social networks. Our results lead us to these findings:

---

[1] Power law $p_k \sim k^{-\alpha}$ where α is usually in the range of $[2,3]$.
[2] Defined in section 4.

- Five of the six attack strategies behave similarly for all different classes of networks, the exception being targeted attack on nodes.
- Clustering coefficient has no effect on the resilience of networks if netoworks with high edge-node ratio are studied.
- Results show scale free and small world-scale free networks are more fragile to targeted attacks. Targeted attack on edges removes the same number of edges from other classes of networks and the behavior of all classes including random networks remains the same indicating that the different behavior in scale free and small world-scale free networks is due to the large number of edges being removed from the network and not due to the structural organization of the network itself.
- Network generation models used to generate small world, scale free, and small world-scale free networks differ largely from the behavior of real networks in terms of resilience suggesting structural flaws in existing network generation models.

The second experiment studies the resilience of the four real social networks in terms of different attack strategies on nodes, which was found to be more interesting in the previous experiment. The results can be summarized below:

- We observe only minor differences between random and almost random failures for blog, epinions and twitter networks as compared to the author network which demonstrates some differences between the two strategies.
- Attack on Targeted nodes clearly differs from random and almost random failures whereas the author network seems to be the most fragile. The blog, epinions and twitter network demonstrate graceful degradation in performance in terms of size of biggest component.

Rest of the paper is structured as follows: In the next section we discuss several studies pertaining to resilience of different types of networks. Section 3 provides the details of the real world datasets and the networks generated using different network generation models. In section 4, we describe our experimental set up and the metrics used for analysis. Section 5 explains the results obtained and provide findings from the experimentation and finally we conclude in section 6 also giving future research directions.

2 Related Work

One of the earliest studies to demonstrate that scale free networks are more robust against random failures was conducted by [3]. The authors also discuss the vulnerability of scale free networks to targeted attacks. Cohen et al. [7,8] study the resilience of internet under random and targeted attacks on nodes. For the case of random attacks, they conclude that even after 100% removal of nodes, the connectivity of the biggest component remains intact that spans the whole of the network. The authors claim that this condition will remain true for other networks if their connectivity distribution follows power law with power law coefficient less than 3. For the case of targeted attacks, scale free networks are highly sensitive to targeted attacks on nodes as the biggest connected component disintegrates much sooner. Holme et al. [13] study attacks on edges using betweenness centrality where edges with the highest centrality are removed. They show that recalculating betweenness centrality after each deletion is a more effective attack strategy for complex networks. Paul et al. [19] discuss that networks with a given degree distribution may be very resilient to one type of failure or attack but not to another. They determine network design strategies to maximize the robustness of networks to both intentional attacks and random failures keeping the cost of the network constant where cost is measured in terms of network connections. Analytical solutions for site percolation on random graphs with general degree distributions were studied by [6] for a variety of cases such as site and bond percolation. Serrano et al. [22] introduce a framework to analyze percolation properties of random clustered networks and small world-scale free networks. They find that the high number of triads can affect some properties such as the size and resilience of biggest connected component. Wang et al. [24] studied the robustness of scale free networks to random failures from the perspective of network heterogeneity. They examine the relationship of entropy of the degree distribution, minimal connectivity and scaling component obtaining optimal design for scale free networks against random failure. Estrada [9] studied sparse complex networks having high connectivity known as good expansion. Using a graph spectral method, the author introduces a new parameter to measure the good expansion and classify 51 real-world complex networks into four groups with different resilience against targeted node attacks. Wang and Rong[23] analyse the response of scale free networks to different types of attacks on edges during cascading propagation. They used the scale free model [4] and reported that scale free networks are more fragile to attacks on the edges with the lowest loads than the ones with the highest loads. Liu et al. also affirm that scale free networks are highly resilient to random failures. The authors suggest network design guidelines which maximize the network robustness to random and targeted attacks. A comprehensive study conducted by Magnien et al.[16] survey the impact of failures and attacks on Poisson and power law random networks considering the main results of the field acquired. The authors also list new findings which are stated as under:

- Focusing on the random failure of nodes and edges, although previous researchers had predicted completely different behavior for Poisson and power law networks, in practice the differences, are vital but not huge. Our results re-enforce these results specially for the case of social networks.
- The authors also invalidate the explanation that targeted attacks are very efficient on power-law networks because they remove many links, random removal of as many links also result in breakdown of the network.
- Networks with Poisson degree distribution behave similarly in case of random node failures and targeted attacks, it must be noted that their threshold is significantly lower in the second case. This goes against the often claimed assumption that, because all nodes have almost the same degree in a Poisson

network, there is little difference between random node failures and targeted attacks.

Resilience has not been extensively studied for social networks. Moreover, studies focus on networks that are either only scale free or their sizes are not comparable to online social networks readily available around us. Considering the new findings that deviate with the previous results, we get a strong motivation to further investigate resilience of different types of complex networks with a focus on social networks. Our empirical results reaffirm most of these findings of [16] where our focus is on semantically different social networks.

3 Data Sets

We have used four semantically different real world networks which represent social communication of different forms. These are the Political Blog network, Twitter, Epinions and Author network which we are abbreviated as (RN) and are described below.

Political Blog network is a network of hyperlinks between weblogs on US politics, recorded in 2005 by Adamic and Glance[1]. *Twitter* network is one of the most popular online social networks for communication among online users and we have used the dataset extracted by [11]. *Epinions* network is a who-trust-whom online network of a customer analysis website Epinions and the data is downloaded from the stanford website (http://snap.stanford.edu/data/) where it is publicly available. The *Author* network is a co-authorship network where two authors are linked with an edge, if they co-authored a common work(an article, book etc). The dataset is made available by Vladimir Batagelj and Andrej Mrvar: Pajek datasets (http://vlado.fmf.uni-lj.si/pub/networks/data/). For all these networks, we only consider the biggest connected component and treat these networks as simple and undirected. Table 1 shows the number of nodes and edges in these networks along with the edge-node ratio. For each of these real networks, we generated equivalent size networks using four network generation models referred above. The introduction of real data not only allowed us to select realistic edge-node ratio, but also to compare these models with real data.

We have also used four network generation models to represent different types of networks. The small world (SW) model of Watts and Strogatz[25], the scale free (SF) model of Barabasi and Albert[4], the Small world-Scale free (HK) model of Holme and Kim[12] and, the Erdös (RD) model for Random graphs.

The small world model can be tuned to the desired number of nodes and edges by initializing a regular graph where each node has a degree of n. The scale free model can be tuned by the number of edges each new node has in the network where all nodes connect preferentially. Similarly the model for small world-scale free networks can be tuned by the number of nodes each new node connects to, giving us a network with the desired edge-node ratio approximately. A random network is generated using n nodes and m edges where the degree distribution p_k of the network follows a Poisson distribution $p_k = e^{-\lambda}\frac{\lambda^k}{k!}$. The networks we

Table 1. Network Statistics for different social networks

Network	Nodes	Edges	Edge-Node Ratio
Blog	1222	16714	13.6
Twitter	2492	17658	7.0
Epinions	2000	48720	24.3
Author	3621	9461	2.6

generated all had $\lambda > 1$ which signifies that most nodes in the network have a degree close to the mean degree of the network.

For the purpose of experimentation and empirical analysis, we generated 5 artificial networks each for small world, scale free, small world-scale free and random networks equivalent to the 4 different social networks giving us a total of 80 networks. We averaged the readings obtained for these networks although the networks had very little variations with standard deviations of less than 1 in all the cases. Table 2 shows the degree of most connected nodes, clustering coefficients and average path lengths for the generated networks in comparison to real networks.

A clear similarity among all these networks is the low average path length which indicates that on average, nodes in all these networks lie close to each other following the famous 'six degrees of separation' rule. All the real networks are both small world and scale free in nature, the scale free networks have a low clustering coefficient and the degree distribution of small world networks and random networks follow a Poisson distribution with $\lambda > 1$.

4 Experimentation

As described above, we studied resilience considering six attack strategies, three of which are for nodes and three for edges. These are Targeted attack on Nodes, Random failure of Nodes, Almost Random failure of Nodes, Targeted attack on Edges, Random failure of Edges and Almost Random failure of Edges. Each of these strategies is described below:

Targeted Attacks on Nodes and Edges: The attack strategy for targeted removal of nodes removes nodes in decreasing order of their degree (connectivity). This strategy is used by many other researchers[10] for such studies.

To determine targeted edges, we propose a slightly different version from the one used by [10]. The authors removed edges connected to high degree nodes which suits well for networks like scale free networks. Our method is inspired by the concept of funneling in social networks [17] where most connections of a person to other people are usually through a small set of people and connections with one or two famous personalities reduces the distance from all other people in the social network. Thus important edges linking many people would be the ones between high degree people. We assign a weight $W(e_{i,j})$ to all m edges where i,j represents the edge between nodes i and j based on the degree of each node

Table 2. Rd=Random Network, Sw=Small World, Sf=Scale Free, Hk=Holme and Kim Model for small world-scale free networks. Table shows different metrics calculated for the real and artificially generated networks for comparison.

Data Set	Real Network	RD	SW	SF	HK
	Highest Degree of a Node				
Blog	351	46	47	211	321
Twitter	237	27	27	253	319
Epinions	1192	77	72	373	560
Author	102	15	16	201	183
	Clustering Coefficient				
Blog	0.32	0.02	0.56	0.07	0.24
Twitter	0.13	0.005	0.49	0.03	0.27
Epinions	0.27	0.02	0.58	0.08	0.22
Author	0.53	0.001	0.31	0.01	0.42
	Average Path Length				
Blog	2.7	2.5	3.2	2.4	2.2
Twitter	3.4	3.2	4.2	2.9	2.8
Epinions	2.2	2.2	3.0	2.2	2.0
Author	5.31	5.07	6.41	3.4	4.0

using the equation: $W(e_{i,j}) = deg(i) + deg(j)$. Nodes are removed in decreasing order of W in an attempt to remove edges that connect most connected people in the network.

Random Failure of Nodes and Edges: Random removal of nodes and edges is modeled by a series of failures of nodes or edges selected randomly from the network with equal probability.

Almost Random Failure of Nodes and Edges: These attack strategies were described by [10] as more efficient attack strategies in case of scale free networks. Almost random failure of nodes removes randomly selected nodes with degree atleast 2 and almost random failure of edges removes edges between vertices where the degree of each vertex is atleast 2.

Quantifying Resilience of a Network: In order to quantify the resilience of a network, we use the two most commonly applied methods, one measures the number of nodes and the other measures the average path length of the biggest connected component in the network after each attack . The percentage of nodes still connected after an attack provides an estimation of how resilient networks are. Similarly the increase in the average distance from any one node to the other also provides an estimation of how resilient the networks are after each attack. We have studied the effects after every 10% removal of either nodes or edges against the percentage of nodes remaining in the biggest connected component and the average path length of this component.

5 Results and Discussion

Figures (1, 2, 3 and 4) show the results for all the four datasets with different attack strategies on nodes and edges. The first findings are for the cases where we studied targeted attack on edges, random attacks on nodes, random attacks on edges, almost random attack on nodes and almost random attack on edges. For all the these networks, we find that the real networks behave similarly to all 4 classes of networks, small world, scale free, small world-scale free and random networks if the same fraction of nodes or edges are removed as shown in figures. We justify these results based on the idea that increasing the minimum mean connectivity of nodes increases the robustness of networks to targeted and random attacks also discussed by [15]. For all the social networks under consideration, they have very high average connectivity as shown in Table 1. Even for the case of author network which has an average connectivity of 2.6, it is still high as compared to internet networks previously studied in the literature.

Another generic finding is with respect to the clustering coefficients of different networks. Although there are extreme differences in random and social networks, having low values of even 0.001 and high values of around 0.5 (see Table 2) respectively. Still the behavior in terms of resilience remains the same for all these networks. This indicates that the presence or absence of triads does not reflect on the robustness of a network.

The analysis of scale free and small world-scale free networks which are fragile to targeted attacks when compared to small world and random networks is also very interesting. This result is the direct implication of the large number of edges removed from scale free and small world-scale free networks as a result of targeted attack on high degree nodes. Since nodes with very high degree are absent from small world and random networks, the same fraction of edges is not removed

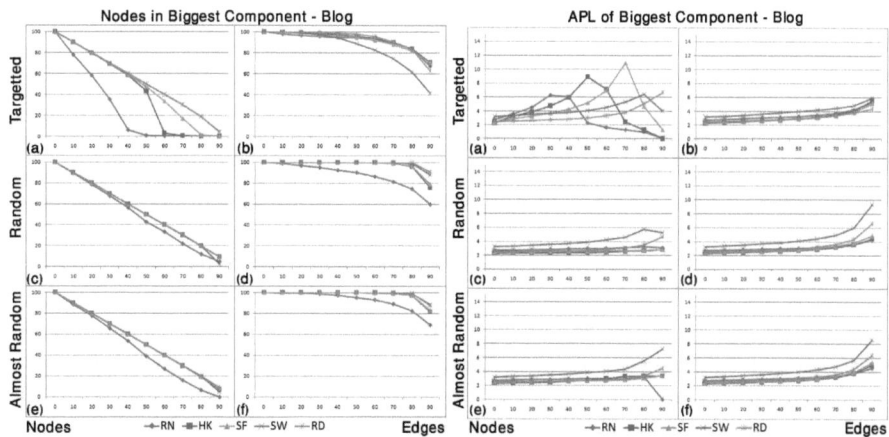

Fig. 1. RN=Blog, HK=Small world-scale free, RD=Random, SF=Scale free, SW=Small world. X-axis: % of nodes (a,c,e) and edges (b,d,f) removed from the network, Y-axis: % of nodes (left) and APL (right) of the biggest connected component.

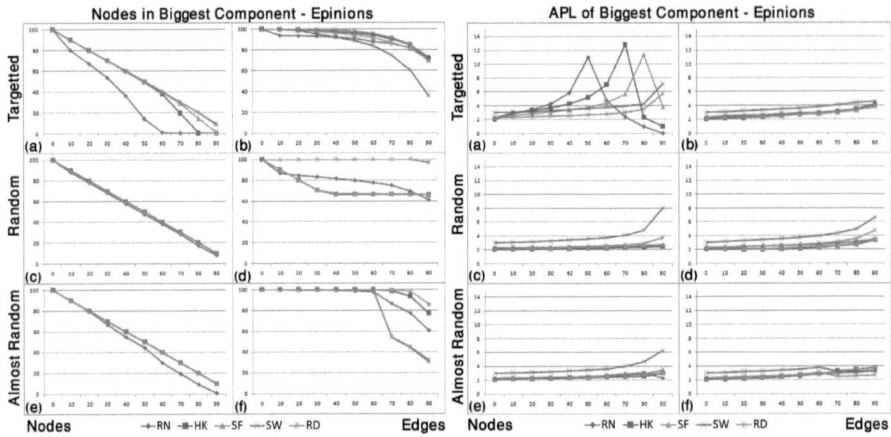

Fig. 2. RN=Epinions, HK=Small world-scale free, RD=Random, SF=Scale free, SW=Small world. X-axis: % of nodes (a,c,e) and edges (b,d,f) removed from the network, Y-axis: % of nodes (left) and APL (right) of the biggest connected component.

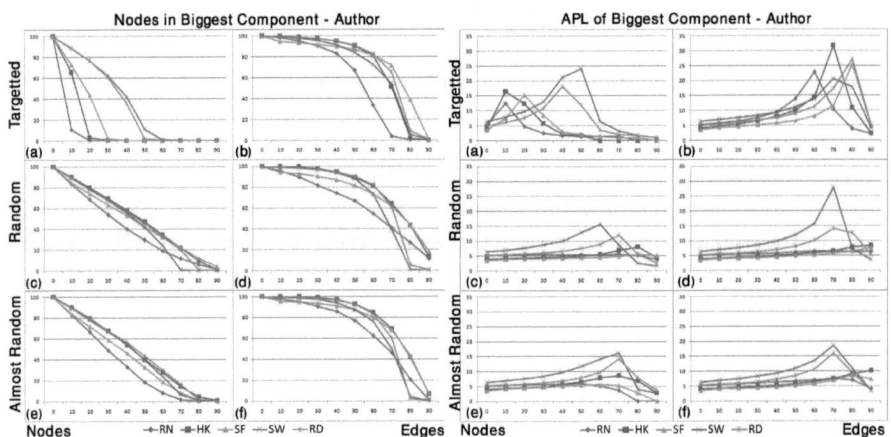

Fig. 3. RN=Author, HK=Small world-scale free, RD=Random, SF=Scale free, SW=Small world. X-axis: % of nodes (a,c,e) and edges (b,d,f) removed from the network, Y-axis: % of nodes (left) and APL (right) of the biggest connected component.

upon removal of high degree nodes and they give an impression that they are more resilient to targeted attack on nodes. The experiment on targeted attack on edges provides a contradiction to this result as equal number of edges are removed from real networks, scale free, small world, small world-scale free and random networks and the results show that the behavior of all these networks is almost the same. This claim is further justified from our results of random attack on edges as, again, they reveal similar behvior for all these classes of networks both in terms of size of biggest connected component and APL.

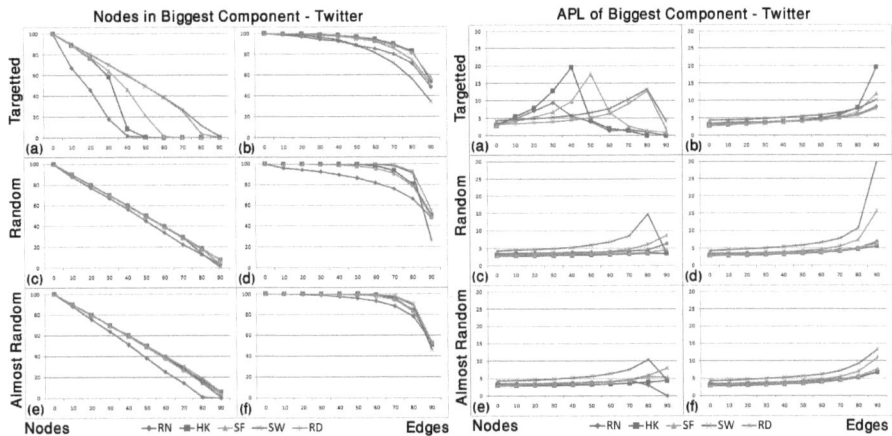

Fig. 4. RN=Twitter, HK=Small world-scale free, RD=Random, SF=Scale free, SW=Small world. X-axis: % of nodes (a,c,e) and edges (b,d,f) removed from the network, Y-axis: % of nodes (left) and APL (right) of the biggest connected component.

Another important result is the behavior of network generation models against the real networks for targeted attack on nodes. All the real world networks disintegrate more quickly than the artificially generated networks for all the four data sets used for experimentation. This observation highlights the fact that network generation models fail to accurately capture all the structural properties of real world networks. This is due to the structural organization of real social networks as compared to the artificially generated networks. In real networks, there is a high percentage of low degree nodes that are connected through high degree nodes only, when these high degree nodes are removed in case of targeted attacks, they immediately become disconnected. On the other hand, artificially generated networks are all based on random connectivity among nodes, and they are not necassarily connected only through high degree nodes, which makes them more resilient when high degree nodes are removed from the network.

We discuss the results for each set of first experiment below:

Targeted Attacks on Nodes: As a general trend, both random and small world networks behave almost similarly for all the datasets. Further more, they are more resilient than small world-scale free (HK) networks and scale free networks (SF). Another important discovery is the behavior of all the real data sets in comparison to the artificially generated networks. Real datasets disintegrate faster than any other model as shown in figures 1(left), 2(left), 3(left) and 4(left) for the case of targeted attacks. The least resilient network is the Author network which disintegrates after 10% highest degree nodes are removed. For the case of Blog and Twitter network, around 40% removal of high degree nodes is sufficient to break the entire network as the size of the biggest component falls below 10%, whereas epinions network falls below 10% after around 50% removal of nodes making it more resilient to targeted attacks. Again the edge-node ratio plays an important role as clearly epinions network has the highest value of 24.3. This, when compared with the Author network with edge-node ration of 2.6 indicates how having more edges nullifies the effects of targeted attacks on networks. The above similarity in the behavoir of networks is further reinforced after looking the behavoir of APL in figures 1(right), 2(right), 3(right) and 4(right) where variations can only be observed in the case of targetted attack on nodes.

Random Failure of Nodes: The behavior of random removal of nodes for all the six cases reveals an interesting similarity specially for the case of generated scale free network, small world network, random network and the real data sets. Particularly for the Blog and Epinions data, almost 100% similar behavior is evident from figure 1(left) and figure 2(left). Twitter and Author networks also show high similarity as shown in figures 3(left) and 4(left). A linear decay is observed in the number of nodes present in the biggest component against linear removal of nodes which suggests that the nodes remain connected even after 90% of the nodes are removed which demonstrates very high resilience for all these networks againt random node failures. The APL of small world networks for all data sets has a slightly higher value indicating minor difference in the empirical values, but the overall behavior and decay pattern is the same for all networks.

Almost Random Failure of Nodes: Just as random failures, almost random failure of nodes demonstrates a high similarity among the different classes of networks and the real networks. Differences can be observed only for the case of author network which has a much lower edge-node ratio. The behavior of the real author network deviates slightly from the other classes of networks. This is contradictory to the results of [10] where they showed that this strategy is more efficient than random failures. The networks used to show these results by [10] had an edge-node ratio of less than 3 and where the networks we use here have a much higher edge-node ratio with the exception of the author network, which has an edge-node ratio of 2.6 and thus we can see differences in the results of random failures and almost random failures in the author network.

Targeted Attacks on Edges: All the networks show an equivalent resilience against targeted attack on edges when compared to random removal of edges. The author network in Figure 3(left) again shows an early breakdown of the biggest component further proving our claim of high mean connectivity being an important reason for resilient structures.

Random Failure of Edges: A slight variation in the resilience can be observed for all the networks. All real networks show a tendency to disintegrate more than the generated networks specially after the removal of 60% edges. Author network is the least resilient case where all the generated and the real networks disintegrate into smaller components after a removal of around 60% edges. Since the author network has the least edge-node ratio (see Table 2), this behavior further proves that other networks show a resilient behavior because of high mean connectivity of the nodes.

Almost Random Failure of Edges: All the dataset behave exactly the same except for the case of epinions network where after the removal of 60% edges result in different pattern. The small world and the random network behave similarly as they are least resilient. The scale free and small world-scale free networks behave similarly being more resilient and epinions network is in between these two behaviors.

The second experiment compares different attack strategies on nodes using four real networks as shown in figure 5. The previous experiment revealed that targeted attack on nodes is the most efficient attack strategy in terms of different classes of networks. The second experiment compares different attack strategies on nodes for different social networks.

The first findings from this experiment are that there are only minor differences in random attacks and almost random attacks when the edge node ratio of the networks is high. Slight differences can be observed for the author network in figure 5(b) which has comparatively low edge-node ratio. This is in contradiction to the results of [10], who studied internet graphs with much less edge-node ratio. Internet graphs are known to have star-like structures where a single node sits (known as hub) in between many other nodes providing efficient connectivity among many nodes. Removing nodes with degree 2 or more unintentionally targets these hubs which in turn results in breakdown of the network. In contrast to this, social networks do not have hubs. Removing nodes with degree 2 or more does not break the network specially for networks with high edge-node ratio because there are many paths that connect a single node, thus making it more resilient to this type of attack.

The second finding is as expected, the effectiveness of targeted attack on nodes as compared to random and almost random failures. The author networks has a low percolation threshold and the network breaks immediately into relatively larger

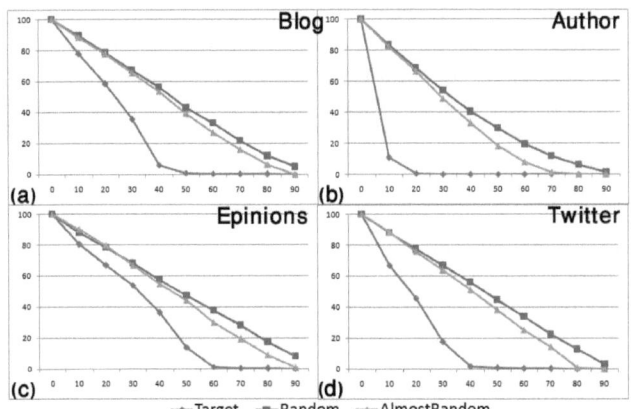

Fig. 5. Comparative analysis of different attack strategies on nodes for the 4 semantically different social networks

size connected components. The Blog, Epinions and Twitter network demonstrate a more graceful degradations with a high percolation threshold as most of the nodes remain connected into a single connected component even after the removal of 40% to 60% high degree nodes.

6 Conclusion

In this paper, we have studied the behavior of small world, scale free and small world-scale free networks in comparison to random and four semantically different social networks. Our results show that that behavior of all these classes of networks remains the same under targeted attack on edges, random attack on nodes and edges, almost random attack on nodes and edges both in terms of size of biggest component and average path length. The behavior of these networks change under targeted attack on nodes. Interesting behavoir was observed on the basis of clustering coefficient and targeted attack on edges. Furthermore structural differences were observed between real social networks and all network generation models. Insignificant differences were observed between random failure of nodes and edges when compared with almost random failures.

We intend to extend this study by incorporating large size social networks. The networks studied are unweighted and undirected, and we intend to analyze the behavior of these networks as well. Another important characteristic of social networks is the temporal dimension which plays an important role in dictating many social processes such as information diffusion and epidemics and we would also like to study resilience for temporal social networks.

References

1. Adamic, L.A., Glance, N.: The political blogosphere and the 2004 u.s. election: divided they blog. In: LinkKDD 2005: Proceedings of the 3rd International Workshop on Link Discovery, pp. 36–43. ACM Press, New York (2005)
2. Ahn, Y.-Y., Han, S., Kwak, H., Moon, S., Jeong, H.: Analysis of topological characteristics of huge online social networking services. In: Proceedings of the 16th International Conference on World Wide Web, WWW 2007, pp. 835–844. ACM, New York (2007)
3. Albert, R., Jeong, H., Barabási, A.-L.: Error and attack tolerance of complex networks. Nature 406, 378–382 (2000)
4. Barabási, A.L., Albert, R.: Emergence of scaling in random networks. Science 286(5439), 509–512 (1999)
5. Broder, A., Kumar, R., Maghoul, F., Raghavan, P., Rajagopalan, S., Stata, R., Tomkins, A., Wiener, J.: Graph structure in the web. Computer Networks 33(1), 309–320 (2000)
6. Callaway, D.S., Newman, M.E., Strogatz, S.H., Watts, D.J.: Network robustness and fragility: Percolation on random graphs. Physical Review Letters 85(25), 5468–5471 (2000)
7. Cohen, R., Erez, K., Ben-Avraham, D., Havlin, S.: Resilience of the internet to random breakdowns. Physical Review Letters 85(21), 4626–4628 (2000)

8. Cohen, R., Erez, K., ben Avraham, D., Havlin, S.: Breakdown of the internet under intentional attack. Physical Review Letters 86, 3682–3685 (2001)
9. Estrada, E.: Network robustness to targeted attacks. the interplay of expansibility and degree distribution. The European Physical Journal B-Condensed Matter and Complex Systems 52(4), 563–574 (2006)
10. Guillaume, J.-L., Latapy, M., Magnien, C.: Comparison of failures and attacks on random and scale-free networks. In: Higashino, T. (ed.) OPODIS 2004. LNCS, vol. 3544, pp. 186–196. Springer, Heidelberg (2005)
11. Hashmi, A., Zaidi, F., Sallaberry, A., Mehmood, T.: Are all social networks structurally similar? In: IEEE/ACM International Conference on Advances in Social Networks Analysis and Mining (ASONAM), pp. 310–314 (2012)
12. Holme, P., Kim, B.J.: Growing scale-free networks with tunable clustering. Physical Review E 65, 026107 (2002)
13. Holme, P., Kim, B.J., Yoon, C.N., Han, S.K.: Attack vulnerability of complex networks. Physical Review E 65(5), 056109 (2002)
14. Iribarren, J.L., Moro, E.: Affinity paths and information diffusion in social networks. Social Networks 33(2), 134–142 (2011)
15. Liu, J.-G., Wang, Z.-T., Dang, Y.-Z.: Optimization of robustness of scale-free network to random and targeted attacks. Modern Physics Letters B 19(16), 785–792 (2005)
16. Magnien, C., Latapy, M., Guillaume, J.-L.: Impact of random failures and attacks on poisson and power-law random networks. ACM Computing Surveys (CSUR) 43(3), 13 (2011)
17. Newman, M.E.J.: Scientific collaboration networks. ii. shortest paths, weighted networks, and centrality. Physical Review E 64(1), 016132 (2001)
18. Newman, M.J., Barabási, A.-L., Watts, D.J.: The Structure and Dynamics of Networks (Princeton Studies in Complexity). Princeton University Press, Princeton (2006)
19. Paul, G., Tanizawa, T., Havlin, S., Stanley, H.: Optimization of robustness of complex networks. The European Physical Journal B-Condensed Matter and Complex Systems 38(2), 187–191 (2004)
20. Prusak, R., Borgatti, S.P.: Supporting knowledge creation and sharing in social networks. Organizational Dynamics 30(2), 100–120 (2001)
21. Reka, A., Barabási: Statistical mechanics of complex networks. Rev. Mod. Phys. 74, 47–97 (2002)
22. Serrano, M.Á., Boguná, M.: Percolation and epidemic thresholds in clustered networks. Physical Review Letters 97(8), 088701 (2006)
23. Wang, G., Zhou, H.: Quantitative logic. Information Sciences 179(3), 226–247 (2009)
24. Wang, L., Du, F., Dai, H.P., Sun, Y.X.: Random pseudofractal scale-free networks with small-world effect. The European Physical Journal B - Condensed Matter and Complex Systems 53, 361–366 (2006)
25. Watts, D.J., Strogatz, S.H.: Collective dynamics of 'small-world' networks. Nature 393, 440–442 (1998)

Changing with Time:
Modelling and Detecting User Lifecycle Periods in Online Community Platforms

Matthew Rowe

School of Computing and Communications, Lancaster University, Lancaster, UK
m.rowe@lancaster.ac.uk

Abstract. In this paper we define the development of a user from *entry* to *churn* as his *lifecycle* that can be divided into discrete stages of development known as *lifecycle periods*. Prior work has examined how social networking site users have developed along isolated dimensions using lexical information [2] and social connections in the context of telecommunications networks [6]. We unify such dimensions by modelling and examining how users develop both socially and lexically, through contrasts of user properties (e.g. time-delimited in-degree distributions) against: (i) prior user properties; and (ii) the community in which the user is interacting. We identify salient traits of user development characterisable in the form of *growth features*, and demonstrate the applicability of such features within a vector space model by outperforming several baselines when detecting the lifecycle period of a given user.

1 Introduction

The modelling, examination and prediction of user development in the context of online community platforms has, to date, been attempted using either lexical information (language used) [2] or '*review expertise*' (topical knowledge) [5], while in the context of telecommunications networks such analyses have been performed using social information [6]. As such, there is a lack of understanding of how users develop throughout their lifecycles (i.e. discrete intervals of user development) along different dimensions (social and lexical indicators), and how users evolve relative to: (i) earlier time periods, and hence their past behaviour and properties; and (ii) the community, and therefore whether users diverge from the community in which they are positioned. Providing such an understanding of how users develop throughout their lifecycles would allow for *churn prediction*, by understanding the lifecycle stage of the user and thus providing incentives to the user for remaining in the community should he be nearing the end of his lifecycle.

In this paper we present a first attempt to model how users develop throughout their lifetimes within three online community platforms: Facebook (groups), SAP Community Network (forums), and ServerFault, along both social and lexical properties and relative to past behaviour and the community as a whole. We demonstrate the applicability of such properties for detecting the lifecycle period that an arbitrary user currently resides in, thereby providing a means for churn prediction. Our contributions in this paper are three-fold: (i) we present a means to model and examine user development along different properties relative to both the community and earlier time periods; (ii) we apply growth rates as features for period detection using a vector space model with varying vector similarity functions; and (iii) we describe an empirical evaluation of the method showing significantly better performance than a random model baseline and the naive Bayes classifier. We begin the paper by describing related work from social network analysis and user evolution, before then describing the datasets used for our work and the means to model user development through lifecycle periods.

2 Related Work

Assessing the social network development of users has been explored in several works: Kairam et al. [4] assessed the dynamics of group formations within the social networking platform Ning finding that the probability of a user joining a group was linked to the number of prior members with whom he had a relationship. Gong et al. [3] inspected the evolution of social networks on Google+ as the platform was growing in memberships, in particular they focused on *social-attribute networks*, finding that the platform exhibited unique growth and characteristics of the networks as more people joined Google+. Chung et al. [1] examined the assortativity (i.e. user interaction extent) of social networks derived from an online community building web site over a ten-year period, and found assortativity to increase with time, while the platform remained dissasortive.

Recent works have examined user evolution throughout their lifecycles within social and telecommunications platforms: Miritello et al. [6] examined the social evolution of users over time in telecommunications networks, finding that as people aged their social circle reduced in size and interaction occurred less. Danescu et al. [2] assessed the lexical dynamics of online community members and how these changed relative to the community, finding that users began their lifecycle within the community by adapting their language to the community but then stopped doing so. McAuley and Leskovec [5] examined how users evolved in their expertise (assuming a monotonic progression) over time in the same beer rating communities as [2], showing that users evolved based on their own '*personal clock*'. In this paper we model and assess how users evolve based on both their social and lexical dynamics, complementing the work of Miritello et al. [6] and Danescu et al. [2], and advance over prior work by contrasting the current state of the user, in terms of his communication behaviour and language, with earlier states and how the community appears at the same time period.

3 Modelling User Lifecycles

We begin this section by defining users' lifecycle periods before then explaining the modelling of user properties and their development over time. For our examination of user lifecycles we used data collected from Facebook, the SAP Community Network (SAP) and Server Fault. Table 1 provides summary statistics of the datasets where we only considered users who had posted more than 40 times within their lifetime on the platform.[1] The Facebook dataset was collected from groups discussing Open University courses, where users talked about their issues with the courses and guidance on studying. The SAP Community Network is a community question answering system related to SAP technologies where users post questions and provide answers related to technical issues. Similarly, Server Fault is a platform that is part of the Stack Overflow question answering site collection[2] where users post questions related to server-related issues. We divided each platform's users up into 80%/20% splits for training (and analysis) and testing, using the former in this section to examine user development and the latter split for our later detection experiments.

Table 1. Statistics of the online community platform datasets

Platform	Time Span	Post Count	User Count
Facebook	[18-08-2007,24-01-2013]	118,432	4,745
SAP	[15-12-2003,20-07-2011]	427,221	32,926
Server Fault	[01-08-2008,31-03-2011]	234,790	33,285

3.1 Defining Lifecycle Periods

In order to examine how users develop over time we needed some means to segment a user's lifetime (i.e. from the first date at which they post to the date of their final post) into discrete intervals. Prior work [6,2,5] has demonstrated the extent to which users develop at their own pace and thus evolve according to their own *'personal clock'* [5]. Hence, for deriving the lifecycle periods of users within the platforms we adopted an activity-slicing approach that divided a user's lifetime into 20 discrete time intervals, emulating the approach in [2], but with an equal proportion of activity within each period. This approach functions as follows: we derive the set of interval tuples ($\{[t_i, t_j]\} \in T$) by first deriving the chunk size (i.e. the number of posts in a single period) for each user, we then sort the posts in ascending date order, before deriving the start and end points of each interval in an incremental manner. This derives the set of time intervals T that are specific to a given user.

3.2 Modelling User Properties

Based on the defined lifecycle periods, we now move on to defining user properties, capturing social and lexical dynamics, and tracking how the properties of users change over time.

---

[1] Choosing 40 posts so that we had at least 2 posts per lifecycle period.
[2] http://stackoverflow.com/

In-degree and Out-degree Distributions. Starting with social dynamics, we assessed the in-degree and out-degree distributions of users: the *in-degree* distribution describes the number of edges that connect to a given user, while the *out-degree* distribution describes the number of edges from the user. As we are dealing with conversation-based platforms for our experiments we can use the *reply-to* graph to construct these edges, where we define an edge connecting to a given user u_i if another user u_j has replied to him. Given our use of lifecycle periods we use the discrete time intervals that constitute $[t,t'] \in T$ to derive the set of users who replied to u_i, defining this set as $\Gamma^{IN}_{[t,t']}$ and the set of users that u_i has replied to within a given time interval: $\Gamma^{OUT}_{[t,t']}$. We then formed a discrete probability distribution that captures the message distribution of repliers to user u_i, using $\Gamma^{IN}_{[t_i,t_j]}$, and user u_i responding to community users, and hence using $\Gamma^{OUT}_{[t_i,t_j]}$. For an arbitrary user $(u_j \in \Gamma^{IN}_{[t_i,t_j]})$ who has contacted user u_i within time segment $[t_i,t_j]$ we define this probability of interaction as follows:[3]

$$Pr(u_j \mid \Gamma^{IN}_{[t,t']}) = \frac{|\{q : p \in P_{u_i}, q \in P_{u_j}, t \leq time(q) < t', q \to p\}|}{\sum_{u_k \in \Gamma^{IN}_{[t,t']}} |\{q : p \in P_{u_i}, q \in P_{u_k}, t \leq time(q) < t', q \to p\}|} \quad (1)$$

For an arbitrary user $(u_j \in \Gamma^{out}_{u_i t})$ who user u_i has contacted within time segment $[t,t']$ we define the probability of interaction using the same formulation as above but looking at outgoing edges from u_i.

Term Distribution. The third user property that we capture is the term distribution of the user, thereby examining the language that he has used. We derive the set of terms that a given user has used by gathering all of his posts within the allotted time interval, removing stop words and then filtering out punctuation We then formed the discrete probability distribution for a user within interval $[t,t']$ based on the conditional probability of term x being used within the time interval. This is produced by defining a multiset containing the set of terms used by a user in a given time period: $x \in C_{[t,t']}$ and a mapping function $\mu : C_{[t,t']} \to \mathbb{N}$ that returns the multiplicity of a given term's usage within the time period. Thus, we defined the conditional probability for term x being used by u_i during $[t,t']$ as: $Pr(x \mid [t,t']) = \mu(x)/\sum_{x' \in C_{[t,t']}} \mu(x')$.

3.3 Analysing User Lifecycles

Given the use of probability distributions to represent the user properties within lifecycle periods we now explain the assessment of these properties: (i) within each lifecycle period; (ii) when comparing one lifecycle period with earlier periods; and (iii) when comparing one lifecycle period with the community platform.

Inspecting Individual Periods (Period Entropy). To analyse the variation in a user's properties within a given lifecycle period we derived the entropy of

---

[3] We use $p \to q$ to denote message q replying to message p.

each probability distribution within each lifecycle period. Entropy describes the amount of variation within a random variable, and therefore provides a useful means to gauge how much a given user is varying: (i) the people with whom he is communicating, and (ii) the terms that he is using within his posts. For each platform (Facebook, SAP, and ServerFault) we derived the entropy of each user in each of his individual lifecycle periods based on the in-degree, out-degree and term distributions. We then recorded the mean of these entropy values over each lifecycle period, thereby providing an assessment of the general changes that users go through. We omitted the plots here for brevity, however it is sufficient to note that users' entropies remained relatively stable throughout their lifecycles and across the three platforms. For the out-degree distribution, the entropy values did appear to increase across the platforms suggesting that users tended to vary the individuals that they were contacting towards the latter stages of their lifetimes.

Historical Contrasts (Period Cross-Entropy). To analyse how users have changed relative to prior lifecycle periods we compared users' in-degree, out-degree and term distributions with *earlier* lifecycle periods through computing the cross-entropy of one probability distribution with respect to another earlier distribution, and then selecting the distribution that minimises cross-entropy. Given a probability distribution (P) formed from one lifecycle period ($[t, t']$), and a probability distribution (Q) from an earlier lifecycle period, we define the cross-entropy between the distributions as: $H(P,Q) = -\sum_x p(x) \log q(x)$

In the same vein as the earlier entropy analysis, we derived the period cross-entropy for each platform's users throughout their lifecycles and then derived the mean cross-entropy for the 20 lifecycle periods. Fig. 1(a), fig. 1(b) and fig. 1(c) present the cross-entropies derived for users' in-degree, out-degree and lexical distributions respectively. We found that for each distribution and each platform cross-entropies reduce throughout users' lifecycles, suggesting that users do not tend to exhibit behaviour that has not been seen previously; i.e. for the in-degree distribution the cross-entropy gauges the extent to which the users who contact a given user at a given lifecycle stage differ from those who have contacted him previously, where a larger value indicates greater divergence. We find that consistently across the platforms, users are contacted by people who have contacted them before and that fewer *novel* users appear. The same is also true for the out-degree distributions: users contact fewer new people than they did before. This is symptomatic of community platforms where, despite new users arriving within the platform, users form sub-communities in which they interact and communicate with the same individuals. Figure 1(c) also demonstrates that users tend to reuse language over time and thus produce a gradually decaying cross-entropy curve.

Community Contrasts (Community Cross-Entropy). To contrast users' properties with the platform's community over the same time interval we took users' in-degree, out-degree and term distributions and compared them with the same distributions derived globally over the same time periods. The global

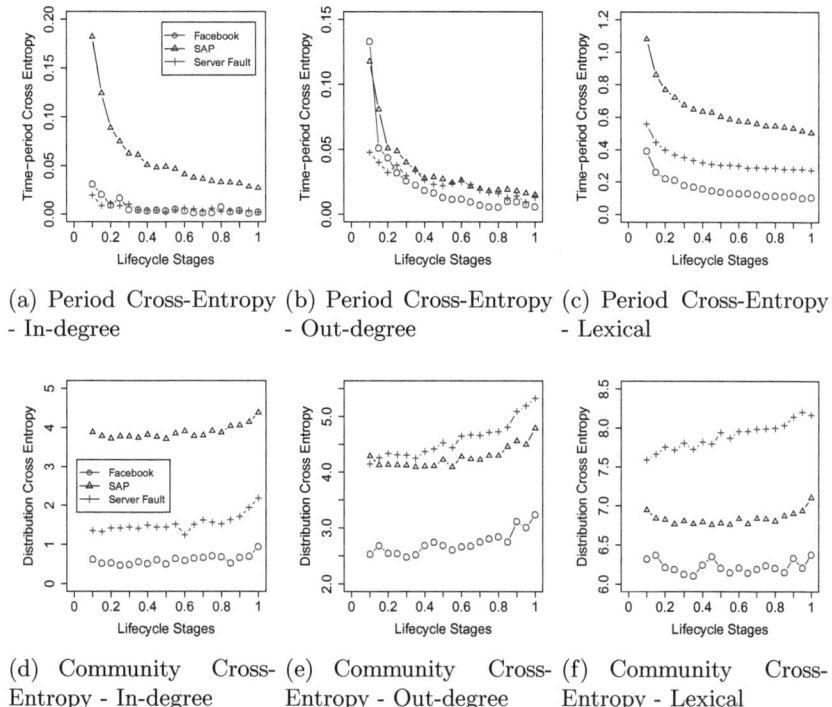

Fig. 1. Changes in user properties throughout lifecycle periods based on: period cross-entropy (Fig. 1(a), 1(b), 1(c)); and community cross-entropy (Fig. 1(d), 1(e), 1(f))

probability distributions were formed using the same means as above for the user-specific distributions, but instead using all posts rather than those by a given user from which to form the edges and term distributions. Therefore, given the discrete probability distribution of a user from a time interval $(P_{[t,t']})$, and the global probability distribution over the same time interval $(Q_{[t,t']})$, we derived the cross-entropy, as above, between the distributions. $(H(P_{[t,t']}, Q_{[t,t']}))$.

As before, we derived the community cross-entropy for each platform's users over their lifetimes and then calculated the mean community cross-entropy for the lifecycle periods. Fig. 1(d), fig. 1(e) and fig. 1(f) present the plots of the cross-entropies for the in-degree, out-degree and term distributions over the lifecycle periods. We find that for all platforms the community cross-entropy of users' in-degree increases over time indicating that a given user tends to diverge in his properties from users of the platform. For instance, for the community cross-entropy of the in-degree distribution the divergence towards later parts of the lifecycle indicates that users who reply to a given user differ from the repliers in the entire community. This complements cross-period findings from above where we see a reduction in cross entropy, thus suggesting that users form sub-communities in which interaction is consistently performed within (i.e. reduction in new users joining). We find a similar effect for the out-degree of the users where divergence from the community is evident towards the latter stages of users' lifecycles. The term distribution demonstrates differing effects however: for Facebook and SAP we find that the community cross-entropy reduces initially before rising again towards the end of the lifecycle, while for Server Fault there is a clear increase in community cross-entropy towards the latter portions of users' lifecycles suggesting that the language used by the users actually tends to diverge from that of the community in a linear manner. This effect is consistent with the findings of Danescu et al. [2] where users adapt their language to the community to begin with, before then diverging towards the end.

4 Lifecycle Period Detection

The above analysis unearthed the development that users go through across the three community platforms, based on different properties and development indicators. We now turn to the problem of detecting the lifecycle period that a given user is in based on his prior development. We characterise this task as a multi-class classification problem in which our goal is to induce a function that returns a user's lifecycle period: $f : \mathbb{R}^n \to S$, where the domain is an n-dimensional feature vector representation of a user's evolution and the co-domain is an integer representation of the lifecycle period ($S = \{6, 7, \ldots, 20\}$).[4]

4.1 Feature Engineering

In the previous section we found that different lifecycle periods can be characterised and thus detected based on the evolution of the user into the lifecycle period: i.e. when inspecting the final lifecycle period of users on each platform in terms of their lexical period cross-entropy, we find that this is less than previous periods, and that the rate of decay has levelled off somewhat. Hence we can use the growth rate from one lifecycle period to the next as information for detecting the lifecycle period of the user. We define this growth rate as follows:[5] $\delta_m(s) = \big(m(s+1) - m(s)\big)/m(s)$, where $m(s)$ denotes a convenience function that returns the measure value (e.g. in-degree period cross-entropy) for the lifecycle period s; thus $\delta_m(s) < 0$ indicates decay, $\delta_m(s) = 0$ indicates no change, and $\delta_m(s) > 0$ indicates growth. The growth rate therefore forms a single *growth feature* in our dataset, and by looking back k lifecycle periods we produce k growth features. Based on the 3 user properties (in-degree, out-degree, and lexical distributions) and 3 development indicators (entropy, period cross-entropy and community cross-entropy) we have a total of 9 measures that the convenience function m can take, and with k growth features derived for each measure ($m \in M$). Our datasets, both for training and testing data, take the following form: $D = \{(\mathbf{x}_1, y_1), (\mathbf{x}_2, y_2), \ldots, (\mathbf{x}_n, y_n)\}$, where $\mathbf{x}$ denotes the

---
[4] We use this integer representation for legibility and consider periods from 6 onwards to provide sufficient training data.
[5] This growth rate is equivalent to proportionate growth rates used in population models.

feature vector of a given user and y denotes the class label (lifecycle period) of the user. Hence the feature vector is a characterisation of the user up until and including the lifecycle period s and describes how the user has developed beforehand. The feature vector contains growth features of the user across different measures: $\mathbf{x} = [\delta_{m_1}(s-1), \ldots, \delta_{m_1}(s-k), \delta_m(s-1), \ldots, \delta_{m_2}(s-k)]$. This format indicates that for both measures (m_1 and m_2) we include k growth features. We maintain the splits mentioned above by using the 80% of analysed users for the training set and the held-out 20% of users for the testing split. For this latter dataset we hid the class labels and detected the label using the below model.

4.2 Vector Space Detection Model

To induce a detection function that can perform multi-class classification we use a vector space representation of classes and their boundaries to identify the most similar, or rather proximal class, to an arbitrary user's feature vector. We define this function as: $f(\mathbf{x}) = \arg\max_{s \in S} sim(\mathbf{x}, s)$, where sim is derived using a given similarity function and thus chooses the class (lifecycle period) that maximises similarity. We vary the similarity function ($sim(\mathbf{x}, s)$) through four measures:

1. *Cosine Similarity:* Measures the cosine of the angle between the user's feature vector $\mathbf{x}$ and the class centroid vector from the training data $\mathbf{p}^s$.
2. *Euclidean Distance:* Measures the distance between the vectors and then takes the reciprocal of this distance to derive the similarity measure, as the reciprocal distance is maximised when the Euclidean distance is minimised.
3. *Mahalanobis Distance:* Accounts for the variance in the class distribution from which the centroid vector is derived by including the covariance matrix Σ^s of the class s: $sim_{mah} = \left((\mathbf{x} - \mathbf{p}^s)\Sigma^s(\mathbf{x} - \mathbf{p}^s)\right)^{-\frac{1}{2}}$
4. *Spearman Rank Correlation Coefficient:* Measures the extent to which a monotonically increasing or decreasing association exists between the feature vector and class centroid: 1 indicates a strong positive association, -1 indicates a strong negative association.

5 Experiments

5.1 Experimental Setup

For the experiments we tested the performance of different feature sets and similarity functions: for the feature sets we tested in-degree (i.e. in-degree entropy, in-degree period cross-entropy, etc.), out-degree and lexical features, followed by entropy (i.e. in-degree entropy, out-degree entropy, etc.), period cross-entropy and community cross-entropy features, before then combining all features together. We set the value of k, the number of previous periods to derive the growth rates from, to 5, and tested the performance of detecting all lifecycle periods ($S = \{6, 7, \ldots, 20\}$). We applied the classification accuracy measures of precision and recall as macro averages over the classes under inspection ($s \in S$)

and then took the harmonic mean of precision and recall as the f-measure (F1 score), We used two baselines in our experiments: (i) a random guesser model, formed from the probability of success in a single Bernoulli trial per class; and (ii) the naive Bayes classifier. The latter comparing the vector space model against an existing generative model that is regularly used in multi-class classification tasks, and the former measuring performance of each vector space model against the random model using the Matthews correlation coefficient (mcc).

Table 2. F1 scores of the different platforms when detecting the user lifecycle periods using different detection models and feature sets, with the Matthews Correlation Coefficient in parentheses to show improvement over the random model baseline

Platform	Feature Set	Cosine	Euclidean	Mahalanobis	Spearman
Facebook	In-degree	0.677 (0.637)	0.757 (0.730)	0.706 (0.659)	0.672 (0.627)
	Out-degree	0.609 (0.582)	0.751 (0.718)	0.703 (0.665)	0.592 (0.553)
	Lexical	0.653 (0.632)	0.757 (0.730)	0.739 (0.700)	0.629 (0.601)
	Entropy	0.674 (0.618)	0.757 (0.730)	0.676 (0.621)	0.654 (0.602)
	Period Cross Entropy	0.650 (0.590)	**0.774**** (0.746)	0.630 (0.586)	0.647 (0.589)
	Comm' Cross Entropy	0.643 (0.592)	0.760 (0.732)	0.657 (0.610)	0.671 (0.621)
	All	0.676 (0.614)	0.757 (0.730)	0.659 (0.608)	0.686 (0.633)
SAP	In-degree	0.582 (0.520)	0.665 (0.652)	0.426 (0.376)	0.583 (0.527)
	Out-degree	0.597 (0.571)	0.658 (0.647)	0.600 (0.588)	0.574 (0.541)
	Lexical	0.583 (0.521)	0.665 (0.652)	0.431 (0.378)	0.558 (0.499)
	Entropy	0.522 (0.468)	0.665 (0.652)	0.470 (0.418)	0.532 (0.467)
	Period Cross Entropy	0.643 (0.591)	0.656 (0.651)	0.434 (0.377)	0.640 (0.590)
	Comm' Cross Entropy	0.546 (0.497)	**0.708***** (0.677)	0.529 (0.475)	0.520 (0.466)
	All	0.619 (0.565)	0.665 (0.652)	0.423 (0.364)	0.640 (0.590)
Server Fault	In-degree	0.671 (0.631)	0.748 (0.721)	0.718 (0.664)	0.667 (0.619)
	Out-degree	0.635 (0.613)	0.760 (0.727)	0.732 (0.683)	0.608 (0.580)
	Lexical	0.666 (0.631)	0.748 (0.721)	0.711 (0.663)	0.643 (0.595)
	Entropy	0.669 (0.631)	0.748 (0.721)	0.703 (0.637)	0.654 (0.602)
	Period Cross Entropy	0.701 (0.650)	**0.774**** (0.747)	0.622 (0.584)	0.702 (0.660)
	Comm' Cross Entropy	0.650 (0.597)	0.738 (0.710)	0.709 (0.647)	0.651 (0.603)
	All	0.698 (0.637)	0.748 (0.721)	0.706 (0.637)	0.680 (0.632)

Significance codes: p-value < 0.001 *** 0.01 ** 0.05 * 0.1 . 1

5.2 Detection Results

Table 2 shows the detection results across the platforms using different feature sets and similarity functions. For all platforms Euclidean distance performed best with performance variation between the feature sets: period cross-entropy for Facebook and Server Fault; and community cross-entropy for SAP. This suggests that information about the user evolving with respect to his earlier properties and/or the community is sufficient to detect the period of the user. We performed significance testing using the Mann-Whitney test of the difference in performance between the best model from each platform and the next best performing model, and found the differences to all be significant (as indicated in the table). Interestingly, for all of our tested models we significantly outperformed the random model baseline, as indicated by the mcc values within the parentheses. We omitted the performance of the naive Bayes classifier for legibility, however it is sufficient to add that it achieved low F1 scores across all platforms and feature sets (0.092, 0.098 and 0.126 for Facebook, SAP and Server Fault respectively), hence our vector space model significantly ($p < 0.001$) outperforms this second baseline.

6 Conclusions and Future Work

In this paper we presented a means to model the development of users in online community platforms along different properties (in-degree, out-degree, lexical terms) and using different development indicators (period specific entropy, past period cross-entropy, community cross-entropy), finding that the examined platforms show similarities in terms of how their users evolved. Armed with such insights we presented a vector space detection model, with varying vector similarity functions, that used the growth rates of users' properties from earlier lifecycle periods to detect the current lifecycle period of the user. Our evaluation demonstrated the high levels of accuracy that we achieve across the four tested similarity functions - finding minimised Euclidean distance to perform best - and significant improvement in performance over a random model and the naive Bayes classifier. Future work will include exploring alternative classification approaches and examining a more *real world* test setting in which the lifecycle period of the user is to be detected: i.e. given an arbitrary point in time return the lifecycle period of a user. As our work is one of the first to propose a model to detect the lfiecycle period of online community users it has implications on churn prediction: we can detect users in their final lifecycle periods, thereby allowing community managers to take the appropriate action to retain those users (e.g. recommending content to interact with).

References

1. Chung, K.S.K., Piraveenan, M., Uddin, S.: Community evolution and engagement through assortative mixing in online social networks. In: 2012 IEEE/ACM International Conference on Advances in Social Networks Analysis and Mining, pp. 724–725 (2012)
2. Danescu-Niculescu-Mizil, C., West, R., Jurafsky, D., Leskovec, J., Potts, C.: No country for old members: User lifecycle and linguistic change in online communities. In: WWW 2013 (2013)
3. Gong, N.Z., Xu, W., Huang, L., Mittal, P., Stefanov, E., Sekar, V., Song, D.: Evolution of social-attribute networks: Measurements, modeling, and implications using google+. CoRR, abs/1209.0835 (2012)
4. Kairam, S.R., Wang, D.J., Leskovec, J.: The life and death of online groups: predicting group growth and longevity. In: Proceedings of the Fifth ACM International Conference on Web Search and Data Mining, pp. 673–682. ACM (2012)
5. McAuley, J., Leskovec, J.: From amateurs to connoisseurs: Modeling the evolution of user expertise through online reviews. In: Proceedings of World Wide Web Conference (2013)
6. Miritello, G., Lara, R., Cebrián, M., Moro, E.: Limited communication capacity unveils strategies for human interaction. Nature (April 2013)

Metro: Exploring Participation in Public Events

Luca Chiarandini[1,2], Luca Maria Aiello[2], Neil O'Hare[2], and Alejandro Jaimes[2]

[1] Web Research Group
Universitat Pompeu Fabra
Barcelona, Spain
[2] Yahoo! Research
Barcelona, Spain
{chiarluc,alucca,nohare,ajaimes}@yahoo-inc.com

Abstract. The structure of a social network is time-dependent, as relationships between entities change in time. In large networks, static or animated visualizations are often insufficient to capture all the information about the interactions between people over time, which could be captured better by interactive interfaces. We propose a novel system for exploring the interactions of entities over time, and support it with an application that displays interactions of public figures at events.

1 Introduction

In the context of image search, people often query and browse photos of celebrities and public figures [5]. Automated query analysis allows search engines to identify the queried person and display structured information on the result screen (*e.g.* biographical information, birth year, related people, *etc.*). The information is usually an aggregated summary of a person's life and does not allow the user to further explore important events, nor the social interactions of a celebrity.

Representing events in a person's life, and especially social interactions, can help us to gain a better understanding of a person, not just as a standalone entity but as an individual in a social environment. A person at a particular instant of time is not just a set of properties (*e.g.* hair color, job, birth date, *etc.*), but is also defined by the connection to other people.

Displaying the interactions of entities over time is a challenging task because of the conflation of the temporal and relational dimensions. Tools to visualize and explore interactions between entities in time tend to focus either on the structural or the temporal dimension. On one hand, tools to animate *dynamic graphs* [14] [1] can visualize the evolution of the whole set of interactions in the system, but they do not provide a way to explore the history of relations. On the other hand, *timelines* [10] [8] and their variants, such as stacked lines charts and stream graphs [4], foster the exploratory visualization of temporal data by explicitly displaying temporal sequences of events as lines on a reference plane. However, being focused on the representation of temporal information only, the interaction between entities is not easily represented in such displays.

Attempts to produce visualization between these two extremes have been made in the past. Tools for the exploration of genealogical data explicitly represent both time and interactions, but are bound to the visual paradigm of the tree [2]. Visualizations with *metro maps* [13] [12] allow a more generic layout, but relax the constraint on time representation, being more similar to graphs than timelines. *Alluvial diagrams* have been used to represent changes in network structure over time [11]. In such representations, each line is a cluster of entities and one can see how entities move across them in time. In *Metro*, however, the focus is on entities and their interactions rather than on clusters. *TimeNets* is a tool for genealogical data visualization [7]. People's lives are represented on a horizontal timeline as lines spanning from the year of birth to the year of death. Lines of different people join and split correspond to weddings or separations. However, Timenets does not allow exploration by query and is tailored towards genealogical data, where the interactions between people are few and, on average, span a long period of time (e.g. marriage).

In this paper we present *Metro*, a system for exploring social interactions over time, leveraging information about participation in public events, using the paradigm of crossing *life lines* over time. *Metro* provides functions to explore online content in an unconventional yet practical way, allowing complex exploration of the information space by querying, pivoting over people and events, and inspecting the context information around the visualized interactions (photos of events and related people). *Metro* not only allows the user to explore existing social interactions and visualize their temporal characteristics, *i.e.* are they sporadic, periodic or clustered around a particular date. It also helps the user to discover tempo-structural holes in the network, *i.e.* moments in time when links between people are missing. Through interactive search, *Metro* can retrieve people that are connected to a particular person and *not* to another. To the best of our knowledge, this work is the first to present such feature automatically.

We present an example to motivate the *Metro* approach. Consider the case of a person living in different geographical locations. His or her social network could be composed of separate components. This is often due to the geographical distance between the people one knows. If we represent the social network as a graph, we would see the person acting as a hub across communities. Without any further information, we are not able to reconstruct the life of this person, nor the reason why he or she is connecting such heterogeneous communities. By exploiting time and structure jointly, we can understand if the person was interacting simultaneously with multiple communities or if he or she was interacting with a community at a time. Moreover we may be able to observe frequent and time-independent interaction, *i.e.* people with whom he or she interacts across locations (*e.g.* family, long-lasting friends, *etc.*). Displaying participation to events instead of explicit connections allows us to distinguish between currently active and non active relationships.

The methodology we present could naturally adapt to many exploratory tasks, such as co-appearance in online social networks, or authorship of scientific publications. In this demo, we present an application in search and browsing photos of public figures, which is a frequent task in image search [6].

2 Exploring People and Events

Metro is a system to explore interactions among people over time. In the application we present, people are public figures and their interactions are the co-participations

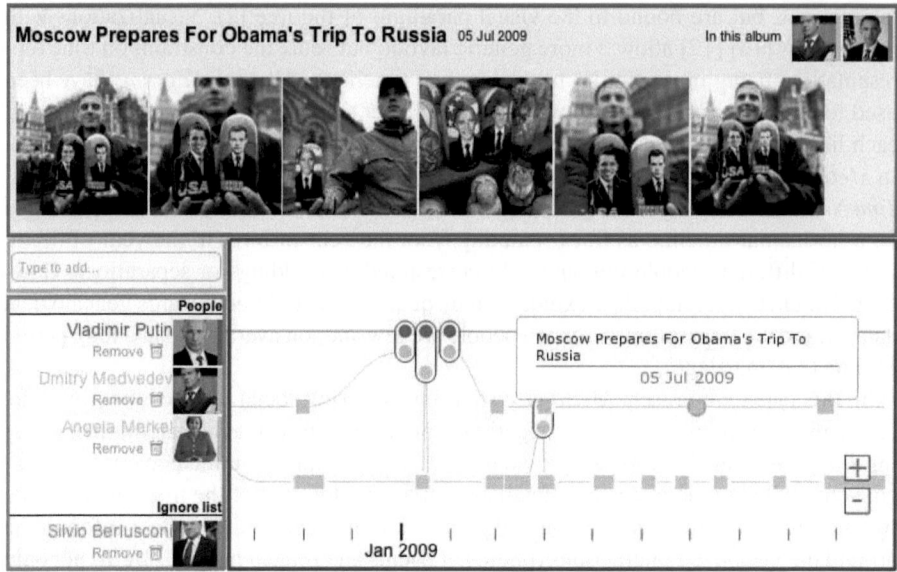

Fig. 1. The interface containing the event bar (red), the search field (yellow), the people and ignore lists (green), and the timeline (blue)

in events. Figure 1 shows a snapshot of the working system. We can see how three politicians interact during 2009. Each person is represented by a horizontal line of the same color as their name on the left. Lines join when the two people appear together in photo. We can see, for example, that Vladimir Putin visits Angela Merkel during mid-January. Along the blue line there is a point surrounded by a gray circle. This is the currently selected point. It refers to the preparation of US president Barack Obama to Russia. The photos of the event are shown in the upper part.

The front-end of this demo has been fully developed in HTML 5 and the back-end uses PHP 5[1] and MySQL[2]. It is powered by the metadata from approximatively 9 million images, taken between the years 2000 to 2011, from a well-known stock photo agency [9]. Photo metadata includes the timestamp, the list of *people* which appear and the id and description of the event the photo relates to (*e.g.* royal wedding, G8 summit, *etc.*). To enhance the visualization with more accurate people description, we crawl the Wikipedia information of all the available person names. Among all the people in our corpus, 78% have a Wikipedia page. The following sections will describe the functionalities currently featured in *Metro*.

2.1 Interface Structure

In this section we describe the structure of the interface. Figure 1 shows the interface of *Metro*; components of the interface are highlighted by different colors to ease the description.

---

[1] http://php.net/
[2] http://www.mysql.com/

(a) The profile of a celebrity.   (b) Recommendation of people.

Fig. 2. The profile of a user and the recommendation box as they appear in the interface

Interactions between people are displayed with intersecting *life lines* on the top of a timeline (bottom right module, highlighted in blue). The horizontal axis represents time and the vertical one the social relations. Participation in events are represented as points on the person's life line. When multiple people attend the same event, the points are grouped together, placed on the topmost free row and enclosed in a black border. To minimize line crossings when many lines are present, we use a greedy algorithm to order them so that lines of people who appear together often are drawn close to each other. The timeline can be explored by horizontally zooming or scrolling.

Hovering on a point opens a text-box with a short event description, while clicking on it loads related pictures in the *event bar* (red box in the figure), together with the full event description and the list of other people attending the event.

On the left in the green box is the *people list*, which contains the list of currently displayed people. Clicking on a person's photo displays the biographical information, extracted from Wikipedia (see Figure 2(a)). People in the *people list* can be dragged down to the *ignore list*. The timeline only shows events in which at least a person in the people list appears and *no* person in the ignore list appears.

2.2 Recommendation Algorithm

The people list can be expanded by searching for a person's name in the search field on top (yellow). If no query is typed, clicking on the search field opens a box with a *recommendation list* of the people who mostly co-occurred with the people already present in the list (see Figure 2(b)). The recommendation algorithm takes in input two disjoint sets of people, P_+ (the current people list) and P_- (the ignore list), and the current time frame on the timeline $t_1 \leq t_2$. The suggested people should be tightly connected to the ones in P_+ but not to the ones in P_- in the given time frame.

The score assigned to people for ranking is computed as follows. First, let $c : \mathbb{R} \times \|P\| \times \|P\| \mapsto \mathbb{R}$ be the co-occurrence function, where $c(t, p_k, p_h)$ returns the number of photos taken at time t in which both p_k and p_h appear. Second, the time-constrained co-occurrence function $\bar{c}(p_1, p_2) = \int_{t_1}^{t_2} c(t, p_1, p_2) \, dt$ is created to quantify how often people co-occur in a specified time interval. Finally, the person score p, which is used to rank people for the recommendation list, is computed as:

$$score(p) = \frac{\prod_{p_+ \in P_+} \bar{c}(p, p_+)}{\prod_{p_- \in P_-} \bar{c}(p, p_-)}. \tag{1}$$

An example of recommendation is shown in Figure 2(b). In this case $p_+ = \{$ *Barack Obama* $\}$ and $p_- = \{$ *Michelle Obama* $\}$. The recommended person, the vice-president of the United States *Joseph Biden*, appears often with Barack Obama but not with Michelle.

2.3 Functionalities

Unlike previous work, *Metro* allows users to explore the history interactions over time at different granularities and across several dimensions. This is done by means of the following functions.

Search People. The social space is explored by searching for a person's name. When adding new life lines in the interface, the intersections in common events are dynamically adapted.

Slice over Social or Time Dimension. Display all events for a person in the timeline or all the attendees at an event in the event bar. The slicing works also for group of people, when more than one are selected.

Context Exploration. The interaction between public figures during events is contextualized by the content displayed in the event bar. Pictures related to the events are shown, together with the full set of attendees.

Pivoting. People related to the ones displayed and to the current time frame are recommended with the algorithm described above, allowing the user to pivot from one person to another based on their past co-appearances. Moreover, event attendees shown in the event bar can be added to the people list. The iteration of this process allows to move smoothly through the space and find related entities [3].

3 Conclusions and Future Work

Metro is a system for explore people's participation in public events. The interface jointly represents social interactions and the temporal dimension, and allows the user to browse through people using either. This rich set of features enables an effective way to explore the information space that could be adapted to different domains. As future steps, we plan to include functionalities to reduce and aggregate the information displayed, by clustering similar entities.

Acknowledgments. This research is partially supported by European Community's Seventh Framework Programme FP7/2007-2013 under the ARCOMEM and Social Sensor projects, by the Spanish Centre for the Development of Industrial Technology under the CENIT program, project CEN-20101037 "Social Media", and by Grant TIN2009-14560-C03-01 of the Ministry of Science and Innovation of Spain.

References

1. Bastian, M., Heymann, S., Jacomy, M.: Gephi: An open source software for exploring and manipulating networks. In: ICWSM: International AAAI Conference on Weblogs and Social Media (2009)
2. Bezerianos, A., Dragicevic, P., Fekete, J.D., Bae, J., Watson, B.: Geneaquilts: A system for exploring large genealogies. IEEE Transactions on Visualization and Computer Graphics 16(6), 1073–1081 (2010)
3. Dörk, M., Riche, N.H., Ramos, G., Dumais, S.T.: Pivotpaths: Strolling through faceted information spaces. IEEE Trans. Vis. Comput. Graph 18(12), 2709–2718 (2012)
4. Havre, S., Hetzler, B., Nowell, L.: Themeriver: Visualizing theme changes over time. In: IEEE Symposium on Information Visualization, InfoVis 2000, pp. 115–123 (2000)
5. Holub, A., Moreels, P., Perona, P.: Unsupervised clustering for google searches of celebrity images. In: 8th IEEE International Conference on Automatic Face & Gesture Recognition, FG 2008, pp. 1–8 (2008)
6. Jansen, B.J.: Searching for Digital Images on the Web. Journal of Documentation 64(1), 81–101 (2008)
7. Kim, N.W., Card, S.K., Heer, J.: Tracing genealogical data with timenets. In: Proceedings of the International Conference on Advanced Visual Interfaces, pp. 241–248. ACM (2010)
8. Ogawa, M., Ma, K.L.: Software evolution storylines. In: Proceedings of the 5th International Symposium on Software Visualization, pp. 35–42. ACM (2010)
9. O'Hare, N., Aiello, L.M., Jaimes, A.: Predicting participants in public events using stock photos. In: ACM Multimedia (2012)
10. Plaisant, C., Milash, B., Rose, A., Widoff, S., Shneiderman, B.: Lifelines: visualizing personal histories. In: Proceedings of the SIGCHI Conference on Human Factors in Computing Systems: Common Ground, pp. 221–227. ACM (1996)
11. Rosvall, M., Bergstrom, C.T.: Mapping Change in Large Networks. PLoS ONE 5(1), e8694+ (2010)
12. Shahaf, D., Guestrin, C., Horvitz, E.: Metro maps of science. In: Proceedings of the 18th ACM SIGKDD International Conference on Knowledge Discovery and Data Mining, pp. 1122–1130. ACM (2012)
13. Shahaf, D., Guestrin, C., Horvitz, E.: Trains of thought: Generating information maps. In: Proceedings of the 21st International Conference on World Wide Web, pp. 899–908. ACM (2012)
14. Yee, K.-P., Fisher, D., Dhamija, R., Hearst, M.: Animated exploration of dynamic graphs with radial layout. Presented at IEEE Symposium on Information Visualization (2001)

Follow My Friends This Friday!
An Analysis of Human-Generated Friendship Recommendations

Ruth Garcia Gavilanes, Neil O'Hare, Luca Maria Aiello, and Alejandro Jaimes

Yahoo! Research, Barcelona

Abstract. Online social networks support users in a wide range of activities, such as sharing information and making recommendations. In Twitter, the hashtag *#ff*, or *#followfriday*, arose as a popular convention for users to create contact recommendations for others. Hitherto, there has not been any quantitative study of the effect of such human-generated recommendations. This paper is the first study of a large-scale corpus of human friendship recommendations based on such hashtags, using a large corpus of recommendations gathered over a 24 week period and involving a set of nearly 6 million users. We show that these explicit recommendations have a measurable effect on the process of link creation, increasing the chance of link creation between two and three times on average, compared with a recommendation-free scenario. Also, ties created after such recommendations have up to 6% more longevity than other Twitter ties. Finally, we build a supervised system to rank user-generated recommendations, surfacing the most valuable ones with high precision (0.52 MAP), and we find that features describing users and the relationships between them, are discriminative for this task.

1 Introduction

Social media services have emerged as platforms on which people express opinions, obtain information about topics of interest (e.g. sports, fashion, etc.), discover breaking news, and receive updates from their friends, contacts, and their favorite celebrities. Most social media sites allow users to set up a network of connections (e.g., friends, contacts, celebrities) from which they can receive information. In some networks, such as Twitter or Google+, connections need not be reciprocal, and any user is free to *follow* any other user with a public profile, to be able to see their posts or status updates.

Since users are allowed to follow people they do not know, an important question is who *else* they should follow, in particular people who might be sources for the type of information they are interested in. In response to this need, *Follow Friday* emerged in 2009 as a spontaneous behavior from the Twitter user base, inspired by a blog post of an influential blogger[1]: users post tweets with the *#followfriday* or *#ff* hashtag, and include the usernames of the users they wish to recommend. As the name suggests, by convention these recommendations are made on Fridays. The key idea behind *Follow*

---

[1] http://mashable.com/2009/03/06/twitter-followfriday/

Friday is that people you already follow should be able to suggest new contacts that you will be interested in following.

In 2009 and 2010, in particular, the popularity of these hashtags on Twitter rose considerably, up to the point that the Twitter hashtags *#followfriday* and *#ff* were among the most popular hashtags observed in several large-scale Twitter corpora [25,20].

Although Twitter now has an automatic recommender system for contacts, the analysis of the dynamics of the *Follow Friday* phenomenon is interesting from multiple perspectives. From the angle of complex systems analysis, measuring the effect that collective recommendation processes have in driving the connectivity choices of individuals is very valuable to quantify the ability of a system to self-organize. Additionally, our analysis identifies features that are most predictive of tie formation in a peer-to-peer link recommendation process. This is useful on the one hand to alleviate the information overload of users receiving recommendations from their peers, by identifying the 'strongest' recommendations among hundreds or even thousands, and on the other hand to improve the design of automatic contact recommendation algorithms.

In this paper we focus on the dynamics of *Follow Friday* as a form of broadcast recommendations, making the following main contributions:

- We analyse for the first time the dynamics of a large-scale human-driven recommendation system and, by comparing it with two baseline conditions, we measure its impact on the process of follower-link creation. We find that recommended users have a chance of being followed that is roughly two or three times higher than a recommendation-free scenario. We also measure how long the recommendation effect lasts, as well as the effect of repeated recommendations and the longevity of the accepted recommendations (i.e. how long these follower links persist).
- We develop a recommender system for ranking the human-generated recommendations received by a user. We evaluate this system against a corpus of known 'accepted' recommendations, identifying the features that are more predictive of link creation. Our recommender achieves a MAP of around 0.52, which is extremely high given the sparsity of the link recommendation problem. To the best of our knowledge, this is the first friend recommender system built and evaluated on human created recommendations.

The rest of this paper is organized as follows. In the next section we summarize related work, followed in Section 3 by a description of the dataset and a summary of key terminology. In Section 4 we analyze the *Follow Friday* phenomenon along a number of dimensions, and quantify the extent to which it has a real effect on users' *following* behavior. We then, in Section 5, propose and evaluate a recommender system for ranking a user's received *Follow Friday* recommendations. Finally, we conclude the paper in Section 6.

2 Related Work

The study of user-generated recommendations based on *Follow Friday* tags lies between two streams of research on recommender systems: recommendation based on user-generated content and social link recommendation.

A number of studies have been done on friendship recommendations in the context of Twitter, for instance Hannon et al. [11] compared collaborative filtering and content-based recommendation for the purpose of link recommendation on Twitter. Garcia Esparza et al. [9] presented a movie recommendation system that extracts information from a Twitter-like microblog platform for movie reviews. They profile 537 users and 1080 movies according to words and tags, and offer content-based and collaborative-filtering recommendations. Several aspects of user profiles have been studied for recommendations, for example Abel et al. [1] propose a methodology for modeling Twitter user profiles to support personalized news recommendation. They compare profiles constructed from the complete long-term user history with profiles based only on users' most recent tweets.

The task of predicting link formation (or deletion) in social graphs is one of the major challenges in the area of link mining, and has been well studied in the last decade [17,19,13]. Approaches have been proposed based on attributes of the nodes [13], structural graph features [26,18], or both [2]. Unlike most of the work on link prediction that tries to predict future links in balanced sets of positive and negative samples, we are interested in a variant of link prediction, namely *link recommendation*, that is strictly user-centered and aims to provide a list of contacts to a user with the objective of maximizing the acceptance rate. Due to its inherent sparsity, this problem is more difficult than general prediction, and it has received little attention so far [4]. Previous studies also investigated what are the most predictive network and profile features for link formation in Twitter [12].

Despite the previous work in the area, we are not aware of any other attempt at characterizing human-generated recommendations and to leverage them to provide automatic contact suggestions. We also quantify the power of different features in predicting the formation of new links, not just considering structural or profile features of the user accounts, but focusing also on features that are descriptive of the human-driven recommendation process, such as the characterization of the relationship between the different human parties involved: the user who produces the recommendation, the one who receives it, and the one who is recommended.

3 A Dataset of Broadcast Friend Recommendations

Twitter is a social media platform on which users post 140-character messages called *tweets*. Users can follow other users and get notified with the tweets they post. For convenience, we will refer to the follower-followee relationship as a friendship relationship, although strictly speaking this relation only occasionally represents a true friendship: follower links are often not reciprocated [7] and often the followee can be an organization or a celebrity. So, in this sense, it is more correct to think of the followee as an information channel whose updates the user may be interested in subscribing to. In Twitter, a *hashtag* is any sequence of characters, without whitespace, preceded by the # symbol, and a 'mention' consists of any Twitter username preceded by the symbol '@'. We define *Follow Friday* recommendations as broadcast mentions of usernames in tweets containing the hashtag *#followfriday* or *#ff* (case-insensitive)[2]. So, for

---

[2] We use the term *Follow Friday* to refer to the use of either of these *Follow Friday* hashtags.

Table 1. Unique # of Receivers, Recommenders, Recommended Users, Recommendation Instances, and Accepted and Rejected Recommendations

	Total
Initial seed set	55,000
Receivers	21,270
Recommenders	589,844
Recommended Users	3,261,133
Recommendation Instances	59,055,205
Accepted Recommendation Instances	354,687
Rejected Recommendation Instances	58,700,518

example, the tweet "`#followfriday @Lula and @Obama for being such great leaders`" recommends people to follow the Twitter users `Lula` and `Obama`.

In March 2011, using the Twitter stream API, we randomly selected a *seed set* of 55,000 users. To remove profiles that are unlikely to be legitimate or active, we follow the approach of Lee et al. [15] and exclude users who have more than 1000, or less than 100, followers or followees. This filter also excludes celebrities, who usually do not interact with other users [14]. We monitored the evolution of the seed users' followees over time by collecting snapshots of the seed users' contact networks during a 24 week period from March 24th, 2011 to September 5th, 2011. The snapshots were taken twice a week, every Thursday and Monday yielding a total of 48 network snapshots. This choice is motivated by the fact that, although the recommendations are mostly broadcast on Fridays (76%), there is still a non-negligible amount of recommendations broadcast on Saturday (14%) and Sunday (3%), therefore Thursday and Monday snapshots can describe the status of the network right before and right after the recommendation takes place.

In the remainder of this paper, we use the following terminology:

- **Receivers (Rcv).** Users from the initial seed set who accepted at least one *Follow Friday* recommendation at any time during the 24 week period.
- **Recommenders (Rdr).** The followees of the *receivers (Rcv)* who made at least one *Follow Friday* recommendation during the 24 week period.
- **Recommended users (Rdd).** The users mentioned after the *Follow Friday* hashtag in the messages of the *recommenders* (Rdr).
- **Recommendation Instance (Rec).** The tuple $\langle rdd, rdr, rcv, w \rangle$ identifying an instance of a recommended user, made by a recommender, and exposed to a specific receiver in a given week (w). We use lowercase letters to identify elements in the actors sets (e.g., $rdr \in Rdr$)
- **Acceptance.** We consider a *recommendation instance* made at time t to be *accepted* if its receiver becomes a follower of the recommended user between time t and time $t + \Delta$. Unless stated otherwise, the Δ considered is one week. Although we use the term *acceptance*, we cannot be sure about the causal relation between recommendation and acceptance (see discussion in Section 4.1).
- **Rejection.** We consider a recommendation instance made at time t to be *rejected* if the receiver does not follow the recommended user between time t and $t + \Delta$. Recommended users who are already followees of the rcv are not considered in the analysis.

Table 1 summarizes the quantities of followers, receivers, recommenders, recommended users and recommendation instances in our dataset.

4 Analysis of Broadcast Recommendations

During the 24 weeks captured by our dataset, we have a total of 144,180 *unique* new followees, from 354,687 *Follow Friday* accepted *recommendation instances*: this means that, on average, for accepted *recommendation instances*, the *receiver* got recommendations to follow the same *recommended user* from 2 distinct *recommenders*. Table 2 shows the acceptance rate (the number of accepted recommendations divided by the number recommendation instances) for recommendations under various conditions where one of the actors involved in the *recommendation instance* mentioned one of the others in the previous week (using the '@username' convention). The first column indicates the direction of the mention and the users involved. The case of $rdr \to rdd$ (recommender mentions recommended) involves all recommendations since this is a necessary condition of a *Follow Friday* recommendation.

We can see that, overall, the acceptance rate is very low at 0.006 (i.e. 0.6% of recommendations are accepted), which is to be expected, since the recommendations are broadcast, as opposed to being personalized, and may not even have been seen by the receiver. When one of the actors mentions another, the acceptance rate tends to increase, which is expected, since these mentions are indicators of an active relationship. When either the *recommended user* (rdd) or the *receiver* (rcv) mention each other, the acceptance rates are roughly 10 times higher than the average (10% to 14% of recommendations accepted), which is not surprising since it shows that there is already a connection between these two users who form the new link. Note that while the acceptance rate for these particular cases is relatively high, the volume is low, indicating that these cases of pre-existing relationships are not typical of *Follow Friday* recommendation acceptances.

4.1 Effect of #FF Recommendation

Since *Follow Friday* is a spontaneous recommendation phenomenon, the first question that arises is whether it has an actual impact on the creation of new follower links, and to what extent. In complex social systems, determining the causes of observed evolutionary phenomena is a very challenging task, due to the intrinsic difficulty in disentangling all the factors that produce the events observed in a-posteriori data-driven studies [23]. Even when controlled experiments are performed [3,5], it is very difficult to know with absolute certainty which factors trigger the observed dynamics.

In our case, the inclusion of a new *recommended user* in the followee list cannot be interpreted directly as a cause-effect sequence, since the adoption may be driven by factors that are not related with the recommendation itself, such as unobserved on-line interactions or even exogenous events. Nevertheless, when sufficiently extensive temporal data is available, it is possible to compare the evolution of the system under different conditions, or null models [6,22], to understand if the the target factor has an effect, distinguishable from the other conditions, on the evolution of the system.

Table 2. Acceptance Rates for *Follow Friday* Recommendations, under various conditions where the users mention each other in the preceding week. For example 'rdd→rcv' indicates that the recommended mentioned the receiver, and 'rdd↔rcv' indicates that the recommended and receiver both mentioned each other. (rdd→rdr is omitted because it is identical to rdd↔rdr in this dataset: by definition, the recommender mentions the recommended for all recommendations.)

Mentions	Volume	Proportion	Acceptance Rate
rdr→rdd	59,055,205	1.000	0.006
rdd↔rdr	4,667,056	0.079	0.009
rcv→rdr	9,071,311	0.154	0.010
rdr→rcv	9,199,224	0.156	0.011
rdr↔rcv	6,242,059	0.106	0.012
rcv→rdd	205,447	0.003	0.095
rdd→rcv	238,822	0.004	0.097
rcv↔rdd	76,482	0.001	0.145

Specifically, we measure the added value of *Follow Friday* by comparing the acceptance rate of *#ff* or *#followfriday* recommendations with two alternative conditions:

(a) **Implicit recommendation model**: all usernames mentioned in any tweet received by users in the receiver set (*Rcv*) are considered implicit recommendations, based on the assumption that being exposed to the names of some users may increase the probability of adopting them as new followees. The implicit recommendations we consider are all mentions appearing in non-*#ff* tweets during the week before the target week, and that never previously appeared as an explicit *#ff* recommendation (for the same *receiver*) in the 24 week sample.

(b) **Unobserved recommendation model**: for this model, we assume that, due to unobserved factors, the contacts recommended through *#ff* hashtags would have been added by the *Rcv* set even in absence of any explicit *#ff* recommendation. These unobserved factors could include, for example, the rising popularity of the recommended user or relevance of the topics discussed by the recommended user to external breaking events. To model this condition, we apply a temporal shift: for the set of *#ff* recommendations made at time t, we measure their acceptance rate at time $t-1$, before the actual recommendation is made, i.e. we measure the acceptance rate in a situation where the external conditions are similar (one week previously), but where no *Follow Friday* recommendation has been made. To keep this model separate from the implicit one, we exclude cases where the receiver received implicit recommendations, up to time $t-1$, for the same recommended user.

The difference in the acceptance rate between the three models, depicted in Figure 1, shows that *#ff* recommendations lead users to follow a higher proportion of contacts compared to models in which *#ff* is not considered. Apart from an outlier at week 1, the margin between the *#ff* model and the two alternative conditions is large, with *#ff* having an acceptance rate always between two and three times that of the others. By disentangling the role of the presence of the *#ff* tag from other factors that play an important role in the creation of social links, mainly homophily, the comparison with

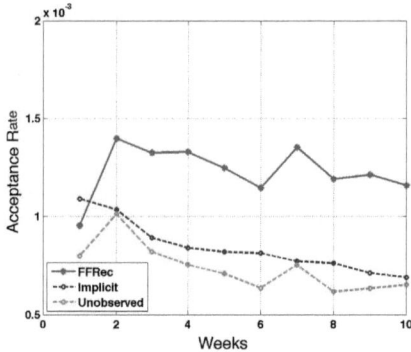

Fig. 1. The acceptance rate for *Follow Friday* recommendations in different weeks, compared with implicit and unobserved recommendation models

these alternative conditions provides strong evidence that the recommendation has an effect on the probability of link creation.

Whereas homophily may have a role in the selection of a recommended profile among other recommended ones, it seems not to be the main reason for the recommendation acceptance itself. Since the unobserved condition is simulated by performing a one week temporal shift, if we assume that the homophily effect between two users is not likely to change drastically in this one week time frame, then if the probability of acceptance is mainly determined by homophily, the *#ff* and *unobserved* conditions would have similar acceptance rates. The fact that this is not the case is, we believe, strong evidence that the *#ff* recommendation, and not purely the similarity between the profiles, drives the creation of the new link. Of course, there may be cases where the homophily effect changes drastically over the one week time shift, but it seems unlikely that this would explain all of the recommendation acceptances in this very large corpus.

A slight decreasing trend in acceptance rates over time is observed for all conditions, most likely due to the effect of the residual signal of explicit and implicit recommendations from the previous weeks (i.e. due to recommendations made before week 1 of our study, which we have no information about). To further verify this hypothesis, we measure how much the effect of an implicit or explicit recommendation lasts in time by computing the acceptance rate n weeks after the recommendation is made. To do so, we split our 24-week sample in half and observe the percentage of recommendations (from the first 12 weeks) that receivers followed up to 12 weeks after the recommendation was made. We do not consider cases where the recommendation was repeated after the week of the initial recommendation. Figure 2 reveals that the likelihood of subscribing to a recommended profile extends over several weeks and, after an initial substantial drop, fades slowly. We observe that the probability does not seem to stabilize even after 12 weeks. Even though the scenario in which a user remembers a *Follow Friday* recommendation after several weeks is unlikely (especially if the recommendation has not been repeated), the probability decay is evident. The reasons behind such a long-lasting decay are difficult to find, since over such a large time scale many other interconnected events co-occur in the network's evolution. We argue that the effect of

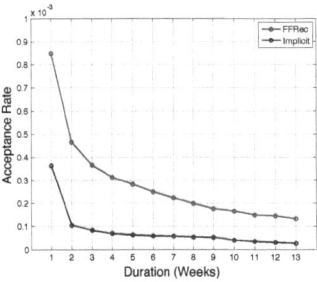
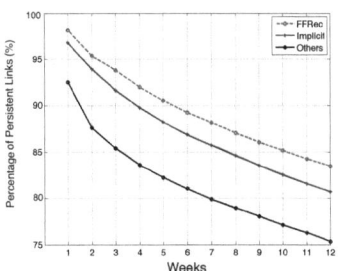

Fig. 2. Acceptance rates n weeks after a recommendation is made

Fig. 3. Longevity of accepted recommendations

the *#ff* recommendation may introduce a perturbation in the network structure that may lead to delayed adoptions.

For instance, a user who received a recommendation before, but did not accept it, may create the link later because other users in his neighboring network accepted it, leading to new opportunities for social triangle closure [16].

To go beyond the acceptance rate of recommendation, we now look at the longevity of the new social ties created as a consequence recommendations. Figure 3 shows the percentage of acceptances that were still in the receivers network after n weeks. The curve labeled as *Others* represents all the users that were followed for reasons not related to the conditions considered in this study. After 12 weeks, we can see that 83% of *#ff* links are still in the receiver's network, versus 80% of links that follow implicit recommendations, and 76% of other follower links. This is an important finding in an environment such as Twitter where social ties have been observed to be very volatile [13].

4.2 Repeated Recommendations

In social sites such as Twitter, it is likely that a single broadcast Tweet may not be seen by many of a user's followers. Repeated recommendations, therefore, are likely to increase the likelihood of a recommendation being accepted, because the follower is more likely to see the recommendation at least once, and because repeated viewings of the recommendation may reinforce it. Figure 4 (a) plots the acceptance rate against recommendation repetitions, where repetitions are counted as recommendations received previously by a user within the time frame covered by the corpus. We consider two cases: when the recommendation is in the form of a *Follow Friday* recommendation only, and when there are only implicit recommendations. The results show that repeated recommendations make a significant difference. We can also see that it takes many implicit recommendations to have a similar effect as even a single *Follow Friday* recommendation, with 15 implicit recommendation having a similar acceptance rate as 1 *Follow Friday* recommendation.

Figure 4 (b) plots the acceptance rate versus the number of distinct recommenders who recommend the same recommended user a receiver, and it shows a similar increase in the acceptance rate as the number of distinct recommenders increases, but with a bigger gap between the *Follow Friday* recommendations and the implicit model.

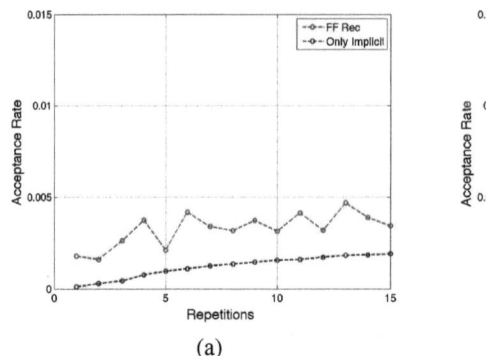 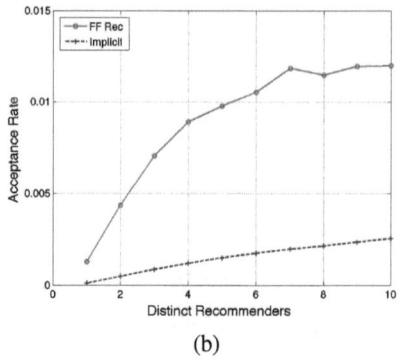

Fig. 4. The effect of repeated recommendations on the acceptance rate. (a) The number of repeated recommendations vs acceptance rate. (b) The number of distinct recommenders vs acceptance rate.

5 Recommender System

In the broadcast recommendation setting given by *Follow Friday*, users are exposed to a large number of friend recommendations every week. In a situation of information overload, the 'good' recommendations are likely to get lost among noisy ones, therefore automated methods are needed to detect the most valuable recommendations. We envision a scenario where all recommendations received by a user in a given week are ranked such that the good recommendations are at the top of the ranking. This essentially corresponds to providing recommender service built on top of the human-generated recommendation system.

In the following, we verify that it is possible to rank Twitter friendship recommendations and surface the most valuable ones, and we evaluate the utility of various features for this task. Secondly, by analysing the predictive value of different features for ranking recommendations, we supplement the analysis of the previous section, giving further insight into features that can predict the creation of a link after a recommendation is made.

5.1 Features for Ranking Recommendations

For each recommendation instance $\langle rdd, rdr, rcv, w \rangle$ we calculate a number of features, and group them into 3 main types: *user-*, *relation-*, and *format-based*.

User-Based Features. These features describe an individual Twitter user, whether it be a *receiver*, a *recommender* or a *recommended user*. We identify two types of user-based features, attention-based and activity-based:

(a) *Attention-Based features* are related to the level of attention given to the user by other users. We measure *popularity* ($followers/(followers + followees)$), the number of times the user has been *mentioned*, the number of people mentioning

the user, the number of times the user has been recommended with a *Follow Friday* hashtag, and the number of distinct recommenders.
(b) *Activity Based features* describe the level of activity of a user. We count the number of *new followees* of a given user for a given week, the *average tweets per day* of the user (over the entire history of the account), the number of *recommendations accepted* by receivers, and the number of distinct recommenders the a receiver has accepted recommendations from. Finally, we also count *mentions*, the number of distinct Twitter accounts mentioned by the user.

Relation-Based Features. These features describe the relation between pairs of users in the $\langle rdd, rdr, rcv \rangle$ triple, based on either profile similarity or communication patterns.

(a) *Communication-based features* describe the level of communication between two users. *Conversational mentions* count the number of times a user mentions another user, and is calculated separately for each pair of users involved in a recommendation. The *number of Follow Friday recommendations*, (the number of times a user recommended another user) is calculated for each pair of users. We also count the number of previous acceptances between the *receiver* and the *recommender*, based on *Follow Friday* recommendations and on *implicit* recommendations. Last, we measure the friendship duration between the receiver and recommender by number of weeks.
(b) *Similarity-based features* describe the similarity of users. Separate *content-based similarity* features calculate similarity between all the tweets of two users, hashtags only, mentions only, and urls only. All these features use the weighted Jaccard similarity coefficient, as in Sudhof et al. [24]. *Geograpical similarity* is a binary feature, set to 1 if actors are in the same country or 0 if not. The location is parsed from the users's declared location using the Yahoo! PlaceMaker API[3].

Format-Based Features. These features describe a recommendation with information of the profile of the users, based either on the context or the format of the recommendation.

(a) The *repetitions* counts the number times the recommendation has been repeated in a *receiver's* timeline, the number of distinct *recommenders* who made the recommendation, and the number of previous weeks in which the recommendation was received.
(b) *Context* features describe the format or the context of the tweets containing the recommendation. We consider the *day of week* on which the recommendation was made, we record whether the recommendation was made in a retweet or not, the number of other users appearing *together* with the recommended user in the *Follow Friday* tweet(s), and the *length of recommendation tweet* (the number of tokens in the tweet, excluding *#ff* hashtags and *@mentions*). Finally, we count the number of urls in the recommendation tweets. Since many of these measures can have more than one value for a given user (i.e. they receive the same recommendation from different people) we calculate both the maximum and minimum for all of them.

---

[3] http://developer.yahoo.com/geo/placemaker/

Most of the features are calculated over a temporal window prior to the recommendation. For all such features, we calculate two versions of the feature: (1) based on the one week period prior to the recommendation (to capture recent activity, similarity, etc), (2) based on all previous weeks in the corpus (to capture longer-term activity).

5.2 Evaluation Methodology

We consider all the unique *Follow Friday recommendation instances* that a given receiver is exposed to at week t and rank them with the aim of putting the 'best' recommendations at the top of the ranking. For the recommender, we set $\Delta = 2$, meaning that we consider the acceptances within two weeks of the recommendation, based on the ground truth of known acceptances. Recommendations accepted after two weeks are considered as not accepted.

Using the *acceptance* information as a ground truth, we evaluate our method by its ability to place the accepted recommendations towards the top of the ranking.

We split the data into training and test sets based on time, with data from weeks 1 to 16 used for training, and weeks 17 to 23 used for testing. We do not test against data from week 24, as we do not have details of the evolution of the followee network one week later.

We use two distinct methods to rank the *Follow Friday* recommendations received by a user in a given week: (1) based on a linear combination of the normalized scores from each feature, and (2) using the confidence score from a supervised classifier trained to classify recommendations as acceptances or rejections. To build the supervised classifier, we take a random subset of recommendations from the training set, ensuring that this subset contains a balanced set of acceptances and rejections. We train a binary classifier on this data using the Rotation Forest algorithm [21] as implemented in the WEKA library [10]. The Rotation Forest method constructs an ensemble of decision trees using random subspaces and principal components transformation applied to the input data [21]. For the linear combination of features, we normalize each list of recommendations by dividing by the feature's maximum value.

We do not normalize the similarity features based on the weighted Jaccard index, since those features are already normalized.

Since a receiver can receive many recommendations in a given week, and can accept one or more of them, we evaluate our various ranking approaches using the standard Information Retrieval measure *Mean Average Precision* (MAP). MAP evaluates a ranking by averaging the precision at the rank position where each relevant item is retrieved [8]. In the evaluation of friendship recommendation, an accepted recommendation is analogous to a relevant item, and a recommendation that is not accepted is non-relevant.

5.3 Results

Table 3 shows performance of the Rotation Forest classifier, compared against the linear combination and a random baseline. The linear combination performs very poorly, while the Rotation Forest gives encouraging performance, with a MAP of almost 0.5, showing that machine learning approaches can give good results for this task.

Table 3. Recommendation Mean Average Precision using all features

Ranking	MAP
Rotation Forest	0.4959
Linear Combination	0.0565
Random	0.0368

Table 4. Recommendation performance for subsets of features (Rotation Forest)

Features	MAP
All	0.4959
User-based	0.0741
Relation-based	0.3976
Format-based	0.0615
User + Relation	0.5176
User + Format	0.0790
Relation + Format	0.3787

In Table 4, we show the results when using various subsets of features, according to the grouping of features described in Section 5.1. The *relation-based* features are the most discriminative for friend recommendation, while the *format-based* features are not useful at all, and always harm performance. Finally, the *user-based* features, while they do not perform particularly well on their own, improve performance when combined them with the *relation-based* features. Overall, the best performing set of features is *user-based + relation-based* (i.e. ignoring the format-based features), with a MAP of almost 0.52.

Due to space we do not show detailed results for individual features, but the single best performing individual feature is the previous behaviour of the *receiver* in accepting recommendations from the *recommender*. Other relation-based features based on similarity (not communication) are also important, however, and the results in Table 4 show that optimal performance is achieved when we also consider user-based features.

6 Conclusions

In this paper, we describe the first study of the *Follow Friday* phenomenon, which aims to better understand the dynamics of a large scale collective process of human-generated link recommendations, and to understand the features and conditions that may predict the creation of new social links.

Furthermore, in contrast with other studies of link prediction in social media, we use a direct and reliable ground-truth of acceptances and rejections, based on real user behavior. We compare acceptance rates of *Follow Friday* recommendations with baseline conditions where (a) another user is mentioned, without being explicitly recommended, and (b) we simulate a condition where there is no observed (explicit or implicit) recommendation made via Twitter. Through this comparison, we show that explicit *Follow*

Friday recommendations have a large, measurable, effect on who users choose to follow on Twitter. We also show that the effect of a recommendation (explicit or implicit) lingers for a number of weeks, that repeating recommendations has a strong effect, and that ties formed after *Follow Friday* recommendations tend to have more longevity than other ties, an important finding in Twitter, where social ties are quite volatile.

To surface more valuable recommendations above others, we propose an automated recommender system based on a number of features, which we group into three distinct categories: *user-based*, *relation-based* and *format-based*. We show that the most discriminative features for friendship recommendation are those features based on communication and similarity between users. In particular, past behavior in following recommendations coming from a given recommender is the most predictive feature of future recommendation acceptance.

Acknowledgments. This research is partially supported by European Community's Seventh Framework Programme FP7/2007-2013 under the ARCOMEM and Social Sensor projects, by the Spanish Centre for the Development of Industrial Technology under the CENIT program, project CEN-20101037 "Social Media", and by Grant TIN2009-14560-C03-01 of the Ministry of Science and Innovation of Spain.

References

1. Abel, F., Gao, Q., Houben, G.-J., Tao, K.: Analyzing user modeling on twitter for personalized news recommendations. In: Konstan, J.A., Conejo, R., Marzo, J.L., Oliver, N. (eds.) UMAP 2011. LNCS, vol. 6787, pp. 1–12. Springer, Heidelberg (2011)
2. Aiello, L.M., Barrat, A., Schifanella, R., Cattuto, C., Markines, B., Menczer, F.: Friendship prediction and homophily in social media. ACM Trans. Web 6 (June 2012)
3. Aiello, L.M., Deplano, M., Schifanella, R., Ruffo, G.: People are Strange when you're a Stranger: Impact and Influence of Bots on Social Networks. In: Proceedings of the 6th AAAI International Conference on Weblogs and Social Media, ICWSM (2012)
4. Backstrom, L., Leskovec, J.: Supervised random walks: predicting and recommending links in social networks. In: Proceedings of the Fourth ACM International Conference on Web Search and Data Mining, WSDM (2011)
5. Bakshy, E., Rosenn, I., Marlow, C., Adamic, L.: The role of social networks in information diffusion. In: Proceedings of the 21st International Conference on World Wide Web, WWW 2012. ACM, New York (2012)
6. Barrat, A., Barthlemy, M., Vespignani, A.: Dynamical Processes on Complex Networks, 1st edn. Cambridge University Press, New York (2008)
7. Cha, M., Haddadi, H., Benevenuto, F., Krishna Gummadi, P.: Measuring User Influence in Twitter: The Million Follower Fallacy. In: Proceedings of the 4th International AAAI Conference on Weblogs and Social Media, ICWSM (2010)
8. Croft, B., Metzler, D., Strohman, T.: Search Engines: Information Retrieval in Practice, 1st edn. Addison-Wesley Publishing Company, USA (2009)
9. Esparza, S.G., O'Mahony, M.P., Smyth, B.: On the real-time web as a source of recommendation knowledge. In: Proceedings of the Fourth ACM Conference on Recommender Systems, RecSys (2010)
10. Gewehr, J.E., Szugat, M., Zimmer, R.: BioWeka - extending the Weka framework for bioinformatics. In: Bioinformatics/Computer Applications in the Biosciences (2007)

11. Hannon, J., Bennett, M., Smyth, B.: Recommending twitter users to follow using content and collaborative filtering approaches. In: Proceedings of the Fourth ACM Conference on Recommender Systems, RecSys 2010. ACM, New York (2010)
12. Hutto, C.J., Yardi, S., Gilbert, E.: A longitudinal study of follow predictors on twitter. In: Proceedings of the SIGCHI Conference on Human Factors in Computing Systems, CHI 2013, pp. 821–830. ACM, New York (2013)
13. Kwak, H., Chun, H., Moon, S.: Fragile online relationship: a first look at unfollow dynamics in twitter. In: Proceedings of the 2011 ACM Annual Conference on Human Factors in Computing Systems, CHI (2011)
14. Kwak, H., Lee, C., Park, H., Moon, S.: What is Twitter, A Social Network or a News Media? In: Proc. 19th ACM International Conference on World Wide Web, WWW (2010)
15. Lee, K., Eoff, B., Caverlee, J.: Seven months with the devils: A long-term study of content polluters on twitter. In: Proceedings of the Fifth International Conference on Weblogs and Social Media (ICWSM), AAAI
16. Leskovec, J., Backstrom, L., Kumar, R., Tomkins, A.: Microscopic evolution of social networks. In: Proceedings of the 14th ACM SIGKDD International Conference on Knowledge Discovery and Data Mining (KDD). ACM, New York (2008)
17. Liben-Nowell, D., Kleinberg, J.: The link prediction problem for social networks. In: Proceedings of the Twelfth International Conference on Information and Knowledge Management, CIKM 2003, pp. 556–559. ACM, New York (2003)
18. Lu, L., Jin, C.-H., Zhou, T.: Effective and efficient similarity index for link prediction of complex networks. arXiv:0905.3558 (2009)
19. Lü, L., Zhou, T.: Link prediction in complex networks: A survey. Physica A (2011)
20. Petrović, S., Osborne, M., Lavrenko, V.: Rt to win! predicting message propagation in twitter. In: Proceedings of the Fifth AAAI International Conference on Weblogs and Social Media (ICWSM)
21. Rodríguez, J.J., Kuncheva, L.I., Alonso, C.J.: Rotation forest: A new classifer ensemble method. IEEE Trans. Pattern Analysis and Machine Intelligence (2006)
22. Schifanella, R., Barrat, A., Cattuto, C., Markines, B., Menczer, F.: Folks in folksonomies: social link prediction from shared metadata. In: Proceedings of the Third ACM International Conference on Web search and Data Mining, WSDM (2010)
23. Shalizi, C.R., Thomas, A.C.: Homophily and contagion are generically confounded in observational social network studies. Sociological Methods & Research (2011)
24. Sudhof, M.: Politics, twitter, and information discovery: Using content and link structures to cluster users based on issue framing 11 (2012)
25. Suh, B., Hong, L., Pirolli, P., Chi, E.H.: Want to be retweeted? large scale analytics on factors impacting retweet in twitter network. In: SocialCom/PASSAT 2010 (2010)
26. Zhou, T., Lu, L., Zhang, Y.-C.: Predicting Missing Links Via Local Information. European Physical Journal B. Special Issue: The Physics Approach to Risk: Agent-Based Models and Networks 71(4) (2009)

A Divide-and-Conquer Approach for Crowdsourced Data Enumeration

Hideto Aoki and Atsuyuki Morishima

University of Tsukuba
{aoki,mori}@slis.tsukuba.ac.jp

Abstract. Crowdsourced data enumeration, in which the Web crowd is requested to enumerate data items within a specified range, is important in many Web applications such as hotel reviews. This paper presents a processing method for crowdsourced data enumeration on microtask-based crowdsourcing platforms. A general approach to achieving a high recall in data enumeration is to apply the divide-and-conquer principle. However, how to apply the principle to data enumeration on microtask-based crowdsourcing platforms is not trivial. The proposed method is unique in that the workers join the process of generating smaller tasks in a divide-and-conquer fashion, and the programmer does not need to provide many microtasks in advance. This paper explains the method, provides theoretical results to show the method works well with microtask-based platforms, and explains our experimental results that suggest the proposed method can achieve higher recalls and produces appropriate tasks for microtask-based crowdsourcing.

1 Introduction

Crowdsourced data enumeration is a form of human-powered search, in which humans are requested to enter data items satisfying a given condition. Compared to other forms of human-powered searches, crowdsourced data enumeration is characterized by large output volume, and recall is of particular concern. Crowdsourced data enumeration is a key component in many Web applications, because it is often the case that algorithms for extracting data items from Web content are difficult to implement, and that data items do not exist in machine-readable storages. For example, Wikipedia terms and hotel names in hotel review Web sites [10] are collected by crowdsourced data enumeration.

This paper presents a processing method for crowdsourced data enumeration on microtask-based crowdsourcing platforms (such as Amazon's mechanical Turk). Such platforms attract much attention today. In microtask-based crowdsourcing, *requesters* insert microtasks to a task pool on the crowdsourcing platform, and *workers* perform the microtasks in the task pool. A simple approach to data enumeration on a microtask-based crowdsourcing platform is just to provide a simple microtask that requests the crowd to enter data items (Figure 1). This approach works well for general human-powered searches for finding one or more query results; however, it is unsuitable for crowdsourced data enumeration

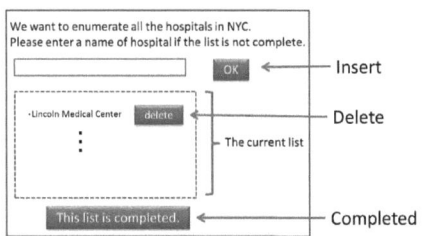
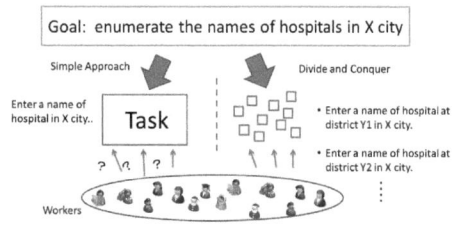

Fig. 1. A task for data enumeration

Fig. 2. Data enumeration by a set of smaller tasks

in which recall is important. For example, this simple approach will generate low recall of hospital names in a large city, because workers cannot easily identify the enumerated data items or missing items in such large enumerated datasets.

A common approach to solve this kind of problem is to apply the *divide-and-conquer* principle. In the microtask-based crowdsourcing framework, we can provide a set of different microtasks, where each task requests workers to enter data items within smaller ranges. For example, when enumerating the names of hospitals in a city, the microtasks can request workers to enumerate hospitals in different districts (Figure 2). However, several questions arise. First, who defines the set of microtasks, given that the task sets differ among application domains? Second, can we guarantee that the set of microtasks achieves complete data enumeration?

The contributions of this paper are twofold:

(1) The divide-and-conquer data enumeration in the microtask-based crowdsourcing framework. To our knowledge, this paper is the first to present the application of the divide-and-conquer principle to data enumeration in the microtask-based framework. The method is unique in that workers join the process of gradually generating smaller tasks for enumerating data items in a divide-and-conquer fashion; therefore the programmer does not need to provide a specific set of microtasks in advance. The underlying idea is that workers who perform the task for data enumeration have some knowledge on the application domain and can assist the division of data enumeration into smaller problems while performing enumeration. As mentioned, how to apply the divide-and-conquer principle to the microtask-based crowdsourcing is nontrivial. To achieve this, the proposed method takes as input a *task-generation plan* that summarizes the hierarchical structure of the application domain. This plan, together with the data entered by the workers, is used to generate smaller tasks.

Note that the divide-and-conquer principle is not a magic wand; there are cases in which it is difficult to divide the problem into smaller problems. We note that our proposed method is a generalization of the simple approach and does not enforce us to divide problems.

(2) Evaluation. The proposed method is theoretically and empirically evaluated. First, we theoretically show that the proposed method is correct, in the sense that it outputs the same set of data items as the simple approach if workers perform the tasks appropriately. Second, we show the result of an experiment using a set of real data to confirm the intuition that the divide-and-conquer approach raises the recall. In addition, we argue that the proposed method is highly appropriate for microtask-based crowdsourcing, because the average time for performing tasks for the proposed method is shorter and the tasks can be performed in parallel.

The remainder of the paper is as follows. Section 2 explains related work. Section 3 presents the microtask-based crowdsourcing framework and discusses the simple approach in the framework. Sections 4 and 5 introduce and evaluate the proposed method, respectively. Section 6 is the summary.

2 Related Work

Previous researches have addressed how to process a variety of human-powered operations. For example, several papers [7] [8] [2] [9] addressed how to efficiently process human-powered database operations such as filtering, joins, and sorts. Among these, human-powered filtering is closely related to human-powered data enumeration, since both operations enumerate data items that satisfy a given condition. However, filtering assumes that the result is a subset of data items stored in the machine; this assumption is absent in crowdsourced data enumeration (more generally, in human-powered searching).

As mentioned in the introduction, if the result of a human-powered search requires no recall, and a worker in the crowd knows some data items, a simple task will suffice. However, if recall is important and the result contains many data items, more sophisticated strategies are required to process the operation.

Crowdsourced data enumeration has been considered in several previous works. In [6], we first proposed that workers could be requested to join the divide-and-conquer process and demonstrated that it is feasible. Trushkowsky et al. [11] adopted species estimation techniques from statistics and biology literature to manage the execution of enumeration queries. In particular, they developed methods that can be used to estimate the completeness of data enumeration, which is not considered in our approach. Instead, we directly ask workers whether the data enumeration is completed or not, because in our approach, each task deals with only a part of data enumeration that is small enough for workers to handle. An interesting future work would be the combination of our method with such estimation techniques.

Kulkarni et al. [4] applied the divide-and-conquer approach to crowdsourcing in a more general setting and presented an experimental evaluation. They found that it is difficult for the crowd to appropriately divide the problem without an appropriate guidance. Our approach is consistent with their finding in that our task generation plan serves as a guidance for generating smaller tasks and our experimental results show that the proposed method worked well.

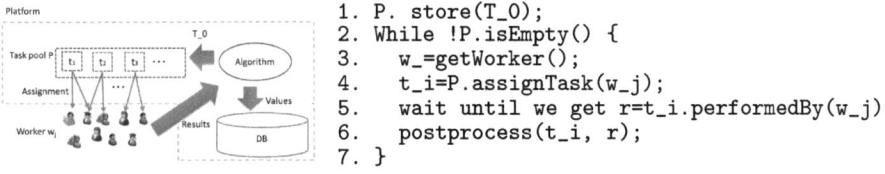

Fig. 3. MCF overview

Fig. 4. MCF algorithm

```
1. P.store(T_0);
2. While !P.isEmpty() {
3.    w_=getWorker();
4.    t_i=P.assignTask(w_j);
5.    wait until we get r=t_i.performedBy(w_j);
6.    postprocess(t_i, r);
7. }
```

Another important issue in the data-centric crowdsourcing is data quality. Our proposed technique can be combined with existing techniques for improving data quality. For example, many crowdsourcing adopt majority voting [7], a technique that relies on the law-of-large-numbers. Another approach is a coordination game [5] [1], in which rational workers give appropriate values. [3] provides a game-theoretic analysis of games with a purpose for obtaining data, and shows that a simple change of the incentive structure can affect the obtained data.

3 Microtask-Based Crowdsourcing

This section first models a framework for microtask-based crowdsourcing and then explains a simple approach to data enumeration in the framework.

3.1 Microtask-Based Crowdsourcing Framework

The modeled framework, referred to as MCF, provides an abstraction of widely used microtask-based crowdsourcing services. In MCF, *workers* perform *microtasks* (abbreviated to tasks) that exist in a *task pool* (Figure 3). Here, we use t_i, w_j, and P to denote a task, a worker and the task pool, respectively. Figure 1 illustrates an example of a task. Each task in P is not deleted unless removed explicitly. Therefore, a task in P can be performed by more than one worker.

Figure 4 shows the MCF algorithm. In Line 1, the task pool P is assigned an initial set T_0 of microtasks. Then, while P contains a task (Line 2), we obtain a worker (Line 3), assign her a task (Line 4), and (if the task is completed) postprocess the task (Line 6). During postprocessing, the task result is stored in the database. For simplicity, the code waits until each task has been performed before assigning other tasks to workers. However, an event-driven version of the code, in which tasks are assigned to other workers without waiting, is a possible extension. The discussions in this paper remain valid in both versions.

3.2 Task Representation

In MCF, each task t_i is an instance of a *task class*. A task class provides a template for instantiating tasks with specified parameters. For example, let *EntryTask(item_type, scope)* be a task class. Then, *EntryTask(*`ResearchTopic`, `"University of Tsukuba"`*)* and *EntryTask(*`Hospital`, `"NYC"`*)* are tasks (instances) of the task class. Here, *item_type* is the type of data items, and

```
postprocess(t, r){
    Let r be the result of t:TaskEntry(item_type, scope)
    switch r.pressed_button {
    case insert:
        DB.insert(item_type, r.dataitem);
        break;
    case delete:
        DB.delete(item_type, r.dataitem);
        break;
    case completed:
        P.remove(t);
    }
}
```

Fig. 5. Postprocess for a simple data enumeration

scope denotes the scope of data items to be enumerated. For example, *EntryTask(*Hospital*, "*NYC"*)* is a task to accept the names of hospitals in NYC.

How each task is presented to workers is defined by the task class. For example, Figure 1 is a screen shot of the task *EntryTask(*Hospital*, "*NYC"*)*, a task of *EntryTask* class. For each task of the *EntryTask* class, workers are required to perform one of the followings:

- Insert: Enter the name of a hospital that does not exist in the current list. Then, click the "Insert" button.
- Delete: Select a name that is incorrect or duplicate in the current list. Then, click the "Delete" button.
- Completed: Click the "Completed" button if a worker assesses the current list as completed.

4 Data Enumeration in the Microtask-Based Crowdsourcing Framework

We first define the concept of data enumeration and implement a simple crowdsourced data enumeration in MCF.

Definition 1. Let *Items(item_type, scope)* be the complete set of data items of *item_type* in *scope*. Then, we define data enumeration of *item_type* in *scope* as an operation to enumerate all items in *Items(item_type, scope)*.

For example, the result of data enumeration of research topics (the item type) at University of Tsukuba (the scope) is *Items(*Research_topic*, "*University of Tsukuba"*)*. □

In MCF, we can implement a simple crowdsourced data enumeration (S-DE) to compute *Items(Research_topic, "Univ of Tsukuba")* as follows.

1. Set $\{EntryTask(\text{Research\_topic, "Univ of Tsukuba"})\}$ to T_0.
2. Implement P.assignTask(w) by a function that always returns *EntryTask(*Research_topic*, "*Univ of Tsukuba"*)* (i.e., the only task in P).
3. Implement postprocess(t, r) by the code shown in Figure 5. In this code, DB.insert(item_type, r.dataitem) stores the data item entered for the task into the relation whose name is item_type (e.g., ResearchTopic).

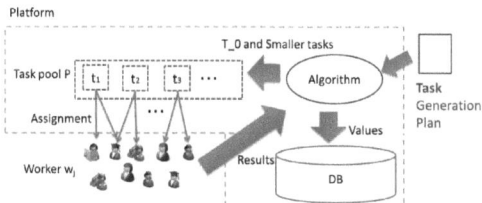
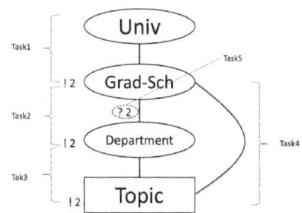

Fig. 6. How DC-DE works in MCF

Fig. 7. Illustration of a task generation plan

4.1 Problems

The simple approach is appropriate provided that recall is unimportant. However, when recall is important and the result is large, this approach fails for the following reasons. First, a worker entering data (using the Insert button) may not readily identify missing items in the current list or suggest an appropriate data item to raise recall. Second, a worker clicking the Delete button may not observe inappropriate data items in the current list. Finally, a worker clicking the Completed button may fail to determine whether the task is completed if many data items exist.

5 Proposed Method

In this section, we propose a method called DC-DE. The points of DC-DE are two-fold. First, it applies the divide-and-conquer approach to crowdsourced data enumeration and uses the smaller tasks for data enumeration. Second, workers not only enumerate data items, but also join the divide-and-conquer process.

Smaller Tasks for Data Enumeration. To explain the smaller tasks generated by DC-DE, we introduce the concept of *covers* among scopes.

Definition 2. Given *item_type* and one of its scope s, let $s_1, \ldots s_n$ be other scopes for *item_type*. Then, we say that the set $\{s_1, \ldots s_n\}$ covers s on *item_type* if the following holds.

$$Items(item\_type, s) = Items(item\_type, s_1) \cup \ldots \cup Items(item\_type, s_n).$$

If set $\{s_1, s_2 \ldots s_n\}$ covers s on *item_type*, the divide and conquer approach can be implemented by setting $T_0 = \{$ *EntryTask(item_type, s_1)*, ..., *EntryTask(item_type, s_n)*$\}$. For example, if we know that University of Tsukuba consists of Research Institutes A, B, and others, T_0 would be $\{$ *EntryTask(*ResearchTopic*, "Institute A")*, *EntryTask(*ResearchTopic*, "Institute B")* ,...$\}$. Note that each task in T_0 is a task to enumerate data items in a more limited scope and the number of the enumerated data items is smaller. Therefore, it is expected that the recall becomes higher than S-DE.

```
[Univ{University of Tsukuba}<uname>,
    Grad-Sch[!2, *]<gname>,
    Department[!2, ?2]<dname>,
    *Topic[!2]<topic>]
]
```

Fig. 8. Description of a task generation plan

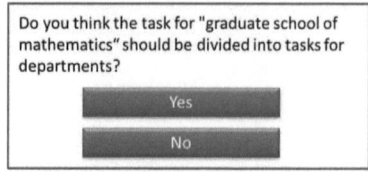

Fig. 9. Example of a DivisionTask

Divide and Conquer by the Crowd. The problem is that it is not easy to provide the set of tasks for the divide-and-conquer approach. We need the domain knowledge to generate the set of tasks. Each domain has a different set of tasks and it is impossible to provide tasks for all domains in advance. The idea underlying DC-DE is to have workers join the divide-and-conquer process, because workers who enumerate data items are expected to have knowledge on the domain and can help the divide-and-conquer process.

The question is how to implement such a mechanism in MCF. Our idea is to have the algorithm take as an additional input a *task generation plan*, which gives a summary of the hierarchical structure of the application domain. Figure 6 illustrates how DC-DE generates smaller tasks with a given task-generation plan in the MCF. First, DC-DE takes as input a task generation plan (explained in Section 5.1). Then, DC-DE matches the given task generation plan with the results of performed microtasks to generate and insert smaller tasks into the task pool (Section 5.2). The smaller tasks are then assigned to workers. Note that the programmer does not need to provide the actual set of smaller tasks in advance.

5.1 Task Generation Plan

A task generation plan specifies how to generate smaller tasks during data enumeration. Essentially, it explains a hierarchy structure for the cover relationships among ranges. For example, the research topics of a university can be covered by the union of all research topics of research institutes of the university.

Figure 7 illustrates what is described in a task generation plan. The plan describes the hierarchical structure for the cover relationships among research topics of a university. Note that the shape of the Topic node is different from those of the others. We call the node in the shape the *terminal node*, which represents the type of data items to be enumerated. Therefore, the plan is used to generate small tasks to enumerate research topics.

In Figure 7, each edge between nodes corresponds to a data entry task for different data item types. Because the plan has four edges, it describes that the following four types of tasks will be generated in the process. (Task 1) Tasks for enumerating graduate schools in the range of each university. (Task 2) Tasks for enumerating departments in the range of each graduate school. (Task 3) Tasks for enumerating research topics in the range of each department. (Task 4) Tasks for enumerating research topics in the range of each graduate school.

Associated to each node is the number of *completion agreements* (!x), which means that we decide that the task is completed if x workers agree on the completion. For example, "!2" associated to the department node in Figure 7 states that we decide that each task for enumerating departments is completed if two workers agree on the completion about the same state of the current list.

Associated to some of the edges is the number of *generation agreements* (?y), which means that we decide to generate the task if y workers agree on the generation. For example, "?2" associated to the edge corresponding to Task 2 (in Figure 7) states that we decide to generate Task 2 if two workers agree on the generation. Each generation agreement is performed with a *scope-division task* (explained below), not with *EntryTask* for accepting data items.

The reason that we introduced the generation agreement is that the hierarchy of the cover relationships is not necessarily uniform in some applications. For example, it is possible that the graduate school of physics has departments but the graduate school of social science has no departments. In that case, if we generate Task 3 for all graduate schools, we would generate inappropriate tasks for graduate school of social science.

Notation. Figure 8 is a task generation plan encoding the information of Figure 7. A task generation plan is described as a list $[n_1, n_2, \ldots, n_m]$ of nodes. The node list represents the sequence of nodes from the root (Univ) to the leaf (Topic). The prefix * of Topic node states that Topic is the terminal node of the plan and the plan is used to generate smaller tasks to enumerate research topics. In the square brackets following each node, we describe (1) the number of completion agreements (!x), (2) the number of generation agreements (?y), and (3) whether the node has a direct edge to the terminal node (*).

Additional information included in the task generation plan, not illustrated in Figure 7, is described below.

1. We write the initial value for the top-most node in {...}. For example, "University of Tsukuba" is the initial value for Univ node.
2. We write in < ... > a relational attribute name to store the task results. The name of the relation for the relational attribute is the same as the node name. For example, <gname> following the Grad-Sch node states that the task result is stored in the gname attribute of Relation Grad-Sch.

Generated Tasks. Given a task generation plan, DC-DE generates two types of tasks.

(1) Data Entry Task (EntryTask) In *EntryTask(item_type, scope)* (Figure 1), workers perform one of the followings:

- (*Insert*) Enter a data item of *item_type* in the specified *scope*.
- (*Delete*) Delete a data item in the current list if it is incorrect or a duplicate.
- (*Completed*) Decide whether all data items in the specified *scope* have been entered in the current list.

(2) Scope-Division Task (DivisionTask) A scope-division task, denoted by *DivisionTask(scope, n_i)*, asks a worker to decide whether *scope* should be divided into scopes in terms of n_i. Figure 9 is a screenshot of *DivisionTask(*"Mathematics", Department*)*, which asks a worker whether the task for the graduate school of mathematics can be divided into tasks for departments. The worker chooses one of the two buttons to answer whether it can be divided or not.

5.2 Task Generation Algorithm

This section explains how DC-DE generates smaller tasks with a given task generation plan $[n_1\{v_1\}, n_2, \ldots, n_m]$. The algorithm can be implemented in The MCF shown in Figure 4. Therefore, it can be applied to any microtask frameworks. The details of how to implement DC-DE in MCF are as follows.

1. Let T_0 be $\{EntryTask(n_2, v_1)\}$ where v_1 is the initial value for the topmost node n_1 and n_2 is the child of n_1. For example, given the task generation plan shown in Figure 8, $t_0 = \{EntryTask($Grad-Sch, "Tsukuba University"$)\}$.
2. Implement P.AssignTask(w) by a function that assigns w to the oldest task inserted into P. The assignment policy performs the width-first traversal of the hierarchical structure of scopes.
3. Implement postprocess(t, r) by the code shown in Figure 10. The function handles the following four cases:

   [Case 1: t is an EntryTask and the pressed button is "Insert"]
   The result value of the task is inserted into the database (Line 6). Then, it generates a new task for the child node using the result value if the node is not the terminal node n_m. Then, it generates a new task for the child node using the result value if the node is not the terminal node n_m. For example, assume that the current task corresponds to the edge $n_1 - n_2$ and the result value for the task is v_2. Then, the child task is $EntryTask(n_3, v_2)$ that corresponds to $n_2 - n_3$. There are two cases: If the task (*i.e.*, the edge $n_2 - n_3$) is not associated with the number of task generation agreements (?y) in the plan, the task is immediately inserted into P (Line 8). Otherwise, $DivisionTask(v_2, n_3)$ is generated and inserted into P (Line 10). The task will be used to ask workers whether the original task can be divided into a smaller set of tasks in terms of n_3.

   [Case2: t is an EntryTask and the pressed button is "Delete"] We remove the chosen value ($r.dataitem$) from the relation n_i in the database (Line 13). We also remove the tasks generated for the value (Line 14).

   [Case 3: t is an EntryTask and the pressed button is "Completed"] We remove the task from P if the number of workers who pressed the button reaches the necessary number of agreements specified by $!x$ (Line 17).

   [Case 4: t is a DivisionTask] If the number of workers who pressed "Yes" in performing $DivisionTask(r.data\_item, n_i)$ reaches the necessary number of agreements (?y), insert $EntryTask(n_i, r.dataitem)$ into P (Line 22).

```
1. postprocess(t, r){
2. Let [n_1{scope}, ..., item_type] be the task generation plan.
3.   If t is EntryTask(n_i, scope) {
4.     switch(r.pressed_button) {
5.       case Insert:                    // Case 1
6.         DB.insert(n_i, r.dataitem);
7.         if ((n_{i+1}.?y==0) && (n_i != n_m)){
8.           P.insert(EntryTask(n_{i+1}, r.dataitem));
9.         }else if (n_i != n_m){
10.          P.insert(DivisionTask(n_{i+1}, r.dataitem));
11.        }
12.       case Delete:                   // Case 2
13.         DB.delete(n_i, r.dataitem);
14.         P.delete(EntryTask(n_{i+1}, r.dataitem));
15.       case completed:                // Case 3
16.         if(completion is agreed) {
17.           P.delete(t);
18.         }
19.    }
20.  } else if t is DivisionTask(n_i, r.dataitem) { // Case 4
21.       if (scope division is agreed) {
22.         P.insert(EntryTask(n_i, r.dataitem));
23.       }
24.  }
25. }
```

Fig. 10. Postprocess for DC-DE

6 Evaluation

This section evaluates the DC-DE to answer several important questions. The first and most important question is whether DC-DE is a correct algorithm for applying the divide-and-conquer principle to data enumeration in MCF. To answer this question, we theoretically assess whether DC-DE is equivalent to S-DE, i.e., whether DC-DE and S-DE return the same set of data items if workers perform tasks appropriately. We also discuss the tradeoffs between the quality of results and number of agreements specified in the task generation plan. The second question is whether DC-DE actually raises recall, although it is intuitively obvious that the divide-and-conquer approach is effective to raise recall. The third question is whether DC-DE generates appropriate tasks for microtask-based crowdsourcing. To answer the second and third questions, we conducted an experiment with a real set of data on the Web. The results showed that (1) the divide-and-conquer approach can raise recall, and that (2) the divide-and-conquer approach suits for microtask-based crowdsourcing, in the sense that it generates smaller tasks and workers can perform those tasks within a smaller average time.

6.1 Theoretical Analysis

Theorem 1. Given an appropriate task generation plan for *scope* and *item_type*, the DC-DE algorithm outputs $Items(item\_type, scope)$ if workers give correct values in performing tasks.

Proof. Figure 11 shows DC-DE algorithm with assertions (conditions) Q, R and S for the proof. Let V be the set of values stored in the database as values to be included in the result. We prove that if the precondition Q holds when we start the loop in Figure 11, the postcondition R holds when the algorithm stops. Here, Q and R are as follows:

(1) Precondition Q: $P = \{EntryTask(n\_2, scope)\}$
(2) Postcondition R: V contains $Items(item\_type, scope)$.

To prove this, we introduce the invariant S that holds at the beginning and ends of the each iteration of the loop.

(3) S: $Items(item\_type, scope) \subseteq \bigcup_{t_i \in P} Items(item\_type, t_i.scope) \cup V$

Note that V is empty until a task for the terminal node has been performed. In short, S states that it is guaranteed that an appropriate set of tasks exists in the task pool P to enumerate the items that are not included in V. If we prove that S holds at each iteration, we can say that R holds when the algorithm stops, because $S \land !P.isEmpty$ means that all of the appropriate tasks in the task pool were successfully completed.

Let S holds at the beginning of the iteration. In the code, the postprocess function that called in Line 6 is the only part to affect S, because the tasks are inserted and deleted only in the function (Figure 10). We ignore Case 2 (See Section 5.2) because we assume that workers do not make a mistake. In Cases 1 and 4, some tasks are added to P and no task is removed. Therefore, the operation does not spoil S. In Case 3, a task is removed from the task pool. However, the task is removed only if workers agree on the completion of the task. If the task was generated for the terminal node, this means that the items covered by the task were inserted into V. Otherwise, this means that the smaller set of tasks that covers the range of the removed task exists in P. In both cases, the operation preserves S.

The loop always ends because the number of tasks in P eventually becomes zero. As discussed in the previous paragraph, Cases 1 and 4 adds tasks to P. However, there is a limit to the number of the tasks if the number of data items is finite, because the number of generated tasks is essentially the number of the internal nodes of a tree generated by the hierarchical structure of the given task generation plan. Because the postprocess either adds or deletes tasks to/from P in any case, the number of tasks in P eventually becomes zero. □

```
1.    P.store(T_0);
//* Q ∧ S *//
2.    While !P.isEmpty() {
//* S ∧ !P.isEmpty() *//
3.        w=getWorker();
4.        t_i=P.assignTask(w);
5.        wait until we get r=t_i.resultOfTaskBy(w);
6.        postprocess(t_i, r);
7.    }
//* S ∧ P.isEmpty → R *//
```

Fig. 11. DC-DE algorithm with assertions

An important feature of DC-DE is that workers join the process of making smaller tasks. Therefore, it is a concern how often they generate inappropriate tasks for data enumeration. Obviously, the following theorem holds.

Theorem 2. Given a task t and a task generation plan p, let $correct(t)$ be a predicate to state that t is a correct task. Also, let N be the smallest number of agreements specified in p. Then, the necessary condition for $correct(t)$ to hold for every t generated with p is that workers do not give incorrect answers N times successively to the same task whose result is used to generate t. □

This theorem states that it is important to set the number of agreements appropriately. Our proposed method is tolerant to incorrect tasks to some extent, because the EntryTask allows workers to delete incorrect data items caused by incorrect tasks. However, it causes increase of the cost due to performing unnecessary tasks, especially if the tasks are generated for the nodes close to the root in the task generation plan. As shown later, although determining the best number for the agreements is beyond of the scope of the paper, our experimental results show that even with two agreements, the number of inappropriate tasks generated was small.

6.2 Experimental Settings

For the experiment, we constructed a dedicated microtask-based crowdsourcing platform that implements MCF. Note that our purpose is not to conduct extensive experiments to choose best parameters but to check if we can obtain the results to support the intuition that the divide-and-conquer approach works well for handling data enumeration. The details are as follows.

Data Items to Enumerate. We chose a set of public swimming pools in Niigata Prefecture, Japan, as the data items for data enumeration, i.e., workers enumerate all data items in *Items(*`public_swimming_pool`*, "*`Niigata Prefecture`*")*.

The reason we chose the data items is that we knew that there are no Web page to provide the complete list of the data items at the time we conducted the experiment. This is important because workers can use Web search engines in performing tasks.

Before conducting the experiment, the authors carefully searched the Web for the public swimming pools to make the correct set of data items. The number of pools in the correct set was 91.

Workers. We made two groups of workers, each of which has six workers. Workers in Groups A and B performed the tasks of S-DE and DC-DE, respectively. Communication among workers during the experiment was not permitted.

Tasks. In S-DE, we used *EntryTask(*`SwimmingPool`*, "*`NiigataPrefacture`*")* as the task. In DC-DE, we used the task generation plan shown in Figure 12. For a fair comparison, we required an agreement by two workers to determine whether data enumeration was completed even for S-DE.

```
[ prefecture {Niigata Prefecture},
  Area[!2]<area_name>,
  Municipality[!2,*]<municipality_name>,
  Ward[!2,?2]<ward_name>,
  *Public_Pool[!2]<Public_Pool_name>
]
```

Fig. 12. Task-generation plan for the experiment

Table 1. Recall and precision

	recall	precision	F-measure
S-DE	0.65	0.97	0.78
DC-DE	0.78	0.92	0.85

Table 2. Number of performed tasks

	EntryTask			DivisionTask	Total
	insert	completed	delete		
S-DE	82	5	17	0	104
DC-DE	179	141	44	64	428
S-DE'	91	2	0	0	91
DC-DE'	133	86	0	60	279

6.3 Results of the Experiment

(1) Recall and precision. Table 1 shows the recall and precision in the experiment. As expected, the recall of DC-DE was higher than that of S-DE. The precision is slightly reduced, but it is natural because recall and precision is negatively correlated in general. We carefully examined the log file to find the reason why precision is slightly less in DC-DE. Then, we find that two workers in Group A were mainly responsible for enumerating data items, while the remaining four checked whether the data items in the list were appropriate. This suggests that there are many factors to affect the precision.

(2) Total number of tasks. The first and second line in Table 2 shows the numbers of performed tasks in the experiment. Because workers sometimes enter and delete incorrect data items, the numbers are larger than the ideal numbers of performed tasks (S-DE' (Line 3) and DC-DE' (Line 4)), in which we assume that workers take no mistake. We found that in 33/44 cases where "Delete" is clicked, workers clicked the button to delete the data items they entered just before or to delete the data items for terminal nodes. This means that the mistakes did not produce many inappropriate tasks. The number of tasks in which workers clicked the "completed" button is much larger in DC-DE than in S-DE because at each level we need the agreement of completion in DC-DE while workers need to press the button only when the list is completed in S-DE. Note that many tasks generated by DC-DE can be performed in parallel so that although the number of tasks for DC-DE is larger than that for S-DE, we can complete all tasks for DC-DE faster if we have many workers.

(3) Elapsed time for performing tasks. The required time for each task is an important factor of successful crowdsourcing. Figure 13 shows the distribution of elapsed times required for completing tasks. The average of the elapsed times of S-DE and DC-DE were 97 and 58 seconds, respectively. In DC-DE, 92% of tasks were completed in 180 seconds. The result confirmed that the divide-and-conquer principle is effective for generating microtasks appropriate for crowdsourcing.

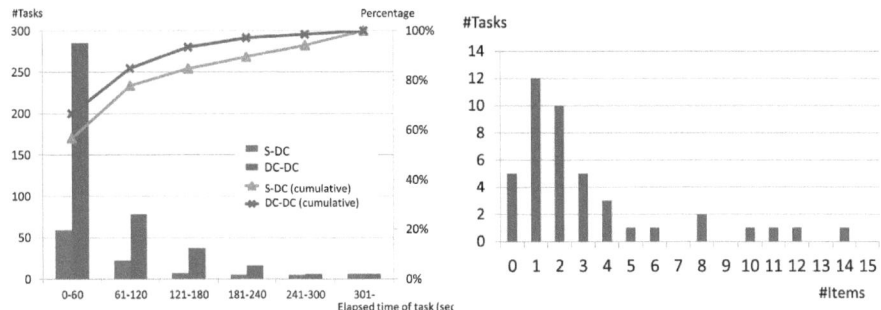

Fig. 13. Distribution of elapsed times for performing tasks

Fig. 14. Distribution of tasks for having items

Figure 14 shows the distribution of the sizes of complete lists for each task in DC-DE. The average is 3.1, while the size of the complete list is 91 in S-DE. This is one of the reasons why the average time for DC-DE is shorter.

If we assume that we have only one worker to perform the tasks and simply sum up the elapsed times of all tasks, we need 10,043 seconds for S-DE, while 24,672 seconds for DC-DE. Thus the simple sum of all elapsed times of DC-DE is 2.5 times higher than that of S-DE. However, DC-DE naturally introduces parallelism in performing tasks because many of the generated tasks are independent. Therefore, we can exploit the parallelism if we have many workers. For example, we estimate that if we have more than 30 workers, all tasks of DC-DE would be completed in 3,652 seconds, about 2.7 times faster than those of S-DE.

7 Summary

This paper proposed a processing method for applying the divide-and-conquer principle to data enumeration on microtask-based crowdsourcing platforms. The difficulty is how to provide a number of small tasks for implementing data enumeration. Our proposed method is unique in that the workers join the process of generating smaller tasks in a divide-and-conquer fashion; therefore, the programmer does not need to provide many microtasks in advance. This paper has presented the method, provided theoretical results that prove the method works well on microtask-based platforms, and explained experimental results that confirmed the proposed method can achieve higher recalls and produce appropriate tasks for microtask-based crowdsourcing.

Future work includes the development of a system to suggest task generation plans, the analysis of the behavior of the proposed method when combined with different incentive structures, and the incorporation of a variety of techniques to improve data quality into our proposed framework.

Acknowledgements. The authors are grateful to Prof. Sugimoto, Prof. Sakaguchi and Prof. Nagamori for the discussion in seminars, and to the contributors of Crowd4U, whose names are partially listed at http://crowd4u.org. This research was partially supported by PRESTO from the Japan Science and Technology Agency, and by the Grant-in-Aid for Scientific Research (#25240012) from MEXT, Japan.

References

1. ESP Game, http://www.gwap.com/gwap/gamesPreview/espgame/
2. Franklin, M.J., Kossmann, D., Kraska, T., Ramesh, S., Xin, R.: CrowdDB: answering queries with crowdsourcing. In: SIGMOD 2011, pp. 61–72 (2011)
3. Jain, S., Parkes, D.C.: A Game-Theoretic Analysis of Games with a Purpose. In: Papadimitriou, C., Zhang, S. (eds.) WINE 2008. LNCS, vol. 5385, pp. 342–350. Springer, Heidelberg (2008)
4. Kulkarni, A.P., Can, M., Hartmann, B.: Collaboratively crowdsourcing workflows with turkomatic. In: CSCW 2012, pp. 1003–1012 (2012)
5. Morishima, A., Shinagawa, N., Mochizuki, S.: The Power of Integrated Abstraction for Data-Centric Human/Machine Computations. In: VLDS 2011, pp. 7–10 (2011)
6. Morishima, A., Shinagawa, N., Mitsuishi, T., Aoki, H., Fukusumi, S.: CyLog/Crowd4U: A Declarative Platform for Complex Data-centric Crowdsourcing. PVLDB 5(12), 1918–1921 (2012)
7. Marcus, A., Wu, E., Karger, D.R., Madden, S., Miller, R.C.: Human-powered Sorts and Joins. PVLDB 5(1), 13–24 (2011)
8. Marcus, A., Wu, E., Madden, S., Miller, R.C.: Crowdsourced Databases: Query Processing with People. In: CIDR 2011, pp. 211–214 (2011)
9. Parameswaran, A.G., Polyzotis, N.: Answering Queries using Humans, Algorithms and Databases. In: CIDR 2011, pp. 160–166 (2011)
10. Tripadvisor, http://www.tripadvisor.com/
11. Trushkowsky, B., Kraska, T., Franklin, M.J., Sarkar, P.: Crowdsourced enumeration queries. In: ICDE 2013, pp. 673–684 (2013)

Social Listening for Customer Acquisition

Juan Du, Biying Tan, Feida Zhu, and Ee-Peng Lim

School of Information Systems,
Singapore Management University
{juandu,biying.tan.2012,fdzhu,eplim}@smu.edu.sg

Abstract. Social network analysis has received much attention from corporations recently. Corporations are trying to utilize social media platforms such as Twitter, Facebook and Sina Weibo to expand their own markets. Our system is an online tool to assist these corporations to 1) find potential customers, and 2) track a list of users by specific events from social networks. We employ both textual and network information, and thus produce a keyword-based relevance score for each user in pre-defined dimensions, which indicates the probability of the adoption of a product. Based on the score and its trend, out tool is able to pick up the potential customers for different kinds of products, such as suits which are time-insensitive and diapers which are time-sensitive. In order to detect the scenario of purchasing products as gifts, we filter the user network and only consider the off-line close friend network. In addition, we could track users in a more flexible way. Despite the pre-defined dimensions, our tool is also able to track users by customized events and catch those who mention the event at an early stage.

1 Introduction

According to the statistics provided by the Official Twitter blog[1] on March 21, 2013, the social network Twitter has over 200 million active users creating over 400 million Tweets each day. Social networks has become the main form of media for online users to express their opinions and connect with other users. According to [2], these social media data represents a vault of precious information that could be used by companies for increasing their sales revenue. For example, a user who tweeted about his new born baby on Twitter is more likely to buy diapers. Her friends who saw the posts might purchase diapers as gifts for her as well. This suggests that the tweet's content as well as the social relationships are useful for identifying potential customers. The identification should be early before potential customers purchase the product.

There are some existing tools to expand the markets by social networks such as "Twitter Business"[2], a tool developed by Twitter. However, the relationship they considered is the whole network structure rather than *off-line friends*. We

---

[1] https://blog.twitter.com/company
[2] https://business.twitter.com/

distinguish our system[3] from the others by focusing on the adoption of product for a user through his off-line close friends. By utilizing off-line network, we are able to catch the user intension to buy something as gifts and influence by his close friends. Our system provides the analyses based on two social networks: the English version is based on *Twitter* while the Chinese one is based on *Sina weibo*. Our tool is able to detect potential customers as well as providing a feature to track a set of users via specific events found in social media. We employ both *textual information* and *relationships* from social networks and produce a relevance score for each user in pre-defined product dimensions, such as cars with a set of weighted words related to cars. The relevance score depends on the user's volume of discussion on the product such that higher relevance score suggests that the user is more likely to adopt the product. Our system gives the relevance score based on the user's whole tweets to catch some loyal customers. And we also provide trends which shows the evolution and occurrence of events, such as birth of a baby from a user. In addition, we filter the network and puts more focus on the off-line friends by the algorithms proposed in [4]. The relevance scores of a user's close friends will also influence his. Moreover, in order to track the users in a more flexible way, our system is able to follow a list of users and define customized topics. Once the user mentions the topic, he will be picked up.

By an aggregated analysis on the tweets and networks, we are able to find the users who intend to adopt a product. In this paper we give an overview of the architecture of our system and its major modules including a briefing on the algorithms in Section 2. We then demonstrate a case in Section 3.

2 Architecture and Algorithm

Figure 1 shows that our system composes of two main parts. In the back-end part, the tool fetches the social media data and processes the data by calculating the users' relevance score to the product and their off-line friend network. While the front-end provides analysis and visualization based on the scores and networks produced by the back-end. The visualization includes ranking of the user's relevance score, detail profiles of a selected user, off-line social network of a target user, and the event-triggered user tracking. The company could directly market their products to the users by posting tweets to them.

2.1 Data Collection and Processing

There are many social networks like Twitter, Sina Weibo and Foursquare. We first choose one of them as our dataset and collect the timelines, user profiles, and networks by the API they provided.

---

[3] http://research.larc.smu.edu.sg/pa/home.php

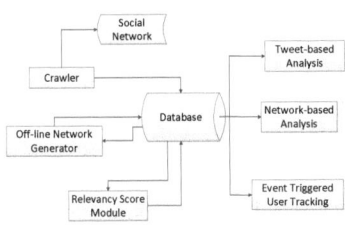

Fig. 1. System overview

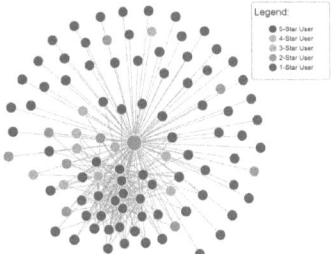

Fig. 2. An example of a user's social network with relevance score

Relevance Score to Business. In Information Retrieval, the basic idea for measuring the relevance of a document to a query has hinged on whether or not a query term is present within a document. A document that mentions a query term more often has more to do with that query and therefore should receive a higher score [3]. We adopt this idea to calculate the relevance score of a user to a business.

We define business related dimensions before processing. For each dimension D, we compose a dictionary as $\{t_1 = w_1, t_2 = w_2, ..., t_n = w_n\}$, where t_i is the term and w_i is the corresponding representative weight of t_i. To take the car-insurance as an example, its dimension could be formed as $\{crash = 0.8, gasoline = 0.5, ..., carparking = 0.3\}$. The relevance score $R(tw)$ of a tweet tw to D is determined as

$$R(tw) = \sum_{t_i \in D} w_i * o_i, \qquad (1)$$

, where o_i indicates the occurrence of t_i in tw.

And therefore, the relevance score of a user to a dimension D in period $[start, end]$ is produced by $\sum_{tw_i \in [start, end]} R(tw_i)$.

To build up D, we employ the ideas in [1] to extract the keywords automatically. Based on a small set of seed terms t_i which is selected at first, some other representative co-occurring terms are then extracted from tweets. We add the new terms and iterate the selection until the number of terms reach a threshold α. And then, w_i is measured as n_{rt}/n, where n_{rt} is the number of dimension-relevant tweets and n is the total number of tweets containing t_i. The terms with w_i under threshold β will be filtered out.

Off-line Friend Network. Since the user will be influenced more by his close friends and may buy something as gifts to their friends, we generate the off-line network by the algorithm proposed in [4]. The algorithm utilizes the structural network and applies random walk with restart iteratively to produce the closeness score between users. For the details, please refer to [4].

 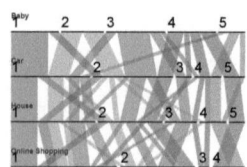 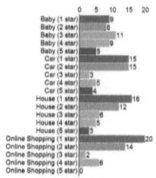

Fig. 3. Radar chart to represent a user's relevance score in different dimensions

Fig. 4. Joint relevance distribution in different dimensions among user's network

Fig. 5. Relevance distribution in different dimensions among user's network

2.2 Social Network Marketing

Tweet-Based Analysis. The tweets that the user sent out directly indicate the possibility of the user to buy a product. If a user posted a lot of tweets about fashions, then he is more likely to buy clothes and shoes. For each user, a radar chart for the relevance score on different dimensions is shown as Figure 3. The radar chart illustrates the user's topic distribution in the pre-defined dimensions. The tweets related to the dimensions will also been provided as word cloud. When advertising, the tweets contribute to eliminating the distance between customers and salesmen.

Network-Based Analysis. Despite of what the user said, the behaviours of other users will also affect his decision to buy a product or not. Just as a famous marketing phenomenon, which is "Since a user care more about what his neighbour thinks than what Google thinks"[2], so the influence of his social network contributes to detecting potential customers. Actually, the influence in the network is two-way. If the target user is a loyal customer, then the neighbors in his network is much easily to be advertised with success. In addition, if the probability of the user's close friends to buy a product is high, and thus the intension of him to purchase becomes higher as well. And consequently, we filter the user's network and only retrieve the off-line friends network. We generate a relevance score for the user by averaging his off-line close friends' scores.

Static and Dynamic Relevance. In fact, some topics are long-term like fashion while others are short-term such as new born baby. And consequently, the measurement of the user's possibility to purchase the product will be decided accordingly. In order to cope with various products, our system provides not only the user's overall relevance score by all his tweets, but also the score trends by splitting the tweets into different periods. To take the topic "baby" as an example, if the user posted a tweet as *"Listen up. I am PREGNANT."*. Even though the relevance score to the baby dimension is not large since the number of tweets related to baby is small, they will also be found by score trends.

Event-Triggered Marketing and User Tracking. A user's relevance score in different dimensions is preprocessed, which could meet the demands of detecting potential customers. However, we still have two problems need to solve.

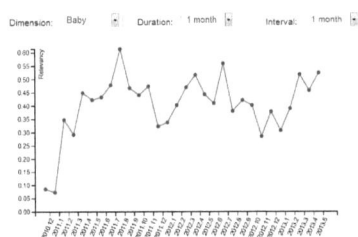 

Fig. 6. The user's relevance trends in a dimension

Fig. 7. The user's word cloud in a dimension according to his tweets

- How to track the user by customized events?
  We pre-define a weighted word dictionary for each product dimension to produce the relevance score. However, in case of some new events, our system is also able to allow system users to customize their own tracking events.
- How to find the user at the early stage?
  Based on the customized events, our system will continuously check whether the tracked users triggers the events. As long as the crawler collects the latest data, once the user triggers the events, he will be highlight immediately and the check frequency is decided accordingly.

3 Demonstration Cases

In this section, we demonstrated a case as a baby-insurance salesman. We could search the top baby related users from "Social Network Marketing". Under the tabs "Rank by User Relevancy", "Rank by Relevancy Growth Rate" and "Rank by Close Friend Relevancy", the static individual relevancy, the dynamic individual relevance trend and the dynamic community relevance are shown separately. If we are willing to advertise the insurance, we just need to choose the users and then send out a tweet to them by "Contact Clients". We are able to view the detail profile and the social network of a user as well. The profile page includes user's basic information provided in social networks. Besides, the radar chart for the user in different dimensions are shown as Figure 3. The relevance trend to the baby dimension is represented in Figure 6, where a sudden burst could be found on 2011.02. By viewing the relevant word cloud and tweets as Figure 7 represents, we could eliminate the gaps between the salesman and the user. He could know what the user mentioned in this dimension. Furthermore, in the social network page, the user's 1-hop relationship is shown as Figure 2, where different colors indicate different relevance scores. Red is the highest while blue is the lowest. The yellow bordered dots stand for the user's off-line friends. Analyses for the network such as joint distribution and relevance score distribution in different dimensions are shown in Figure 4 and Figure 5 respectively. In the "Event Trigger Marketing" page, after entering the salesman's login ID "1", we could view his own tracking users. Manually adding some events such as "milk powder" with keywords "kid, milk, powder, buy". Once a follow user mentions the event, he will be highlighted as Figure 8 shows. The refresh frequency could be adjust accordingly.

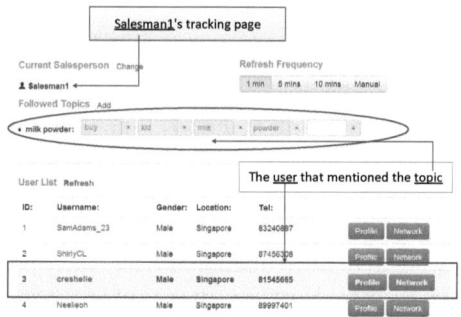

Fig. 8. Tracking users by customized events or topics. (This is just an example and the information like "Tel." is not real.)

4 Conclusions

Our system is an online tool to help corporations to expand the market from social networks. We analysis the users by their own relevance, relevance trends and close friends' relevance score for different types of products. Our system is also able to track users by customized events. In conclusion, we could assist corporations to find and reach potential customers on social networks with ease.

Acknowledgments. This research is partially supported by the Pinnacle Lab for Business, Consumer and Social Insights, a Ping An-SMU Joint Research Lab.

References

[1] Clark, M., Kim, Y., Kruschwitz, U., Song, D., Albakour, D., Dignum, S., Beresi, U.C., Fasli, M., De Roeck, A.: Automatically structuring domain knowledge from text: An overview of current research. Information Processing & Management 48(3), 552–568 (2012)
[2] Kaplan, A.M., Haenlein, M.: Users of the world, unite! the challenges and opportunities of social media. Business Horizons 53(1), 59–68 (2010)
[3] Manning, C.D., Raghavan, P., Schütze, H.: Introduction to information retrieval, vol. 1. Cambridge University Press, Cambridge (2008)
[4] Xie, W., Li, C., Zhu, F., Lim, E., Gong, X.: When a friend in twitter is a friend in life. In: Proceedings of the 4th International Conference on Web Science, pp. 493–496. ACM (2012)

Passive Participation in Communities of Practice: Scope and Motivations

Azi Lev-On[1] and Odelia Adler[2]

[1] School of Communication, Ariel University, Ariel, Israel
azilevon@gmail.com
[2] Department of Communication, Ben-Gurion University, Beer Sheva, Israel
odeliadler@gmail.com

Abstract. In spite of its prevalence in online social media platforms, there have been very few studies of passive participation. The current study uses interviews to understand the motivations for passive participation in online communities of the Israeli Ministry of Social Services. In addition to the motivations commonly found when not participating "more actively", such as concerns of criticism, lack of need, lack of motivation and technological concerns, users stated two additional and secondary reasons for not contributing content in the communities: concern that their posts would not be addressed (i.e. lack of reciprocation), and reasons relating to the graphical interface of the communities.

Keywords: communication, online communities, communities of practice, participation, passive participation, lurking.

1 Theoretical Background

Studies of online communities commonly distinguish between two main types of participants – "posters", i.e. *active* participants who initiate discussions and respond to posts, and "lurkers", *passive* participants who exclusively or almost always read content created by others without contributing content of their own [1-19].

Already from the term "passive participants" one can discern that their profile of activity integrates participation on one hand, with passivity on the other. The two, obviously, sit uneasily with each other. One way to reconcile "passivity" with "participation" is by conceptualizing online activity as composed of a range of activities carried out by community members, where "low frequency posters" and "complete" lurking are near the end or at the end of the spectrum [6].

Passive participation is an important phenomenon in social media environments. Studies consistently demonstrate that passive participants, or "lurkers", tend to be the majority of the population in the forums studied, sometimes consisting of no less than 90% of community members [6,7,10,16,25].

The literature addressing passive participation focuses on reasons originating from lack of need, fear of criticism, concerns about personal skills, technological obstacles and the dynamics of community interaction:

1. A primary reason often stated by passive participants is the lack of need to contribute, i.e. reading and/or browsing satisfies the passive participants' needs. For example, 53.9% of the participants in the study conducted by Preece, Nonnecke, & Andrews (2004) stated this as a primary reason for non-contribution. [16]
2. Shyness and uncomfortable feelings of public participation [10,16].
3. Lack of willingness to be exposed to others and risk one's reputation [6,13,16].
4. Concerns of criticism [12].
5. Language issues, such as spelling and grammatical errors [10,11,16].
6. Learning about the group and the rules and norms that govern its interactions [10,16].
7. Concern that information provided is not important or relevant enough for other community members [10,12,14,16]
8. Concern for misleading others as a result of posting inaccurate or incomplete information [12,16].
9. Limited identification with the community and its members [9,10,13].
10. Lack of familiarity with online communication technologies, and consequently lack of interest in understanding the ways it can be capitalized on to assist community members [9,16].
11. Lack of access or inferior access to technology, for example to new computers and fast Internet connectivity [9].
12. Embededdness in a culture that does not focus on sharing, and/or focuses on competitiveness instead [12,24].
13. A final concern is that in over-crowded communities with much activity, users may think that their input is redundant and may not want to generate further information overload [3,4,6,10,16].

Some refer to passive participants as "lurkers", "free riders" or individuals who carry out "social loafing" [13,15]. Indeed, if online communities are composed only of such free-riders who prefer to benefit from other people's work without contributing themselves, the community would not be able to function as a scaffolding for knowledge creation, sharing and circulation and may eventually disintegrate.

Yet, others argue that passive participation should not be understood as a dysfunction, and at times offering quite a few advantages for online communities. Passive participants can spread important information about the community, thus assisting in publicizing it [4,19]. Their participation can also become more active as they grow more comfortable within the community [6,19]. Moreover, large scale passive participation can prevent information overload [4,6,7]. Additionally, due to the large number of passive participants exposed to messages posted to the communities, marketers have come to realize their significant consumption potential, and oftentimes target them for marketing efforts and publications [6,7,11].

According to such views, passive participants are not necessarily "free riders." Rather, they do not contribute for a variety of reasons not necessarily as a result of a deliberative effort to benefit from the work of others. They can also contribute to the community in various ways other than posting to the community. Thus, Crawford (2009) suggests to replace the word "lurkers" with "listeners" [8], and Zhang and Storck offer the term "peripheral participants" [17].

Studies demonstrate that active and passive participants differ in significant respects. For the active participants, the online community is often perceived as a public place in which friends and colleagues meet and thoughts are shared. For passive participants, the community is more often perceived as an instrument for obtaining specific information [5,6,18]. Online participation patterns frequently reflect offline participation patterns; for example, people who are more cautious and concerned offline, generally lean towards passive participation online [13].

1.1 Research Setting: Social Workers' Online Communities of Practice

The current study takes a fresh look at passive participation through analysis of social workers' online communities of practice established by the Israeli Ministry of Social Affairs and Social Services.

The communities of practice studied here represent a unique case in Israel, whereby a governmental ministry established online forums to enable interaction between its employees and the broader community of practitioners. Such communities may have many advantages in terms of exposing local knowledge, improving knowledge circulation and even supporting professional acquaintance and creating solidarity between practitioners [1,2]. At the same time, they can disseminate employees' open criticism of their supervisors, damage working relations within the office, and generate criticism of the employing agency's work and routines, among other risks. As a result, governmental ministries, in general, tend to avoid platforms for such interaction among employees [20]. Nonetheless, with the rise of social networking platforms, an increasing number of Israeli governmental bodies are offering more opportunities for direct online interactions amongst their own employees, as well as between employees and citizens. The communities of practice that are studied here were established in 2006, and can be seen as pioneers of this phenomenon.

Since its establishment in 2006 some 9,000 members have enrolled into one or more of the 31 online communities with topics ranging from Domestic Violence, Adoption, Juvenile Delinquency, Mental Disabilities, and the like, to discuss ideas, regulations and ethics. The communication in the forums is identified, as members use their real names. The forum is based on a standard web platform, where discussion is threaded and last comments appear first, pushing down earlier comments.

2 Methodology

The paper is a part of a larger project analyzing the content of online communities of practice within the Israeli Ministry of Social Affairs and Social Services, as well as the perceived effects on members.

A twofold research methodology was applied for this purpose. First, we undertook a content analysis of all the available materials – 7,248 posts through mid-2012. Second, 71 semi-structured interviews were conducted with community members. Based on data received from the Ministry of Social Affairs and Social Services, members were sampled according to their levels of engagement – number of logins to the community and number of times they contributed content. Interviews were conducted by five interviewers across Israel, with the average length of an Interview around 45 minutes. Interviews were transcribed and analyzed using a thematic-interpretive method [26]. The findings in the paper below are largely based on these interviews.

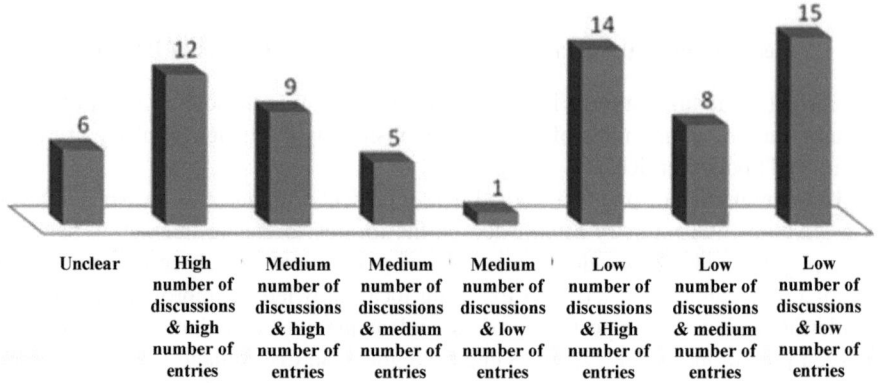

Fig. 1. Interviewees' participation patterns

3   Passive Participation in Social Workers' Communities of Practice: Findings

Within the communities of the Ministry of Social Services, active participation can take three forms:

1. Posting an article/ document in the community content library.
2. Responding to a post initiated by another user
3. Initiating a discussion [1].

An analysis of data demonstrates that 76% of community members have never initiated a discussion or responded to an existing discussion. 21% of members have initiated a discussion or responded to an existing discussion between one to ten times. Thus, in these communities, passive participation is a frequent phenomenon, as documented in previous literature.

In general, community members who did not post can be divided into two kinds: those who read the content but do not post and those who do not enter the community and hence do not read the content. As noted above, interviewees were sampled according to clusters of participation patterns, with the aim of interviewing those who are either active or passive, and with varying levels of activity. Most of the passive participants interviewed are not only subscribed to the communities but also visit them from time to time. The interviews assisted in identifying factors leading to negligible or lack of content contribution.

3.1 Insecurity and Concerns of Criticism

Insecurity was the primary reason for passive participation stated in the interviews, along with concern from direct or overt criticism of other community members, whether supervisors or peers [10,12]. This concern is of course associated with the fact that participation in the community is identified using the members' real name, and cannot take place anonymously.

For example, SH, who is a member in the "community work" community, does not enter the community much and nearly refrains from posting (28 entries, 2 comments). When asked for the reasons, she replies: "because it's online!... I mean, if you want to write there, then you have to expose yourself, you need to know to accept criticism... I mean, you cannot control the reactions". Similarly, Y (20 entries, 4 comments) says: "I am concerned with the exposure... it's a kind of - risk, because everyone reads it, so you constantly need to be politically correct". A stated: "you always ask yourself, who is reads it and what would she think? And how would she react?"

The concern for criticism can be divided into two reoccurring thoughts. First, concern for reaction and criticism of supervisors and community managers. Such criticism can jeopardize their reputation and even endanger the continued employment of the poster. Many community members argued that they prefer not to jeopardize themselves by writing things that run against community policies or norms. In addition, they feared they would be perceived as non-professional by other members. Oftentimes interviewees said they were afraid that managers would prefer not discussing a certain theme generating a "chilling effect" leading to refrainment from writing about controversial issues.

An additional concern was based on the expected reactions of the "peer group" – colleagues and peers with the same professional level who may dislike nonconformist and challenging posts, and may therefore think negatively about the poster.

A few interviewees described criticism by peers as a potentially offensive situation, which may prevent them from writing in the near future. They also described the need to be absolutely certain of what they have to say prior to posting in the communities. The need for total control and certainty also comes up in the interviews with Z and ZB, both passive participants, whose number of entries into the communities is large, yet the number of posts nears zero. Z (170 entries, 2 posts): "If I feel an urgent need to respond and I'm certain about the content- I'll react. But I do not initiate discussions." ZB (85 entries, 0 posts): "I did my master's thesis about treatment of women who were victims of violence. The paper was good and I got an A for it, and still I did not upload it to the community. It's as if I did not feel comfortable to... I was uncertain about how this would contribute beyond the knowledge that people already have, and what would happen if they would not find it interesting... I think this is a consequence of insecurity".

The possible reaction of other community members is sometimes considered as a "threat". Interviewees often justified their passive participation by fear they would be misunderstood and would then have to explain and justify their positions. For example, K said:

I don't like to write so much... when you write, the experience is of something more deliberate and responsible... should I use this sentence, or this word, or that word... if I get misunderstood, can this generate a thread of reaction... and then you have to intervene and explain what you meant and what you did not mean".

Some of the interviewees "dare" to write, but are cautious about providing their personal opinion. A, a social worker in the Youth Law community, argues:
I know I'll be eaten alive, because employees often like to just accept what others say and not think about what we should do. What I have to say I say in small forums or face to face, because I do not like to be eaten alive... I will maintain a low profile about some of the things... it's hard to break the conventions...after I retire I'll say everything I have to say.

In addition, A gives a few more reasons that lead individuals to refrain from writing such as insecurity, concerns from exposure, lack of experience and more:
Not everyone would like everyone to know what they think, people cherish anonymity... some graduated from universities, and yet they are shy, some do not like to lose control, prefer to do what's asked from them and not think outside the box... some are too young or lack experience ".

T provides a different kind of concern for criticism:
I have a problem, I have learning disabilities, and when I write the letters escape me, it is especially difficult on the computer [...] if the problem was not so severe... I would have made many more comments... It is problematic, everyone would ask, 'who is the retard who writes like that'..."

Lastly, IK described concerns for the reactions of his colleagues at work:
I'm afraid to expose myself. I'm the kind of person who thinks twice about every word he says. Internet posts have a much bigger audience than what I can see. I have no problem talking face to face, but it's difficult for me to address large crowds, even invisible [...] I would like to be able to write more but I'm closed. Many times I have thoughts about things that I could post on the forum, but it never materializes."

Concerns for criticism seem to be the predominant reason for passive participation. These concerns exist in offline communities as well [20]. However, online the information is visible to everyone and is "there to stay", with past comments about posts made a long time ago "resurfacing" years after they were made. When concerns about criticism are present this generates an extra "chilling effect".

3.2 Lack of Time and Motivation

An almost absolute majority of interviewees mentioned lack of time and motivation- as a critical reason for passive participation. Posting in the forum is perceived as much more time consuming than writing. With the busy daily schedule of social workers, finding time to write in professional communities seems to be very demanding.

SY, a probation officer, agreed to be interviewed although she believed she will not contribute much to the study due to the lack of her activity in the communities: "there's an article that I opened from the community, I got it in my email. Most of the chances are that I will not read it. I would really like to, but I do not have the time".

SL says she is passive due to time constraints: "The number of times I uploaded stuff or reacted in the community are very limited. I think it's mostly an issue of time... There are quite a few instances of things I see and I was happy to respond to and even think it was appropriate if I only had the time".

SR is a member of four of the communities of practice, and demonstrates a big difference between the large number of entries (722), and the low number of posts (6). In regards to this gap, she says:

In our work we are out there in the field a lot, do not have access to computers... It's a matter of time, because we have lots of meetings, and when we sit on the computer it is in order to answer letters, respond to urgent things, update diaries. So it's a matter of priorities... If I have 200 emails on average when I arrive, then the community of practice is not a first priority, because there are lots of other things. I think it's a shame, I wish I could be more involved and more active.

AS: "if I had more time I could have thought harder about the content and respond more... the way things are, I do it superficially and it's a shame... I mostly browse through the text and don't get deeper".

A related reason stated by some is that the time spent on the professional communities comes at the expense of free time at home. SH: "After long hours of work, I cannot come home and continue my work by taking time to assist a colleague with her issues... it's definitely not considered a part of my job. Not that I don't want to help her but, once more, I have a limited amount of hours so if I do it from my home voluntarily it will blur the boundaries between home and work and I do not know if I like this ... I prefer my work to remain [in my workplace] rather than migrate to my home."

Many of the interviewees think that reading posts from the communities should be done as a part of their work hours and not as a part of their private time. But some, like A, says that "I read my mail at home and not from my work, this way I can read things peacefully. I cannot get that from work. Still, these are strictly professional material". YK presents the opposite view: "It's good to have a library of content, but I never used it because I don't have time, I come home from work tired and I have no energies to read, I prefer being with my family while at home."

A (member of 12 forums, but only 49 entries and 0 posts) argues that there is not enough active participation in the forum due to work pressure: "Social workers have no time to sit and read. If the service's management had allocated an hour weekly in which all employees must sit on the forum and create a discourse that would encourage me more."

A small number of interviewees refer to their lack of active participation as a kind of laziness. Maybe just a few think this is the true reason why they do not participate, but possibly just a few are sincere and/or self-conscious enough to point this reason out. For example LA, who entered the community many times (263) but hardly wrote (3) says: " I think it is wonderful to post, you just need energies and willingness to make an investment, and this was never too high on my priority list… it's a kind of laziness". CH Writes:

> *Maybe it's laziness. I once nearly responded to a story about an attack against social workers who were not defended enough, and we had a case like this in our department. They touched on it in the forum, and I thought I had to react, but it … Such laziness, it's always late at night until I get to that, so I left it.*

3.3   Concern from Possible Lack of Feedback and Reciprocation

A third reason for passive participation, which came up in a few of the interviews, involved concerns about the *absence* of members' reactions, i.e. in order for community members to contribute, they need to feel that their contribution is important for the group as expressed by replies by other group members [14,21-23].

S enters the communities quite frequently (145 times) but has only posted three times in six years. She explained that once she uploaded a file she invested a lot of effort in to describe a project she worked on through the institution. She waited in vain for comments that did not come. Therefore, this reason for passive participation does not involve concern from direct criticism, rather from the lack of *any* reaction. An absolute lack of reaction for S meant that the project was not significant or interesting enough, and the reciprocity, an important aspect in maintaining relationships online, according to previous literature , was gravely missed [21-23].

Y describes another such disappointment: "Two states in the US now legalized Cannabis. It's a bombshell in our field! I posted it, and nothing happened. Nothing…" One of the community managers who was interviewed, said that every time an item is uploaded to the community, she hopes that "this item does not remain orphan!" she describes her conflict on whether to wait for members' reactions- which may not happen, or to respond herself, in which case "this will be the end of the discussion, because no one would react after I did". According to these interviews, members' reactions support active participation, they are like "fuel" for further posting. The members whose comments get plenty of reactions feel that their comments are useful, while those whose comments remain orphan may feel like R, who says: "I like very much to receive feedback, but when there are no reactions you say to yourself, 'oh, I probably said something stupid'".

R., who works with blind and vision-impaired individuals, reports that he wanted to contribute to the community and uploaded large amounts of materials which got no replies. He says: "for a long time I reported about the daily lives of social workers, but I realized at some point, that no one cared! … I assume that just like myself, my colleagues in Haifa or Jerusalem or other places are also very busy with their daily activities, and perhaps instead of a bonus, my contribution seem like a burden to them… I saw that I throw rocks but no one want to pick them, so at some point I just quit…" Similarly, L (99 entries, 7 posts) says: "once you see no reaction it gives you a sense of: OK, so I'm not going to upload any more materials".

3.4 Reasons Related to Status and Expertise

Visibility of professional hierarchy within the various communities was another reason provided for passive participation by members of both high and low seniority, .

D, who is in charge of a mental institution, presents her concerns: "when you are in a certain position within the professional hierarchy you have to be very cautious with what you write… people do not take it as your personal opinion, it is your personal opinion in this specific hat worn in the forum. I think that because of this I think twice about everything I write in difference places… people give it various meanings because I'm an administrator. Personally, I find it upsetting that people in high positions write their opinions when it does not correspond to the position of the ministry".

D demonstrates a unique case in which the organizational hierarchy limits the posting of high-status members. But in general, such concerns were stated more by rank-and-file or new employees. For example, G is a social worker who recently moved from handling addictions to patients with special needs, and therefore moved to a new online community. G thinks that expertise is an important determinant of the scope of participation: "In the first few years I did not feel confident enough to upload materials… maybe now I would feel more confident to post. When you are new to the field, you are more interested in reading".

Similarly, ZB, has worked in the field for five years, yet still feels like a "beginner":

There are people who write there with so much experience. For example, X, who is very active, has thirty or forty years of experience. So, I feel like I need to learn more… plus, I know that what I post is exposed to many people, some of whom are administrators in various positions, and what I write may have consequences…

Another manager of one of the communities said that forums have their "hard core" participants, and sometimes the new members in the community are afraid to post. She says that she wants to bring up topics for discussion, and is very curious what new members think about them, "if they dare to write". S, with an experience of a year and half, says: "if this does not touch me enough, I will not comment… I do not have enough experience and self-confidence… I'm sure other people will do it".

3.5 Sense of Redundancy

R: "I haven't seen anything that provoked me to initiate or respond personally. There are some interesting things but… I had nothing to add from my knowledge or experience".

Redundancy was one of the reasons mentioned for passive participating [14], especially if someone else already responded, and even if not – "somebody else would do it". Studies show that when a participant asks for help or advice in an online community, people spend more effort to reply in a situation where they know that others will not do it [15,25]. Although the tendency to diffuse responsibility and rely on others exists in offline environments as well, it is easier to do so in online communities. This is especially true when participants tend to be passive in general and when the person asking for assistance may not know they are in the forum, and certainly do not expect a reaction from them. Thirteen such members have been interviewed for the current study.

SZ: "someone else will write it at some point…"

ZB: "The things I could have said are, like, being said by others".

D: "Actually I see that enough people respond so I enjoy reading the responses and sometimes I feel it unnecessary to give further opinion".

N, who subscribes to four communities and mainly enters one of them- 99 entries per year, with zero comments, finds it difficult to point at the reasons for passive participation, which is obvious and comfortable for her. Eventually she replies: "thinking about it, maybe I haven't felt until now that I have something to write… maybe when I feel the urgency I will write".

M is a member in five communities:

I think maybe once in my six years in the forum I've consulted people and got replies, and once or twice I replied to people who asked for assistance ... because I see that people who ask questions get their responses already before I have a chance to offer my comments. And if I have nothing to add, I will not add, and usually there are always people that respond... I do not think I have something really important to add to the world at such a level that makes contribution absolutely necessary...

BB: *"There are social workers there who have worked in the community for years. I do not feel I could contribute anything new... The ability to express oneself and request assistance- that's what I cherish in the forum."*

G also did not find it necessary to respond even once: *"I am not one of those people that write 'way to go! What a beauty!' or all these things-it's not me. Don't feel the need to do this."*

In line with the literature, the interviews demonstrate that sometimes community members do not contribute content because of a feeling that their information is not important enough [10,12]. In addition, it was found that people tend not to respond if they assume that someone else would answer the question, especially if a reply has already been written, they feel exempt even if they have something to add. If what they wanted to say has already been said, then there is no point for them to write again. They don't see their own personal addition a contribution to sharpening the point.

3.6 Dislike of Technology

Some of the social workers interviewed, especially the older ones, very frequently stated that it was difficult for them to "connect" to the virtual platforms because face-to-face meetings were irreplaceable and computer-mediation generates a "cold" and "alienated" environment.

S, who is a passive participant in the *Juvenile Delinquency* community and does not read nor post herself stated:
> *I am not a technological person, I don't chat or post comments. I am a person who needs to speak [with the other party]. Being in front of a computer is difficult for me...I keep saying I am a social worker! Let me talk to people, not to computers.*

TS, a member of the Addictions community, says,
> *Maybe this is because of my advanced age, but I really don't understand how one can have a sense of belonging to something on the Internet...let's talk about work issues, why do we need the forum for that? Why do we need the Internet to get closer to one another as human beings...why those emails? Pick up the phone – let's talk as humans.*

D, a member of a community concerning Mental Disabilities, notes the need for an unmediated human connection: "I am interested in some response, some verbal interaction. I am not able to fully connect to the electronic world..." R argues that "[online interaction does] not replace friendship. It's not instead of acquaintance. It cannot replace the peer discourse we sometimes generate amongst ourselves."

Z, from the Family Courts community, states that, "I handle myself much better in a human, not text-based environment... [online contact is] not like sitting with people and carrying out a dialogue, getting support and feeling the 'softness.'" Interviewees complained that the online connection hinders their ability to express emotions, grant and receive support – essential elements in this profession. For example, D notes: "I think that in our world, most people don't get support through this media...we are people persons, and I don't think that the Internet can do the job..." The online forum seems to provide an inferior arena in particular for expressing emotions as compared to face-to-face conversations.

M, a member of the Juvenile Delinquency and Addictions communities who posts frequently, expresses similar concerns. "Generally, I would not go to some Internet forum to ask a question even if I get an immediate online response...this is not my medium."

S, from the Addictions community, mentioned how
> *This medium has many features that are inferior to face-to-face conversation. I need a little more of the personal contact, the intonation that allows you to fully understand what was said and reduce misunderstandings [...] I prefer to get my support from someone I am personally acquainted with. It's much better to talk to people that I know than [to those that] I don't know.*

3.7 The Visual Aspect- "Not Facebook-ish Enough"

Whereas some community members, especially the older ones, were concerned that the community environment is too technological, others, especially the younger and more "digital native" expressed the opposite concern – that it is not savvy and up-to-date.

The Ministry of Social Affairs is portrayed in both the public eye and to younger social workers as gray and anachronistic. Some said that this feeling is manifest also by looking at the community interface, one they consider dry and with a boring look-and-feel. In the colorful and lively interactivity Facebook era, the communities of practice of the Ministry of Social Affairs seem to be "lagging behind", and require an upgrade, eventually bringing more contribution by members.

The communities of practice do not only exist within the forums rather include mailing lists, Facebook groups and more. The more appealing look-and-feel competing social networks is an additional reason stated by a few for not contributing content to the communities. For example, R states: "now I joined the professional Facebook group ... It is very different from what the Ministry of Social Affairs operates. People have all kinds of groups and advertisers are there. I think they are more open and modern... and this makes it also easier for me to participate". SN: "the site is grey and depressing, it does not give you a feeling like you want to enter. It is b-o-r-i-n-g! It's no fun to enter."

Many of the interviewees repeated the arguments that the visual component can assist or alternately hamper the degree of participation on the site. Y: "the forum is very uninviting... if you do not have the patience to look for what you need, you give up and leave. It could have been different, because I see it in other open forums where people participate more... It does not happen in the community, it does not invite participation".

4 Discussion and Conclusions

Passive participation is a common phenomenon in communities of practice, and in online forums in general. The study of the motivations for passive participation is fascinating because it is often unconscious. Often interviewees had difficulty to pinpoint exact reasons for their lack of participation, but nevertheless some reasons repeated themselves. From the interviews it was possible to discern a number of main motives for this mode of participation, most of them consistent with the academic literature.

This study demonstrates that lack of anonymity is critical in understanding participation patterns. The main motives for passive participation as raised in this study were related to concerns about the reputation of the author, concerns from others in the organizational hierarchy as well as from one's peers. Some of these concerns could have been minimized if writing was anonymous. Identification has its advantages but also disadvantages and community managers must weigh its pros and cons when deciding policies regarding anonymity and identification.

Some members are afraid to put their reputation at risk, especially when they know that their superiors could perceive their posts negatively. Therefore, they may self-censor and avoid posting altogether. In light of this, it is appropriate to examine how positive comments from the community and in particular from managers can encourage participation.

The findings demonstrate that passive participation stems from several additional reasons, such as technological concerns and concerns that the post is redundant, both as dominant reasons for lack of contribution. Even more, as the interviewees argued, social workers have very little free time and their work is very demanding [24]. Therefore, presumably if some time was dedicated during working hours for the use of the online communities, this might have generated greater participation.

Two reasons for passively participating in the community that did not come up in past studies came up in the interviews. First, the visual component of the communities can support or impair participation. The site, its colors and the ease of use of the interface can encourage the participation and contribution of content.

An additional reason for the absence of contribution which arose in the interviews yet not in past studies, is the concern for lack of reciprocation. As a person who reads a magazine, the passive participant can internalize the informational benefit that the community provides without exposing himself. However, by not posting, s/he is losing the response, support and appreciation that may be expressed in the forum. Because they contribute less, their expectations from the community are usually lower than those of the frequent participants [6,11,18]. Arguably, members should be encouraged to reply to posts and generate motives for the few who do participate in order to entice them to continue to do so [6,7,11]. Future studies should continue to research the links between reciprocation and appearance on one hand, and participation in communities of practice on the other.

References

1. Fein, T.: Online Communities of Practice in the Social Services: A Tool for Sharing and Knowledge Circulation between Employees. MA Thesis, Hebrew University of Jerusa-lem (2011)
2. Cook-Craig, P.G., Sabah, Y.: The role of virtual communities of practice in supporting collaborative learning among social workers. British Journal of Social Work 39(4), 725–739 (2009)
3. Meier, A.: Offering social support via the internet: A case study of an online support group for social workers. Journal of Technology in Human Services 17(2-3), 237–266 (2000)
4. Nonnecke, B., Preece, J.: Lurker demographics: Counting the silent. In: Proceedings of the SIGCHI Conference on Human Factors in Computing Systems, The Hague, The Netherlands, pp. 73–80. ACM Press, New York (2000)
5. Sassenberg, K.: Common bond and common identity groups on the Internet: Attachment and normative behavior in on-topic and off-topic chats. Group Dynamics: Theory, Research, and Practice 6, 27–37 (2002)
6. Ridings, C., Gefen, D., Arinze, B.: Psychological barriers: Lurker and poster motivation and behavior in online communities. Communications of AIS 18(16) (2006)
7. Rafaeli, S., Ravid, G., Soroka, V.: De-lurking in virtual communities: A social communication network approach to measuring the effects of social and cultural capital. In: Proceedings of the 2004 Hawaii International Conference on System Sciences, HICSS 37 (2004)
8. Crawford, K.: Following You: Disciplines of Listening in Social Media. Continuum 23(4), 525–535 (2009)

9. Gray, B.: Informal learning in an online community of practice. Journal of Distance Education 19(1), 20–35 (2004)
10. Nonnecke, B., Preece, J.: Why lurkers lurk. Paper Presented at the Americas Conference on Information Systems, Boston (2001)
11. Nonnecke, B., Preece, J., Andrews, D.: What lurkers and posters think of each other. Paper presented at the HICSS-37, Maui, Hawai (2004)
12. Ardichvili, A., Page, V., Wentling, T.: Motivation and barriers to participation on virtual knowledge-sharing communities of practice. Journal of Knowledge Management 7(1), 64–77 (2003)
13. Cheshire, C., Antin, J.: None of us is as lazy as all of us: Social Intelligence and Loafing in Information Pools. Information. Communication & Society 13(4), 537–555 (2010)
14. Beenen, G., Ling, K., Wang, X., Chang, K., Frankowski, D., Resnick, P., et al.: Using social psychology to motivate contributions to online communities. In: CSCW 2004: Proceedings of the ACM Conference On Computer Supported Cooperative Work. ACM Press, New York (2004)
15. Blair, C.A., Thompson, L.F., Wuensch, K.L.: Electronic helping behavior: The virtual presence of others makes a difference. Basic and Applied Social Psychology 27, 171–178 (2005)
16. Preece, J., Nonnecke, B., Andrews, D.: The top 5 reasons for lurking: Improving community experiences for everyone. Computers in Human Behavior 20, 2 (2004)
17. Zhang, W., Storck, J.: Peripheral members in online communities. In: AMCIS 2001 Proceeding, Paper 117 (2001), http://aisel.aisnet.org/amci2001/117
18. White, D.S., Le Cornu, A.: Visitors and Residents: A New Typology for Online Engagement. First Monday 16(9), 5 (2011)
19. Yeow, A., Johnson, S., Faraj, S.: Lurking: Legitimate or illegitimate peripheral participation? In: ICIS Proceedings, pp. 967–982 (2006)
20. Haber, K.: Organizational Culture Obstacles for Implementing Open Governance Policy. Jerusalem: Israel Democracy Institute (2013)
21. Davenport, E., Hall, H.: Organizational knowledge and communities of practice. Annual Review of Information Science and Technology 36, 171–227 (2002)
22. Ipe, M.: Knowledge sharing in organizations: A conceptual Framework. Human Resource Development Review 2(4), 337–359 (2003)
23. Meier, A.: An online stress management support group for social workers. Journal of Technology in Human Services 20, 107–132 (2002)
24. Ardichvili, A., Maurer, M., Li, W., Wentling, T., Stuedemann, R.: Cultural influences on knowledge sharing through online communities of practice. Journal of Knowledge Management 10(1), 94–107 (2006)
25. Sproull, L., Conley, C.A., Moon, J.Y.: Prosocial behavior on the Net. In: Amichai-Hamburger, Y. (ed.) The Social Net: The Social Psychology of the Internet, pp. 139–161. Oxford University Press, Oxford (2005)
26. Strauss, A.L., Corbin, J.M.: Basics of Qualitative Research: Techniques and Procedures for Developing Grounded Theory. Sage Publications, Thousand Oaks (1998)

An Ontology-Based Approach to Sentiment Classification of Mixed Opinions in Online Restaurant Reviews

Hea-Jin Kim and Min Song

Department of Library and Information Science, Yonsei University, Seoul, Korea
{erin.hj.kim,min.song}@yonsei.ac.kr

Abstract. Consumers review other consumer's opinion and experience of the quality of various products before making purchase. Automatic sentiment analysis of WOM in the form of user product reviews, blog posts and comments in online forum can support strategies in areas such as search engines, recommender systems, and market research and benefit to both consumers and sellers. The ontology-based approach designed in this work aims to investigate how to detect and classify mixed positive and negative opinions by interpreting with an ontology containing opinion information on terms. Our research question is whether disinterested subjectivity scores of sentiment ontology are pertinent to sentiment orientations not affected by reviewer's linguistic bias. The experimental results adopting opinion lexical resource achieve better and more stable performance in F-measure.

Keywords: opinion mining, sentiment analysis, ontology-based approach, SentiWordNet.

1 Introduction

With the rapid growth of the Internet and social network services consumers now review other consumer's opinion and experience of the quality of various products before making purchase. It is obvious for the field of business, as well, to take care of consumer's opinion of products or services that a company provides because online product reviews, posted online by consumers are other forms of word of mouth (WOM) communication [1-2]. Sentiments inside user reviews play a vital role in consumer decision making. Automatic sentiment analysis of WOM in the form of user product reviews, blog posts and comments in online forum can support strategies in areas such as search engines, recommender systems, and market research and benefit to both consumers and sellers [3-4]. However Thelwall et al. [5] argue that traditional approaches adopting machine-learning algorithms to sentiment analysis are problematic for several reasons; human-coded data and domain-dependency. Human-coded data means a set of texts assigned sentiment ratings by authors (i.e., consumers, reviewers, or readers) and these subjective texts are as an input to train in most machine-learning sentiment analysis algorithms [6-10]. Domain-dependency refers to the identification of expression topics in sentiment detection. Frequency-based features from texts for learning predictors are likely to be skewed to domain topics, for instance books, electronics, hotels, and restaurants and so on [11-14]. We will present a domain-independent and impartial ontology-based approach in orientations prediction in this paper.

Furthermore, traditional sentiment analysis mainly focuses on polarity of non-informal texts: positive or negative (e.g., [7, 10, 15-16]). However, mixed positive and negative sentiment often is observed in online communication [17] and multi points rating (e.g., one to five 'stars') from peers is more helpful for potential customers to make purchase decision [12].

The goal of the present paper is to investigate how to detect and classify mixed positive and negative opinions by interpreting with an ontology containing opinion information on terms. Our research question is whether disinterested subjectivity scores of sentiment ontology are pertinent to sentiment orientations not affected by reviewer's linguistic bias from individual's abstract language use [52]. To scrutinize this question, we exploit SentiWordNet 3.0[1] [18] and map subjectivity scores to opinion words in natural informal texts with Part-of-Speech (POS) information. Among scale of one to five ratings, it is obvious that score of 5 or 1 is as very positive or as very negative (or strongly recommended or strongly not recommended), the rest of the ratings, however, are mixed opinions. The most difficult task is neutral opinion, i.e., score of 3 in 5-stars rating. This score is as an ambiguous opinion to recommend, so reviews assigned three stars were removed from the corpus for the evaluation [34]. Therefore our study represents how well detect mixed opinions without reviewer's linguistic bias by a lexical induction approach.

This paper is structured as follows: we discuss related work in section 2. Section 3 describes our datasets and methodology. Then in Section 4, we analyze the experiment results followed by discussion and conclusion in Section 5.

2 Related Work

In this literature review, we discuss related opinion mining, known as sentiment, and/or subjectivity analysis.

A majority of sentiment or opinion analysis focuses on sentiments polarity and specialized features. Machine-learning algorithms are commonly used in detection of sentiments and features from informal texts like words [6, 8, 15, 20]. In terms of identification of sentiment orientations in sentence-level, it is important to figure out subjective words containing reviewer's opinion. Ghose and Ipeirotis [19] considered objective information in texts which means production descriptions by a seller, and subjective information by a customer which is very personal descriptions of the product. Then they proposed two ranking mechanisms; a manufacturer-oriented and a consumer-oriented. Dave et al. [15] used substitutions for product's name, unique or certain product type words in order to improve performance but yielded lower results. Jin et al. [21] ignored sentences describing products without reviewer's opinions and expressing opinions on another product model regarded as non-effective opinion

---

[1] http://sentiwordnet.isti.cnr.it/

sentences. These previous studies discovered subjectivity based on term frequencies the classifiers have observed from the training data. This approach lacks a semantic structure of words.

To consider relating to meanings in language, Turney [22] used two reference words "excellent" for positive estimate and "poor" for negative estimate of orientation in extracted phases. He implemented pointwise mutual information (PMI) to calculate the semantic distance between reference word and extracted phrase and yielded sentiment orientation of it. Hatzivassiloglou and McKeown [23] demonstrated prediction of sentiment orientations of linguistic link type based on conjunctions between adjectives like 'but' or 'and' in text. They assumed conjunctions constrain the orientation of conjoined adjectives and clustered the adjectives into two different orientations, i.e., positive or negative. The results gained 82% accuracy in prediction of same or different orientation between conjoined adjectives. A lexical resource like WordNet [24] was not chosen in their paper because they appealed lexical resources are not explicitly helpful to identify semantic orientation information, while a number of researchers have utilized WordNet with part-of-speech tagging techniques in order to extract opinion words [10, 25-27].

A human-labored ontology-based approach on opinion mining has some drawbacks [3], for instance, time-consuming effort from manually created opinion lexicons [6, 29], or annotator bias from human-assigned sentiment annotations [17, 23]. SentiWordNet, introduced by Esuli andSebastiani [28], a lexical resource derived from the synsets (set of synonyms) of WordNet to support sentiment classification and opinion mining application terms built on semi-automated method annotating all WordNet synsets. Each synset is connected to three sentiment scores, i.e., *pos, neg,* and *obj* according to how positive, negative, and objective terms are in the range of [0.0, 1.0] and their sum is 1.0 for each synset. Therefore SentiWordNet as graded lexical resource can provide detailed information of emotional nuances. Because of short history of SentiWordNet, there has rarely been evaluative works of SentiWordNet though, most of related studies reported outperformance against the traditional sentiment approaches [3, 31-36]. Our paper presents a lexical approach mapping with opinion information in SentiWordNet to user reviews to detect sentiment orientations.

An alternative approach has used topic models techniques as a fine-grained sentiments analysis [11, 37-38]. Some readers may want to find more detailed opinions for subtopics, for instance, price in electronic product review domain. Titov and McDonald [11] and Mei et al. [37] attempted to solve this problem by topic models and explored associated topics for sentiment extraction. Prediction of thumbs-up is not confined to products or services but also scientific articles. Wang and Blei [38] developed a recommendation algorithm to scholarly papers in bibliography sharing service community. They used latent factor models to predict ratings of never rated items by users which mean newly published articles, based on the predictive model they built on training sets which are previously rated items by users. In terms of a psychological approach, LIWC (Linquistic Inquery and Word Count) software[2] and the Japanese software proposed by Ptaszynski et al. [53] were attempted to detect emotions at an arousal level such as anger, anxiety, sadness and so on.

---

[2] www.liwc.net

As growing popularity of microblogging services like Twitter, sentiment analysis on Twitter data has been recently focused [39-42].

3 Methodology

3.1 Datasets

To conduct our experiment, we crawled a set of customer reviews from *yelp.com*[3]. The site provides Internet rating and review services including a directory service and review site with social networking features [43]. We collected total of 11,598 restaurant reviews with the 5-stars rating of restaurant given by the reviewer. The rating that a reviewer allocates to a review is denoted by a number of stars on a scale of 1 to 5 which is regarded as five categories of opinion in our experiment; 'strongly negative', 'mildly negative', 'neutral', 'mildly positive', and 'strongly positive', respectively.

We observed that our collected datasets have imbalanced proportions in five categories of opinion which is 592 texts in strongly negative, 900 in mildly negative, 1,827 in neutral, 4,045 in mildly positive, and 4,234 in strongly positive. To avoid undersampling results [44-45], we randomly selected 2,992 reviews. The final distribution of the reviews is 600 each (except for strongly negative opinion) for each class in the 5-classes categorization. Descriptive statistics of our final datasets is shown in Table 1.

Table 1. Descriptive statistics based on five categories of opinion

Opinion category	Number of texts	Mean sentences	Mean words
Strongly negative (s_neg)	592	12.8	86.5
Mildly negative (m_neg)	600	12.0	83.7
Neural (neutral)	600	11.3	80.1
Mildly positive (m_pos)	600	10.6	74.4
Strongly positive (s_pos)	600	10.3	71.4
Total	2,992	10.8	75.6

3.2 Proposed Approach

The first step is for preprocessing collected user reviews. This is done by using traditional text mining process, which contains removal of stop words, tokenizing, part-of-speech tagging, and lemmatizing. We do not use any negation detection techniques instead of excluding negatives like 'not', 'no', 'nor', 'only', or 'few' in stop words list.

---

[3] www.yelp.com

The second step is term selection and weighting. We set feature selection based on document frequency (*df*) and term frequency (*tf*). Terms on lower document frequencies below 5 were ignored [30] when feature selection and term frequencies were calculated for each cleaned term. We used *tf* and normalized *tf* with logarithm (*logtf*) as baseline term weighting. We then set different features set sizes using the top-down percentile features from 10 to 100 percentages.

The last step is to adjust base term frequencies from the mapping *swn_score* (formula 1). A word of SentiWordNet normally has several senses. To identify sentiment orientation of each word, an overall score [3, 31], first sense value [32], maximum value [33], or average score [34-35] is calculated for terms found with POS tagging. Unlike previous studies, we aggregated every value of each term for positive and negative, and subtracted negative value from positive value; hence, we get a negative score for negative orientation terms, according to the formula of *SentiWordNet score* for *term T*:

$$SWN\_score(t) = \sum_i^n Pos(t) - \sum_i^n Neg(t) \qquad (1)$$

By doing this, our term weights are extended to negative value from positive. We expect these stretched values would be distinguished from non-opinion words.

We exploited part-of-speech information after preprocessing of our datasets in opinion mining. Then we mapped our selected terms to SentiWordNet. In mapping process, we multiplied *swn_score* and *term frequency* weights of a matched term in order to adjust term weights, otherwise left a genuine *term frequency* weight for non-matched terms to entries of SentiWordNet.

The adjusted term weights for each document provides the input for Support Vector Machine (SVM) classifier implemented in the Weka 3.7.9[4] package. We trained and tested LibSVM wrapper in Weka on our datasets. LibSVM [46] is a popular Java library for building SVM classifiers and a wrapper called LibSVM, original code by Yasser EL-Manzalawy (= WLSVM) [47] has been provided by Weka. The classifier has 5 targets classes (s_neg, m_neg, neutral, m_pos, and s_pos).

We performed 10-fold cross validation with 10 different features set sizes (10, 20,…100% of all features). Figure 1 depicts the workflow of our approach.

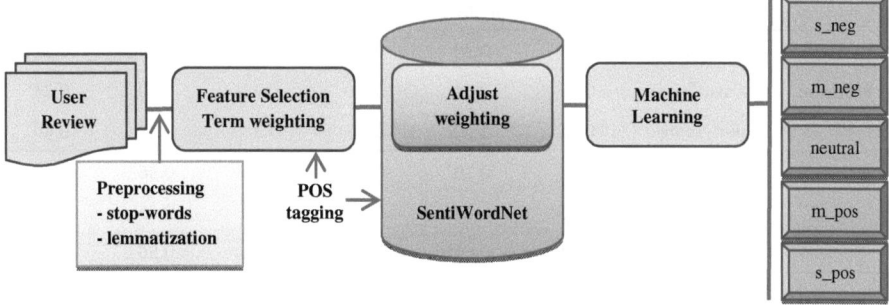

Fig. 1. Workflow of sentiment classification

---

[4] http://www.cs.waikato.ac.nz/ml/weka/

As the measure for the evaluation, we use classification accuracy and *F*-measure. One of the most effectiveness measures to classification performance is the accuracy, which is the proportion of the number of correctly allocated instances to the total number of instances, calculated from a confusion matrix [48]. *F*-measure is a single measure combining recall and precision scores. Recall is the proportion of the corrected allocated cases to the total cases correct. And precision is the proportion of the corrected allocated cases to the total cases [50-51]. Accuracy in each class has the same value as recall in text categorization evaluation.

$$accuracy = \frac{the\ number\ of\ cases\ correctly\ allocated\ to\ the\ class}{the\ total\ number\ of\ cases\ of\ that\ class} \quad (2)$$

$$F\_measure = \frac{2(precision \times recall)}{precision + recall} \quad (3)$$

3.2.1 Experiments

Before we fulfill our experiment, we wanted to test performance *swn_scores* of each term. We extracted verbs (v), adjectives (a), and adverbs (r) POS tags information from each document as opinion indicators [3] and compared predictivity between genuine opinion indicator values (*var_tf*) and ontology-derived values (*swn_score*) weighting of each term. Table 2 shows the results of pre-experiments. *var_tf* means simple term frequencies of verbs, adjectives, and adverbs, *swn*tf* means revised term frequencies multiplied by *swn_score*, and *swn_score* as standalone calculated values by formula (1) from SentiWordNet. Except for two cases (*var_tf* for s_neg in precision and s_pos in recall), adjusted term frequency (*swn*tf*) and *swn_score* alone showed better performance in results of precision, recall, and *F*-measure. So we were confident that implementation of lexicon resource into sentiment predictions are valuable. We moved on to the next step.

Table 2. Results of precision, recall, and *F*-measure in pre-experiment

	Term weight	s_neg	m_neg	neutal	m_pos	s_pos	avg
	var_tf	0.64	0.30	0.29	0.13	0.24	0.32
Precision	swn*tf	0.42	0.34	0.31	0.28	0.29	0.33
	swn_score	0.41	1.00	0.26	0.38	0.27	0.46
	var_tf	0.01	0.10	0.16	0.02	0.96	0.25
Recall	swn*tf	0.30	0.12	0.25	0.25	0.65	0.31
	swn_score	0.21	0.00	0.61	0.01	0.56	0.28
	var_tf	0.02	0.15	0.21	0.04	0.39	0.16
F-measure	swn*tf	0.35	0.17	0.28	0.26	0.41	0.29
	swn_score	0.28	0.00	0.36	0.02	0.36	0.21

To this end, we compare the proposed method with the traditional term weighting techniques in opinion mining. *base_tf* means a simple term frequency and *base_logtf* is a normalized term frequency. Those two serve as baseline whereas *rev_tf* and *rev_logtf* are calculated as revised weights mapping from a lexical resource, SentiWordNet.

4 Results

The main goal of the evaluation is to explore orientations prediction of mixed positive and negative opinions without reviewer's linguistic bias by a lexical induction approach.

Figures 2 to 6 show the experimental results of these aforementioned approaches in each opinion category with different features set sizes, as selected using the top-percentile features.

Strongly negative sentiment
In Figure 2, the SVM classifier using the adjusted term weights set as feature vectors are significantly more accurate than baseline on all data sets in the category of strongly negative sentiment. Overall, *rev_tf* showed the best performance, whereas the performance of traditional methods decreased in both accuracy and *F*-measure. The performance of *rev_tf* is noticeable because *rev_tf*, a modified term frequency based on an opinion lexicon, infers negative orientation very well over different features set sizes. Among four different term weights, *rev_tf* yielded the highest accuracy of 59.0 % at threshold of the 10th percentile features set. Note that we did not deal with any negation processing like previous research [3, 5, 25].

Mildly negative sentiment
In mildly negative orientation, the prediction performance is poor as shown in Figure 3. Unlike above very strong negativeness, baseline outperforms our ontology-based method in both accuracy and *F*-measure. Particularly, *rev_logtf* weight makes a weak performance from one half percentile features set, and does *base_tf* in strongly negative sentiment.

Neutral sentiment
Figure 4 shows comparison of the results of neutral sentiment prediction. *base_logtf* tends to opposite to *rev_logtf* as the size of features set increases. In accuracy, the performance gaps between *base_logtf* (38.0%) and the proposed *rev_logtf* (31.2%) is insignificant (6.8% at the 10th percentile features set) whereas the gap becomes 43.5% at the very last percentile using all terms as the feature selection set. This implies there are many common words, whether containing opinions or not, distributed in 3 stars rated informal texts, and these terms may share emotions either negative or positive. Thus, the classifier could train the inputs as negative or positive class. In this

category, we observe the semi-supervised sentiment scores bear sentimental difference which is confirmed by [28] and return to alleviate human linguistic bias. Actually this phenomenon prevail over five opinion classes regardless of competiveness. Either *rev_tf* or *rev_logtf* shows the stable performance over all features set sizes in *F*-measure as shown in Figures 2 to 6. This indicates that the proposed technique suggests unbiased weights than the existing methods.

Mildly positive sentiment
In mildly positive sentiment, term weights with SentiWordNet outperform baseline approaches both in accuracy and *F*-measure. In Figure 5, especially *rev_tf* achieves a higher and more stable performance over the all features set size. Table 3 shows specific figures in prediction performance in *F*-score at threshold of every 10th percentile features set size. To be specific, the results of *F*-score show the improvement from 2.2% to 19.6% in revised term frequency (*rev_tf*) and from 3.7% to 14.8% in revised normalized term frequency (*rev_logtf*).

Table 3. Experimental results on mildly positive sentiment at threshold of every 10th percentile features set in *F*-measure

features set	10th	20th	30th	40th	50th	60th	70th	80th	90th	all
base_tf	0.30	0.25	0.23	0.19	0.17	0.15	0.11	0.09	0.08	0.07
rev_tf	0.32	0.33	0.31	0.30	0.29	0.29	0.28	0.27	0.27	0.26
% increase	2.2	7.6	7.8	10.5	12.0	14.5	17.0	17.8	18.6	19.6
base_logtf	0.28	0.21	0.17	0.13	0.11	0.11	0.10	0.10	0.10	0.09
rev_logtf	0.32	0.32	0.30	0.28	0.26	0.25	0.22	0.21	0.20	0.18
% increase	3.7	10.8	12.6	14.8	14.4	13.7	12.4	10.8	10.5	9.2

Strongly positive sentiment
Unfortunately, ontology modified weights are degraded in accuracy in very strong positiveness prediction than baseline as shown Figure 6. Nevertheless, *F*-values are very close. This means that baseline methods show very low precision, while our technique achieves the better precision because *F*-score cares both recall and precision [50]. In Table 4, a traditional approaches yield lower precision results on the contrary to higher accuracy in the category of strongly positive sentiment predictions as shown in Figure 6. The precisions of baseline trade off against accuracy [51], our proposed weights using SentiWordNet, however, kept consistencies in both accuracy and precision, on the other hand, outperformed in precision as shown in Table 5.

Table 4. Accuracy and precision results on strongly positive sentiment at threshold of every 10th percentile features set

features set		10th	20th	30th	40th	50th	60th	70th	80th	90th	all
base_tf	accuracy	0.79	0.85	0.87	0.90	0.92	0.93	0.93	0.94	0.95	0.95
	precision	0.38	0.33	0.30	0.29	0.28	0.27	0.27	0.26	0.26	0.25
base_logtf	accuracy	0.77	0.86	0.90	0.92	0.94	0.96	0.97	0.98	0.98	0.98
	precision	0.38	0.32	0.29	0.27	0.25	0.24	0.23	0.23	0.23	0.23
rev_tf	accuracy	0.61	0.62	0.62	0.61	0.62	0.62	0.63	0.64	0.65	0.67
	precision	0.42	0.39	0.37	0.35	0.34	0.32	0.32	0.31	0.30	0.30
rev_logtf	accuracy	0.55	0.54	0.53	0.54	0.54	0.54	0.55	0.56	0.56	0.56
	precision	0.42	0.36	0.34	0.32	0.31	0.30	0.29	0.29	0.29	0.28

Table 5. Evaluating strongly positive sentiment predictions with precision at threshold of every 10th percentile features set

features set	10th	20th	30th	40th	50th	60th	70th	80th	90th	all
base_tf	0.38	0.33	0.30	0.29	0.28	0.27	0.27	0.26	0.26	0.25
rev_tf	0.42	0.39	0.37	0.35	0.34	0.32	0.32	0.31	0.30	0.30
% increase	4.1	5.7	6.7	5.9	5.8	5.1	4.9	4.6	4.6	4.6
base_logtf	0.38	0.32	0.29	0.27	0.25	0.24	0.23	0.23	0.23	0.23
rev_logtf	0.42	0.36	0.34	0.32	0.31	0.30	0.29	0.29	0.29	0.28
% increase	3.3	4.0	4.9	5.2	5.7	5.9	6.1	6.4	6.5	5.6

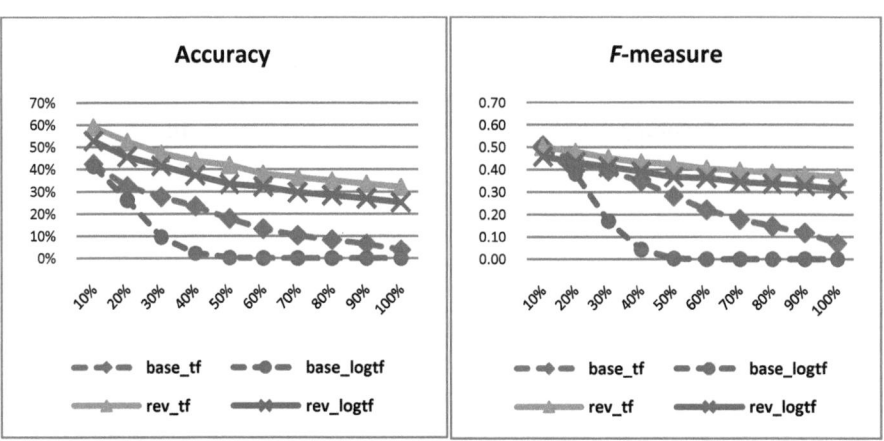

Fig. 2. Strongly negative sentiment classification accuracy and *F*-measure

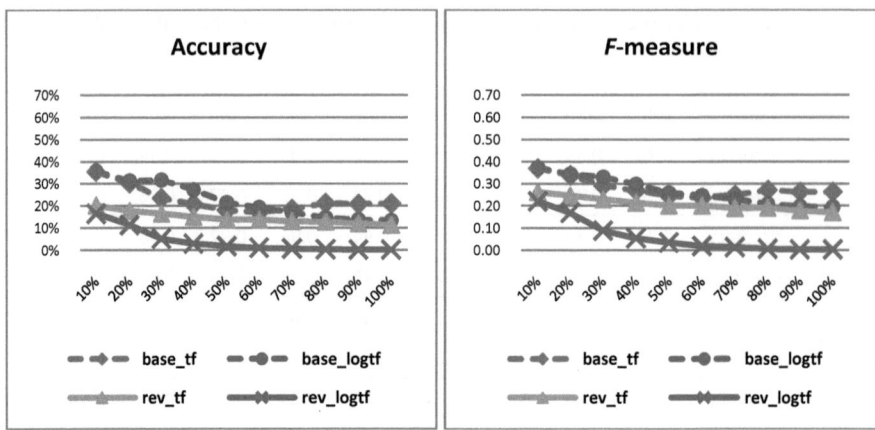

Fig. 3. Mildly negative sentiment classification accuracy and *F*-measure

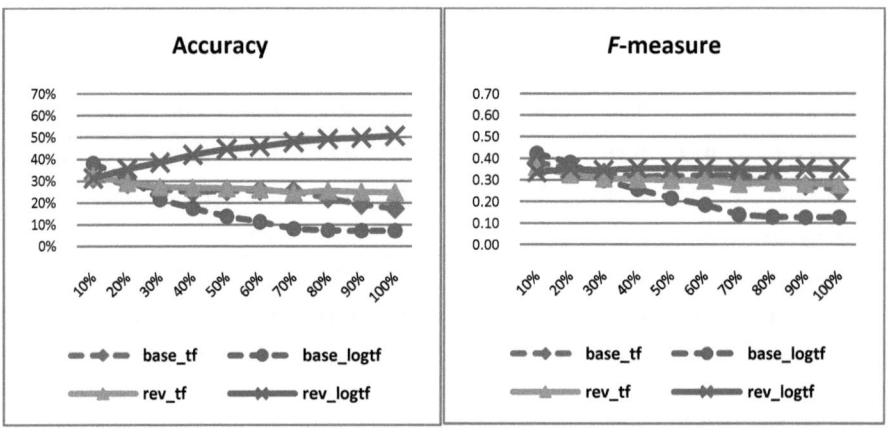

Fig. 4. Neural sentiment classification accuracy and *F*-measure

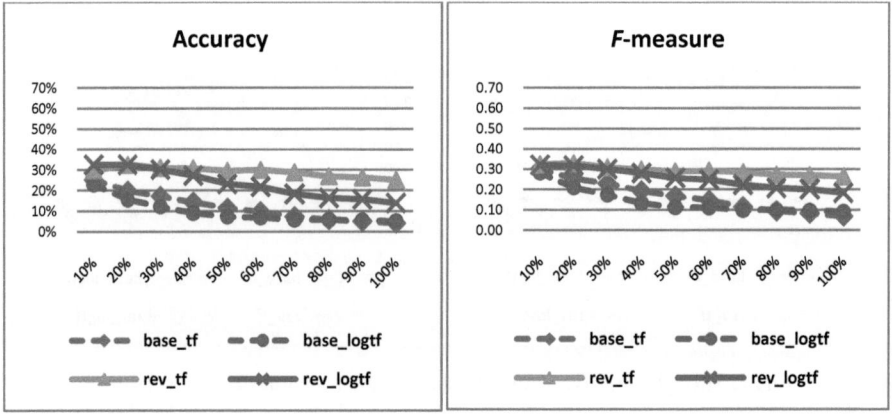

Fig. 5. Mildly positive sentiment classification accuracy and *F*-measure

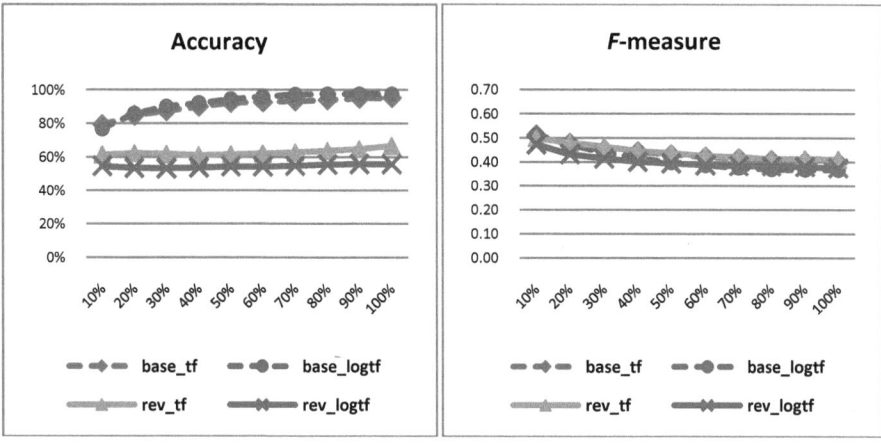

Fig. 6. Strongly positive sentiment classification accuracy and F-measure

5 Discussion and Conclusion

In this paper, we presented an opinion classification technique of mixed opinions without reviewer's linguistic bias by a lexical induction approach. The experimental results show that our ontology-based approach on opinion mining achieves the more stable classification performance than baseline approaches in F-measure values. In addition, the proposed method detects mixed positive and negative opinions, i.e., neutral sentiment in an effective manner. Even though our proposed method achieves the best performance in the category of strongly negative orientation, it yields lower prediction in mildly negative opinion. Given these mixed results our sentiment ontology-based approach still shows some promising results in several ways.

Firstly, our approach does not require prior domain knowledge and human labors. While the previous studies attempted to elaborate domain subjects with orientations prediction [11, 14, 19], our approach simply adopts pre-defined lexical resource. This empowers our technique to apply to cross- or multiple domains without prior knowledge and item features. Secondly, our method can avoid human biases stemmed from reviewer's habitual writing patterns by using machine-learned annotations. Lastly, we observe that in our approach, the rule of feature selection (removal of unique terms up to 90% or more) does not affect categorization average precision [30].

As the future study, we plan to incorporate semantic relations in SentiWordNet into our weighting scheme. SentiWordNet values for each term are calculated by sum of total scores for positive and negative each and subtraction between them. But, we do not utilize semantic relations available in SentiWordNet. Our research method is domain-independent prediction, and one further research direction may focus on compatibility of the proposed technique over heterogeneous domains.

Acknowledgments. This work was supported by the National Research Foundation of Korea Grant funded by the Korean Government (NRF-2012-2012S1A3A2033291).

References

1. Chevalier, J.A., Mayzlin, D.: The Effect of Word of Mouth on Sales: Online Book Reviews. No. w10148. National Bureau of Economic Research (2003)
2. Chen, Y., Xie, J.: Online Consumer Review: Word-of-Mouth as a New Element of Marketing Communication Mix. Manage. Sci. 54(3), 477–491 (2008)
3. Ohana, B., Tierney, B.: Sentiment Classification of Reviews Using SentiWordNet. In: 9th IT & T Conference, p. 13 (2009)
4. Ye, Q., Shi, W., Li, Y.: Sentiment Classification for Movie Reviews in Chinese by Improved Semantic Oriented Approach. In: 39th Annual Hawaii International Conference on System Sciences, HICSS 2006, vol. 3, p. 53b. IEEE (2006)
5. Thelwall, M., Buckley, K., Paltoglou, G.: Sentiment Strength Detection for the Social Web. Journal of the American Society for Information Science and Technology 63(1), 163–173 (2012)
6. Pang, B., Lee, L., Vaithyanathan, S.: Thumbs Up?: Sentiment Classification Using Machine Learning Techniques. In: ACL 2002 Conference on Empirical Methods in Natural Language Processing, vol. 10, pp. 79–86. Association for Computational Linguistics (2002)
7. Cui, H., Mittal, V., Datar, M.: Comparative Experiments on Sentiment Classification for Online Product Reviews. In: AAAI, vol. 6, pp. 1265–1270 (2006)
8. Ye, Q., Zhang, Z., Law, R.: Sentiment Classification of Online Reviews to Travel Destinations by Supervised Machine Learning Approaches. Expert Systems with Applications 36(3), 6527–6535 (2009)
9. Li, S., Wang, Z., Zhou, G., Lee, S.Y.M.: Semi-Supervised Learning for Imbalanced Sentiment Classification. In: 22nd International Joint Conference on Artificial Intelligence, vol. 3, pp. 1826–1831. AAAI Press (2011)
10. Hu, M., Liu, B.: Mining and Summarizing Customer Reviews. In: 10th ACM SIGKDD, pp. 168–177. ACM (2004)
11. Titov, I., McDonald, R.: Modeling Online Reviews with Multi-Grain Topic Models. In: 17th International Conference on World Wide Web, pp. 111–120. ACM (2008)
12. Li, F., Liu, N., Jin, H., Zhao, K., Yang, Q., Zhu, X.: Incorporating Reviewer and Product Information for Review Rating Prediction. In: 22nd International Joint Conference on Artificial Intelligence, vol. 3, pp. 1820–1825. AAAI Press (2011)
13. de Albornoz, J.C., Plaza, L., Gervás, P., Díaz, A.: A Joint Model of Feature Mining and Sentiment Analysis for Product Review Rating. In: Clough, P., Foley, C., Gurrin, C., Jones, G.J.F., Kraaij, W., Lee, H., Mudoch, V. (eds.) ECIR 2011. LNCS, vol. 6611, pp. 55–66. Springer, Heidelberg (2011)
14. Pan, S.J., Ni, X., Sun, J.T., Yang, Q., Chen, Z.: Cross-Domain Sentiment Classification via Spectral Feature Alignment. In: 19th International Conference on World Wide Web, pp. 751–760. ACM (2010)
15. Dave, K., Lawrence, S., Pennock, D.M.: Mining the Peanut Gallery: Opinion Extraction and Semantic Classification of Product Reviews. In: 12th International Conference on World Wide Web, pp. 519–528. ACM (2003)
16. Wilson, T., Wiebe, J., Hoffmann, P.: Recognizing Contextual Polarity in Phrase-Level Sentiment Analysis. In: Human Language Technology and Empirical Methods in Natural Language Processing, pp. 347–354. Association for Computational Linguistics (2005)
17. Thelwall, M., Buckley, K., Paltoglou, G., Cai, D., Kappas, A.: Sentiment Strength Detection in Short Informal Text. Journal of the American Society for Information Science and Technology 61(12), 2544–2558 (2010)

18. Baccianella, S., Esuli, A., Sebastiani, F.: SentiWordNet 3.0: An Enhanced Lexical Resource for Sentiment Analysis and Opinion Mining. In: LREC, vol. 10, pp. 2200–2204 (2010)
19. Ghose, A., Ipeirotis, P.G.: Designing Novel Review Ranking Systems: Predicting the Usefulness and Impact of Reviews. In: 9th International Conference on Electronic Commerce, pp. 303–310. ACM (2007)
20. Hu, X., Downie, J.S., West, K., Ehmann, A.: Mining Music Reviews: Promising Preliminary Results. In: 6th International Symposium on Music Information Retrieval (2005)
21. Jin, W., Ho, H.H., Srihari, R.K.: OpinionMiner: A Novel Machine Learning System for Web Opinion Mining and Extraction. In: 15th ACM SIGKDD, pp. 1195–1204. ACM (2009)
22. Turney, P.D.: Thumbs Up or Thumbs Down?: Semantic Orientation Applied to Unsupervised Classification of Reviews. In: 40th Annual Meeting on Association for Computational Linguistics, pp. 417–424. Association for Computational Linguistics (2002)
23. Hatzivassiloglou, V., McKeown, K.R.: Predicting the Semantic Orientation of Adjectives. In: 35th Annual Meeting of the Association for Computational Linguistics and 8th Conference of the European Chapter of the Association for Computational Linguistics, pp. 174–181. Association for Computational Linguistics (1997)
24. Miller, G.A.: WordNet: ALexical Database for English. Communications of the ACM 38(11), 39–41 (1995)
25. Zhuang, L., Jing, F., Zhu, X.Y.: Movie Review Mining and Summarization. In: 15th ACM International Conference on Information and Knowledge Management, pp. 43–50. ACM (2006)
26. Ding, X., Liu, B., Yu, P.S.: A Holistic Lexicon-Based Approach to Opinion Mining. In: International Conference on Web Search and Web Data Mining, pp. 231–240. ACM (2008)
27. Andreevskaia, A., Bergler, S.: Mining WordNet for a Fuzzy Sentiment: Sentiment Tag Extraction from WordNet Glosses. In: EACL, pp. 209–216 (2006)
28. Esuli, A., Sebastiani, F.: SentiWordNet: A Publicly Available Lexical Resource for Opinion mining. In: LREC, vol. 6, pp. 417–422 (2006)
29. Taboada, M., Brooke, J., Tofiloski, M., Voll, K., Stede, M.: Lexicon-Based Methods for Sentiment Analysis. Computational Linguistics 37(2), 267–307 (2011)
30. Yang, Y., Pedersen, J.O.: A Comparative Study on Feature Selection in Text Categorization. In: ICML, vol. 97, pp. 412–420 (1997)
31. Saggiona, H., Funk, A.: Interpreting SentiWordNet for Opinion Classification. In: 7th Conference on International Language Resources and Evaluation, LREC 2010 (2010)
32. Hung, C., Lin, H.: Using Objective Words in SentiWordNet to Improve Sentiment Classification for Word of Mouth. IEEE Intelligent Systems 28(2), 47–54 (2013)
33. Abulaish, M., Jahiruddin, Doja, M.N., Ahmad, T.: Feature and Opinion Mining for Customer Review Summarization. In: Chaudhury, S., Mitra, S., Murthy, C.A., Sastry, P.S., Pal, S.K. (eds.) PReMI 2009. LNCS, vol. 5909, pp. 219–224. Springer, Heidelberg (2009)
34. Denecke, K.: Using SentiWordNet for Multilingual Sentiment Analysis. In: IEEE 24th International Conference on Data Engineering Workshop, ICDEW 2008, pp. 507–512. IEEE (2008)
35. Hamouda, A., Rohaim, M.: Reviews Classification Using SentiWordNetLexicon. In: World Congress on Computer Science and Information Technology (2011)
36. Peñalver-Martínez, I., Valencia-García, R., García-Sánchez, F.: Ontology-Guided Approach to Feature-Based Opinion Mining. In: Muñoz, R., Montoyo, A., Métais, E. (eds.) NLDB 2011. LNCS, vol. 6716, pp. 193–200. Springer, Heidelberg (2011)

37. Mei, Q., Ling, X., Wondra, M., Su, H., Zhai, C.: Topic Sentiment Mixture: Modeling facets and Opinions in Weblogs. In: 16th International Conference on World Wide Web, pp. 171–180. ACM (2007)
38. Wang, C., Blei, D.M.: Collaborative Topic Modeling for Recommending Scientific Articles. In: 17th ACM SIGKDD, pp. 448–456. ACM (2011)
39. Go, A., Bhayani, R., Huang, L.: Twitter Sentiment Classification Using Distant Supervision. CS224N Project Report, pp. 1-12 (2009)
40. Jiang, L., Yu, M., Zhou, M., Liu, X., Zhao, T.: Target-Dependent Twitter Sentiment Classification. In: ACL, pp. 151–160 (2011)
41. Kouloumpis, E., Wilson, T., Moore, J.: Twitter Sentiment Analysis: The Good the Bad and the OMG! In: ICWSM (2011)
42. Barbosa, L., Feng, J.: Robust Sentiment Detection on Twitter from Biased and Noisy Data. In: 23rd International Conference on Computational Linguistics: Posters, pp. 36–44. Association for Computational Linguistics (2010)
43. Wikipedia.org, http://en.wikipedia.org/wiki/Yelp.com
44. Akbani, R., Kwek, S., Japkowicz, N.: Applying Support Vector Machines to Imbalanced Datasets. In: Boulicaut, J.-F., Esposito, F., Giannotti, F., Pedreschi, D. (eds.) ECML 2004. LNCS (LNAI), vol. 3201, pp. 39–50. Springer, Heidelberg (2004)
45. Wu, G., Chang, E.Y.: Class-Boundary Alignment for Imbalanced Dataset Learning. In: ICML 2003 Workshop on Learning from Imbalanced Data Sets II, pp. 49–56 (2003)
46. Chang, C.C., Lin, C.J.: LIBSVM: ALibrary for Support Vector Machines. ACM Transactions on Intelligent Systems and Technology (TIST) 2(3), 27 (2011)
47. EL-Manzalawy, Y., Honavar, V.: WLSVM: Integrating LibSVM into Weka Environment (2005), http://www.cs.iastate.edu/~yasser/wlsvm
48. Foody, G.M.: Status of Land Cover Classification Accuracy Assessment. Remote Sensing of Environment 80(1), 185–201 (2002)
49. Yang, Y., Pedersen, J.O.: A Comparative Study on Feature Selection in Text Categorization. In: ICML, vol. 97, pp. 412–420 (1997)
50. van Rijsbergen, C.J.: Information Retrieval, 2nd edn. Butterworth (1979)
51. Musicant, D.R., Kumar, V., Ozgur, A.: Optimizing F-Measure with Support Vector Machines. In: FLAIRS Conference, pp. 356–360 (2003)
52. Douglas, K.M., Sutton, R.M., Wilkin, K.: Could You Mind YourLanguage? An Investigation of Communicators' Ability to Inhibit Linguistic Bias. Journal of Language and Social Psychology 27(2), 123–139 (2008)
53. Ptaszynski, M., Dybala, P., Shi, W., Rzepka, R.: Contextual Affect Analysis: A Aystem for Verification of Emotion Appropriateness Supported with Contextual Valence Shifters. Int. J. Biometrics 2(2), 134–154 (2010)

A Novel Social Event Recommendation Method Based on Social and Collaborative Friendships

Yu-Chun Sun and Chien Chin Chen

Department of Information Management,
National Taiwan University, Taipei, Taiwan
{r00725005,patonchen}@ntu.edu.tw

Abstract. Many social network sites (SNSs) provide social event functions to facilitate user interactions. However, it is difficult for users to find interesting events among the huge number posted on such sites. In this paper, we investigate the problem and propose a social event recommendation method that exploits user's social and collaborative friendships to recommend events of interest. As events are one-and-only items, their ratings are not available until they are over. Hence, traditional recommendation methods are incapable of event recommendation because they need sufficient ratings to generate recommendations. Instead of using ratings, we analyze the behavior patterns of social network users to measure their social and collaborative friendships. The friendships are aggregated to identify the acquaintances of a user and events relevant to the preferences of the acquaintances and the user are recommended. The results of experiments show that the proposed method is effective and it outperforms many well-known recommendation methods.

Keywords: social network, recommendation systems, friendship analysis.

1 Introduction

With the rapid growth in mobile computing applications, many mobile services have been developed to help people exchange information. Social network sites (SNSs), such as Facebook, Google+, and Meetup[1], are popular mobile services that provide social functions to enable people to define their social networks and share information. Because SNSs are readily available through various mobile devices, people use the functions frequently to update friends with the latest news and information. Consequently, SNSs contain a great deal of valuable information that can be used to learn about a person's status and behavior [1]. Many SNSs provide event functions that enable users to raise or attend a social event. Basically, a social event groups SNS users together to achieve a specific goal at a certain time. Examples of social events are seminars, reunions, and group buying auctions. Invariably, event functions receive a great deal of attention because they enrich people's social relationships. For instance, more than 16 million social events are created on

---

[1] {facebook,plus.google,meetup}.com

Facebook each month[2]. While the functions enhance SNSs significantly, users often have difficulty finding events of interest. This is because many events created on SNSs keep popping that most of them fade away quickly from users' attentions. To provide better social event functions, recommendation systems must suggest events that are relevant to users' interests.

Recommendation systems constitute a practical research topic in the field of e-commerce. Methods like collaborative filtering [2] and content-based filtering [3] have been developed to recommend items of interest to users. Specifically, collaborative filtering utilizes the ratings of items to compute the similarity between items (or users). The similarity is then used to predict a relevance score of an item for a user. Items with high relevance scores are similar to the preferences of the user and are therefore recommended. Instead of using the item ratings, content-based filtering determines the similarity based on item descriptions. Items whose descriptions are similar to those rated by a user are considered relevant and recommended to the user. While both methods have been used successfully by several e-commerce sites like Amazon[3], they may not be appropriate for social event recommendations. This is because events are one-and-only items [4] that are only valid for a short time (i.e., several hours or few days) and ratings are not available until the events are over. Hence, it is difficult for collaborative filtering methods to collect enough ratings to make effective event recommendations. Meanwhile, content-based methods tend to recommend events that are similar to those a user has attended, rather than new events that may be of interest to the user.

Social influence is a key driver in motivating users to participate in a social event [5]. In other words, a user is more likely to attend an event if several of his/her friends are interested in it. In terms of SNSs, a user's friends may be defined as persons he/she interacts with frequently. In this paper, we propose a social event recommendation method that analyzes the behavior of users in SNSs to identify potential acquaintances. Specifically, we aggregate explicit, implicit, and collaborative friendships to identify a user's acquaintances. An event that attracts many acquaintances is recommended if its topic is relevant to the user's preferences. To evaluate the proposed event recommendation method, we collected 342 social events and 21101 event attendance records from Facebook and Meetup, and examined the effect of the above friendship aggregation. The experiment results demonstrate that the proposed method outperforms a number of well-known recommendation methods in terms of the recall rate.

2 Related Work

The social event recommendation task is affected by the cold start problem because of an event's one-and-only property [4]. That is, social events are only valid for a very short time and ratings are not usually available until the events are over.

---

[2] Facebook Newsroom: Products. http://newsroom.fb.com/Products
[3] http://www.amazon.com/

Consequently, it is difficult for recommendation systems to collect enough ratings to make effective event recommendations. Cornelis et al. [6] introduced a fuzzy method that integrates a content-based approach with collaborative filtering to remedy the above ratings sparseness problem. However, the authors did not validate their event recommendation method on a real-world dataset. Klamma et al. [7] integrated social network analysis techniques into a user-based collaborative filtering method to recommend academic events. The method examines researchers' participation networks to measure the similarity between their research interests. Then, researchers with similar interests are considered neighbors and recommendations are generated by suggesting events favored by them. However, the events studied in [7] are annual conferences, which do not possess the one-and-only property. In other words, the ratings of previous conferences can be used to avoid the rating sparseness problem. Mincov et al. [8] developed a dimension reduction method that extracts salient topics from event descriptions and utilizes the RankSVM formula to determine appropriate weights for the topics. Then, the topics and the corresponding weights are used to rank upcoming events. Daly and Geyer [9] established an interesting event management service that considers the location of events when making recommendations. They observed that attendees at an event are sometimes from nearby locations. Thus, they developed a location-based recommendation method to recommend local events to the target user. Dooms et al. [10] collected implicit feedback about events (e.g., event browsing or sharing an event on Facebook) as input for event recommendation algorithms. Their experiment results show that implicit feedback is useful for recommending events that are relevant to users' interests.

To summarize, many event recommendation methods exploit the content of events to recommend events that are relevant to users' interests. In addition to the content, we investigate the behavior of SNS users to recommend social events.

3 The Proposed Event Recommendation Method

Our event recommendation method is comprised of two key components, *acquaintance identification* and *recommendation generation*. The acquaintance identification process analyzes SNS users' behavior to learn their social and collaborative friendships. The friendships are then aggregated to identify acquaintances of the target user. In the recommendation generation process, events that the acquaintances and the target user have attended are examined to compute a recommendation score for an upcoming event. Then, upcoming events are ranked based on their recommendation scores and top-ranked events are suggested to the target user.

3.1 Acquaintance Identification

We first define the symbols used in the event recommendation method. Let $U = \{u_1, u_2, ..., u_N\}$ be a set of users in a social network site, and let $E = \{e_1, e_2, ..., e_M\}$ be all past events. E_n denotes all the events that user u_n has ever attended. Our objective is to recommend an upcoming event e_f that is relevant to the target user's preferences.

Social Friendships: In [11], Chen et al. showed that like-minded people prefer similar items. Hence, events favored by the target user's acquaintances should be recommended. In SNSs, acquaintances can be the friends that are made explicitly through user-friendly interfaces. However, few users are willing to disclose their friend lists due to privacy concerns. Here, we define $S_e(u_m,u_n)$ and $S_i(u_m,u_n)$ to measure, respectively, the explicit and implicit social friendships between u_m and u_n. The value of $S_e(u_m,u_n)$ is equal to 1 if u_n is on u_m's friend list; otherwise, it is 0. The implicit friendship $S_i(u_m,u_n)$ is defined as follows:

$$S_i(u_m,u_n) = (\frac{comm(u_m,u_n)}{\max_{u_i,u_j \in U} comm(u_i,u_j)} + \frac{like(u_m,u_n)}{\max_{u_i,u_j \in U} like(u_i,u_j)})/2, \quad (1)$$

where $comm(u_m,u_n)$ and $like(u_m,u_n)$ return, respectively, the number of comments and the number of 'likes' u_m left on u_n's posts. Simply, $S_i(u_m,u_n)$ is the normalized frequency of the comments and 'likes' u_m left on u_n's posts. It ranges from 0 to 1. The higher the value of $S_i(u_m,u_n)$, the greater the likelihood that u_m and u_n are friends because they interacted heavily on the social network site.

Collaborative Friendships: We also examine users' attendances at events (i.e., E_n) to measure their collaborative friendships. In the user-based collaborative filtering approach, users have similar tastes if their rating behavior is similar. We assume that SNS users who frequently attend events together would prefer similar upcoming events. Intuitively, the Jaccard coefficient [12], which measures the degree of overlap between the events attended by users, can be used to measure collaborative friendships. However, new events pop up on SNSs constantly. At any one time, there are numerous events, but users only attend a few of them. In such situations, the Jaccard coefficient underestimates the collaborative friendships between users. To address the problem, we extend the Jaccard coefficient by considering the content of events as follows:

$$C(u_m,u_n) = \sum_{e_i \in E_m, e_j \in E_n} sim(e_i,e_j)/(|E_m| \times |E_n|), \quad (2)$$

where $C(u_m,u_n)$ denotes the collaborative friendship between u_m and u_n; and $|E_m|$ indicates the number of the events u_m has ever attended. The function $sim(e_i,e_j)$ returns the content similarity between events e_i and e_j. We examine the titles and textual descriptions of events to measure their content similarity. If users attended similar events, we say they have a strong collaborative friendship. Here, we exploit the bag-of-words content similarity approach [13] to compute $sim(e_i,e_j)$. The approach represents an event as a high dimensional term vector where each dimension denotes a term extracted from the contents of previous events. We utilize the well-known TF-IDF [13] scheme to determine the weight of a term in a term vector. Once events are represented as weighted term vectors, we use the following cosine function to measure their content similarity.

$$sim(e_i, e_j) = \frac{\sum_{t \in T} w_{t,i} \times w_{t,j}}{\sqrt{\sum_{t \in T} w_{t,i} \times w_{t,i}} \times \sqrt{\sum_{t \in T} w_{t,j} \times w_{t,j}}}, \quad (3)$$

where $w_{t,i}$ is the weight of term t in e_i's term vector and T is the set of terms extracted from E. The value of $sim(e_i,e_j)$ is within the range [0,1]. The larger the value of $sim(e_i,e_j)$, the greater will be the content similarity between e_i and e_j.

Acquaintance Selection: Finally, we aggregate the collaborative and social friendships to measure the strength of friendships between users as follows:

$$F(u_m, u_n) = \alpha C(u_m, u_n) + (1-\alpha)[\beta S_i(u_m, u_n) + (1-\beta)S_e(u_m, u_n)], \quad (4)$$

where $0 \leq F(u_m,u_n) \leq 1$ and indicates the strength of the friendship between u_m and u_n. The parameters α and β weight the influence of the collaborative friendship and the implicit social friendship respectively. They also range from 0 to 1. To identify the acquaintances of u_m, we rank all other users based on $F(u_m,u_n)$. Then, the top H users are selected as the acquaintances of u_m.

3.2 Recommendation Generation

We use the following function to compute the recommendation score of an upcoming event e_f for a target user u_m.

$$R(u_m, e_f) = (1-\gamma) \frac{\sum_{e_i \in E_m} sim(e_i, e_f)}{|E_m|} + \gamma \frac{\sum_{u_n \in U_m} (\frac{\sum_{e_i \in E_n} sim(e_i, e_f)}{|E_n|} \times F(u_m, u_n))}{\sum_{u_n \in U_m} F(u_m, u_n)}, \quad (5)$$

where U_m denotes the top H acquaintances of u_m. The first part of the function measures the relevance of e_f to all the events u_m has ever attended. The second part computes the relevance of e_f to all events attended by the acquaintances of u_m. Parameter γ, which ranges from 0 to 1, controls the weight of the acquaintance relevance. In the next section, we investigate its effect on event recommendations. The value of $R(u_m,e_f)$ is within the range [0,1]. We rank upcoming events based on their recommendation scores and select the top-ranked events to compile an event recommendation list for u_m.

4 Experiments

4.1 Dataset and Performance Metrics

We collected events from Facebook and Meetup to evaluate our method's performance. Facebook provides a group function that helps people with common interests form communities. Users in a group can initiate or join events for a certain purpose. We collected 1925 users, 148 events, and 6476 event attendance records of a group-buying group from February 22, 2012 to November 22, 2012. The users initiate or join events to buy discounted products together or to share shipping fees. Meetup is a social network site that facilitates offline group meetings all over the world. Similar to Facebook, users in Meetup form communities and are also encouraged to create or join events (called "meetups") in the communities. We collected 9430 users, 194 events, and 14625 event attendance records of a mobile entrepreneur community from October 7, 2010 to October 7, 2012. Due to privacy concerns, no explicit social friendship information about users is available in the two datasets. Nevertheless, the users constantly interact with each other by commenting on, and expressing their feelings about, one another's posts. Therefore, we set parameter β at 1 to analyze the interactions and measure the users' implicit social friendships.

To assess our method's performance, we utilize the conventional leave-one-out procedure [14] to evaluate our method's performance. For each user, we evaluate the event recommendation performance over multiple runs. Each run treats an attended event as an upcoming event e_f and generates an event recommendation list based on all the information prior to e_f. In other words, the events, posts, and likes that occurred after e_f are excluded from the recommendation process. There are eight events on the recommendation list, as suggested by Bollen et al. [15]. The list is deemed successful if e_f is on it. The results of all the evaluation runs are then averaged to obtain the recommendation performance. The evaluation metrics are the overall recall and user recall rates, which measure the fraction of successful recommendations made for all events and all users respectively.

4.2   System Component Evaluation

First, we examine parameter H, which controls the number of acquaintances when event recommendations are generated. In this evaluation, we set H at 5, 10, 15, 25, 30, 35, 40, and 45. To evaluate the true effect of H, parameters α and γ are set at 0.5 and 0.5 respectively.

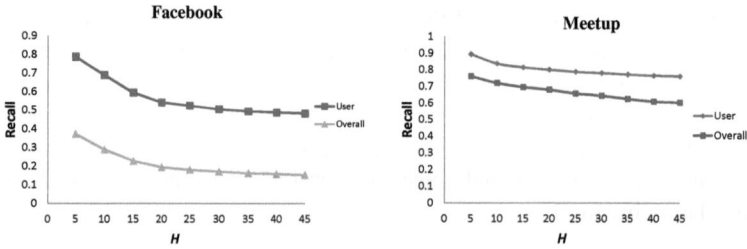

Fig. 1. The effect of parameter H on the system performance of Facebook and Meetup

As shown in Fig. 1, the overall recall and user recall decrease as H increases. This is because we rank acquaintances according to the strength of their friendships. If H is large, the friendships will be weaker than those under a small H. Consequently, the recommendations generated by a large H are prone to error. It is noteworthy that $H = 5$ produces the best recommendation performance. The result suggests that a small number of acquaintances is enough to make successful recommendations. While SNS users enjoy making friends, few of them are really close and influential. Because setting H at 5 yields the best recall rate, we utilize the setting in the following experiments.

Next, we examine the effect of parameter α, which determines the weight of collaborative friendships in the acquaintance identification process. Parameter α is set between 0 and 1, and increased in increments of 0.25. Under $\alpha = 0$, only interactions between users (i.e., social friendships) are considered when ranking their friendships. Hence, the selected acquaintances are people that frequently interact with the target user. By contrast, under $\alpha = 1$, all the selected acquaintances have similar tastes, and the events they attend are very similar to those attended by the target user. As shown in Fig. 2, the recall rates under $\alpha = 0$ are inferior in both datasets. For instance, in the Meetup dataset, only 6% of the evaluated events are successfully recommended under $\alpha = 0$. However, the recall rates improve dramatically when $\alpha > 0$. The results indicate that, when making event recommendations, acquaintances with similar tastes are more useful than acquaintances that interact frequently. We speculate that most SNS interactions are just for fun and therefore meaningless. By contrast, the collaborative friendship is informative because users generally think carefully before joining an event (especially group-buying events because joining them means spending money). It is noteworthy that the performances under $\alpha = 0.25$, 0.5, and 1 are quite close. The results demonstrate that our method is robust; that is, useful recommendations are always generated when collaborative friendships between users are considered. We utilize the setting $\alpha = 0.25$ in the following experiments.

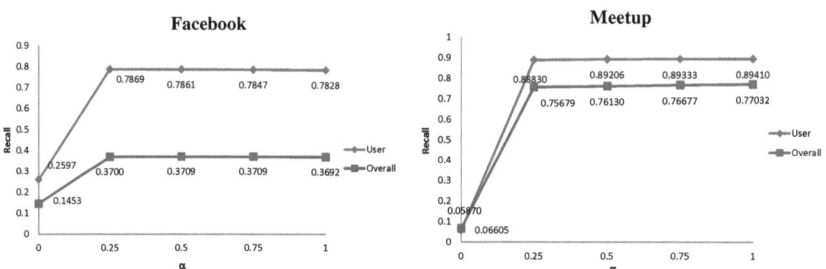

Fig. 2. The effect of parameter α on the system performance

Finally, we examine parameter γ, which controls the weight of the identified acquaintances when computing the recommendation score for an upcoming event. A large γ tends to suggest events relevant to the preferences of the identified acquaintances; while a small γ recommends events similar to those attended by the target user. As shown in Fig. 3, the recall rate decreases as the value of γ is reduced. In other words, users do not join events that are similar to those they have attended previously. For instance, users would not join group-buying events if the products are similar to items they bought at previous events. Similarly, users join meetups to learn something new. Therefore, Meetup users would not join events that are similar to those they have attended before. Consequently, a small γ value yields an inferior event recommendation performance. The proposed method achieves a superior recommendation performance by setting γ at 1. The result demonstrates that the identified acquaintances are representative and can be used to make effective event recommendations. We utilize the setting in the following experiments.

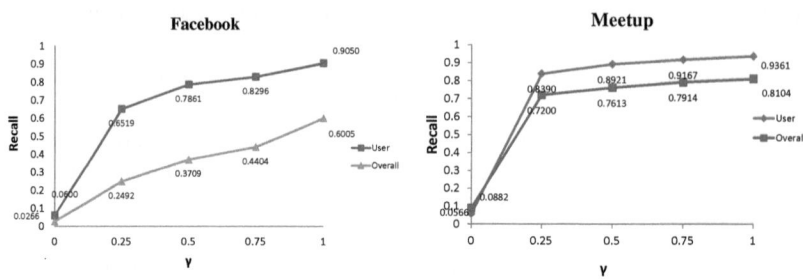

Fig. 3. The effect of parameter γ on the system performance

4.3 Comparison with Other Recommendation Methods

Next, we compare our method with four well-known recommendation methods: the user-based collaborative filtering (CF) method [16]; the singular value decomposition (SVD) method [17]; the content-based (CB) method [3]; and the hybrid method (CB+CF) [10], which compiles a recommendation list by combining the top four events produced by the CB and CF methods. To assess the effectiveness of the compared methods, the performance of a naïve method, which compiles a recommendation list by randomly selecting available events, is taken as the baseline. To ensure that the comparison is fair, the leave-one-out procedure (used in the previous experiment) is employed again. The length of a recommendation list is eight. Figures 4 and 5 show the performance results of the compared methods.

Except CB, the compared methods outperform the baseline method. The CB method yields low recall rates. The result corresponds with the finding of the previous experiment, i.e., users do not prefer events similar to those they have attended before. The recommendation results of CF are comparable to those of the proposed method under a large H; otherwise, our method outperforms the compared methods significantly. To summarize, our method explores the behavior of users to identify representative acquaintances. Then, the preferences of those acquaintances are utilized to recommend events of interest to the target user. Consequently, our method achieves the best event recommendation performance.

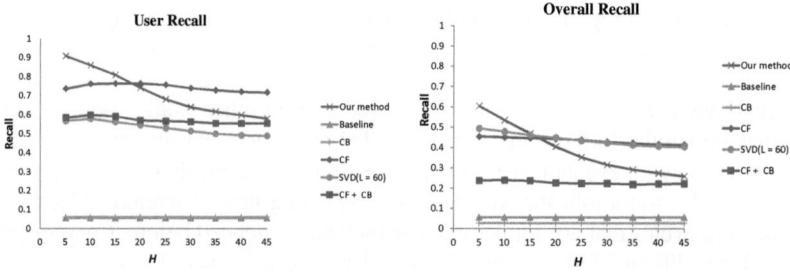

Fig. 4. The recall rates of the compared methods on the Facebook Dataset

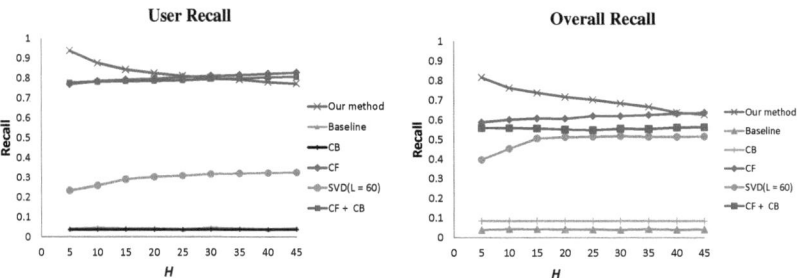

Fig. 5. The recall rates of the compared methods on the Meetup Dataset

5 Concluding Remarks

Recommending social events to SNS users is difficult due to the one-and-only property of events. In this paper, we have proposed an event recommendation method that analyzes the behavior of SNS users to identify their social and collaborative friendships. The friendships are then combined to identify a user's acquaintances so that upcoming events relevant to the preferences of the acquaintances and the user can be recommended. The results of experiments on two real-word datasets show that the proposed method resolves the rating sparseness problem of social events and identifies representative acquaintances of users. Thus, the method is effective, and it outperforms well-known recommendation methods.

Acknowledgements. This research was supported in part by NSC 100-2628-E-002-037-MY3 from the National Science Council, Republic of China.

References

1. Boyd, D.M., Ellison, N.B.: Social Network Sites: Definition, History, and Scholarship. Journal of Computer-Mediated Communication 13, 210–230 (2007)
2. Su, X., Khoshgoftaar, T.M.: A Survey of Collaborative Filtering Techniques. In: Advances in Artificial Intelligence (2009)
3. Pazzani, M.J., Billsus, D.: Content-Based Recommendation Systems. In: Brusilovsky, P., Kobsa, A., Nejdl, W. (eds.) Adaptive Web 2007. LNCS, vol. 4321, pp. 325–341. Springer, Heidelberg (2007)
4. Guo, X., Zhang, G., Chew, E., Burdon, S.: A hybrid recommendation approach for one-and-only items. In: Zhang, S., Jarvis, R.A. (eds.) AI 2005. LNCS (LNAI), vol. 3809, pp. 457–466. Springer, Heidelberg (2005)
5. Backstrom, L., Huttenlocher, D., Kleinberg, J., Lan, X.: Group formation in large social networks: membership, growth, and evolution. In: Proceedings of the 12th ACM SIGKDD International Conference on Knowledge Discovery and Data Mining, pp. 44–54. ACM, New York (2006)
6. Cornelis, C., Guo, X., Lu, J., Zhang, G.: A Fuzzy Relational Approach to Event Recommendation. In: Proceedings of the Indian International Conference on Artificial Intelligence, Pune, India, pp. 2231–2242 (2005)

7. Klamma, R., Cuong, P.M., Cao, Y.: You Never Walk Alone: Recommending Academic Events Based on Social Network Analysis. In: Zhou, J. (ed.) Complex 2009. LNICST, vol. 4, pp. 657–670. Springer, Heidelberg (2009)
8. Minkov, E., Charrow, B., Ledlie, J., Teller, S., Jaakkola, T.: Collaborative future event recommendation. In: Proceedings of the 19th ACM International Conference on Information and Knowledge Management, pp. 819–828. ACM, New York (2010)
9. Daly, E.M., Geyer, W.: Effective event discovery: using location and social information for scoping event recommendations. In: Proceedings of the Fifth ACM Conference on Recommender Systems, pp. 277–280. ACM, New York (2011)
10. Dooms, S., Pessemier, T.D., Martens, L.: A user-centric evaluation of recommender algorithms for an event recommendation system. In: Proceeding of the RecSys 2011: Workshop on Human Decision Making in Recommender Systems and User-Centric Evaluation of Recommender Systems and Their Interfaces, pp. 67–73. Ghent University, Department of Information technology, Chicago, IL, USA (2011)
11. Chen, C.C., Wan, Y.-H., Chung, M.-C., Sun, Y.-C.: An effective recommendation method for cold start new users using trust and distrust networks. Information Sciences 224, 19–36 (2013)
12. Manning, C.D., Schütze, H.: Foundations of Statistical Natural Language Processing. The MIT Press (1999)
13. Salton, G., Buckley, C.: Term-weighting approaches in automatic text retrieval. Information Processing & Management 24, 513–523 (1988)
14. Massa, P., Avesani, P.: Trust-aware recommender systems. In: Proceedings of the 2007 ACM Conference on Recommender Systems, pp. 17–24. ACM, New York (2007)
15. Bollen, D., Knijnenburg, B.P., Willemsen, M.C., Graus, M.: Understanding choice overload in recommender systems. In: Proceedings of the Fourth ACM Conference on Recommender Systems, pp. 63–70. ACM, New York (2010)
16. Candillier, L., Meyer, F., Fessant, F.: Designing Specific Weighted Similarity Measures to Improve Collaborative Filtering Systems. In: Perner, P. (ed.) ICDM 2008. LNCS (LNAI), vol. 5077, pp. 242–255. Springer, Heidelberg (2008)
17. Sarwar, B., Karypis, G., Konstan, J., Riedl, J.: Application of dimensionality reduction in recommender systems–a case study. In: Proceedings of the ACM WebKDD Web Mining for E-Commerce Workshop (2000)

Factors That Influence Social Networking Service Private Information Disclosure at Diverse Openness and Scopes

Basilisa Mvungi and Mizuho Iwaihara

Graduate School of Information, Production and System, Waseda University

Abstract. In this paper, we present findings about factors that influence private information disclosure activities in social networking service (SNS). Our study is based on two data sets: (1) Survey Data, responses to a questionnaire consisting of items such as users' privacy settings, motivations of using SNS (motivations), and risk awareness, and (2) Public Data, consisting of user profiles that were opened to the public. Openness score is calculated from the number of disclosed items. We study influential factors and their rankings at varying openness scores and disclosure scopes. Our findings reveal that, gender, profile photo, certain motivations, and risk awareness highly affect private information disclosure activities. However the ranking of influential factors is not uniform. Gender and profile photo have greater influence, however their influence becomes lower and loses significance as openness is getting higher, falling behind motivations and number of friends. We also observe consistent tendencies between both data.

Keywords: Information disclosure, motivations of using SNS, privacy, risk awareness, social networking service.

1 Introduction

Among popular social networking services (SNSs), Facebook (FB) has become of interest to scholars because of tremendous increase in the number of active users recently and richness of information provided on their profile pages. In spite of its popularity, none of previous studies measure factors that have higher contribution to information disclosure behavior, instead they looked at contribution of few factors [1,2,3]. In addition, most of information disclosure questions were based on yes/no responses instead of looking at precise levels of information disclosure. Users in FB may provide private information but limit to themselves or selected individuals who are considered safe. Effect of openness level of information disclosure has not been studied.

In this paper, we investigate factors that influence private information disclosure activities for FB users. Our study is based on (1) Survey Data, based on questionnaire, asking users about their privacy settings, motivations of using SNS (motivations), number of friends, and risk awareness, and (2) Public Data consisting of user profiles opened to the public. We use FB because it has detailed privacy settings that are suitable for our study. However, our study can also be applied to other SNSs, having the following features: (a) promoting expansion of friend networks and (b) *access control* for each profile attribute that can be specified by *privacy settings*. Survey Data is important because we can obtain latent factors which can only be acquired through user responses like motivations and risk awareness. Also, Survey Data capture intermediate disclosure scopes such as friends of friends. We aim to detect which factors are affecting private information disclosure based on (a) *opened* particular attributes, (b) *openness score*, which is the total number of attributes disclosed for representing amount of overall disclosed information. The importance of this study is that influential factors and their rankings at varying openness scores and disclosure scopes are verified statistically. Wider implication of this study is that SNS providers, conscious SNS users, and research communities will be aware of situations which users are encouraged or discouraged to share their private information, leading to development of more active and/or comfortable SNSs. In addition, researchers and marketers who are interested in profile attributes of users can realize which observable factors shall be given priority in promoting user activities and information sharing in an SNS.

The rest of the paper is organized as follows. In Section 2, we will review related work. In Section 3, we describe our research methodologies. Methods used for our study will be explained in Section 4. In Section 5, we will present our results. In Section 6, we will present a summary in the form of discussion, and Section 7 is a conclusion.

2   Related Work

Previous work [3,4] mentioned that the following factors influence information revelation: benefits of selectively revealing information outweigh the cost of privacy invasion, influence of peer pressure and "herding" behavior, ignorance about implication of personal information disclosure, and trust in participants and host sites. Only few factors were statistically verified. As part of their work, [3] looked at relationship between trust in other members and SNS as independent variables and information sharing as dependent variable. They found that weak significant positive relationship between the variables in FB and MySpace. [1] studied self-disclosure patterns (amount of disclosed information, number of friends, and number of disclosed attributes) of users as a function of age and gender. They found that females of the "Moj Mir" Russian SNS disclose more information till age of 46 where males begin to disclose more information. In addition, younger people have more friends than older people. In spite of the similarity of some of variables used, their objective, methodology and SNS used is different from ours. Comparison of personal and contact formation disclosure between genders was studied in [5] and found that percentage of females of the FB who disclosed contact information was significantly less than males. Our methodology will statistically verify whether gender affects information disclosure. [2] found that being males in FB caused them to more actively disclose MobilePhone and Address than females. Also younger people disclose PoliticalView, RomanticStatus, SexualOrientation, MobilePhone, and Classes more than older people. Similar methodology and SNS was used, however, factors they used (demography, audience, and general privacy) are few and slightly different from ours.

We have observed from previous work that, only few factors were studied their relationships to information disclosure. Motivations have been studied by many researchers, however they relate with other factors [6,7,8] that are beyond the scope of this study. It has also been studied as a separate variable [9,10,11] and their results reveal that members use FB more to interact with friends than acquire new friends. The relationship between motivations and private information disclosure has not been studied. In this paper we include a new factor *risk awareness* which indicates how much degree a user is familiar with online/offline security risks, and we study how risk awareness can affect private information disclosure. We also include number of friends as factor to predict information disclosure. A larger number of friends mean exposure of private information to larger audience, indicating a larger privacy risk.

3 Openness Score and Explanatory Variables

For each data set, we set a *scope threshold*. If a user open his/her attribute value beyond scope threshold, we regard that he/she *opened* this attribute. *Openness* is general notion of how user discloses information. *Openness score (OS)* is the number of his/her open attributes, where we assume an equal weight to each attribute. FB has five levels of *disclosure scopes* which are 1) only me, 2) selected individuals, 3) friends 4) friends of friends, and 5) everyone. For Survey Data, we set its *scope threshold* at between 3) friends and 4) friends of friends. This threshold is intended to illuminate whether the user has an intention to reveal information to friends of friends and beyond, or just to share with his/her friends. For Public Data, we set its *scope threshold* at between 4) friends of friends and 5) everyone, which is the only threshold we can capture for public data.

Let n be the number of attributes for each data set. We introduce a threshold to classify users by openness scores with parameter $k \in \{1,2,3,....,n\}$. For each k we classify users by whether or not a user's *openness score (OS)* is no less than k, and investigate factors that influence at each k through binary logistic regression.

We use the following explanatory variables. In Survey Data, the independent variables are Gender, Age, Number of friends in logarithmic form (LogFriends), ProfilePhoto, Motivations of using SNS items (Table 2), and Risk awareness items (Table 3), while the dependent variables are dichotomous variables on whether profile attributes are open or not, and openness score within thresholds. In Public Data, the independent variables are Gender and LogFriends while the dependent variable is openness score within thresholds. Independent variables, such as Motivations and Risk awareness, are missing from Public Data because they acquire users' response.

4 Method

4.1 Data Collection Method Used for Public Data

We collected FB users' profiles that are accessible from the Internet. We generated random user ID numbers and collected user profiles that were valid, having published friendlists. The number of collected user profiles is 49,281. In each user profile, we check whether the following list of attribute values are provided to the public:

Address, Bio and favorite quotations, Birthday, Current work, Email, Interested in and looking for, Location, Religious and political views, Relationship, Websites, Gender, and Number of friends. Users disclosing friend lists are considered.

4.2 Data Collection Method Used for Survey Data

Participants

A total of 276 participants with age range 20 to 59 agreed to participate in our survey. Females were 134 while males were 142. As compensation participants received a complimentary copy of possible social network risks and recommendation for safe practice in their email. Participants were recruited in six ways: invitation posted on Foreign Student Division Board and on an international house board, mailing lists, visiting close members who have FB account to their houses and giving them a questionnaire to participate, posting an invitation on several FB group walls and giving links to members who were interested in participating, recruited by a Japanese company, and chatting with online FB members and inviting them to participate by giving a link to a survey for those who showed interest. The link was opened for three weeks.

Survey Design

The survey questionnaire contained fourteen questions, which were FB related questions (motivations, disclosure scopes of some attributes, and the total number of friends). Question regarding risk awareness includes both SNS and non-SNS risk types. We used this technique in order not to prime the participants on the subject matter. The survey also contained the set of demographic questions (gender, age, status and current country of residence). The survey was prepared by using Microsoft Word to create a paper version and was also hosted on GoogleDocs. The survey is available in English and Japanese versions upon request from the authors.

To ensure content validity, pre-tested scales were relied on where possible. All items in questions were anchored using Likert scale except for questions related to age, current country of residents, and the total number of friends. We relied on [9,10,11,12,13,14] to prepare items for Motivations construct. We modified some items where needed. For the purpose of this study, we only considered interactive motivations and activities without lurking. Items for the construct Risk Awareness were developed by the authors on the basis of knowledge acquired from [4,15].

5 Results

Table 2 and Table 3 show the description of the items used and descriptive statistics of Motivations of using SNS and Risk awareness, respectively. Factor analysis could be performed to produce distinctive factors, however we wanted to look at individual items because statistically significant results were more detected at the individual level rather than grouped items.

We use binary logistic regression to examine the influence of each factor on dependent variables. The results are shown in Tables 3 to 5. Our study also aims to detect factors in ranked order; therefore we use Wald's statistics to rank the factors in descending order of influence. For Survey Data we considered five most influential factors for each criterion variable and ignore the rest.

Table 1. Summary statistics for Motivations of using SNS

Symbol	Motivations of Using SNS Items[1]	Mean	SD
Motiv1a	Keep in touch or finding out what an old friend is doing	3.90	1.15
Motiv1b	Reconnect with people you have lost contact with	4.20	.974
Motiv1c	Maintaining relationships with friends you see rarely	4.05	.958
Motiv1d	Contacting people who are very far from you cheaply	3.98	1.13
Motiv1e	Finding out about, joining or organizing events	3.02	1.14
Motiv1f	Creating group/groups	2.72	1.20
Motiv1g	Interact with group/groups (e.g. like, comment, post on group wall)	3.38	1.17
Motiv1h	Meeting new people and make new friends	3.20	1.29
Motiv1i	Meet face-to-face with someone that I learned about through Facebook	2.38	1.15
Motiv1j	Date, flirt	1.78	1.03
Motiv1k	Posting on wall (personal, friend or group wall)	3.11	1.22
Motiv1l	Updating or editing profile	2.88	1.15
Motiv1m	Changing current status	2.87	1.15
Motiv1n	Entertainment (to pass time, not be bored)	3.54	1.15
Motiv1o	All my friends have account/friends suggested it	2.64	1.22
Motiv1p	Commented on photo/post	3.13	1.10

Note: [1]Items ranged from 1 = strongly disagree to 5 = strongly agree   SD(Standard deviation)

Table 2. Summary statistics for Risk awareness

Symbol	Risk awareness Items[1]	Mean	SD
Risk5a	Forgetting to lock the door puts you in rich risk of being robed	3.82	1.60
Risk5b	If someone steal your credit card or has all your credit card information can easily buy things online	3.55	1.73
Risk5c	Using public networked computer and forget to sign out your account e.g. Email can make someone get your private information from your account	4.14	1.42
Risk5d	If you reveal information online like your current location, phone number, home address, birthday and gender you can easily be stalked, embarrassed, blackmailed and someone else can use your identity	3.71	1.45
Risk5e	Your Social Security Number can be detected by strangers if you disclosed zip code, gender and date of birth online (Applicable to U.S.)	2.20	1.86
Risk5f	The information you put public can be extracted by other people without your knowledge	3.72	1.50
Risk5g	The photos posted online e.g. Facebook can be used to identify your anonymous account if the same photo is used	3.36	1.70
Risk5h	Using AIM and become available whenever you are online will lead you to become monitored by strangers	2.79	1.81

Note: [1]Items ranged from 0 = no knowledge to 5 = very high knowledge   SD(Standard deviation)

Table 3. Binary logistic regression between factors and information disclosure of each attribute beyond scope threshold for Surveyed Data

	CurrentLocation	Hometown	Email	MobilePhone	IM	Address	Birthday Full	Birthday Partial	ContSum
LogFriends			2/0.24*						
Age			1/0.8***						2/0.89**
ProfilPhoto	1/8.52****	1/5.99****					1/5.75**		1/5.15****
Gender	3/0.50*	3/0.46*		1/0.02**	1/0.15*	1/0.18*	2/0.38*		1/0.32**
Motiv1a					5/2.29				
Motiv1b								5/0.62	
Motiv1d	5/0.79	2/0.61**							
Motiv1e							3/0.60*	2/1.50*	
Motiv1g						2/2.30*	4/0.72		
Motiv1h	2/1.52**	4/1.44*	4/1.84*						
Motiv1i		5/0.71*			3/0.18*				
Motiv1j									
Motiv1k					4/0.45	4/0.44*			
Motiv1l			5/2.50*	4/6.31*			5/1.52		4/1.80*
Motiv1m									3/0.58*
Risk5c					3/2.71*			3/0.71*	5/1.47
Risk5d				2/0.26*		3/0.51*			
Risk5e	4/0.86			3/0.66*	5/2.81*			4/0.81*	
Risk5f					2/0.52*				
Risk5g							5/0.72		

Notes: [1]Cell value represents: Ranked factors in Walds' value descending order / Exp (B) value. [2]Regression method: Enter. [3]Bold and Italic ($p < .10$). [4]Meaning of abbreviations used for headers in brackets: IM(InstantMessage), ContSum(Openness of contacts), LogFriends(Number of friends in log base 10). [5]* $p<.05$, ** $p<.01$, *** $p<.001$, **** $p<.0001$.

Table 4. Binary logistic regression between factors and trend of openness score for Surveyed Data

Openness score	≥1	≥2	≥3	≥4	≥5	≥6	≥7	≥8
LogFriends								2/2332*
Age							4/0.85	
ProfilePhoto	1/10.13****	1/9.42****	1/7.46****	1/9.97****	1/7.28****	2/7.00**		
Gender	5/0.60	3/0.50*	2/0.47*	2/0.34***	2/0.43*	1/0.23***	2/0.15*	
Motiv1a	4/0.76							4/5.75
Motiv1d	3/0.77			4/0.79	3/0.71	4/0.67		
Motiv1h		2/1.52*	3/1.41*		5/1.34			
Motiv1j	2/1.43							
Motiv1k		5/0.76					1/0.29**	1/0.07*
Motiv1l				5/1.34				

Table 4. (Continued.)

						3/1.76*	5/1.75
Motiv1n						3/1.76*	5/1.75
Motiv1p					4/1.54		
Risk5a			5/0.81			5/1.45	3/2.56*	3/6.53
Risk5b		4/1.31*	4/1.26					
Risk5e								50/0.45
Risk5g				3/0.83				

Notes: [1]Cell value represents: Ranked factors in Walds' value descending order / Exp(B) value. [2]Regression method: Enter. [3]Bold and Italic value (p <.10). [4]Meaning of abbreviations used for headers in brackets: LogFriends(Number of friends in log base 10). [5]* p<.05, ** p<.01, *** p<.001, **** p<.0001.

Table 5. Binary logistic regression between factors and openness score for Public Data

Openness score	≥1	≥2	≥3	≥4	≥5	≥6	≥7
LogFriends	2.262****	1.915****	2.183****	2.926****	3.674****	5.138****	9.401****
Gender	0.753****	0.683****	0.605****	0.524****	.439****	.428****	.404**

[1]The numbers represents the values of exp(B). [2]Meaning of abbreviations used for row headers in brackets: LogFriends(Number of friends in log base 10). [3]** p<.01, **** p<.0001.

6 Discussions

In both Surveyed and Public Data, female users are less likely to disclose their information in FB profile than males, at each of their scope thresholds, confirmed by exp(B) values being less than one (We encode female=0 and male=1). Being females cause them to be less likely to disclose their age, InstantMessage and MobilePhone. Our results are consistent with [2,5] who also observed that females disclose less contact information. However, our results differ from [1] because they observed that females disclose more attributes than males. Contrary to previous studies we further observe that, as openness scores increases, the odds of females to disclose information to the public decrease, in comparison with males. However, Gender loses significance at higher openness scores, as indicated in Table 4. At $OS \geq 7$, Motiv1k surpasses significance over Gender, and at $OS \geq 8$, Gender no longer attains statistical significance of 0.05. Also as Table 5 shows, the product of exp(B) values of LogFriends and Gender increases as openness increases beyond 2, which indicates Gender becomes less significant than Number of friends as openness becomes higher, and at $OS \geq 7$, being female is cancelled by having 40 percent more friends.

We observed that, users who disclose their profile photos to the public tend to disclose other information in their FB's wall. FB users disclosing profile photos to the public are more easily identified by friends than those without photos. So motivations to communicate with known friends and creating new friends are both expected to be positive factors of ProfilePhoto. Our results are consistent with these observations. In Table 3, ProfilePhoto has positive effect on disclosing CurrentLocation, Hometown, and BirthdayFull/Partial. In Table 4, ProfilePhoto is the most influential factor to openness when $1 \leq k \leq 6$. However, when $OS \geq 7$ and $OS \geq 8$, ProfilePhoto loses significance and factors Motiv1k, LogFriends, and Risk5a attain significance. This can be explained as at higher openness levels, most users already publish profile photos, so it is not a decisive factor; rather motivation and activity levels better identify these users.

Number of friends is not a strong predictor for each private information disclosure in Survey Data. The result in Table 3 reveals that users with higher numbers of friends, tend to disclose their Email address less to indirectly connected friends. One possible explanation is that having friends increases the chance of the user's profile to be viewed by indirectly connected friends which is considered to be unsafe. In Table 4, Number of friends becomes a significant factor when $OS \geq 8$. In Table 5, we can see increase of exp(B) values of LogFriends as openness becomes higher, indicating that having more friends is a positive factor of having more open attributes. We also observed that, the mean value of openness is positively correlated with LogFriends. However, the standard deviation of openness increases with LogFriends, indicating that openness level is diversified as LogFriends increases. A possible explanation is, users who are actively using FB tend to create more friends, so Number of Friends is a positive factor of the activity level of the user. Highly active users in FB are also more likely to disclose more attributes in their profiles, but the degree of openness to the public depend on their privacy attitudes, which explains the diversities observed at the higher openness levels.

As seen in Table 3, Age is a negative factor of disclosure of Email, meaning that younger users disclose more contact information than older users. We could say younger users are less sensitive to identity-disclosure risk. This result slightly contradicts with [1] who found that younger "Moj Mir" users disclose more attributes than older users except for Education, Career, Location, and Military.

Risk awareness has certain influence on private information disclosure. In Table 4, eight relationships between Risk awareness types and attributes are having significance level $p< 0.05$. In Table 4, one relationship, namely Risk5a and $OS \geq 7$, has significant level $p<0.05$. In Table 3, six out of eight relationships have exp(B) values below 1.0, indicating that these risk types are negative factors of disclosing the attributes. In these instances, we can say that users of high risk knowledge are actually avoiding exposure of their private information. However, three relationships are rather positive, meaning that exp(B) values are over 1. These relationships are between Risk5c and InstantMessage, and between Risk5e and MobilePhone in Table 3, and between Risk5a and $OS \geq 7$ in Table 4. In these cases, higher knowledge levels of these risk types are rather increasing the chance of disclosing corresponding attributes. It can be said as there exist users (employees and students in our findings) who are well aware of privacy risk; nevertheless they disclose their mobile phone numbers or IM address to a larger scope, indicating their awareness of risk-taking behavior. In Table 3, we observe that profile attributes which are influenced by risk knowledge levels can be categorized as *contact information*, consisting of Email, MobilePhone, and InstantMessage, and non-contact information BirthdayPartial. Exposing such information to unknown people is indeed a risky behavior, so it is reasonable that these attributes are more influenced by risk awareness than other attributes. One explanation of why BirthdayPartial is negatively influenced by risk knowledge levels is that users who disclose BirthdayPartial, are not disclosing BirthdayFull, meaning they are hiding their year of birth and age, indicating their risk-avoiding attitude.

We discovered that motivations to acquire new friends (partially or fully) in Facebook (Motiv1h and Motiv1l in Table 3 and 4, Motiv1e, Motiv1g, and Motiv1j in Table 3, and Motiv1n and Motiv1p in Table 4) are positive factors to openness score. But they are mostly weak. It can be considered as users seeking for new friends are more likely to publish profile attributes to promote themselves. Motivations related to staying in contact with known friends are considered to be negative factor (Motiv1d in Table 3 and Table 4 is a negative factor and significant). Users communicating with known friends already have ways of communicating offline, so they disclose only necessary information to facilitate their SNS motivations. We also found that the motivation of posting on self or others' profile wall (Motiv1k) is a negative factor. One explanation for this is that Motiv1k is to visit others' wall or groups and post, which can be done without disclosing their own profile information, so users who want more privacy can choose this activity. On the contrary, users who want to attract visits and postings from users beyond friends level need to publish more personal information.

7 Conclusion

We investigated factors that influence private information disclosure activities for Facebook users. We used two types of data sets, namely questionnaire-based Survey Data and public profile-based Public Data. These data sets encompass different degrees of disclosure scopes.

Our results revealed that, factors that highly affect private information disclosure activities are gender, profile photo, certain motivations of using SNS, and risk awareness, but their influences vary depending on openness scores and disclosure scopes, as evidenced by variations in influence rankings. Younger age causes more contact information disclosure, indicating more openness of younger users.

Gender and profile photo have greater influence when openness is lower. But at higher openness levels, certain motivations of using SNS and number of friends have greater influence. This suggests that a prediction model for openness should be developed, based on an activity indicator which incorporates number of friends.

We also observed that a consistent result between Surveyed Data and Public Data such that Gender is more influential when openness is lower, but number of friends is more influential when openness is higher. Our future work includes a more detailed user behavior analysis through a Facebook application that can provide analysis results to users.

References

1. Kisilevich, S., Ang, C., Last, M.: Large-scale Analysis of Self-disclosure Patterns among Online Social Network Users. Knowledge and Information Systems 32(3), 609–628 (2011)
2. Tufekci, C.: Can You See Me Now? Audience and Disclosure Regulation in Online Social Network Sites. Journal of Bulletin of Science, Technology, and Society 28(1), 20–36 (2008)

3. Dwyer, C., Hiltz, S.R., Passerini, K.: Trust and Privacy Concern within Social Networking Sites: A comparison of Facebook and MySpace. In: Proceedings of AMCIS 2007, Keystone (2007)
4. Gross, R., Acquist, A.: Information Revelation and Privacy in Online Social Networks. In: Proceedings of the ACM Workshop on Privacy in Electronic Society. Alexandria, Virginia (2005)
5. Taraszow, T., Aristodemou, E., Shitta, G., Laouris, Y., Arsoy, A.: Disclosure of Personal and Contact Information by Young people in Social Networking Sites: An Analysis Using Facebook Profiles as an Example. International Journal of Media and Cultural Politics 6(1), 81–102 (2010)
6. Papacharissi, Z., Mendelson, A.: Toward a New(er) Sociability: Uses, Gratifications and Social Capital on Facebook. In: Papathanassopoulos, S. (ed.) Media Perspectives for the 21st Century. Routledge (2011)
7. Spiliotopoulos, T., Oakley, I.: Understanding Motivations for Facebook Use: Usage Metrics, Network Structure, and Privacy. In: Proceedings of the SIGCHI Conference on Human Factors in Computing Systems, Paris, France, pp. 3287–3296 (2013)
8. Joinson, A.: 'Looking at', 'looking up' or 'Keeping up with' People? Motives and Uses of Facebook. In: Proceeding of the SIGCHI Conference on Human Factors in Computing Systems, pp. 1027–1036. ACM Press, New York (2008)
9. Lampe, C., Ellison, N., Streinfield, C.: A Face (book) in the Crowd: Social Searching vs. Social Browsing. In: Proceeding of the 2006 20th Anniversary Conference on Computer Supported Cooperative Work, pp. 167–170. ACM Press, New York (2006)
10. Lampe, C., Ellison, N., Streinfield, C.: Changes in Use and Perception of Facebook. In: Proceedings of the ACM Conference on Computer Supported Cooperative Work, San Diego, USA, pp. 721–730 (2008)
11. Subrahmanyam, K., Reich, S., Waechter, N., Espinoza, G.: Online and Offline Social Networks: Use of Social Networking Sites by Emerging Adults. Journal of Applied Developmental Psychology 29(6), 420–433 (2008)
12. Tuunainen, V., Pitkanen, O., Hovi, M.: Users' Awareness of Privacy on Online Social Networking Sites. In: Proceedings of the 22nd BLED eConference, Bled, Slovenia, pp. 1–16 (2009)
13. Govani, T., Pashley, H.: Student Awareness of Privacy Implications when Using Facebook (2012), http://lorrie.cranor.org/courses/fa05/tubzhlp (unpublished manuscript retrieved on September 3, 2012)
14. Tiffany, A., Yevdokiya, A., Yermolayeva, S.L.: College Students' Social Networking Experience on Facebook. Journal of Applied Developmental Psychology 30(3), 227–238 (2009)
15. Hasib, A.: Threats of Online Social Network. Seminar on Internetworking, Helsinki University of Technology, Helsinki, Finland (2009)

An Approach to Building High-Quality Tag Hierarchies from Crowdsourced Taxonomic Tag Pairs

Fahad Almoqhim, David E. Millard, and Nigel Shadbolt

Electronics and Computer Science, University of Southampton, Southampton, United Kingdom
{fibm1e09,dem,nrs}@ecs.soton.ac.uk

Abstract. Building taxonomies for web content is costly. An alternative is to allow users to create folksonomies, collective social classifications. However, folksonomies lack structure and their use for searching and browsing is limited. Current approaches for acquiring latent hierarchical structures from folksonomies have had limited success. We explore whether asking users for tag pairs, rather than individual tags, can increase the quality of derived tag hierarchies. We measure the usability cost, and in particular cognitive effort required to create tag pairs rather than individual tags. Our results show that when applied to tag pairs a hierarchy creation algorithm (Heymann-Benz) has superior performance than when applied to individual tags, and with little impact on usability. However, the resulting hierarchies lack richness, and could be seen as less expressive than those derived from individual tags. This indicates that expressivity, not usability, is the limiting factor for collective tagging approaches aimed at crowdsourcing taxonomies.

Keywords: Folksonomies, Taxonomies, Collective Intelligence, Social Information Processing, Social Metadata, Tag similarities.

1 Introduction

One of the essential principles behind the success of Web 2.0 applications is to harness the power of Collective Intelligence (CI) [1]. Collaborative tagging is one of the most successful examples of the power of CI for constructing and organizing knowledge in the Web. Tagging is a process that allows individuals to freely assign tags to a web object or resource, whereas folksonomy (a set of user, tag, resource triples) is the result of that process [2].

In recent years, folksonomies have emerged as an alternative to traditional classifications of organizing information [3]. However, they share the inconsistent structure problem that is inherited from uncontrolled vocabularies, which causes many problems like ambiguity, homonymy (same spelling but different meanings), and synonymy (terms have the same meaning) [4,5]. As a result, many researchers have focused on resolving this problem by proposing approaches for acquiring latent hierarchical structures from folksonomies and building tag hierarchies [6,7,8]. Building tag hierarchies from folksonomies can be useful in different tasks, like improving content retrieval [9], building lightweight ontologies [10] and enriching knowledge bases [11].

However, current approaches to automatic tag hierarchy construction come with limitations [12,13], such as suffering from the "generality-popularity" problem or the limited coverage of the existing knowledge resources. In this research, rather than propose a new algorithm for analyzing folksonomies we seek to explore whether a slight change in the tagging process itself could improve the resulting tag hierarchies. The new tagging approach takes the form of an "is-a" relationship, where users should type two related tags; i.e. Tag1 is a tag for the resource and Tag2 is a generalization of Tag1. The research hypothesis of this paper is that this simple relationship (Tag1 is-a Tag2) can be gained with low user interaction cost and provides higher quality tag hierarchies, compared to ones constructed from flat tags.

2 Related Work

Recently there have been several promising approaches proposed for building tag hierarchies from folksonomies. These approaches can be seen in two directions: First, **knowledge resources based approaches,** which aim to discover the meaning of tags and their relationships by using some knowledge resources, like WordNet and online ontologies. However, such resources are limited and they can only handle standard terms [12]. Second, **clustering techniques based approaches**. First pair-wise tag similarities are computed and then divided into groups based on these similarities. After that, pair-wise group similarities are computed and then merged as one until all tags are in the same group. For example, Heymann and Garcia-Molinay [6] propose an extensible algorithm that automatically builds tag hierarchies from folksonomies, extracted from Delicious and CiteULike. Their claim is that the tag with the highest centrality is the most general tag thus it should be merged with the hierarchy before others. Benz et al. [8] improved Heymann's algorithm by applying tag co-occurrence as the similarity measure and the degree centrality as the generality measure.

C. Schmitz et al [14] adopted the theory of association rule mining to analyze and structure folksonomies from Delicious. P. Schmitz [15] adapted the work of [16] to propose a subsumption-based model for constructing tag hierarchical relations from Flickr. Plangprasopchok et al. [7] adapted affinity propagation introduced by Frey & Dueck [17] to construct deeper and denser tag hierarchies from folksonomies. Yet Strohmaier et al. [3] showed that generality-based approaches of tag hierarchy, with degree centrality as generality measure and co-occurrence as similarity measure, e.g. [8] show a superior performance compared to probabilistic models, e.g. [7].

Although several approaches based on clustering techniques have been tried to structure folksonomies, they come with limitations [12,13]. These include the suffering from the "generality-popularity" problem. For example, Plangprasopchok and Lerman [18] found, on Flickr, that the number of photos tagged with "car" are ten times as many as that tagged with "automobile". By applying clustering techniques, "car" is likely to have higher centrality, and thus it will be more general than "automobile". Therefore, while tag statistics are an important source for constructing tag hierarchies, they are not enough evidence to discover concept hierarchies.

The experiment in this paper aims to explore whether a key reason for these limitations is that the current tagging approach, flat tags, does not provide a source of enough semantic evidence for building high-quality tag hierarchies. Rather we propose a slight change to the current tagging approach to benefit more from the power of CI by moving from collective folksonomies to collective taxonomic tag pairs.

3 Tag Hierarchies Leaning from Taxonomic Tag Pairs

In the proposed 'tag pairs' approach (Fig. 1), the user is required to tag the resource in the form of an "is-a" relationship, where Tag1 (the left box) is a tag for the resource and Tag2 (the right box) is a generalization of Tag1. For example, "Tower of London" is a "tower", or "tower" is a "building". The users can tag as much as they want for each resource in this way. Although this tag pairs approach shares some of the issues of single tags, such as spelling errors, it also provides additional semantics between tags. The algorithm we have adopted in our work (Table 1) is an extension of Benz's algorithm [8], which itself is an extension of Heymann's algorithm [6].

Table 1. Pseudo-code for the proposed algorithm

***Input**: user-generated terms (tag pairs)* , ***Output**: tag hierarchy*

1. Filter the tag pairs {tag1, tag2} by an occurrence threshold *occ*.
2. Order the tag pairs in descending order by generality (measured by degree centrality in the tag2– tag2 co-occurrence network).
3. Starting from the most general tag2, as the root node, and append tag1 as a less general term underneath tag2.
4. add all tags tag$2i$ subsequently to an evolving tree structure:
    (a) Calculate the similarities (using the co-occurrence weights as similarity measure) between the current tag tag$2i$ and each tag currently present in the hierarchy and add the current tag tag$2i$ as a child to its most similar tag *tag_sim*.
    (b) If tag$2i$ is very general (determined by a generality threshold *min_gen*) or no sufficiently similar tag exists (determined by a similarity threshold *min_sim*), append tag$2i$ underneath the root node of the hierarchy.
    (c) Append tag$1i$ as a less general term underneath tag tag$2i$.
5. Apply a post-processing to the resulting hierarchy by re-inserting orphaned tags underneath the root node in order to create a balanced representation. The re-insertion is done based on step 4.

The algorithm is affected by various parameters, including: occurrence threshold *occ* (the number of tag occurrences); similarity threshold *min_sim* (the number of tag co-occurrences with another tag); and generality threshold *min_gen* (the number of tag co-occurrences with other tags). Empirical experiments were performed to optimize these parameters. By incorporating tag pairs this variation of Benz's algorithm reduces the reliance on co-occurrence to create relationships, as nearly a half of the resulting tag hierarchy is created directly by users.

4 Empirical Study

Since we are proposing a new tagging approach, our experiment must look at two distinct aspects. Firstly its usability, in terms of efficiency, effectiveness and satisfaction, and secondly its performance in building high-quality tag hierarchies, in terms of semantics and structure. The technique will be successful if it increases the quality of tag hierarchy structure and semantics without significant impacting the ease of use.

To test the proposed tagging approach and collect data for executing the empirical study, we created the TagTree System. It is a web-based prototype which allows participants to tag some online resources by using the two tagging approaches (tag pairs and flat tags). The TagTree System consists of four main components:

- **User Interface:** The user interface describes to users how to use the tag pairs approach, with an example, and requires them to tag five resources with the tag pairs approach and another five with the flat tags approach (Fig. 1). To give a fair balance between the two tagging approaches, each of them is used first by half of the participants before they swap to the second approach.
- **Tag Content Recording:** This component records the tag content, including: user ID (user session), tags and time spent for each tagging action by the user.
- **Tags Normalising:** Before hierarchy construction, tags are passed to the normalisation process for clearing, e.g.: *Letters Lower-case* and *Non-English Deleting*.
- **Tag Hierarchy Constructing:** This component uses the proposed algorithm to construct tag hierarchies from the tags.

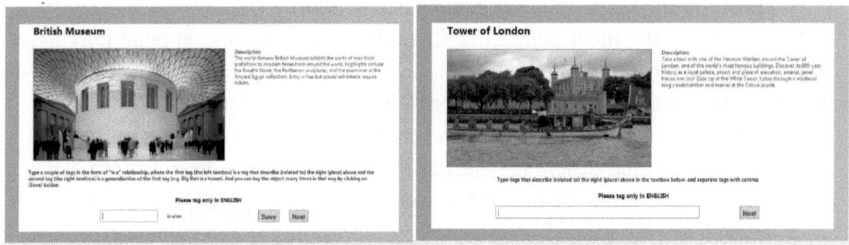

Fig. 1. The tag pairs (left) and the flat tags (right) tagging approaches

The Top 10 London Attractions[1], elected by visitlondon.com, were selected to be the resources used in the TagTree system for their popularity. The link of the TagTree system was sent to many people to take part in the study through email and social networks. After performing the normalization process, the dataset contains: 215 users, 333 tag pairs and 550 individual tags.

4.1 Evaluation Methodology

Evaluating taxonomy construction is a major challenge since there is not an approved evaluation dataset [3], nor an agreed methodology in the literature [19]. However, this

---

[1] http://www.visitlondon.com/things-to-do/sightseeing/tourist-attraction/top-ten-attractions

subsection proposes a broad evaluation process to evaluate two things: 1) The quality of tag hierarchies constructed from our tag pairs approach, compared to tag hierarchies constructed from flat tags (Evaluation metrics: 1, 3 and 4). 2) The usability of the tag pairs approach compared to the flat tags approach, in terms of efficiency, effectiveness and satisfaction (Evaluation metric: 2). The proposed evaluation process consists of four evaluation metrics as follows:

Evaluation metric 1: Evaluation by Human Assessment (subjective). We chose a simple but effective approach, used by [3], for evaluating the consistency of our tag hierarchy relations. Each direct taxonomic pair (t1, t2) from the tag hierarchy is extracted and manually judged as a relation of either: "same as", "kind of/part of", "somehow related", "not related", or "unclear". The idea behind this approach is that a better tag hierarchy will have a higher percentage of pairs being judged as "kind of" or "part of", and a lower percentage of pairs being judged as "not related" or "unclear".

Evaluation metric 2: Usability Evaluation (subjective). We conducted an online survey based on the System Usability Scale (SUS) [20], a Likert scale questionnaire of 10 items that is a standardized tool and has been used and verified in many domains [21]. The survey yields a single score, from 0 to 100. Bangor et al. found that a product with SUS scores below 50 will mostly have usability difficulties, whereas scores between 70-89, though promising, do not assure high acceptance of usability [22].

Evaluation metric 3: Evaluation against Reference Taxonomy (objective). Two researchers, in the field of Semantic Web and Knowledge Engineering, were asked to create appropriate reference taxonomy of the experiment domain (Fig. 2). To perform the comparison between a produced taxonomy (PT) and reference taxonomy (RT), Dellschaft and Staab propose two measures: taxonomic precision (*tp*) and taxonomic recall (*tr*) for comparing concept hierarchies [23]. The main idea is to compare the positions of two common concepts (*c*) in both hierarchies (local measure), and then to compare the two whole hierarchies (global measure). The local measure of taxonomic precision (*tp*) and taxonomic recall (*tr*) are defined, respectively, as follows:

$$tp(c, PT, RT) = \frac{|ce(c,PT) \cap ce(c,RT)|}{|ce(c,PT)|} \quad (1) \qquad tr(c, PT, RT) = \frac{|ce(c,PT) \cap ce(c,RT)|}{|ce(c,RT)|} \quad (2)$$

Where (*ce*) is characteristic excerpts that contain the ancestors (super-concepts) and descendants (sub-concepts) of the concept which are present in both hierarchies. The global measure of taxonomic precision (TP) is defined, as follows:

$$TP(PT, RT) = \frac{1}{|Cp \cap Cr|} \sum_{c \in Cp \cap Cr} tp(c, PT, RT) \quad (3)$$

Where Cp is the concepts set in the produced taxonomy, and Cr is the concepts set of the reference taxonomy. To give an overall overview, taxonomic F-measure (TF) is computed as the harmonic mean of taxonomic precision and recall as follows:

$$TF(RT, PT) = \frac{2.TP(RT,PT).TR(RT,PT)}{TP(RT,PT)+TR(RT,PT)} \quad (4)$$

Evaluation metric 4: Structural Evaluation (objective). It considers that a better tag hierarchy is a bushier and deeper hierarchy. To perform this evaluation, Plangprasopchok et al. [7] introduce a simple measure known as Area Under Tree (AUT). To compute AUT for a hierarchy, the distribution of nodes numbers in each level is computed first, and then the area under the distribution is calculated.

5 Results and Analysis

Two data sets were extracted from the TagTree system. The first one was collected by the tag pairs approach, while the second one was collected by the flat tags approach. In the experiment, three tag hierarchies (Fig. 2) are produced as follows: 1) **H1**: By using the tag pairs algorithm and the *tag pairs* dataset. 2) **H2**: By using the Benz's algorithm and the *flat tags* dataset. 3) **H3**: By using the Benz's algorithm and using the *tag pairs* dataset in which {tag1 is-a tag2} is considered as flat tags, i.e. ignoring the "is-a" relations.

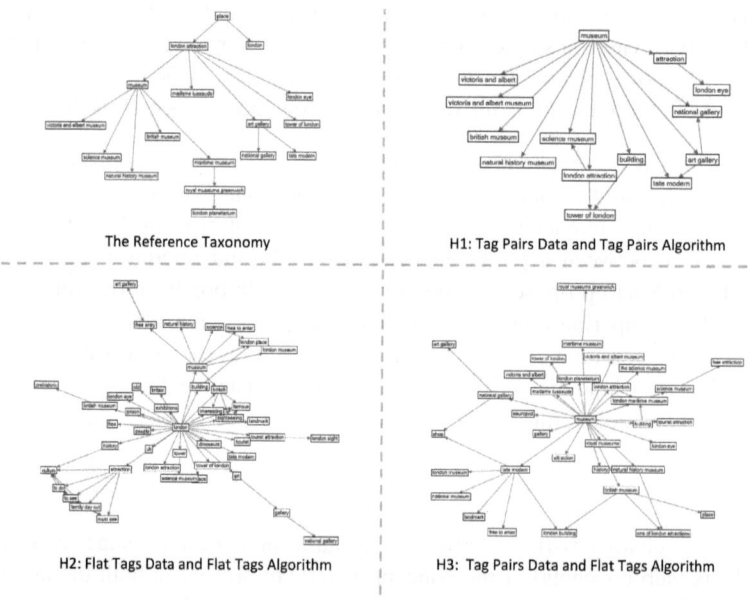

Fig. 2. The tag hierarchies used and produced in the experiment

5.1 Results of Semantic Evaluation

Fig. 3 shows the results of the semantic evaluation against the reference taxonomy, in terms of taxonomic precision (TP), taxonomic recall (TR) and taxonomic F-measure (TF). More similarity between a tag hierarchy and the reference taxonomy indicates that tag hierarchy has a higher quality.

The first observation that can be drawn from these empirical results is that there is a significant difference between H1 and H2. Our proposed extended algorithm yields tag taxonomies from our proposed tagging approach that is more similar to the reference taxonomy with taxonomic F-measure (TF) equal to 70.16%. Another important observation is that the quality of H3 is much better than the quality of H2, although both have been constructed by the same process (Benz's algorithm). However, H3 is built from tags originally collected from the tag pairs approach. This confirms our expectation and validates our research hypothesis that to make a small change to the current tagging approach can make a big change to the quality of the knowledge structure that can be built. Fig 4 shows there is a large difference between the percentages of pairs being judged as "is-a" in H1 and others. Also, all the pairs in H1 are related. On the other hand, H2 is the worst since it has the lowest portion of "is-a" relation and the highest portion of "not related" relation between pairs. Furthermore, similar to the observation in Fig. 3, the quality of H3 is much better than the quality of H2.

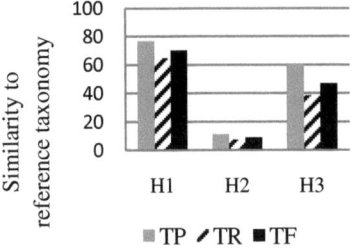

Fig. 3. Results of semantic evaluation against reference taxonomy

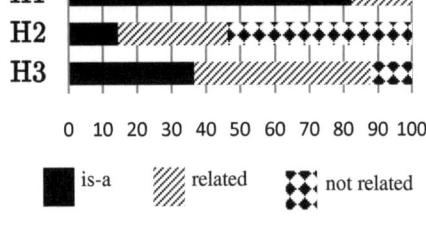

Fig. 4. Results of semantic evaluation by human assessment

5.2 Results of Structural Evaluation

Fig. 5 shows the results of AUT on the three produced tag hierarchies. H2 yields the highest AUT result, which indicates H2 is bushier and deeper than the others, whereas H1 yields the lowest AUT result. However, according to the previous results, H2 has a very small degree (TF=8.83%) of similarity to the reference taxonomy and also a big amount (53.62%) of "not related" tags pairs. This indicates that H2 has many noisy tags, while the proposed tagging approach and algorithm succeed in avoiding them. Ideally, it is a better to have an approach that generates both high quality and expressive tag hierarchies. While our approach succeeded in tackling the lack of consistent structure in folksonomies, it generated a less expressive hierarchy.

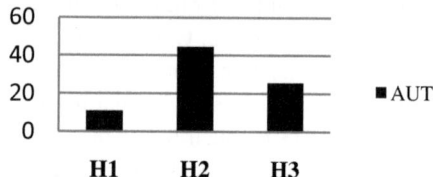

Fig. 5. Results of structural evaluation (AUT)

5.3 Results of Usability Evaluation

The results of this evaluation indicate that the average of SUS for the flat tags approach is 71.3%, with a standard deviation of 19.57, whereas the new approach obtains 54.6%, with a standard deviation of 16.22. First of all, this results show that the new approach is marginal acceptable since its average SUS score is over 50%. SUS scores are affected by the user experience by 15-16% between users who have "never" and "extensive" experience of the approach [24]. Consequently, the incipient SUS score of the tag pairs approach my get better over the time.

To measure the efficiency of the tag pairs approach compared to the flat tags approach, the time spent for each tagging action by users is recorded. The tagging action (ta) for the tag pairs approach means a pair of tags typed by the user, whereas for the flat tags approach means one tag or more typed by the user. The average time spent for using the tag pairs approach is 44.88 sec/ta, and 22.44 sec/tag. In contrast, the average time spent for using the flat tags approach is 71.90 sec/ta, and 36.37 sec/tag. This is a surprising result, as we expected the additional cognitive load of creating a tag pair to increase time taken, whereas our results show that users created a tag pair (containing two tags) in only slightly more time than it takes to generate a single tag.

6 Discussion and Conclusion

Current approaches to automatic tag hierarchy construction are limited, such as suffering from the "generality-popularity" problem or the limited coverage of the existing knowledge resources. Therefore, we proposed a slight change to the current tagging approach to cope with the lack of a consistent structure in folksonomies, and raise their semantic quality, whilst keeping the interaction cost of the process down. Our aim was to see if collecting tag pairs resulted in better quality hierarchy structure and semantics while minimizing the cost to usability.

The evaluation results of the produced tag hierarchies have shown that on the one hand the proposed tagging approach and algorithm have a superior performance in building high quality tag hierarchies when compared to ones built from the flat tags approach, but on the other hand they are not as rich, and therefore could be seen as less expressive. This problem might be caused by one or both of the following two reasons: First, the small size of the tagging resources and dataset in our experiment (we might expect to see a power law [25] in tag occurrence, and therefore the more tags gathered the longer the tail of rare tags), and second, the inability of the tag pairs approach to capture the intermediate concepts between the high levels and leaves of the hierarchy. To solve the first problem, there is a need to run the experiment with a system that can motivate a larger number of participants for a longer time. And to solve the second problem, the approach itself need to be improved, e.g. adding other relations between tag pairs or asking users some specific questions to encourage them to provide intermediate concepts. However, this has the risk of making the approach complicated and losing the simplicity and flexibility features of folksonomies.

In terms of usability, the results have shown that the tag pairs approach is marginal acceptable. Users were even able to complete the task by using the tag pairs approach in quicker way compared to the flat tags approach. Although the tag pairs approach succeeded in avoiding noisy tags, it seemed to restrict users in their choice of tags, leading them to particular key taxonomic relations (this may a reason behind the surprisingly low average time taken to create tag pairs). This meant that the tags produced by the tag pairs approach had lower quantity and diversity than with flat tags.

Our results therefore indicate that expressivity, not usability, is the limiting factor for collective tagging approaches aimed at crowdsourcing taxonomies.

References

1. O'Reilly, T.: What is web 2.0: design patterns and business models for the next generation of software (2005), http://oreilly.com/web2/archive/what-is-web-20.html (June 20, 2013)
2. Vander Wal, T.: Folksonomy Coinage and Definition (2007), http://vanderwal.net/folksonomy.html
3. Strohmaier, M., Helic, D., Benz, D., Körner, C., Kern, R.: Evaluation of Folksonomy Induction Algorithms. ACM Transactions on Intelligent Systems and Technology 3(4), Article 74 (2012)
4. Golder, S., Huberman, B.: Usage patterns of collaborative tagging systems. Journal of Information Science 32(2), 198–208 (2006)
5. Guy, M., Tonkin, E.: Tidying up tags. D-Lib Magazine 12(1) (January 2006) ISSN 1082-9873
6. Heymann, P., Garcia-Molinay, H.: Collaborative Creation of Communal Hierarchical Taxonomies in Social Tagging Systems. InfoLab Technical Report, Stanford (2006)
7. Plangprasopchok, A., Lerman, K., Getoor, L.: From saplings to a tree: Integrating structured metadata via relational affinity propagation. In: Proceedings of the AAAI Workshop on Statistical Relational AI, Menlo Park, CA, USA (2010)
8. Benz, D., Hotho, A., Stutzer, S.: Semantics made by you and me: Self-emerging ontologies cancapture the diversity of shared knowledge. In: 2nd Web Science Conference (WebSci 2010), Raleigh, NC, USA (2010)
9. Laniado, D., Eynard, D., Colombetti, M.: Using WordNet to turn a folksonomy into a hierarchy of concepts. In: 4th Italian Semantic Web Workshop: Semantic Web Application and Perspectives, Bari, Italy, pp. 192–201 (2007)
10. Mika, P.: Ontologies are us: A unified model of social networks and semantics. Web Semantics: Science, Services and Agents on the World Wide Web 5(1), 5–15 (2007)

11. Zheng, H., Wu, X., Yu, Y.: Enriching WordNet with Folksonomies. In: Washio, T., Suzuki, E., Ting, K.M., Inokuchi, A. (eds.) PAKDD 2008. LNCS (LNAI), vol. 5012, pp. 1075–1080. Springer, Heidelberg (2008)
12. Lin, H., Davis, J.: Computational and crowdsourcing methods for extracting ontological structure from folksonomy. In: Aroyo, L., Antoniou, G., Hyvönen, E., ten Teije, A., Stuckenschmidt, H., Cabral, L., Tudorache, T. (eds.) ESWC 2010, Part II. LNCS, vol. 6089, pp. 472–477. Springer, Heidelberg (2010)
13. Solskinnsbakk, G., Gulla, J.: Mining tag similarity in folksonomies. In: 3rd International Workshop on Search and Mining User-generated Contents (SMUC 2011), Glasgow, Scotland, pp. 53–60 (2011)
14. Schmitz, C., Hotho, A., Jäschke, R., Stumme, G.: Mining association rules in folksonomies. In : 10th IFCS Conference: Studies in Classification, Data Analysis and Knowledge Organization, Ljubljana, Slovenia, pp.261-270 (2006)
15. Schmitz, P.: Inducing ontology from flickr tags. In: Collaborative Web Tagging Workshop at WWW2006, Edinburgh, Scotland (2006)
16. Sanderson, M., Croft, B.: Deriving concept hierarchies from text. In: 22nd ACM Conference of the Special Interest Group in Information Retrieval, Berkeley, California, USA, pp. 206–213 (1999)
17. Frey, B., Dueck, D.: Clustering by passing messages between data points. Science 315(5814), 972–976 (2007)
18. Plangprasopchok, A., Lerman, K.: Constructing Folksonomies from User-Specified Relations on Flickr. In: 18th International World Wide Web Conference, Madrid, Spain, pp. 781–790 (2009)
19. Andrews, P., Pane, J.: Sense induction in folksonomies: a review. Artificial Intelligence Review, 1-28 (2012)
20. Brooke, J.: SUS-A quick and dirty usability scale. Usability Evaluation in Industry, 189–194 (1996)
21. Tullis, T., Stetson, J.: A comparison of questionnaires for assessing website usability. In: Usability Professional Association Conference, Minneapolis, USA, pp. 1–12 (2004)
22. Bangor, A., Kortum, P., Miller, J.: An empirical evaluation of the system usability scale. International Journal of Human-Computer Interaction 24(6), 574–594 (2008)
23. Dellschaft, K., Staab, S.: On how to perform a gold standard based evaluation of ontology learning. In: Cruz, I., Decker, S., Allemang, D., Preist, C., Schwabe, D., Mika, P., Uschold, M., Aroyo, L.M. (eds.) ISWC 2006. LNCS, vol. 4273, pp. 228–241. Springer, Heidelberg (2006)
24. McLellan, S., Muddimer, A., Peres, S.: The Effect of Experience on System Usability Scale Ratings. Journal of Usability Studies 7(2), 56–67 (2012)
25. Halpin, H., Robu, V., Shepherd, H.: The Complex Dynamics of Collaborative Tagging. In: 6th International Conference on the World Wide Web, Banff, Canada, pp. 211–220 (2007)

Automating Credibility Assessment of Arabic News

Mohamed Hammad and Elsayed Hemayed

Cairo University, Giza, Egypt
ibra@hammadian.com, hemayed@ieee.org

Abstract. During the past few years internet has witnessed a massive increase of Arabic language users. Accompanied with this increase in the number of users is an increase in e-publishing. However, necessary laws and regulations are not yet available to control the credibility of e-published content[1]. Furthermore, many political conflicts have risen after the Arab Spring. All of this led to an increasing demand for assessing the credibility of news in general and e-news in particular.

In this work, we present a system for automating credibility assessment of a news article based on two of the most important and most frequently violated criteria; (i) Does the news article indicate the source of its information? (ii) Does the news article indicate the time of occurrence of the reported event? For each of the chosen criteria, we build a classification model to classify a news article as either violating the criteria or not. News articles previously evaluated by MCE Watch (a manual service for news credibility assessment) are used in building and evaluation of our model. Experimental evaluations show that our model has accuracy that exceeds 82% for both criteria.

Keywords: Arabic language, credibility, machine learning, natural language processing, news.

1 Introduction

Internet has become one of the main sources of our daily life news, hundreds or thousands of news articles, videos, blog posts, microblog posts, RSS or social network feeds are published daily. This big chunk of info disseminated by governments, organizations, or persons impacts heavily different aspects of our daily life including but not limited to aspects of health, education, society, economics, politics, and religion.

During the past few years internet has witnessed a massive increase of Arabic language users with a total of 65,365,400 users as of 2010 and a growth rate of 2501.2% placing the Arabic language as the seventh most used language among all languages used on the internet and the first with respect to the growth rate. Egypt is placed on top of the Arabic speaking countries in terms of the number of users with an estimate of 21,691,776 users recorded in December, 2011 [11].

---

[1] Without violating freedom of speech.

From the statistics presented above, users in Arabic speaking countries are increasingly switching to consuming electronic content from social networks, news sites, web blogs ... etc. Accordingly, publishers are marching in the same direction. However, this switch is not yet accompanied by necessary laws and regulations that control e-publishing.

Nowadays and after more than two years from Egypt's January 2011 revolution, political conflicts are leaving no place in the media for an ordinary news consumer to evaluate the credibility of perceived information. This raises many questions about how technology can help an ordinary user evaluate the credibility of a piece of information before placing many (probably life changing) decisions on unreliable news.

As a first step towards answering the above question, a group of media experts created a service called MCE Watch [10]. A user of the service can submit a link of a news article, one of the experts analyzes the submitted article for potential credibility issues and replies to the user stating whether the article is credible. Credibility of news articles is assessed according to preset criteria published on the site, these criteria are listed in table 1. If an article was judged as not credible then a reasoning accompanies this judgment listing each of the violated criteria.

Table 1. Credibility evaluation criteria and points corresponding to each one

Alias	Criterion	Points
indicates_how_info_got	News article indicates how information was got	0.5
correct_info	News article doesn't contain incorrect or incomplete information	2
correct_news	News article doesn't report incorrect news	6
correct_photos	News article doesn't contain incorrect or manipulated photos	2
correct_order	Chronological order of information in the news article is correct	2
correct_video	Attached video is not conflicting with the article's text	2
correct_title	News article doesn't contain a misleading title	2
no_old_info	No old information posted as new	3
has_source	News article indicates the source's identity	2
correct_numbers	News article doesn't contain inaccurate numbers or statistics	2
unbiased	News article is unbiased	2
answers_how	News article answers 'how?'	0.5
answers_why	News article answers 'why?'	0.5
answers_where	News article answers 'where?'	1
answers_when	News article answers 'when?'	1
answers_who	News article answers 'who?'	4

MCE Watch provides credibility assessment services for articles from a predefined set of sources, these news sources are some of the most widely spread e-publishers in Egypt. The credibility of a source is updated once a new article from this source is assessed. Each source is assigned a pool of 500 points, once an article is judged as not credible, points corresponding to the violated criteria are subtracted from the pool. The pool of points is reset to 500 points monthly. The credibility of the source at any point of time corresponds to the percentage of remaining points.

Operating such a service manually has some limitations: (i) It is expensive to scale in terms of the number of requests to be served daily. (ii) There is a delay accompanied with the response that is variable according to the time of the request and the load on the service. Requests done at night are usually answered in the next morning. More users requesting assessment at the same time implies more delay.

The above issues motivated us to take a step towards automation of this process. In this paper we present a system that automates Arabic news credibility measurement. We crawl news articles previously labeled by MCE Watch and analyze them. Based on this analysis we choose two of the most violated and most important criteria to automate their measurement. Next, we build a classifier for each of the chosen criteria using data from MCE Watch for training and evaluation.

The rest of the paper is organized as follows. In section 2 we present the related work. The architecture and details of our system are presented in section 3. Section 4 presents an analysis of the dataset and the details and results of the conducted experiments. Finally, section 5 presents our conclusions.

2 Related Work

Measuring credibility is not something new, Gaziano and McGrath [6] showed that credibility of news in television and newspapers can be measured by multiple factors among them is being fair, unbiased, trustworthy, complete, factual, and accurate.

Automating information credibility measurement was considered by previous research for different sources of information on the internet. For webpages, Akamine et al. [2] used appearance information such as number of sentences, number of images, advertisements, presence of contact address, and privacy policy to assess the credibility of a website. Xu et al. [14] relied on trust features of a website to indicate its credibility.

For blogs, Weerkamp and de Rijke [13] used post-level features such as capitalization and spelling and blog-level features such as spamminess and comments to assess a blog's credibility. Juffinger et al. [8] relied on the similarity of blog posts with verified content which is an available news corpus. Al-Eidan et al. [3] combined blog-level and post-level features from Weerkamp and de Rijke [13] together with the similarity with verified content from Juffinger et al. [8] to assess the credibility of Arabic blogs. They considered content from Aljazeera[2] and Saudi Press[3] as verified.

For microblogs *e.g. Twitter*, credibility assessment has witnessed an increased attention from researchers during the past couple of years. Approaches use message based or content based features such as length of tweet, number of hashtags, lack or presence of inappropriate words, sentiment polarity in a tweet, similarity with verified content, presence of urls, retweets, special characters, emoticons,

---

[2] http://www.aljazeera.net/
[3] http://www.spa.gov.sa/

and @ mentions. Other approaches use source based or user based features such as number of followers, number of followings, ratio between followers and followings, if the account is verified by Twitter.com, age of user, and screen name. Al-Eidan et al. [4] combined message and source based features to classify Arabic tweets into three credibility classes, low, moderate or high. Gupta and Kumaraguru [7] used message-based features and user-based features to rank tweets by credibility during high impact events.

For news articles, Zhang et al. [15] and Kawai et al. [9] detect the sentiment bias for a query topic in a news website. Zhang et al. [15] detects sentiment bias for a query topic on four dimensions $Joy \Leftrightarrow Sadness$, $Acceptance \Leftrightarrow Disgust$, $Anticipation \Leftrightarrow Surprise$, and $Fear \Leftrightarrow Anger$. This bias is then visualized for users to help in credibility assessment. Kawai et al. [9] calculate sentiment on four different dimensions $Bright \Leftrightarrow Dark$, $Acceptance \Leftrightarrow Rejection$, $Relaxation \Leftrightarrow Strain$, and $Anger \Leftrightarrow Fear$.

As seen from the above, most research in automating credibility measurement used domain specific features to indicate credibility but none tries to extract content features from the articles which can be used to signal different credibility aspects. Furthermore, research in automating credibility measurement for Arabic content can be concluded in the work done by Al-Eidan [3, 4] for blogs and Twitter; both of them used news content from specific sources as verified content which doesn't fit our purpose of evaluating credibility for news content itself. In addition, the work done by Zhang et al. [15] and Kawai et al. [9] analyzes sentiment in news articles and visualizes it without taking credibility decisions, leaving that to the users. Accordingly and to the best of our knowledge noone has tackled the problem the same way we do in this paper.

3 Methodology

In this section we present a detailed description of our system. Figure 1 shows a high level block diagram of our system's architecture. Each of the upcoming subsections presents details for a block in this diagram.

The system starts by downloading and parsing data from MCE Watch – the next subsection presents details of this stage. The result of this stage is a labeled data set of news articles; each article is labeled as either credible or not and if not credible a list of the violations of the criteria in table 1 is compiled.

The distribution of non-credible articles among different credibility criteria is presented in table 2. As can be seen, *indicates_how_info_got*, *has_source*, and *answers_when* are the most frequently violated criteria. Together, *has_source* and *answers_when* are violated in 80.64% of the non-credible articles. If we added *indicates_how_info_got* to them we get a coverage of 91.5%. We choose to automate *has_source* and *answers_when* only since we currently have enough training to build a model for them. Although, we also have data for *indicates_how_info_got* but we leave it out because (i) *indicates_how_info_got* is assigned low weight of 0.5 points vs 1 and 2 for *has_source* and *answers_when*, respectively (ii) *indicates_how_info_got* and *has_source* are highly correlated, since both are tightly

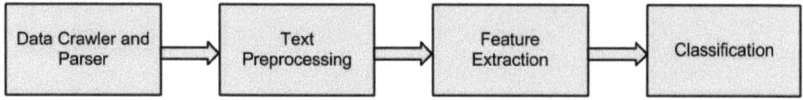

Fig. 1. System Architecture

coupled to the source, the later indicates if the source was explicitly mentioned in the article while the former indicates how did the source provide the writer of the article by information. In terms of numbers, 90.28% of the articles that violate *has_source* also violate *indicates_how_info_got* (iii) Our manual analysis of the collected data show that this signal has very high degree of noise.

Table 2. Distribution of non-credible articles among different credibility criteria

Criterion	Percentage %[4]
indicates_how_info_got	**49.67**
correct_info	2.22
correct_news	0.45
correct_photos	16.73
correct_order	0.01
correct_video	0.42
correct_title	3.51
no_old_info	0.04
has_source	**24.83**
correct_numbers	0.00
unbiased	2.28
answers_how	1.15
answers_why	2.32
answers_where	0.79
answers_when	**69.69**
answers_who	3.49

For each of the two criteria we build a binary text classifier. Typically, we perform preprocessing on the text, feature extraction, training and classification. In subsections [3.2-3.4] we discuss the details of each of these steps. We discuss details of preprocessing in the subsection following that for feature extraction since we base some of our preprocessing on the features.

3.1 Data Crawler and Parser

Figure 2 presents a detailed architecture of this stage. First, we crawl MCE Watch for news articles judged between December 2012 and April 2013. Next,

---

[4] Percentage doesn't sum to 100% because an article can violate more than one criteria at the same time.

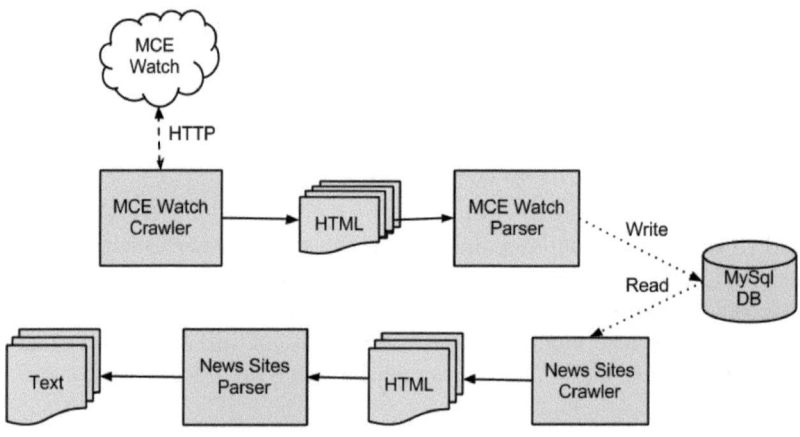

Fig. 2. System Architecture: Data crawler and parser

we parse the HTML from MCE Watch to populate a database with URLs of news articles, the name of the news site for each article, judgment of MCE Watch; credible or not and the list of violations in case of a non-credible article. After that we download the news articles and parse each one extracting the title and text.

3.2 Feature Extraction

In this subsection we present the features for each of the chosen criteria. We start by features for *has_when* followed by features for *has_source_info*.

3.2.1 Does the Article Answer 'When'?

We model the answer to this question as a binary text classification problem. We classify news articles into one of two classes; **HasWhen** or **NoWhen**. Articles of **HasWhen** class doesn't violate *answers_when* criteria while articles of **NoWhen** do violate it. We use a combination of features to represent each article:

- *N-GRAMS(NG)*: we run experiments with combinations of word n-grams with $n \leq 3$. We use term frequencies without inverse document frequency weighting (IDF).
- *TIME(T)*: this is a binary feature indicating whether any word from our manually created lexicon[5] of time-bearing words exists in the article. The lexicon consists of all weekdays, today and yesterday.
- *TIME-DISTANCE(TD)*: this is a real-valued feature. It is calculated as the distance of the time-bearing word found by *TIME* relative to the beginning of the article. It is transformed by an exponential function to the

---

[5] The idea of using a manually crafted lexicon was effective in other domains as well [1].

range $[1, e^1]$, Eq. (1). Our analysis indicates that time-bearing words that refer to the time of the reported event tend to occur early on in the article.

$$TD = \begin{cases} e^{(1-\frac{T_i+1}{|D|})} & \text{if } T = 1 \\ 1 & \text{if } T = 0 \end{cases} \quad (1)$$

where T_i is the index of a time-bearing word in a vector of the article words and $|D|$ is the number of words in the article.
- *TIME-CONTEXT(TC)*: this is a real-valued feature. It is calculated as the minimum absolute distance between the time-bearing word found by *TIME* and any of the words in a manually created lexicon of contexts (LC), Eq. (2). Contexts include conferences, tweets, Facebook updates, meetings, TV shows, telephone calls, and statements. We only search for context words in a frame of $\pm \epsilon$.

$$TC = \begin{cases} \min_{T_i-\epsilon < j < T_i+\epsilon} \frac{|T_i-j|}{\epsilon} & \text{if } T = 1 \text{ and } j \text{ in } LC \\ 1 & \text{otherwise} \end{cases} \quad (2)$$

where T_i is the index of time-bearing word in a vector of the article words. In our experiments, we choose $\epsilon = 10$.

3.2.2 Does the Article Has 'Source Info'?

We again model the answer to this question as a binary text classification problem. We classify news articles into one of two classes; **HasSourceInfo** or **NoSourceInfo**. Articles of **HasSourceInfo** class doesn't violate *has_source* criteria while articles of **NoSourceInfo** violate it. We use a combination of features to represent each article:

- *N-GRAMS*: we run experiments with combinations of word n-grams with $n \leq 3$. We use term frequencies without IDF weighting.
- *SOURCE*: this is a binary feature indicating whether the news article is referring to one of the popular news agencies as a source.
- *VERB*: this is a binary feature indicating whether a verb from a manually created lexicon of verbs exists. All verbs indicate that someone said something.

3.3 Preprocessing

For both tasks, we perform basic normalization [5] of Arabic text. This normalization involves transformation of one form of a letter to another form mostly because both forms are erroneously used interchangeably:

---

[6] We use Buckwalter transliteration in this paper.

- We transform all forms of *alef*: "إ($>^6$), إ(<), آ(|)" to "ا(A)".
- We transform all forms of *alef layyena*: "ى(Y) to "ي(y)"
- We remove all Arabic diacritics. We also remove all punctuation and digits.

We tokenize each news article into a vector of words and discard tokens of length < 3. In case of *has_source*, we add two processing steps on the tokens:

- If a token starts with "ال(Al)" – meaning "the", we strip it. Like, "الأحزاب (Al>HzAb)" – meaning "the parties" → "أحزاب(>HzAb)" – meaning "parties".
- If a token starts with "و(w)" – meaning "and", it is commonly written as the first letter of the following word as in "وقال"(wqAl) – meaning "and he said". We remove it giving "قال(qAl)" – meaning "he said".

We don't do the above for *"answers_when"* because all Arabic weekdays are commonly preceded by *"The"* and when stripped can mean a different thing leading to ambiguity. For example, "الأحد(Al>Hd)" – meaning "Sunday", when stripped gives "أحد(>Hd)" – meaning "Someone". "الإثنين(Al<vnyn)" – meaning "Monday" gives "إثنين(<vnyn)" – meaning the number "Two". Similarly, for "الخميس(Alxmys)" and "الجمعة(AljmEp)" – meaning "Thursday" and "Friday" respectively, when stripped gives "خميس(xmys)" and "جمعة(jmEp)" that correspond to common Egyptian male names.

3.4 Classification

We use SVM with linear kernel from Python's Scikit-learn package [12]. Literature suggests that it does well on text classification tasks. We set c to 1. All feature vectors are L2 normalized.

4 Results and Evaluation

In this section we describe the details of our experiments and discuss their results. We conducted three experiments; the first one presents classification results for *answers_when* criteria. It presents results for the different combinations of features presented in section 3.2.1. The second experiment presents classification results for *has_source* criteria. It presents results for the different combinations of features presented in section 3.2.2. The third and last experiment compares the credibility scores we assign to different news sites to the ones assigned by MCE Watch.

In the next subsection we present statistics and details of our data set. After that we go into the details of our experiments.

4.1 Data Description

A total of 9358 articles were crawled; 2606 articles judged as credible and 6752 articles judged as not credible by violating at least one of the credibility criteria listed in table 1. The dataset is obviously unbalanced with 72.15% of the articles being not credible and 27.84% being credible. The main reason behind this is the way the service operates; users submit articles for credibility assessment when in doubt. However, our assumption is that this imbalance is not a good representative of the general case, we try to overcome this in our experiments.

The articles belong to different news categories; *Politics*, *Accidents*, *Economics*, *Arts*, and *Sports*. This diversity in news categories poses a challenge to our analysis but in the mean time it tests the ability of our model to generalize. One more challenge introduced by the data set is that some articles are written in Modern Standard Arabic(MSA) and others are in the Egyptian dialect.

Since labeled Data was crawled from MCE Watch, the labeling is done by media experts but this didn't involve annotating every article by multiple annotators and filtering conflicting annotations. Given that multiple experts provide the service it is common to find very similar cases where reasoning differs. It is also common to find cases that were judged and reasoned based on some well known incidents at the time of judgment like public speeches, TV shows ... etc. without having a reference to the incident in the news article.

Length of articles in our news corpus is not uniform, the shortest article has 158 characters (the article has no body just a title) and the longest one has 45423 characters with average length of 1980.2 characters.

We divide our data into 80% training set and 20% test set. During development (*DEV*), we perform 5-fold cross validation where we train with 4 folds and test on the 5th, we report the average accuracy of all 5 folds. During testing (*TEST*), we use the best performing settings on DEV to train with all the training set and report accuracy on the test set.

We conduct experiments to show the effectiveness of each of our designed features and their combinations. For all experiments we compare our results against the baseline which is the majority class.

4.2 Results for "When?"

For this problem, we have a total of 9358 articles divided among the two classes. For **HasWhen** class, we have a total of 4652 articles; 2606 credible articles and 2046 non-credible articles but doesn't violate *answers_when* criteria – we augmented the credible articles by this set of non-credible articles to overcome class imbalance (35.64% → 49.71%). For **NoWhen** class, we have a total of 4706 articles that violate *answers_when* criteria.

First we try all combinations of N-GRAMS for $n \leq 3$ and determine the best setting on DEV. From table 3, unigrams + bigrams together give the highest accuracy of 78.99% on DEV, which is 28.7% higher than the baseline. The same setting also gives the highest accuracy of 77.12% on TEST which is 26.83%

Table 3. answers_when accuracy(%) on both DEV and TEST using all combinations of N-GRAMS $n \leq 3$

	DEV	TEST	Baseline
n = 1	78.16	76.27	50.29
n = 2	71.71	68.00	50.29
n = 3	61.44	57.6	50.29
n = 1, 2	**78.99**	**77.12**	50.29
n = 2, 3	71.29	66.67	50.29
n = 1, 2, 3	78.98	76.91	50.29

higher than the baseline. Next, we show results for combinations of the other features with the best setting of N-GRAMS i.e. n = 1 and 2.

From table 4, the domain dependent feature *TIME* gives an accuracy of 85.18% on DEV that is 34.89% higher than the baseline and 6.19% higher than N-GRAMS. It performs even better on TEST with a slight accuracy increase of 0.1%. The right part of table 4 displays results of using the combination of *TIME-DISTANCE* and *TIME-CONTEXT* features which performs 0.7% and 2.19% higher than *TIME* for DEV and TEST, respectively. Combining these two features with N-GRAMS – by appending them to the L2 normalized N-GRAMS feature vectors and re-normalizing – gives our best setting on DEV that exceeds the baseline by 36.02%. Although this setting is 36.64% higher than the baseline for TEST, it is -0.53% less than *TIME-DISTANCE + TIME-CONTEXT*.

Table 4. answers_when accuracy(%) on both DEV and TEST

	NG (n = 1, 2)	TIME	TD + TC	NG (n = 1, 2) + TD + TC
DEV	78.99	85.18	85.88	**86.31**
TEST	77.12	85.28	**87.47**	86.93
Baseline	50.29	50.29	50.29	50.29

4.3 Results for "Source Info"

For this problem we have a total of 4283 articles distributed among two classes. Class **HasSourceInfo** has a total of 2606 articles and class **NoSourceInfo** has a total of 1677 articles.

Once more we try all combinations of N-GRAMS for $n \leq 3$ and determine the best setting on DEV. From table 5, unigrams give the highest accuracy on DEV. It exceeds the baseline by 23.43%. Although unigrams + bigrams together give an accuracy that exceeds unigrams alone by 0.34% on TEST, we still use the setting performing best on DEV i.e. unigrams in our next experiments.

Table 6 shows accuracies for domain dependent features. Although, best results for domain dependent features *VERB + SOURCE* exceeds baseline by 19.15% and 14.4% on DEV and TEST respectively, it is still lower than results of unigrams by 4.27% for DEV and 7.34% for TEST. Combining both of them

Table 5. has_source accuracy(%) on both DEV and TEST using all combinations of N-GRAMS $n \leq 3$

	DEV	TEST	Baseline
n = 1	84.23	82.54	60.8
n = 2	81.42	79.16	60.8
n = 3	76.37	73.57	60.8
n = 1, 2	83.91	**82.89**	60.8
n = 2, 3	81.19	79.28	60.8
n = 1, 2, 3	83.82	82.77	60.8

with unigrams gives our best setting on both DEV and TEST with accuracies exceeding baseline by 24.1% and 22.2%, respectively.

As can be seen, domain dependent features for *"has_source"* weren't as good as their corresponding for *"answers_when"*. We attribute this to the lack of **NoSourceInfo** training examples; we had a total of 1341 examples which is slightly more than 35% of what we had for **NoWhen**.

Table 6. has_source accuracy(%) on both DEV and TEST

	NG (n = 1)	VERB	VERB + SOURCE	NG + VERB + SOURCE
DEV	84.23	78.88	79.96	**84.9**
TEST	82.54	75.55	75.2	83
Baseline	60.8	60.8	60.8	60.8

4.4 Credibility Score Per News Site

In this last experiment we compute the credibility score of each news site in the same way as MCE Watch for all news articles evaluated in April 2013. Basically, for each news site we classify all articles from this site according to *has_source* and *answers_when* criteria. For all articles violating a criteria we subtract the corresponding points of the violated criteria from the monthly pool of 500 points assigned to each news site. The percentage of remaining points corresponds to the credibility score of the news site, Eq. (3).

$$Score_i = \frac{500 - (P_{has\_source} \times NS_i + P_{answers\_when} \times NW_i)}{500} \times 100\% \quad (3)$$

Where

- $Score_i$: the credibility score assigned to news site i at the end of April 2013.
- P_{has_source}: the points assigned to *has_source*. From table 1, its value is 2.
- $P_{answers\_when}$: the points assigned to *answers_when*. From table 1, its value is 1.

- NS_i: the number of articles from source i that violate *has_source* criteria at the end of April 2013.
- NW_i: the number of articles from source i that violate *answers_when* criteria at the end of April 2013.

Figure 3 shows the score calculated by MCE Watch versus the score calculated based on our model. The mean absolute error (MAE) is 17.56%. Two factors contribute to this error (i) the error introduced by our classification (ii) and the error due to the criteria we don't consider. To get a closer picture, we adjust the scores assigned by MCE Watch by adding back points deducted due to criteria that we don't consider. Figure 4 shows the comparison between our scores and the adjusted MCE Watch scores, it is clear that the gap between the scores is much less. Consequently, MAE drops significantly to 2.55%. We note a couple of issues here: (i) The larger portion of the error is introduced by classification errors for *has_source*. This is very clear for youm7.com, shorouknews.com, rassd.com, fj-p.com, and elfagr.org with highest absolute error in the range [4, 6.4] (ii) The drop in MAE is not due to our model accuracy alone but there is still a

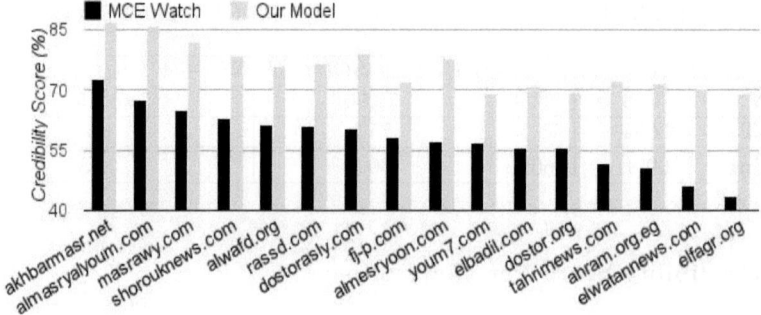

Fig. 3. Credibility score of each news site measured by MCE Watch vs. score based on our model

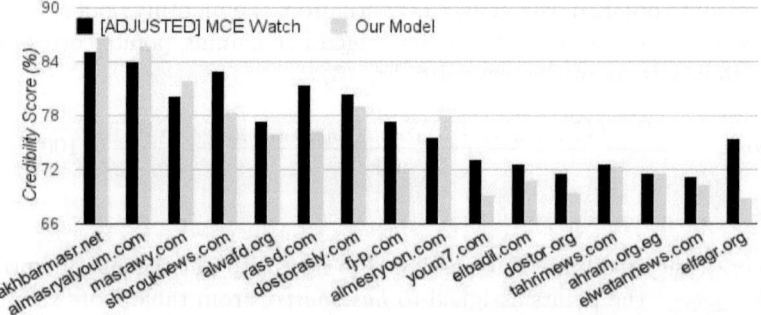

Fig. 4. Adjusted credibility score of each news site measured by MCE Watch vs. score based on our model

hidden error that contributes towards this drop; this happens when erroneously classified articles contribute positively in penalizing (or not penalizing) a news site. However, this can be used as a high level indicator for a news site credibility.

5 Conclusion

In this paper we presented a new approach to automate Arabic news credibility measurement. Our approach is a first step towards automating a manual solution for the same problem presented by MCE Watch. The solution depends on evaluating each news article on a preset collection of criteria developed by media experts. Criteria are evaluated according to the text content of the news article. To automate this, we crawled news articles previously labeled by MCE Watch. Then, we picked two criteria of the most violated ones, and modeled each one as a separate binary classification problem dividing the crawled data into training and test sets.

The first criterion we chose is whether the article answers the question "When?". Using SVM and a combination of unigrams, bigrams and domain dependent features we were able to reach an accuracy of 86.93% on a total of 1875 articles representing 20% of the crawled data. Our baseline is 50.29% which is the majority class.

The second criterion we chose is whether the news article mentions the *"Source Info"*. Using SVM, unigrams and domain dependent features we got an accuracy of 83% on a total of 859 articles representing 20% of all the data we have for this problem and 9.1% of the crawled data. Our baseline is 60.8% representing the majority class.

As we previously mentioned, this is a first step towards automation so there is plenty to be done in the future. Currently, we are looking into collecting more data to automate the rest of the criteria as well as enhance the models for the chosen ones. Also looking into how the state-of-art automatic feature extractors can help leverage our work. Next steps involve building a website that automatically scores news articles using our model.

References

1. Abdul-Mageed, M., Diab, M.T., Korayem, M.: Subjectivity and sentiment analysis of modern standard arabic. In: ACL (Short Papers), pp. 587–591 (2011)
2. Akamine, S., Kato, Y., Inui, K., Kurohashi, S.: Using appearance information for web information credibility analysis. In: Second International Symposium on Universal Communication, ISUC 2008, pp. 363–365. IEEE (2008)
3. Al-Eidan, R.M.B., Al-Khalifa, H.S., Al-Salman, A.S.: Towards the measurement of arabic weblogs credibility automatically. In: Proceedings of the 11th International Conference on Information Integration and Web-based Applications & Services, pp. 618–622. ACM (2009)
4. Al-Eidan, R.M.B., Al-Khalifa, H.S., Al-Salman, A.S.: Measuring the credibility of arabic text content in twitter. In: 2010 Fifth International Conference on Digital Information Management (ICDIM), pp. 285–291. IEEE (2010)

5. Darwish, K., Magdy, W., Mourad, A.: Language processing for arabic microblog retrieval. In: Proceedings of the 21st ACM International Conference on Information and Knowledge Management, pp. 2427–2430. ACM (2012)
6. Gaziano, C., McGrath, K.: Measuring the concept of credibility. Journalism Quarterly 63(3), 451–462 (1986)
7. Gupta, A., Kumaraguru, P.: Credibility ranking of tweets during high impact events. In: Proceedings of the 1st Workshop on Privacy and Security in Online Social Media, p. 2. ACM (2012)
8. Juffinger, A., Granitzer, M., Lex, E.: Blog credibility ranking by exploiting verified content. In: Proceedings of the 3rd Workshop on Information Credibility on the Web, pp. 51–58. ACM (2009)
9. Kawai, Y., Fujita, Y., Kumamoto, T., Jianwei, J., Tanaka, K.: Using a sentiment map for visualizing credibility of news sites on the web. In: Proceedings of the 2nd ACM Workshop on Information Credibility on the Web, pp. 53–58. ACM (2008)
10. MCE Watch: Media Credibility in Egypt, http://www.mcewatch.com (accessed April 20, 2013)
11. Miniwatts Marketing Group: Internet World Stats, http://www.internetworldstats.com (accessed April 20, 2013)
12. Pedregosa, F., Varoquaux, G., Gramfort, A., Michel, V., Thirion, B., Grisel, O., Blondel, M., Prettenhofer, P., Weiss, R., Dubourg, V., Vanderplas, J., Passos, A., Cournapeau, D., Brucher, M., Perrot, M., Duchesnay, E.: Scikit-learn: Machine learning in Python. Journal of Machine Learning Research 12, 2825–2830 (2011)
13. Weerkamp, W., de Rijke, M.: Credibility-inspired ranking for blog post retrieval. Information Retrieval 15(3-4), 243–277 (2012)
14. Xu, J., Yang, X., Wang, L.: Evaluation method of information credibility based on the trust features of web page. In: 2011 Eighth Web Information Systems and Applications Conference (WISA), pp. 69–72. IEEE (2011)
15. Zhang, J., Kawai, Y., Nakajima, S., Matsumoto, Y., Tanaka, K.: Sentiment bias detection in support of news credibility judgment. In: 2011 44th Hawaii International Conference on System Sciences (HICSS), pp. 1–10. IEEE (2011)

Polarity Detection of Foursquare Tips

Felipe Moraes, Marisa Vasconcelos, Patrick Prado, Daniel Dalip,
Jussara M. Almeida, and Marcos Gonçalves

Universidade Federal de Minas Gerais, Brazil
{felipemoraes,marisav,patrickprado,hasan,jussara,mgolcalves}@dcc.ufmg.br

Abstract. In location-based social networks, such as Foursquare, users may post tips with their opinions about visited places. Tips may directly impact the behavior of future visitors, providing valuable feedback to business owners. Sentiment or polarity detection has attracted great attention due to its vast applicability in opinion summarization, ranking or recommendation. However, the automatic detection of polarity of tips faces challenges due to their short sizes and informal content. This paper presents an empirical study of supervised and unsupervised techniques to detect the polarity of Foursquare tips. We evaluate the effectiveness of four methods on two sets of tips, finding that a simpler lexicon-based approach, which does not require costly manual labeling, can be as effective as state-of-the-art supervised methods. We also find that a hybrid approach that combines all considered methods by means of stacking does not significantly outperform the best individual method.

Keywords: Web 2.0 applications, Sentiment Analysis, Micro-reviews.

1 Introduction

The widespread use of *smartphones* with geolocation technologies like GPS (*Global Positioning System*) and the increasing interest in social networks led to the appearance of location-based social networks (LBSNs), such as Foursquare, as well as the use of geolocation services by other networks like Google Plus and Instagram.[1] On Foursquare, the currently most popular LBSN, users may share not only their locations, by checking in at venues, but also their opinions about those places, by writing micro-reviews or tips.

Tips are short and informal texts containing opinions about any aspect related to the target venue. For instance, a tip left at a restaurant may contain a recommendation or a complaint about a specific dish or service offered. Moreover, users may *like* a tip in sign of agreement with or interest in its content. Thus, tips and likes foster interactions among users, and provide valuable feedback to business owners to improve the quality of their products and services.

This paper analyzes methods for polarity or sentiment detection of Foursquare tips. In general terms, polarity detection aims at determining the attitude of a

---

[1] http://foursquare.com/, http://plus.google.com/
and https://instagram.com/

speaker (or writer) towards some topic, classifying it as, for example, positive or negative. Automatic polarity detection has several applications, including opinion summarization in online reviews [1] and real time monitoring of people's opinions [2]. The polarity detection of Foursquare tips could be used to summarize the sentiment of users towards a specific place (venue), providing quick feedback to venue owners from their potential customers, and assisting other users when choosing places to visit. However, tips have inherent characteristics that bring challenges to polarity detection: they are typically very short (limited to 200 characters), contain very informal content and often slangs and expressions (e.g., "coool!!"), which are hard to analyze and make their polarity unclear.

Existing polarity detection techniques are grouped into *supervised* and *unsupervised* methods. In the former, automatic classifiers are learned from previously labeled examples [1],[3–7], whereas the latter often relies on lists of positive and negative words (lexicons), using the polarity of each term to classify a text [3],[4],[6],[7]. Although supervised methods have been effectively used for polarity detection in "traditional" environments (e.g., long texts) [8], their efficacy strongly depends on the availability of reliable labeled examples for training. As these examples are often labeled by people, they are subject to errors caused by misinterpretation of the text. The number of training examples also affects classification accuracy, as ideally they should cover as many scenarios as possible [9], which implies high costs for building training sets. In contrast, lexicon-based methods do not require any training and can be applied in various contexts and applications. However, there are not many lexicons suitable to all application domains, as the sentiment of some terms may depend on the topic domain [9]. Moreover, there is no consensus as to which of the two approaches behaves best in short texts like tips [4],[7],[6].

This paper analyzes alternative techniques for automatic polarity detection of Foursquare tips, namely three supervised classifiers - Naïve Bayes, Maximum Entropy and Support Vector Machine [10], and one unsupervised method based on the SentiWordNet lexicon [11]. Our study is based on two sets of tips: one was manually labeled (positive or negative) by volunteers, while the polarity of tips in the other was inferred from emoticons. Our experimental results show that the unsupervised approach produces average Macro-F1 results that are statistically tied to those of the best supervised method - Naïve Bayes - in both datasets, without the cost of labeling. We also find that the unsupervised method is at least as good as, if not better than, Naïve Bayes to detect positive tips, particularly in terms of F1 and recall, whereas the latter produces slightly better results for negative tips. Finally, we also evaluated a hybrid classifier that combines all techniques using stacking, finding that it leads to no further gains over the best individual method, possibly due to the large agreement among the techniques.

2 Related Work

Existing polarity detection solutions can be grouped into supervised machine learning based methods and unsupervised lexicon based methods. Regardless of the technique employed, most previous studies target long texts, often long reviews. For example, Pang *et al.* [1] evaluated three supervised classification algorithms - Naïve Bayes, Maximum Entropy and SVM - in detecting the polarity of movie reviews, representing each review as a bag-of-words based on unigrams and bigrams. Ohana *et al.* [3], instead, proposed two approaches to use the SentiWordNet lexicon for the same task: (1) using the sum of scores of positive and negative words in the text and taking the highest score as the polarity of the text, and (2) using the scores as features to train an SVM classifier. We here consider an unsupervised method similar to the first approach, representing each tip as a bag-of-words with TF-IDF (product of the term frequency by the inverse document frequency) weights. We also evaluate a hybrid approach that combines supervised and unsupervised methods, similarly to the aforementioned second approach [3], but that goes beyond by combining the predictions of the supervised methods along with the scores of the lexicon for the polarity detection.

Recently, the polarity detection of short texts has attracted attention, with focus mainly on Twitter [5],[4],[6]. In [5], the authors used features extracted from the textual content and related to the social context of the tweet's author (e.g., polarity of the followers' messages) to detect its polarity. They also exploited the presence of positive or negative emoticons to determine the polarity of a tweet, as we do here to build one of our datasets of tips.

We are aware of only two previous studies of polarity of Foursquare tips [12],[13]. In [12], the authors proposed a lexicon-based method that relies on SentiWordNet to build classifiers, and evaluated them using a dataset collected from both Yelp and Foursquare. Although the datasets are different - in particular Yelp reviews tend to be much longer than Foursquare tips - the results reported in [12] are worse than those obtained with our unsupervised method, perhaps due to the way they estimated polarity scores. The other previous effort also used SentiWordNet to detect the polarity of Foursquare tips, with the goal of building a location recommendation method [13]. Thus, they did not evaluate the effectiveness of the detection methods considered, which is our goal.

A few studies compared supervised and unsupervised approaches to classify the polarity of short texts. In [4], the authors showed that the supervised methods analyzed in [1] may be applied to short tweets, comparing their accuracy with that of an unsupervised lexicon-based method.[2] Bermingham *et al.* [7] concluded that the supervised SVM and multinomial Naïve Bayes classifiers outperform an unsupervised method based on SentiWordNet for tweets and micro-reviews in Blippr. In contrast, Paltoglou *et al.* [6] proposed an unsupervised method based on the LIWC lexicon [14] to detect the polarity of *tweets* and user comments, showing that their method outperforms the Naïve Bayes, Maximum Entropy and SVM classifiers. In sum, there is no consensus as to the best approach for short texts, particularly for Foursquare tips, as previous efforts on that context focused only on lexicon based approaches. This work aims at contributing to this discussion, focusing on detecting the polarity of tips, a noisy environment, but rich in information. To our knowledge, this is the first work that provides a deeper investigation of alternative polarity detection methods in this context.

---

[2] http://twitrratr.com/

3 Overview of Datasets

We here use two datasets of tips, which are random samples of a larger dataset containing around 10 million tips posted by 13 million users, collected between August and October 2011. To build our datasets, we considered only tips posted in venues with English as the official language,[3] since the tools used by the unsupervised method to determine polarity are constrained to the English language.

To build our first dataset, referred to as **manually labeled dataset**, we randomly selected 1,250 tips to be labeled by 15 volunteers. Each tip was analyzed by 3 volunteers. Each volunteer received 250 tips along with information about the venue (i.e., name, category) where each tip was posted, and was asked to label the content of each tip as positive, negative or neutral. There was agreement among at least two of the volunteers in 94% of the tips. The remaining 6% of tips were discarded due to lack of any agreement. The result of this manual classification was: 57.78% of tips were classified as positive, 15.64% as negative, and 26.58% as neutral. As in [5],[1], we focus on classifying tips as either positive or negative, and thus disregard neutral tips.

The second dataset, referred to as **emoticon based dataset**, was built from a sample of 3,512 tips in English containing at least one emoticon. Emoticons may serve as noisy labels [4], as some texts are not easily classified, such as those that express sarcasm. As in [5],[4], we assume that positive emoticons, notably ':)', '(:', ':-)', '(-:', ':)', ':D', '=D', indicate positive tips, whereas ':(', '):', ':-(', ')-:', ': (', ') :' indicate negative tips.

For both datasets, we considered only tips with at least one word in SentiWordNet to be able to apply our unsupervised method, which caused the removal of only a small fraction of tips (at most 1.6%) from both datasets. In sum, our manually labeled dataset consists of 669 positive and 182 negative tips, and our emoticons based dataset contains 3014 positive and 440 negative tips. Note that both datasets are very unbalanced, reflecting a general trend towards users writing positive tips more often (at least among those written in English).

4 Polarity Detection

This section presents the techniques used to automatically detect the polarity (positive or negative) of Foursquare tips, starting with the supervised methods (Section 4.1), and then introducing our unsupervised approach (Section 4.2).

4.1 Supervised Methods

Given a training set with instances (tips) represented by various features and previously labeled in interest classes (polarities), a supervised method "learns" a model, which can then be applied to classify unlabeled data (test set) into the given classes. We analyze three state-of-the-art text classifiers - Naïve Bayes (NB), Maximum Entropy (ME) and Support Vector Machine (SVM) [10].

---
[3] http://en.wikipedia.org/wiki/List_of_official_languages

Naïve Bayes is a probabilistic classifier that makes use of Bayes' theorem to infer the probability that a new document (or tip) belongs to each defined class (polarity). It has been applied to various applications such as spam filtering, disease diagnosis, and classification of text polarity [1],[7]. We used the multinomial version of this classifier, which is more adequate for text classification [15], where the probability of a class is parameterized by a multinomial distribution.

The main disadvantage of Naïve Bayes is the assumption of independence between the features exploited by the classifier. Maximum Entropy does not make such assumption. Instead, to estimate the probability distribution, it assumes that, without external knowledge, the distribution should be as uniform as possible, and thus have maximal entropy. The training data is then used to derive a set of constraints that represent the class-specific expectations for the distribution. An improved iterative scaling algorithm is then used to find the maximum entropy distribution without violating the given constraints.

Finally, Support Vector Machines try to find the best hyperplane, defined in the feature space, that separates with maximum distance (margin) the training instances of the two classes (positive and negative). We use a linear kernel, since the number of instances is smaller than the number of features, which is common for textual classification and usually produces a linearly separable problem. We here used the implementations of Naïve Bayes and Maximum Entropy provided in *scikit-learn*[4], and the SVM implementation available in the LIBSVM package.[5]

To apply these algorithms, we modeled each tip as a bag-of-words, removing first the stopwords, as in [1]. However, instead of using the presence/absence of unigrams in the tip as features, we used the representation proposed in [5]: each tip t is modeled as a vector $p_1, ...p_n$, where p_i is the frequency of a term i in tip t and normalized by the frequency of term i in the training set (i.e., TF-IDF). Preliminary experiments showed that this representation gives better results than those obtained using unigrams or bigrams. The values of p_i were used as features exploited by the classification algorithms.

4.2 Unsupervised Method

Supervised methods require previously labeled instances (training) for the development of the classifiers. Unsupervised techniques minimize this need by directly exploiting the contents of the text, often relying on lexicons.

Opinion lexicons are sets of words that express some form of positive or negative feeling (e.g., "amazing" or "bad"). These lists are largely used by polarity detection methods [3], which typically "count" the number of positive or negative words found in a piece of text. Unlike the training sets exploited by supervised techniques, which are typically application-specific, lexicons may be generic enough to be useful for multiple applications in different contexts. Thus, the cost of building the lexicon may be amortized over a larger number of applications and investigations. One aspect to consider when choosing a lexicon is the

---
[4] http://scikit-learn.org/
[5] http://www.csie.ntu.edu.tw/~cjlin/LIBSVM/

number of terms included or its *coverage*, which may impact the effectiveness of methods that use the lexicon [3]. Many previously proposed unsupervised methods [3],[16],[7] make use of two specific lexicons: SentiWordNet [11] and LIWC [14]. We here chose SentiWordNet due to its larger coverage.

SentiWordNet is a lexical resource for opinion mining derived from WordNet[17]. WordNet is an English lexicon which groups nouns, verbs, adjectives and adverbs into synonym sets, or *synsets*, each expressing a distinct concept.[6] The SentiWordNet combines three scores - positive, negative and neutral - for every WordNet *synset*, representing the positive, negative or neutral feeling or sentiment associated with that word. These scores are normalized between 0 and 1, so that they add up to 1. Our unsupervised approach to classify a given tip as positive or negative consists of the following steps:

1. Part-Of-Speech and Stemming[7]**:** Each word in the tip is first associated with a single grammatical category, such as an adjective, adverb, verb and pronoun, using a part-of-speech parser, and then converted to its canonical form (e.g., verbs are kept in their infinitive form).

2. Treatment of negative terms: the polarity of a word may be influenced if a negative term (e.g., *not*) precedes it. To handle this scenario, we build a dependency tree that models the grammatical relations of each word or phrase of the tip, and used it to identify the words that are influenced by a negative term. These words have their positive and negative SentiWordNet scores exchanged.

3. Sense of a word: a word in SentiWordNet may have multiple *synsets* associated with the same part-of-speech. Thus, we considered the mean of the scores of all the *synsets* associated with the part of speech of the word.

4. Tip Polarity: we assign positive, negative and neutral scores to the tip, each one computed as the average of the corresponding scores of the *synsets* of all words of the tip that were found in SentiWordNet. A tip is considered as having a positive polarity if its final positive score is higher than the negative one, and vice-versa. In case of tie, the polarity of the tip is considered as undefined, and the tip is discarded from our evaluation.

5 Experimental Results

We evaluated the four polarity detection methods on our two datasets (Section 3) using a 5-fold cross-validation: each dataset was divided into 5 folds, 4 folds were used as training set, and 1 fold as test set. The training set was used only by the supervised methods to "learn" the classification models. In particular, best parameter values were determined by performing cross-validation within the training sets. All methods were evaluated only on the test sets. In order to compensate for the large class imbalance, we applied a commonly used technique of undersampling, in which the smallest class determines the number of instances of each class used for training. Thus, for each round of the cross-validation we produced 5 random samples for each of the training classes, and repeated the

---

[6] Every possible meaning of the same word corresponds to a different *synset*.
[7] We use a tool from http://www-nlp.stanford.edu/software/corenlp.shtml

Table 1. Results of Polarity Detection Methods

Metric	Method	Manually labeled dataset			Emoticon based dataset		
		Positive class	Negative class	Average	Positive class	Negative class	Average
Precision	NB	**0.9173±0.0067**	0.4333±0.0180	**0.6753±0.0103**	**0.9393±0.0033**	0.2457±0.0110	0.5925±0.0056
	ME	0.9017±0.0092	0.3950±0.0206	0.6484±0.0120	0.9270±0.0040	0.2193±0.0098	0.5731±0.0050
	SVM	**0.9097±0.0096**	0.4169±0.0221	0.6633±0.0127	**0.9399±0.0040**	0.2394±0.0133	0.5896±0.0065
	Lexicon	0.8861±0.0121	**0.4846±0.0390**	**0.6853±0.0189**	0.9139±0.0054	**0.2416±0.0127**	0.5778±0.0070
	Hybrid	**0.9179±0.0082**	0.4381±0.0231	**0.6780±0.0119**	**0.9365±0.0037**	0.2437±0.0134	0.5901±0.0066
Recall	NB	0.7311±0.0126	**0.7547±0.0241**	**0.7429±0.0119**	0.6888±0.0078	**0.6936±0.0181**	0.6912±0.0087
	ME	0.7015±0.0201	0.7124±0.0364	0.7070±0.0146	0.6670±0.0078	0.6400±0.0156	0.6535±0.0085
	SVM	0.7176±0.0270	**0.7302±0.0387**	0.7239±0.0157	0.6698±0.0227	**0.7028±0.0266**	0.6863±0.0096
	Lexicon	**0.8183±0.0180**	0.6159±0.0276	0.7171±0.0154	**0.7651±0.0055**	0.5101±0.0247	0.6376±0.0122
	Hybrid	0.7335±0.0227	**0.7521±0.0338**	**0.7428±0.0132**	0.6900±0.0206	**0.6768±0.0248**	0.6834±0.0900
F1	NB	0.8133±0.0083	**0.5496±0.0190**	**0.6814±0.0116**	0.7946±0.0051	**0.3623±0.0133**	0.5785±0.0078
	ME	0.7879±0,0116	0.5058±0.0222	0.6469±0.0135	0.7756±0.0055	0.3261±0.0118	0,5508±0.0075
	SVM	0.8003±0.0166	**0.5278±0.0226**	0.6640±0.0157	0.7807±0.0153	**0.3554±0.0150**	0.5681±0.0128
	Lexicon	**0.8502±0.0118**	0.5369±0.0313	**0.6935±0.0187**	**0.8328±0.0040**	0.3271±0.0156	0.5800±0.0086
	Hybrid	0.8141±0.0125	**0.5504±0.0213**	0.6823±0.0140	0.7934±0.0135	**0.3569±0.0152**	0.5752±0.0123

process 5 times. The results discussed here are averages of 25 runs, along with corresponding 95% confidence intervals.

The effectiveness of each method was evaluated in terms of precision, recall and F1. The precision p of a class c is the number of tips correctly classified in class c by the number of tips predicted as c. The recall r of a class c is the number of tips correctly classified in class c by the number of tips in c. The F1 measure is the harmonic mean, $2pr/(p+r)$ between precision p and recall r. We computed precision, recall and F1 for each class (polarity) separately, as well as average values for the two classes.

Table 1 shows the results of each supervised approach - Naïve Bayes (NB), Maximum Entropy (ME), SVM - and the unsupervised method (Lexicon) for both datasets. Best results, including statistical ties, are shown in bold. The significance of these values was tested using a paired t-test considering a confidence interval of 95%.

Regarding precision, the best results for the positive class are produced by the supervised NB and SVM methods, which are statistically tied in both datasets. NB, in particular, produces gains of up to 3.6% and 2.8% over the other methods, on average, in the manually labeled and emoticon based datasets, respectively. For the negative class, in turn, the unsupervised lexicon based method is the best performer, with gains of up to 22.7% over the others in the manually labeled dataset, although it is statistically tied with both SVM and NB in the emoticon based dataset. Note that, in general, precision results are smaller for the negative class, due to the class imbalance which leads to a dominance of the largest (positive) class on the classification results. The differences are larger in the emoticon-based dataset where the imbalance is more severe. Finally, considering average precision, both NB and the lexicon based method are statistically tied as the best solutions for the manually labeled dataset, whereas, in the emoticon based dataset, the supervised SVM and NB produce better results, with statistically significant but small gains of up to 2.5% over the unsupervised method.

In terms of recall of the positive class, the unsupervised approach outperforms all supervised methods, in both datasets, with average gains of up to 11.9% over the best supervised method. However, for the negative class, the best recall is produced by the NB and SVM classifiers, which are statistically tied in both datasets, with gains over the unsupervised method of 22.5% (manually labeled dataset) and 36% (emoticon based dataset). The larger gains in the negative class lead to a superiority of the supervised methods in terms of average recall in both classes: the best performer in both datasets is NB, although SVM appears tied with it as best solution in the emoticon based dataset.

Considering the F1 metric, which combines precision and recall, the lexicon based method produces the best results for positive tips in both datasets. For negative tips, NB and SVM are tied with the lexicon based method as best performers in the manually labeled dataset, whereas in the emoticon based dataset, the two supervised classifiers stand out as the best methods. Overall, NB and the lexicon based methods appear tied as the best methods in both datasets, whereas, in the emoticon based dataset, this tie also includes SVM.

Finally, we also tested a hybrid approach that combines, by means of a stacking method [18], the results of the supervised and the unsupervised methods. In this technique, the predictions of the three supervised methods as well as the scores produced by the unsupervised method are used as input to another supervised classifier (an SVM with a linear kernel, in the present case), which learns how to combine the outputs of all methods (e.g., by assigning proper weights for each single prediction). The results of this hybrid approach are also shown in Table 1. Note that, in both datasets, the results of the hybrid method are, at best, statistically tied with the best performer, although in some cases (e.g., recall of positive class), the results are clearly worse. The lack of improvements from the hybrid method is possibly due to the large agreement among the methods (e.g., the NB and the lexicon based methods produce the same results for 70% of the tips in the manually labeled dataset), which leaves little room for improvement from the stacking approach. Further investigations with other classifiers and alternative strategies to combine multiple methods are required and are left for future work.

Our results can be summarized as follows:

- The unsupervised lexicon based method produces better or statistically tied results, in terms of average F1, when compared to the best supervised methods (NB and SVM) in both datasets. The hybrid method, in turn, does not lead to further improvements.

- The unsupervised method improves the F1 of the positive class in up to 4.8% over the best supervised method. The gains in recall, which is particularly important if one is interested in retrieving most tips of that class, reach 11.9%. Thus, this method should be used when the focus is on *positive tips*.

- If the focus is on the *negative tips*, the best methods are the supervised ones, particularly NB and SVM. In the emoticon based dataset, these supervised methods produce gains in F1 of up to 10.8% over the unsupervised solution. In terms of recall, the gains are even more impressive, reaching 36% in that dataset.

- All methods perform better, in all metrics, in the manually labeled dataset (differences of up to 19.5%), possibly due to a higher level of noise (e.g., sarcasm) and uncertainty in the automatic labeling through emoticons.

These results indicate that, in the specific context of Foursquare tips, the overall effectiveness of the unsupervised lexicon based method is comparable to the best supervised methods (NB and SVM), without the labeling costs associated with the latter. However, the choice of the best method should take the costs and limitations of each approach into account. A supervised method generally requires a costly manual labeling effort. Automatic labeling, for example by exploring emoticons, may be employed. However its effectiveness may be limited by the coverage of emoticons in the tips[8] as well as by higher levels of noise and uncertainty. Unsupervised methods, on the other hand, require the availability of a lexicon for the target language or domain. Moreover, with this type of method, there are constraints in the lexicon coverage of the target domain, which may cause some tips not to be classified. In particular, the method based on the SentiWordNet could not be applied in 1.4% and 1.6% of the tips originally present in the manually labeled and emoticon based datasets, respectively, since none of the words in these tips could be found in the lexicon.[9]

6 Conclusions and Future Work

We have analyzed the effectiveness of three state-of-the-art supervised classifiers - Naïve Bayes, SVM and Maximum Entropy, an unsupervised method that uses the SentiWordNet lexicon as data source, as well as a hybrid approach that combines all four methods by means of stacking, for polarity detection of Foursquare tips. We evaluated all methods in two sets of tips: one manually labeled by volunteers and the other automatically labeled by exploring the presence of emoticons. Our results indicate that, in terms of average F1, the unsupervised method produces better or statistically tied results when compared to the best supervised methods, which are Naïve Bayes and SVM, in both sets. Considering each class separately, we find that although the lexicon based method is better to retrieve positive tips, the Naïve Bayes and SVM classifiers produce great gains, in terms of both F1 and recall, if the focus is on retrieving the negative tips. However, the choice of the best solution should also take the availability of resources and costs associated with each method (training set, lexicon) into account. Finally, we also found that the hybrid approach does not produce significant improvements over the best individual technique, possibly due to the large agreement among the methods.

Future work includes extending our evaluation to other datasets, investigating other strategies to combine multiple methods, and using our methods to build opinion summarization and venue recommendation methods.

---

[8] Only 3.39% of English tips in our dataset of 10 million tips contain emoticons.
[9] These tips were filtered from both datasets (see Section 3).

Acknowledgments. This research is partially funded by the Brazilian National Institute of Science and Technology for the Web (MCT/CNPq/INCT grant number 573871/2008-6), CNPq, CAPES and FAPEMIG.

References

1. Pang, B., Lee, L., Vaithyanathan, S.: Thumbs up? Sentiment Classification using Machine Learning Techniques. In: Proc. EMNLP (2002)
2. Guerra, P., Veloso, A., Meira, W., Almeida, V.: From Bias to Opinion: A Transfer-Learning Approach to Real-Time Sentiment Analysis. In: Proc. SIGKDD (2011)
3. Ohana, B., Tierney, B.: Sentiment classification of reviews using SentiWordNet. In: Proc. of 9th IT & T (2009)
4. Go, A., Bhayani, R., Huang, L.: Twitter Sentiment Classification using Distant Supervision. Technical report, Stanford University (2009)
5. Aisopos, F., Papadakis, G., Tserpes, K., Varvarigou, T.: Content vs. Context for Sentiment Analysis: A Comparative Analysis over Microblogs. In: Proc. HT (2012)
6. Paltoglou, G., Thelwall, M.: Twitter, MySpace, Digg: Unsupervised Sentiment Analysis in Social Media. ACM TIST 3(4) (2012)
7. Bermingham, A., Smeaton, A.: Classifying Sentiment in Microblogs: Is Brevity an Advantage? In: Proc. CIKM (2010)
8. Pang, B., Lee, L.: Opinion Mining and Sentiment Analysis. Foundations and Trends in Information Retrieval 2(1-2) (2008)
9. Lu, Y., Castellanos, M., Dayal, U., Zhai, C.: Automatic Construction of a Context-Aware Sentiment Lexicon: An Optimization Approach. In: Proc. WWW (2011)
10. Pustejovsky, J., Stubbs, A.: Natural Language Annotation for Machine Learning. O'Reilly Media (2012)
11. Esuli, A., Sebastiani, F.: Sentiwordnet: A Publicly Available Lexical Resource for Opinion Mining. In: Proc. LREC (2006)
12. Carlone, D., Ortiz-Arroyo, D.: Semantically Oriented Sentiment Mining in Location-Based Social Network Spaces. In: Christiansen, H., De Tré, G., Yazici, A., Zadrozny, S., Andreasen, T., Larsen, H.L. (eds.) FQAS 2011. LNCS, vol. 7022, pp. 234–245. Springer, Heidelberg (2011)
13. Yang, D., Zhang, D., Yu, Z., Wang, Z.: A sentiment-enhanced personalized location recommendation system. In: Proc. ACM HT (2013)
14. Tausczik, Y., Pennebaker, J.: The Psychological Meaning of Words: LIWC and Computerized Text Analysis Methods. JLS 29(1) (2010)
15. McCallum, A., Nigam, K.: A Comparison of Event Models for Naive Bayes Text Classification. In: Proc. AAAI Workshop Learning for Text Categorization (1998)
16. Hamouda, A., Rohaim, M.: Reviews Classification Using SentiWordNet Lexicon. OJCSIT 2(4) (2011)
17. Miller, G.: WordNet: A Lexical Database for English. Comm. of ACM 38(11) (1995)
18. Dzeroski, S., Zenko, B.: Is Combining Classifiers with Stacking Better than Selecting the Best One? JMLR 54(3) (2004)

The Study of Social Mechanisms of Organization, Boundary Capabilities, and Information System

Shiuann-Shuoh Chen[1], Pei-Yi Chen[2], Min Yu[1,*], and Yu-Wei Chuang[3]

[1] Department of Business Administration, National Central University, No. 300, Jung-da Rd., Jung-li City, Taoyuan 320, Republic of China, Taiwan
kenchen@cc.ncu.edu.tw, m2121374@gmail.com
[2] Department of International Business, Hsin Sheng College of Medical Care and Management, No. 418, Gaoping Sec., Zhongfeng Rd., Longtan Township, Taoyuan 320, Republic of China, Taiwan
peiyi01@ms47.hinet.net
[3] Department of Computer Science and Information Management, Providence University, No. 200 Chung-Chi Rd., Salu Dist., Taichung City 43301, Republic of China, Taiwan
ywchuang@gmail.com

Abstract. Exploring how organizational antecedents affect the boundary capabilities, this study identifies the differing effects for three components of boundary capabilities. The results indicate that the organizational mechanisms associated with the coordination capabilities primarily enhance a team's syntactic transfer, semantic translation, and pragmatic transformation. The organizational mechanisms associated with the socialization capabilities primarily increase a team's semantic translation and pragmatic transformation. The information system primarily enhances the coordination and socialization capabilities. Our findings reveal why the teams may have difficulty managing the levels of syntactic transfer, semantic translation, and pragmatic transformation and vary in their ability to create the value from their boundary capabilities.

Keywords: organization mechanisms, boundary capabilities, knowledge sharing, information system, system capabilities, coordination capabilities, socialization capabilities.

1 Introduction

The fact that most innovation happens at the boundaries between domains has ensured focused attention on effective boundary management as a dominant source of competitive advantage [1]. To survive selection pressures, the firms need to transfer knowledge, translate it, and transform it to the commercial ends. This ability, referred to as the boundary capabilities [1], has emerged as an essential subject in the studies on strategy and organization.

As rich as the scholars are regarding the boundary capabilities, it is surprisingly silent about their richness and multidimensionality. Moreover, while most studies have underscored the competitive edge of boundary capabilities, the organizational antecedents have

---

[*] Corresponding author.

been largely ignored. Even when the organizational antecedents have been considered [2], their connections with the specific types of boundary capabilities have not been examined empirically. For example, Carlile distinguished the boundary capabilities into the syntactic transfer, semantic translation, and pragmatic transformation [1]. He asserted that the firms need to manage these capabilities successfully to generate innovation. Examining the different effects of organizational antecedents on the syntactic transfer, semantic translation, and pragmatic transformation would not only specify how the boundary capabilities can be developed, but also infer why the firms have difficulties in managing the dimensions of boundary capabilities successfully.

Besides, the capacity of communication media to process information has been shown to support the communication and coordination across boundaries [4] [5]. The information system (IS) appears to be related to the coordination and socialization capabilities, little research has empirically explored the causal mechanisms through which IS leads to greater coordination and socialization capabilities. Our study is an attempt to address this issue and therefore refine and extend the comprehension of the link between the combinative and boundary capabilities, and the effects of IS on coordination and socialization capabilities.

We organize this paper as follows: the next section presents a review of theory and hypotheses. The following section shows our methodology. The data analysis and results will appear in section 4. The final section reports the implications and conclusion of our work.

2   Theory and Hypotheses

2.1   Boundary Capabilities

Following Carlile [1], the boundary capabilities categorized into the syntactic transfer, semantic translation, and pragmatic transformation capture the efforts expended in managing knowledge across specialized domains when innovation is a desired outcome. They are a combination of the capacities of common lexicons, meanings, and interests and the abilities of team members involved to use them.

At a syntactic boundary, domain-specific knowledge can be efficiently managed across the boundary when knowledge is transferred according to a common lexicon. At a semantic boundary, developing the common meanings is regarded as a way to address the interpretive differences. At a pragmatic boundary, establishing the common interests affords the key to interest conflicts [1].

2.2   Organizational Mechanisms

Drawing from Jansen and colleagues [2], the organizational mechanisms classified into system, coordination, and socialization capabilities are conceptualized as the common features of combinative capabilities enabling the organizations to synthesize and apply knowledge.

The System capabilities formed by the routinization and formalization establish the patterns of organizational action either through the memory or the standard operational procedure respectively [6]. The formalization captures the extent to which an organization sets its rules and procedures to prevent its employees deviating from established behavior [7], while the routinization reflects how an organization establishes the grammars of action through the individuals' repeated actions to support the complex patterns of interactions between the employees [8].

The formalization enables the team members to efficiently apply a team's codified knowledge through the best practices [9]. By doing so, a team can expose its members to the jargon used in the different functional areas in turn facilitate the shared understanding about the specific terms [10]. Once the shared understanding about the specific terms is in place, there is a greater chance that the group members are able to effectively create a common lexicon, leading to the syntactic transferring knowledge across boundaries [1]. *The measures of formalization* are the situations where the team members had procedures to follow in dealing with any situation, or the organization kept a written record of everyone's performance.

The routinization is embodied by the repetitious behaviors guided by the experiences [12]. It reflects strict patterns of norms and rules intended for imitation, replication, and control [13]. Drawing on Cohen and colleagues' generalization [14], a critical portion of the representations of routines encompasses the memories of individuals for their respective roles, locally shared language, and general language forms such as formal oral codes and pledges. We contend that the more routines a team has, the more likely its members will increase their tendencies to use the common lexicons for effective communication [11]. Therefore, the routinization is instrumental to shape the common lexicons and thereby strengthen the syntactic transferring knowledge across boundaries [1]. *The measures of routinization* include the extent to which the individuals will regard their work as routine, or the volume of repeated tasks from day-to-day.

Hypothesis 1: *The system capabilities will be positively related to the syntactic transfer of boundary management.*

The high level of formalization tends to confine the individuals' thoughts into a frame of reference and in turn provide a common perspective to mitigate the interpretative discrepancies [7]. It consequently leads the members to interpret circumstances with the same path-dependent trajectory of prevailing knowledge and thereby assist the development of shared meanings [2].

A team routinizing tasks is seeking for invariably performing sequences of activities with a few exceptions [15]. Such institutionalization often steers the team members to focus on the areas closely related to the existing knowledge and on what has previously proved useful [2]. In such a case, there is a greater likelihood that the team members are prone to respond to the environmental changes in a shared perspective and interpret the circumstances according to the prevailing norms [16]. Eventually, such processes will facilitate the development of shared meanings underlying the semantic translating knowledge. Therefore, we hypothesize the following:

Hypothesis 2: *The system capabilities will be positively related to the semantic translation of boundary management.*

The coordination capabilities including the cross-functional interfaces, participation in decision making, and job rotation capture the efforts expended in incorporating different sources of expertise and promoting the lateral interaction between the individuals [2].

The cross-functional interfaces contain the liaison personnel, task forces, teams, and so forth [18]. The liaison personnel serve as a knowledge broker to bridge the differences among functions and thus develop a consensus on the meanings of specific terms, symbols, or behaviors [2]. Consequently, such interfaces enable the **common lexicons** needed for the success of transferring knowledge [1]. They also foster the constructive dialogues in turn reconcile the discrepancies in interpreting the tasks [2]. They function as a knowledge translator to interpret the problems in turn enable a shared understanding about the team's goals and thus lead to a shared perspective on the specific issues [1]. In this way, they incorporate the interpretations from diverse functions into **a common meaning** needed for the success of translating knowledge [1]. Furthermore, they encourage the team members to reconsider the value of existing products and to review the combination of components [19]. By doing so, the team members are more willing to negotiate interests and make trade-off with one another, in turn reach a consensus to change the knowledge and interests from their own domains for the shared interests [1] [2]. Therefore, they enable the **common interests** that constitute the transforming knowledge across boundaries [1]. ***The measure of cross-functional interfaces*** is the extent to which the subsidiary used liaison personnel, temporary tasks forces, and permanent teams to coordinate decisions and actions with sister subsidiaries.

The participation in decision making brings the opportunities of oppinion sharing [20]. The participants tend to largely use the general language instead of jargon to increase a shared understanding in turn facilitate the development of **shared language** [10]. A shared understanding also increases their willingness to identify the commonality of each other's notion in turn develop a shared perception [3] [20]. They consequently tend to put similar interpretations on the specific problems and in turn develop the **shared meanings**. Besides, a shared understanding also improves their relations in turn enables the development of shared beliefs inspiring the social climate of trust. Such climate encourages them to believe that a current trade-off will lead to the later **common interests** and thereby increases their tendencies to risk changing their domain-specific knowledge. ***The measures of participation in decision making*** are the situations where the team members have wide latitude in the choice of means to accomplish goals, or the managers are allowed flexibility in getting work done.

Job rotation provides the opportunities to gain the experience and learn the jargon in different fields [20]. Consequently, the team members tend to develop a shared understanding on the specific facts or terms in turn improve their efficiency to create the **common lexicons** [10] [11]. They also tend to develop a shared experience about the projects and thereby increase their tendencies to develop a mutual knowledge, supporting the integration of interpretations [3]. In this way, they can improve their efficiency to create the **common meanings** [1]. The shared experience also improves their relations in turn contributes to the shared beliefs that directly influence the trust [3]. Trustworthy social conditions increase their tendencies to believe that a current trade-off will lead to the later **common interests** and thereby encourage them to risk changing their domain-specific knowledge [21]. ***The measures of job rotation*** include the situations where the team members are regularly rotated between different functions or subunits. This discussion suggests the following hypotheses:

Hypothesis 3: *The coordination capabilities will be positively related to the syntactic transfer of boundary management.*

Hypothesis 4: *The coordination capabilities will be positively related to the semantic translation of boundary management.*

Hypothesis 5: *The coordination capabilities will be positively related to the pragmatic transformation of boundary management.*

The socialization capabilities including the connectedness and socialization tactics capture the efforts expended in developing the unspoken rules, dominant values, and common codes of communication [2].

The connectedness fosters the communication and knowledge exchange in turn contributes to a shared understanding [2], which is proposed to enable the **common meanings and interests,** as explained in the section of coordination capabilities above. *The measures of connectedness* include the situations where the relationships among team members are very close, or the team members understand the personalities of one another.

The socialization tactics has been shown to structure the shared experiences [2], which are also proposed to enable the **common meanings and interests,** as explained in the section of coordination capabilities above. *The measures of* **socialization tactics** include the situations where the new team members are trained by the same program to know the operation of the team, or the senior team members often provide others with many working guides. Formally, we posit that:

Hypothesis 6: The socialization capabilities will be positively related to the semantic translation of boundary management.

Hypothesis 7: The socialization capabilities will be positively related to the pragmatic transformation of boundary management.

2.3 Information System

A burning question for the organizations and information systems (IS) researchers is how IS can support the socialization capabilities in turn improve the team communications for sharing knowledge and developing connection. Such connection has been shown to increase the social capital [23]. IS can also direct joint problem-solving in the process of R&D. Such collaboration enables the identification of relevant concepts underlying the creation of new product and the advancement of R&D process. The cooperative solution therefore enables the cooperation climate.

IS enhances the coordination capabilities for many reasons [24]. First, by making it easier to identify the available resources and providing the visibility of real-time project data, the effective use of project and resource management systems (PRMS) can support the NPD work units to quickly and accurately allocate the resources to the project tasks. Second, the effective use of scheduling and time management functionalities in PRMS helps the NPD managers effectively appoint the NPD workers to relevant tasks and enables them to better monitor the performance of NPD workers. Third, by providing the real-time information on project status and enabling the aggregate project portfolios, the workflow capabilities of PRMS can help the NPD work units become more capable in identifying the synergies among their resources and tasks, better synchronizing their activities, and executing their collective activities in parallel. By enhancing the ability of NPD work units to allocate resources, assign tasks, and synchronize activities, IS can enhance the coordination capabilities.

In addition to improving the communications, IS also provides various media such as the visualization tools, groupware, and video conferencing, helping the team members cope with fine-grained information exchanges and enabling closer cooperation. Friendly relations increase the team members' willingness to share the valuable information and problem solutions as well as reduce the risks and coordination costs, given that the NPD process involves a great deal of complexity and uncertainty [27]. In this way, IS enables the problem solving through the group work when faced with difficulties. As Schulz and Jobe noted [26], the integration of codified knowledge (e.g., schedules, milestones, meeting minutes, and training manuals, organization) facilitates the knowledge exchange in turn support the development of specialized languages, technical lingo, and various expertise [25]. Therefore, we posit that IS can enhance the socialization capabilities.

The measures of IS includes the situations where IS can move remote working beyond telecommuting to include the systems that support the collaboration, group document management, cooperative knowledge management, and so forth, or support the team work and meetings when the employees are distributed throughout the world, and thus reduce the impact of air travel.

Hypothesis 8: IS will be positively related to the coordination capabilities.
Hypothesis 9: IS will be positively related to the socialization capabilities.

3 Methodology

3.1 Data Collection Procedure

In this survey, we restricted our focus to Taiwan's electronic manufacturing firms. Our target respondents were the employees and managers associated with the new product development department or those responsible for related business in the firms. A sample of 221 firms was selected from top 2000 manufacturing enterprises in Taiwan reported by China Credit Information Service, LTD in 2010.

We sent 221 questionnaires by email and fax and received questionnaires from 140 firms, the response rate is about 63.35% and all of them are usable responses. Each respondent was asked to provide their opinions about project A and B because we hope to obtain the responses from projects with different performance and thereby we got two samples from single questionnaire, consequently, there are 280 samples in this study. Preliminary analyses were conducted to provide the information about the characteristics of sample firms in Table 1.

3.2 Measurements

The measurements used in this study were primarily derived from the previous studies and some items were modified to make them applicable to our research purposes. The constructs and items are shown in Appendix A, which is available if needed. All of the items were measured with 7-point Likert scale (1=strongly disagree, 7=strongly agree). Drawing on relevant theories and evidence, the conceptual model is proposed as Figure 1.

Table 1. Profiles of participating firms

	Items	Frequency	Percentage	Chi-	d.f.	P value
Seniority	Below 5 years	94	65.7%	7.721	5	.172
	6~10 years	26	18.2%			
	11~15 years	13	9.1%			
	Over 15 years	7	7.0%			
Number of employees in the company	Below 100 persons	22	15.4%	6.494	5	.261
	101~500 persons	23	16.1%			
	501~1000 persons	12	8.4%			
	1001~3000 persons	24	16.8%			
	3001~5000 persons	11	7.7%%			
	Over 5000 persons	48	33.6%			
Average total sales in recent 3 years (New Taiwan $)	Below 1 billion	32	22.4%	4.581	6	.599
	1.1~5 billions	27	18.9%			
	5.1~10 billions	12	8.4%			
	10.1~30 billions	13	9.1%			
	30.1~50 billions	10	7.0%			
	50.1~100 billions	14	9.8%			
	Over 100.1 billions	32	22.4%			

4 Data Analysis and Results

This study includes the formative constructs such as the system, coordination and socialization capabilities, as well as the reflective constructs like the information system, boundary management knowledge for transferring, translating and transforming, so partial least squares(PLS), a component-based technique for the structural equation modeling, is required. PLS regression was selected to examine both measurement and structural models, allowing the latent factors to be formed as either reflective or formative factors [31] [36] [41].

4.1 Non-response Bias

To examine the possibility of nonresponsive bias, we split the sample into two half-sized subgroups based on the time when each response was received [30] [47]. We then compared the early response group with the late response group on the demographic and project variables such as the seniority, number of employees in the company and average total sales. A Chi-square test was conducted to compare the early and late respondents on the research variables. The responses from the first mailing were 78 questionnaires. The late respondents were 62 questionnaires after a follow-up mailing. The results revealed no significant differences (p>.05) between the early and late respondents) suggesting that non-response bias is not a problem in this study [28]. Table 1 shows the non-response analysis results.

4.2 Analysis of Measurement Model

The indicators include reflective and formative indicators (in table 2).These indicators are not expected to have covariation within the same latent construct and they are causes of, rather than caused by, their latent construct [46]. The reflective indicators and formative indicators require different approaches and criteria for validating the reliability, convergent validity, and discriminant validity [41] [43] [46].

1. Formative Constructs

We reveal the weights of formative indicators linked with the factors for the system capability, coordination capability and socialization capability in Table 2. With the component-based structural equation modeling, the weights are estimated based on the overall model, which provides the useful insight into the meaningfulness of the formative indicator and its relative importance in the context of the homology [44]. To validate the measurement model, three types of validity should be assessed: content validity, convergent validity, and discriminate validity.

For the tests of *convergent and discriminant validity* of the formative independent variable, one possible validation approach is to examine the patterns of correlation between the items and constructs [46]. Diamantopoulos and Winklhofer [39] propose that the formative items should correlate with a "global item that summarizes the essence of the construct" (p. 272). PLS item weights, which indicate the impact of individual formative items [33], can be multiplied by the item values and summed, as noted by Bagozzi & Fornell [29]. In effect, this results in a modified multitrait, multimethod (MTMM) matrix of item-to-construct and item-to-item correlations similar to that analyzed by Bagozzi and Fornell as well as Loch et al. [45]. The resulting matrix, showing the item-to-construct correlations as the grayed out cells, appears as Table 3.

Following the logic of Campbell and Fiske [34], Loch et al. [45] suggest that *the convergent validity* is demonstrated if the items of the same construct correlate significantly with their corresponding composite construct value (item-to-construct correlation) (Campbell & Fiske, 1959; Loch et al., 2003) [34] [45]. This condition has been met, as all items correlated significantly ($p < 0.01$) with their respective construct composite value. *The discriminant validity* can be established if the item-to-construct correlations are higher with each other than with other construct measures and their composite value [45]. This condition is also met. The resulting matrix, showing the item-to-construct correlations as the grayed out cells, appears as Table 3. Assessing the *reliability* is more difficult with the formative measures than with the reflective measures and it is not always possible to accomplish it [39] [46]. In a sense, very high reliability can be undesirable for the formative constructs because the excessive multicollinearity among the formative indicators can destabilize the model [46]. To ensure that the multicollinearity is not a significant issue, we assessed the VIF (variance inflator factor) statistic. If the VIF statistic is greater than 3.3, the conflicting item should be removed as long as the overall content validity of the construct measures is not compromised [38]. For our formative measures, we find the VIF values of both formalization and routinization in the system capability to be 1.187 equivalently, 2.278, 2.491 and 1.335 for the cross-functional interfaces, job rotation and participation in decision making in the coordination capability; while 2.112, 2.951 and 2.025 for the connectedness, preservice training and experiences inheritance in the socialization capability. The results suggest that all indicators have VIF statistics lower than 3.3.

2. Reflective Constructs

The construct validity tests were also conducted for the reflective variables. The factor loadings were examined to ensure that the items loaded cleanly on those constructs to which they were intended to load, and did not cross-load on the constructs to which they should not load [48]. Generally, *the convergent validity* is demonstrated if (1) the item loadings are in excess of 0.70 on their respective factors and (2) the average variance extracted (AVE) for each construct is above 0.50 [42]. As indicted in Table 2, these conditions have been met. Gefen and Straub [42] also contend that *the discriminant validity* is demonstrated if (1) the square root of each construct's AVE is greater than the inter construct correlations in table 4 and (2) the item loadings on their respective constructs are greater than their loadings on other constructs in table 5.

Finally, the *reliability* for the scales was gauged via the composite reliability scores provided in the PLS output. The composite reliability scores equal to or greater than 0.70 are regarded as acceptable in Table 2 [40] [42]. As indicated in Table 2, the composite reliability scores of these reflective variables are acceptable. Our validation results suggest that all reflective measures demonstrated satisfactory reliability and construct validity and all formative measures demonstrated satisfactory construct validity and no significant multicollinearity. Therefore, all of the measures were valid and reliable.

Table 2. Parameter estimates

Factor	Indicator	Mean	St. dev.	Loading	Weight	Cronbach	CR	AVE
System capability(second order factor)	Formative Reflective			0.810	0.605	0.909	0.769	0.848
Formalization (first order factor)	Fom1	4.682	1.513	0.873	0.379			
	Fom2	4.979	1.383	0.912	0.395			
	Fom3	4.489	1.486	0.845	0.365			
	Reflective			0.821	0.622	0.868	0.688	0.773
Routinization (first order factor)	Ro1	4.043	1.512	0.876	0.388			
	Ro2	3.932	1.579	0.866	0.396			
	Ro3	4.636	1.423	0.740	0.428			
Coordination capability(second order factor)	Formative Reflective			0.876	0.414	0.908	0.767	0.845
Cross-functional interfaces (first order factor)	Cr1	4.996	1.513	0.850	0.364			
	Cr2	4.889	1.380	0.925	0.402			
	Cr3	4.557	1.390	0.851	0.376			
Participation in decision making(first order factor)	Reflective			0.728	0.343	0.887	0.723	0.805
	Pa1	4.836	1.389	0.897	0.422			
	Pa2	4.654	1.449	0.849	0.381			
	Pa3	4.371	1.495	0.802	0.372			
	Reflective			0.907	0.428	0.938	0.835	0.901
Job rotation (first order factor)	Jo1	4.918	1.327	0.929	0.374			
	Jo2	4.864	1.387	0.942	0.377			
	Jo3	4.682	1.332	0.869	0.343			
Socialization capability(second order factor)	Formative Reflective			0.866	0.370	0.943	0.846	0.909
Connectedness (first order factor)	Co1	5.361	1.265	0.902	0.360			
	Co2	5.246	1.301	0.940	0.361			
	Co3	4.993	1.306	0.916	0.367			
	Reflective			0.847	0.367	0.908	0.767	0.846
Preservice training (first order factor)	Pr1	4.739	1.476	0.856	0.377			
	Pr2	4.711	1.519	0.912	0.377			
	Pr3	4.850	1.473	0.800	0.416			
	Reflective			0.929	0.398	0.912	0.776	0.855
Experiences inheritance (first order factor)	Ex1	4.886	1.467	0.895	0.389			
	Ex2	4.954	1.379	0.831	0.348			
	Ex3	5.146	1.310	0.899	0.404			

Table 2. (*Continued.*)

Syntactic transferring (first order factor)	Reflective					0.866	0.683	0.767
	TR1	4.589	1.466	0.854	0.452			
	TR2	4.596	1.357	0.860	0.379			
	TR3	5.314	1.158	0.761	0.379			
Semantic translating (first order factor)	Reflective					0.894	0.737	0.815
	TT1	5.046	1.162	0.874	0.428			
	TT2	4.575	1.467	0.832	0.358			
	TT3	4.668	1.250	0.868	0.379			
Pragmatic transforming (first order factor)	Reflective					0.913	0.778	0.858
	TM1	4.654	1.303	0.870	0.388			
	TM2	4.668	1.301	0.897	0.358			
	TM3	4.571	1.334	0.879	0.388			
Information System (first order factor)	Reflective					0.857	0.667	0.746
	Is1	5.164	1.536	0.768	0.414			
	Is2	4.602	1.620	0.876	0.387			
	Is3	3.876	1.799	0.803	0.427			

Table 3. Inter-Item and Item-to Construct correlation Matrix for Formative Constructs

		1	2	3	4	5	6	7	8	9	10	11
1	Formalization	1										
2	Routinization	.330**	1									
3	System capability	.816**	.816**	1								
4	Cross-functional	.753**	.228**	.601**	1							
5	Participation in	.282**	.137*	.257**	.417**	1						
6	Job rotation	.626**	.169**	.488**	.747**	.495**	1					
7	Coordination	.676**	.214**	.545**	.876**	.727**	.906**	1				
8	Connectedness	.562**	.134*	.427**	.637**	.506**	.731**	.750**	1			
9	Training	.714**	.201**	.561**	.644**	.364**	.636**	.663**	.546**	1		
10	Experiences	.597**	.147**	.456**	.683**	.516**	.726**	.770**	.743**	.698**	1	
11	Socialization	.706**	.182**	.544**	.743**	.526**	.793**	.827**	.867**	.845**	.929**	1

*Significant at the .05 level; **significant at the .01 level

Table 4. Reliability and Convergent and Discriminant Validity for Reflective Constructs

	1	2	3	4
1. Information system	0.864			
2. Syntactic transferring	0.614	0.876		
3. Pragmatic transforming	0.445	0.665	0.926	
4. Semantic translating	0.518	0.722	0.757	0.903

Square root of AVE reported along diagonal in bold type.

Table 5. PLS Component-Based Analysis: Cross-Loadings for Reflective Constructs

			1	2	3	4
1	Syntactic transferring	TR1	0.856	0.521	0.561	0.549
		TR2	0.860	0.633	0.524	0.578
		TR3	0.759	0.652	0.564	0.388
2	Semantic translating	TT1	0.711	0.874	0.651	0.473
		TT2	0.566	0.832	0.637	0.48
		TT3	0.569	0.868	0.662	0.381
3	Pragmatic transforming	TM1	0.583	0.704	0.870	0.357
		TM2	0.557	0.637	0.897	0.398
		TM3	0.617	0.658	0.879	0.423
4	Information System	Is1	0.424	0.453	0.346	0.850
		Is2	0.560	0.436	0.385	0.898
		Is3	0.566	0.424	0.397	0.772

4.3 Assessment of Structural Model

The structural model aims to examine the relationship among a set of dependent and independent constructs. In this section, we tested the amount of variance explained and the significance of the relationships. Additionally, a bootstrap re-sampling approach is suggested in order to estimate the precision of the PLS estimates [32] [37]. Following this suggestion, a bootstrap analysis with 500 bootstrap samples and the original 280 cases was performed to examine the significance of the path coefficients [35]. The result of our structural model analysis is presented in Figure 1.

Figure 1 shows the structural model with the coefficients for each path (hypothesized relationship), where the solid and dashed lines indicate a supported and unsupported relationship respectively. All hypothesized relationships are supported, except for the hypothesis of correlation between system capability and semantic translating.

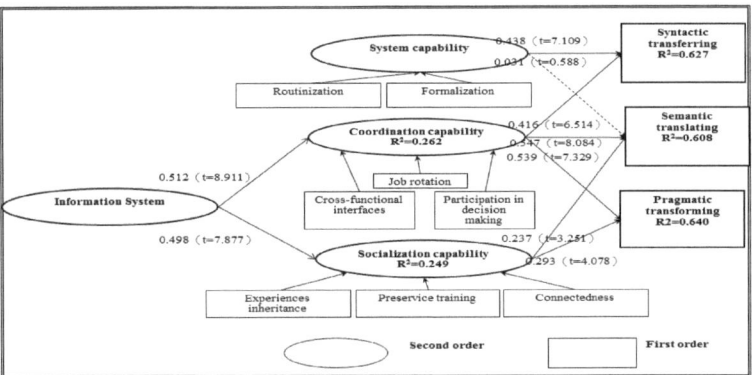

Fig. 1. Result of structural equation model (PLS) analysis

5 Discussion

5.1 Implications

Our study contributes to the research on IS, boundary capabilities and organizational mechanisms in several ways. The information processing and communication capabilities of team members can be enhanced through the information system. Facing more difficult tasks of the new product development project and more difficulties in the clear definition of objective, the more we are going to need the information system in order to enhance the interaction among team members in turn establish the collaborate relationship. Most importantly, our results also reveal that the social mechanisms of organization associated with the combinative capabilities drive a team's boundary capabilities in different ways. The present study contributes to the scholars' understanding as to why certain teams are able to transfer their domain-specific knowledge, but not able to translate and transform it successfully.

Overall, our research indicates that the organizational mechanisms associated with the coordination capabilities primarily enhance a team's syntactic transfer, semantic translation, and pragmatic transformation, while the organizational mechanisms associated with the socialization capabilities primarily enhance a team's semantic translation, and pragmatic transformation. These results reveal that the project teams may differ in their ability to manage the levels of boundary capabilities; follow different developmental paths; and differ in their ability to create the value from their boundary capabilities.

The organizational mechanisms associated with the system capabilities provide somewhat surprising results. Although they enable the syntactic transfer of boundary management as we predicted, they do not contribute to the semantic translation. Firstly, the main reason could be that the common meanings may be formalized to such an extent that impedes the flexible incorporation of newly acquired and existing knowledge. Secondly, the routinization tends to isolate knowledge, to limit the joint dialogues, and to confine the development of new perspective by imposing the existing knowledge.

5.2   Conclusion

To conclude, our study provides an empirically grounded framework simultaneously linking various aspects of organizational mechanisms and their interrelationships to different types of boundary capabilities. This framework shows how teams need to distinctively utilize their varied knowledge resources to achieve different types of boundary capabilities. It also provides a structure for future research probing of more specific questions regarding the knowledge-innovation link.

References

1. Carlile, P.R.: Transferring, Translating, and Transforming: An Integrative Framework for Managing Knowledge across Boundaries. Organization Science 15(5), 555–568 (2004)
2. Jansen, J.J.P., Van Den Bosch, F.A.J., Volberda, H.W.: Managing potential and realized absorptive capacity: how do organizational antecedents matter? Academy of Management Journal 48(6), 999–1015 (2005)
3. Gardner, H.K., Gino, F.: Dynamically integrating knowledge in teams: transforming resources into performance. Academy of Management Journal 55(4), 998–1022 (2012)
4. Dennis, A., Kinney, S.: Testing media richness theory in the new media: The effects of cues, feedback, and task equivocality. Information Systems Research 9, 256–274 (1998)
5. Daft, R.L., Lengel, R.H.: Organizational information requirements, media richness and structural design. Management Science (1986)
6. Van Den Bosch, F.A.J., Volberda, H.W., De Boer, M.: Coevolution of firm absorptive capacity and knowledge environment: Organizational forms and combinative capabilities. Organization Science 10, 551–568 (1999)
7. Weick, K.E.: The social psychology of organizing. Addison-Wesley, Reading (1979)
8. Grant, R.M.: Prospering in dynamically-competitive environments: Organizational capability as knowledge creation. Organization Science 7, 375–387 (1996)

9. Lin, X., Germain, R.: Organizational structure, context, customer orientation, and performance: Lessons from Chinese state-owned enterprises. Strategic Management Journal 24, 1131–1151 (2003)
10. Noe, R.A.: Employee training and development. Irwin McGraw-Hill, Boston (1999)
11. Nahapiet, J., Ghoshal, S.: Social capital, intellectual capital, and the organizational advantage. Academy of Management Review 23, 242–266 (1998)
12. Rerup, C., Feldman, M.S.: Routines as a source of change in organizational schemata: the role of trial-and-error learning. Academy of Management Journal 54(3), 577–610 (2011)
13. Nelson, R.R., Winter, S.J.: An evolutionary theory of economic change. Harvard University Press, Cambridge (1982)
14. Cohen, M.D., Burkhart, R., Dosi, G., Egidi, M., Marengo, L., Warglien, M., Winter, S.: Contemporary issues in research on routines and other recurring action patterns of organizations. Industrial and Corporate Change 5, 653–698 (1996)
15. Galunic, D.C., Rodan, S.: Resource recombinations in the firm: Knowledge structures and the potential for Schumpeterian innovation. Strategic Management Journal 19, 1193–1201 (1998)
16. Subramaniam, Youndt: The influence of intellectual capital on the types of innovative capabilities. Academy of Management Journal 48(3), 450–463 (2005)
17. Matusik, S.F.: An empirical investigation of firm public and private knowledge. Strategic Management Journal 23, 457–467 (2002)
18. Gupta, A.K., Govindarajan, V.: Knowledge flows within multinational corporations. Strategic Management Journal 21, 473–496 (2000)
19. Henderson, R., Cockburn, I.: Measuring competence? Exploring firm effects in pharmaceutical research. Strategic Management Journal 15, 63–84 (1994)
20. Cohen, W., Levinthal, D.: Absorptive capacity: A new perspective on learning and innovation. Administrative Science Quarterly 35, 128–152 (1990)
21. Collins, Smith: Knowledge exchange and combination: the role of human resource practices in the performance of high-technology firms. Academy of Management Journal 49(3), 544–560 (2006)
22. Adler, P.S., Kwon, S.: Social capital: Prospects for a new concept. Academy of Management Review 27, 17–40 (2002)
23. Ahuja, M.K., Carley, K.M.: Network structure in virtual organizations, 757 ed. 10, 741–757 (1999)
24. Bendoly, E., Bharadwaj, A., Bharadwaj, S.: Complementary Drivers of New Product Development Performance: Cross-Functional Coordination, Information System Capability, and Intelligence Quality. Production and Operations Management 21(4), 653–667 (2012)
25. Menon, A., Bharadwaj, S.G., Adidam, P.T., Edison, S.W.: Antecedents and consequences of marketing strategy making: A model and a test. Journal of Marketing 63(2), 18–40 (1999)
26. Schulz, M., Jobe, L.A.: Codification and tacitness as knowledge management strategies. An empirical exploration. Journal of High Technology Management Research 12(1), 139–165 (2000)
27. Uzzi, B., Lancaster, R.: Relational Embeddedness and Learning: The Case of Bank Loan Managers and Their Clients. Management Science 49(4), 383–399 (2003)
28. Armstrong, J.S., Overton, T.S.: Estimating nonresponse bias in mail surveys. Journal of Marketing Research 14(3), 396–402 (1977)
29. Bagozzi, R.P., Fornell, C.: Theoretical Concepts, Measurement, and Meaning. In: Fornell, C. (ed.) A Second Generation of Multivariate Analysis. Praeger, New York (1982)
30. Bailey, K.D.: Methods of Social Research. Free Press, New York (1987)

31. Barclay, D., Higgins, C., Thompson, R.: The partial least squares approach (PLS) to causal modeling, personal computer adoption and use as an illustration. Technology Studies 2(2), 285–309 (1995)
32. Barroso, C., Carrión, G.C., Roldán, J.L.: Applying Maximum Likelihood and PLS on Different Sample Sizes: Studies on SERVQUAL Model and Employee Behavior Model. In: Vinzi, V.E., Chin, W.W., Henseler, J., Wang, H. (eds.) Handbook of Partial Least Squares. Springer Handbooks of Computational Statistics, pp. 427–447 (2010)
33. Bollen, K., Lennox, R.: Conventional Wisdom on Measurement: A Structural Equation Perspective. Psychological Bulletin 110(2), 305–314 (1991)
34. Campbell, D.T., Fiske, D.W.: Convergent and Discriminant Validation by the Multi-Trait–Multi-Method Matrix. Psychological Bulletin 56(2), 81–105 (1959)
35. Chin, W.W.: The partial least squares approach for structural equation modeling. In: Marcoulides, G.A. (ed.) Methodology for Business and Management, pp. 295–336. Lawrence Erlbaum Associates, New Jersey (1998a)
36. Chin, W.W.: The PLS approach to SEM. In: Marcoulides, G.A. (ed.) Modern Methods for Business Research, pp. 295–336. Erlbaum, Mahwah (1998b)
37. Chin, W.W.: How to Write Up and Report PLS Analyses. In: Vinzi, V.E., Chin, W.W., Henseler, J. (eds.) Handbook of Partial Least Squares. Springer Handbooks of Computational Statistics, pp. 655–690. Springer (2010)
38. Diamantopoulos, A., Siguaw, J.A.: Formative versus Reflective Indicators in Organizational Measure Development: A Comparison and Empirical Illustration. British Journal of Management 17(4), 263–282 (2006)
39. Diamantopoulos, A., Winklhofer, H.M.: Index Construction with Formative Indicators: An Alternative to Scale Development. Journal of Marketing Research 38(2), 269–277 (2001)
40. Fornell, C., Larcker, D.F.: Evaluating Structural Equations with Unobservable Variables and Measurement Error. Journal of Marketing Research 18(1), 39–50 (1981)
41. Gefen, D., Rigdon, E.E., Straub, D.: An Update and Extension to SEM Guidelines for Administrative and Social Science Research. MIS Quarterly 35(2), iii-xiv (2011)
42. Gefen, D., Straub, D.: A Practical Guide to Factorial Validity using PLS-Graph: Tutorial and Annotated Example. Communications of the Association for Information Systems 16(25), 91–109 (2005)
43. Gefen, D., Straub, D.W., Boudreau, M.-C.: Structural Equation Modeling and Regression: Guidelines for Research Practice. Communications of the Association for Information Systems 4(7), 1–77 (2000)
44. Klein, R., Rai, A.: Interfirm strategic information flows in logistics supply chain relationships. MIS Quarterly 33(4), 735–762 (2009)
45. Loch, K.D., Straub, D.W., Kamel, S.: Diffusing the Internet in the Arab World: The Role of Social Norms and Technological Culturation. IEEE Transactions on Engineering Management 50(1), 45–63 (2003)
46. Petter, S., Straub, D., Rai, A.: Specifying Formative Constructs in Information Systems Research. MIS Quarterly 31(4), 623–656 (2007)
47. Sivo, S.A., Saunders, C., Chang, Q., Jiang, J.J.: How Low Should You Go? Low Response Rates and the Validity of Inference in IS Questionnaire Research. Journal of the Association for Information Systems 7(8), 351–414 (2006)
48. Straub, D., Boudreau, M.-C., Gefen, D.: Validation Guidelines for IS Positivist Research. Communications of the Association for Information Systems 13(24), 380–427 (2004)

Predicting User's Political Party Using Ideological Stances

Swapna Gottipati[1], Minghui Qiu[1], Liu Yang[1,2], Feida Zhu[1], and Jing Jiang[1]

[1] School of Information Systems, Singapore Management University
[2] School of Software and Microelectronics, Peking University
{swapnag.2010,minghui.qiu.2010,liuyang,fdzhu,jingjiang}@smu.edu.sg

Abstract. Predicting users political party in social media has important impacts on many real world applications such as targeted advertising, recommendation and personalization. Several political research studies on it indicate that political parties' ideological beliefs on sociopolitical issues may influence the users political leaning. In our work, we exploit users' ideological stances on controversial issues to predict political party of online users. We propose a collaborative filtering approach to solve the data sparsity problem of users stances on ideological topics and apply clustering method to group the users with the same party. We evaluated several state-of-the-art methods for party prediction task on debate.org dataset. The experiments show that using ideological stances with Probabilistic Matrix Factorization (PMF) technique achieves a high accuracy of 88.9% at 22.9% data sparsity rate and 80.5% at 70% data sparsity rate on users' party prediction task.

Keywords: Collaborative Filtering, Ideological Stances, Memory-based CF, Model-based CF, Probabilistic Matrix Factorization.

1 Introduction

Social media provides ample opportunities for citizens to participate in the political campaigns through forums, facebook, twitter and debates. These sites provide a testbed for analyzing voters behavior. Among such tasks is political affiliation detection which has been gaining much attention in recent years [1] [2] [3]. This has important consequences in targeted advertising, recommendation and personalization [4].

In our research for party prediction task, we first study American politics and draw inspiration from various political science studies on the behavior of Democrats and Republicans on social and political aspects [5] [6] [7]. These studies demonstrate the fact that the political parties take positions towards critical policies and sociopolitical issues which can ultimately lead to great differences in philosophies and ideal. Subsequently, a citizen leans towards the party that is very close to his ideological beliefs [8]. Table 1 shows the ideological beliefs of the two major parties on major social and political issues[1]. Henceforth, a users'

---
[1] http://www.diffen.com/difference/Democrat_vs_Republican

political affiliation is majorally dependent on his stances on the major social and political issues. We refer to such stances on controversial sociopolitical issues as *Ideological Stances*. For example, a user who supports *abortion* and is against *gun rights* is more likely a *Democrat*. His/her other stances on issues like gay marriage, health care, flat tax, death penalty, etc. can aid in detecting his party affiliation with high accuracy. In this paper, we focus on the problem of how users' ideological stances on the sociopolitical issues impacts their political affiliation and aid in detecting his party. There are other attributes such as gender, income, education, etc., that may be correlated with the party, but preliminary analysis shows the ideological beliefs is more correlated to user politics. Hence the focus of our paper is on investigating the relationship between ideological beliefs and politics.

Table 1. Ideological stances of Republicans and Democrats in US politics

	Death Penalty	Gay Marriage	National Healthcare	Flat Taxes	Gun Rights	Abortion
Republicans	Support (some disagree)	Oppose (some disagree)	Oppose	Support	Support	Oppose (some disagree)
Democrats	Oppose (substantial disagree)	Support (some disagree)	Support	Oppose	Oppose (some disagree)	Support (some disagree)

However, the approach of using ideological stances for party prediction poses two main challenges, i.e. *data collection* and *data sparsity*.

Data collection: Gathering user's ideological stances can be on one hand a trivial problem, where a survey methods can be used or on the other hand very challenging problem, where stances are hidden in user generated content in the form of debates. Ideological belief of a user is exhibited during his/her participation in the forums or debates related to sociopolitical issues [9]. Studies such as [9] [10] can aid in generating the users' ideological stances on the controversial issues. In our work, therefore, we focus on the data sparsity problem.

Data Sparsity: The main challenge we face with the real world data is the sparsity of users' ideological stances on the controversial issues. Not all users may provide their stances on all the issues. In our corpus, the data *sparsity rate*[2] is 22.95% for six controversial issues shown in Table 1. With sparse data, standard clustering techniques would not give satisfactory results. To tackle this problem, we propose collaborative filtering based approach which has been applied successfully for recommendation tasks [11] [12].

In this paper, we present a collaborative filtering approach for predicting users party. We assume that the parties' ideological stances are known, as shown in Table 1. We also assume that users provide stances on some of these controversial issues (incomplete user-ideological stance matrix). For predicting the remaining stances, we use collaborative filtering techniques. Collaborative filtering methods have been used successfully to estimate the user-items rating for the missing

---

[2] The percentage of missing values in a matrix.

ratings in the user-item matrix [13] [11]. Clustering the users based on the ideological beliefs aids in detecting users within the same party. Finally, to label the clusters, the distance between the party ideology and average cluster ideology can be estimated with the standard similarity techniques.

Our Contribution: First, to the best of our knowledge, we propose the first study on the impact of ideological stances on party affiliation of online users. Traditional studies rely on social network structure or text to predict the party affiliation [4] [14]. Whereas, we claim that exploiting the ideological beliefs of users suffice the party prediction task and propose a collaborative filtering method to handle the data sparsity challenges. Second, we design our experiments to evaluate intermediate results on the stance prediction and the party prediction results. Our evaluation results show that PMF achieves a high accuracy of 88.9% over state-of-the-art methods.

The rest of our paper is organized as follows. Section 2 presents related works. Section 3 presents our problem setting and followed by our solution in Section 4. We describe experiments in Section 5. Section 6 concludes the paper.

2 Related Work

User Profiling. Party prediction is among many user profiling studies which examine users' interests, gender, age, geo-localization, and other characteristics of the user profile. [15] [2] proposed supervised approach for gender prediction. Other similar approaches are taken for age prediction on social networks [3] and location of origin prediction in twitter [14]. Aggregating social activity data from multiple social networks to study the users' online behavior [1] shows promising results. In this paper, we study the problem of party prediction. In our approach, we also exploits users online behavior but with the focus on the ideology belief correlated with party leaning.

Political Affiliation Prediction. Some studies focussed on discovering political affiliations of informal web-based contents like news articles [16], political speeches [17] and web documents [18][19][20]. Political datasets such as debates and tweets are explored for classifying user stances [10][9] and also for predicting election results [21]. Closer to these studies is subgroup detection [22][23][24][25]. These works exploit content and other corpus specific properties such as hashtags, social networks etc., for predicting tasks. Some studies are motivated with the fact that the users are influenced by the community in the social network [1][2][3]. Such peer influence may impact a user on his/her political leaning. In real situations, a user social network can be sparse and politically opposing users can be friends. This can limit the performance of the existing methods. In our approach, we use ideological beliefs for party prediction and study if it has high impact on prediction task. Other factors such as hashtags, social networks can be a complimentary to our method.

Memory-Based and Model-Based Collaborative Filtering. Memory-based techniques have been proposed for recommendation tasks [26]. However, due to their limitation, model-based techniques have been more popular recently. PMF has been applied on social recommendation [11][27], news article [12] recommendation, relation prediction [13][28] and modeling friendship-interest propagations [29].Inspired by these works, we propose an approach based on PMF for users' party prediction task.

Our work was also inspired by some observations from [14] [4], who exploited corpus specific text properties such as hashtags on twitter data for party prediction. Some hashtags such as; #gay, #dadt, #912 etc., represent the controversial issues. However, in many case the stances of users cannot be captured by hashtags alone and a need of other methods arises to detect the stances. Similar to them, in our work, we exploited the controversial issues, but we used the stances of the users to predict the party affiliation.

3 Problem Setting

To formally define our problem, we first introduce a few basic concepts.

1. Issue: We refer issue to a controversial sociopolitical topics such as like "Abortion" or "Gun Control" or "Gay Marriage" etc., The controversial issues are those that segregates the political parties with great differences in ideals. In our work, we use major issues[3] studied by [5] [6] [9] as shown in Table 1.

2. Stance: Users express their positions as "Support" or "Oppose" to the issues related to sociopolitical context. Such positions are referred to as stances. Stances can be of pro/con/neural. In general, we observe a binary pattern for the issues shown in Table 1.

3. Ideological Stance: Debates on the controversial issues are referred to as ideological debates [9] and a user's stance on such issues is referred to as ideological stance.

The problem space can be formulated as a set of 2 matrices:

4. Party Ideology Matrix: Matrix of political party versus sociopolitical issues, denoted by $\mathcal{P}$, with each cell representing the stance of the political party on a specific issue. This matrix can be generated from Table 1.

5. User Ideology Matrix: Matrix of users versus sociopolitical issues, denoted by $\mathcal{R}$, with each cell representing a user's ideological stance on a specific issue. Table 2 shows a simplified example of a user-stance matrix where users take pro/con positions towards controversial sociopolitical issues.

User Party Prediction: The main task now is to predict the political party of the users who belong to the matrix $\mathcal{R}$. Clustering methods can be applied for grouping users and labeling them using $\mathcal{P}$. However, in real world, this matrix is generally very sparse, since each user will only have positioned themselves for a small percentage of the total number of issues. The challenge of sparse data can degrade the performance of the clustering algorithms. Hence, we propose collaborative filtering based approach to predict the missing user stances and then apply clustering techniques for party prediction. In the next section, we explain the details of collaborative filtering based approach for prediction task.

---

[3] http://www.diffen.com/difference/Democrat_vs_Republican

Table 2. This is an example of a user ideology matrix where each filled cell represents a user's ideological stance for an issue. The collaborative filtering technique attempts to provide a prediction for missing stances.

	Death Penalty	Gay Marriage	National Healthcare	Flat Taxes	Gun Rights	Abortion
User1	Pro	Con	Con	Pro	?	?
User2	Pro	Con	Pro	?	?	Con
User3	?	Pro	?	Pro	Pro	Con
User4	Con	?	Pro	Pro	?	Pro
User5	Con	Con	Con	Pro	Con	?

4 Solution

Our approach takes two step processing for the party prediction. In the first step, we predict missing users' ideological stances in $\mathcal{R}$ using collaborative filtering method. In the second step, we use the predicted ideological matrix to group users using clustering technique and label the groups in a principle manner. We explain the details below.

4.1 Ideological Stance Prediction

Under this formulation, the problem is to predict the values for specific empty cells of $\mathcal{R}$ (i.e. predict a user's stance for an issue). In a typical CF setting, we have a list of n users, **U** and a list of m issues, **I**. Each user, u takes position as pro/con or 1/0 for each issue, i. The task of CF algorithm aims at predicting the missing value, $\hat{r}_{u,i}$ which indicates user u's position likeliness for an issue i. $\bar{r}_u$ denotes mean rating value for user u and $\bar{r}_i$ denotes mean rating value for issue i. CF algorithms can be divided into to main categories: *memory-based* and *model-based* algorithms[30]. We explain the most popular CF algorithms in this section.

Memory-Based CF Algorithms: Memory-based methods utilize the entire user ideology data to calculate the similarity or weight between users or issues and make predictions according to those calculated similarity values. We explore three popular memory-based models: user-based, item-based and slope-one method.

a. User-based: User-based method predicts missing ratings by firstly finding similar users [26]. The similarity between two users, u and v using Pearson correlation is given by Equation 1. Predicted values are computed using Equation 2.

$$sim(u,v) = \frac{\sum_{i \in \mathcal{I}}(r_{u,i} - \bar{r}_u)(r_{v,i} - \bar{r}_v)}{\sqrt{\sum_{i \in \mathcal{I}}(r_{u,i} - \bar{r}_u)^2}\sqrt{\sum_{i \in \mathcal{I}}(r_{v,i} - \bar{r}_v)^2}}, \quad (1)$$

where $i \in \mathcal{I}$ summations over the issues that both the users u and v have rated.

$$\hat{r}_{u,i} = \bar{r}_u + \frac{\sum_{v \in \mathcal{U}}(r_{v,i} - \bar{r}_v)sim(u,v)}{\sum_{v \in \mathcal{U}}|sim(u,v)|} \quad (2)$$

b. Item-based: Item-based method predicts missing ratings by first finding similar items [31]. The similarity between two items, i and j using Pearson correlation is given by Equation 3 and the predicted values are given by Equation 4.

$$sim(i,j) = \frac{\sum_{u \in U}(r_{u,i} - \bar{r}_i)(r_{u,j} - \bar{r}_j)}{\sqrt{\sum_{u \in U}(r_{u,i} - \bar{r}_i)^2}\sqrt{\sum_{u \in U}(r_{u,j} - \bar{r}_j)^2}}, \quad (3)$$

where $u \in U$ denote users who have rated both the items, i and j.

$$\hat{r}_{u,i} = \bar{r}_i + \frac{\sum_{j \in \mathcal{I}}(r_{u,j} - \bar{r}_j)sim(i,j)}{\sum_{j \in \mathcal{I}}|sim(i,j)|} \quad (4)$$

c. Slope-One: The main idea of the Slope One algorithms is to use the difference between User A's ratings of two items X and Y and User B's rating of item X in common to predict User B's unknown rating of item Y [32]. To predict missing values we first compute the average deviation, $dev_{i,j}$ of each pair of items given by Equation 5. The predicted values are given by Equation 6.

$$dev_{i,j} = \frac{\sum_{v \in \mathcal{U}}(r_{v,i} - r_{v,j})}{|\mathcal{U}|} \quad (5)$$

where $u \in \mathcal{U}$ denote users who have rated both the items, i and j.

$$\hat{r}_{u,i} = \frac{\sum_{j \in \mathcal{I}}(dev_{i,j} + r_{u,j})}{|\mathcal{I}|} \quad (6)$$

where $j \in \mathcal{I}$ denote issues for which $dev_{i,j} \neq 0$.

Model-Based CF Algorithms: Different from memory-based algorithms, model based CF techniques utilize user item matrix (the pure stance data in our case) to estimate or learn a model offline to make predictions. Among the model based CF techniques, matrix factorization models have been mostly studied and successfully applied in real recommender systems [33]. The matrix factorization models try to explain the ratings by characterizing both items and users on a latent factor space, such that user-item ratings are based on the inner products of them in the factor space. We describe two popular model based models: Singular Value Decomposition (SVD) and Probabilistic Matrix Factorization (PMF).

a. Singular value decomposition: SVD is a well-established technique for finding latent factors in information retrieval. Conventional SVD is undefined when there are missing entries in the matrix. It is a matrix factorization technique commonly used for producing low-rank approximations [33] [34]. Singular value decomposition for user ideology matrix $\mathcal{R}$, $SVD(\mathcal{R})$ is defined as

$$SVD(\mathcal{R}) = ASV^T \quad (7)$$

where A, S and V are of dimensions $n \times d$, $d \times d$ and $d \times m$. The S diagonal matrix contains singular values, which are positive and always in decreasing order (singular matrix). For low-dimensional representation we retain only $k \ll d$ entries for all matrices. $\hat{\mathcal{R}}_k = A_k.S_k.V_k^T$ is the rank-k matrix that is closest approximation of $\mathcal{R}$. SVD produces a set of uncorrelated eigenvectors used in collaborative filtering process. Each user and issue is represented by its corresponding eigenvector. The process of dimensionality reduction may help user who rated similar issues (but not exactly the same issues) to be mapped into the space spanned by the same eigenvectors. Once the matrix is decomposed, the prediction can be generated by computing the cosine similarities (dot products) between n pseudo-users and m pseudo-issues. The predicted values are given by,

$$\hat{r}_{u,i} = \bar{r}_u + A_k \cdot \sqrt{S_k}^T(u) \cdot \sqrt{S_k} \cdot V_k^T(i) \tag{8}$$

SVD finds A_k, S_k and V_k to obtain a approximation of $\mathcal{R}$, which requires $\mathcal{R}$ matrix to be complete. In real cases, the rating matrix $\mathcal{R}$ is sparse which make the conventional SVD not suitable. A solution approach is to minimize the squared error with the target $\mathcal{R}$ only for the observed entries of the target matrix R. This will result in a difficult non-convex optimization problem as discussed in [35]. Moreover, the SVD method does not scale well with the number of observations and is highly prone to overfitting on sparse data.

b. PMF: Different from SVD, PMF [27] is a probabilistic algorithms that scale linearly with the number of observations and perform well on sparse data. In PMF, we assume that there are K latent factors with which both users and items can be represented. The generative process of the user u and the item i are as follows.

$$p(u|\sigma_U^2) = \mathcal{N}(u|\mathbf{0}, \sigma_U^2 \mathbf{I}), \tag{9}$$
$$p(i|\sigma_I^2) = \mathcal{N}(i|\mathbf{0}, \sigma_I^2 \mathbf{I}), \tag{10}$$

where σ_U^2 and σ_I^2 are two variance parameters for users and items, respectively, $\mathbf{I}$ is the identify matrix, and $\mathcal{N}(\cdot|\mu, \sigma^2)$ is the normal distribution with mean μ and variance σ^2.

Furthermore, the rating score $r_{u,i}$ between user u and item i is generated as:

$$p(r_{u,i}|u, i, \sigma_1^2) = \mathcal{N}(r_{u,i}|g(u^T i), \sigma_1^2), \tag{11}$$

where σ_1^2 is variance parameter and $g(\cdot)$ the logistic function.

In PMF, we seek to minimize the regularized estimation error as follows:

$$\sum_{u \in \mathcal{U}} \sum_{i \in \mathcal{I}} \mathbb{I}(r_{u,i})(r_{u,i} - g(u^T i))^2 + \lambda_U ||\mathcal{U}||_F^2 + \lambda_I ||\mathcal{I}||_F^2, \tag{12}$$

where $\mathbb{I}(s)$ is an indicator function which equals 1 when s is not empty and otherwise 0. $g(\cdot)$ is the logistic function.

To optimize the objective function above, we can perform gradient descent on $\mathcal{U}$ and $\mathcal{I}$ to find a local optimum point. This method is efficient even on large data sets as it does not need to infer the full posterior distribution over the model parameters and observations. Nevertheless, a fully Bayesian treatment of the PMF model can further boost the model performance [36].

After profiling users and items in the latent factor space, to predict a user u's rating $\hat{r}_{u,i}$ on an given item i, we simply take the dot product of the user and item vector: $\hat{r}_{u,i} = g(u^T i)$.

4.2 Party Prediction

Clustering algorithm on the predicted user stance matrix generates groups of users. We propose simple K-means for clustering the users. To label the clusters to the respective parties, we compute the similarity distance between the cluster's average ideology and the party ideology using the similarity distance techniques like Hamming distance or Euclidean.

5 Experiments

In this section we want to answer the following research questions:
  RQ1: How accurate is the ideological stance prediction and by which model?
  RQ2: Which model performs better for stance prediction on sparse data?
  RQ3: Is our proposed approach effective in users' party prediction?
  RQ4: Does ideological belief aids in users' party prediction?

We answer the first two research questions through ideology stance prediction experiments and the last two questions through the party prediction experiments. We first describe the data set we used to evaluate our approach and our evaluation criteria. We next describe our ideological stance prediction experiments and results of collaborative filtering methods. Finally, we describe our party prediction experimental study that shows the effectiveness of our approach in predicting users political party.

5.1 Dataset

Our dataset is constructed by crawling the data from debate.org. We collected 1000 user profiles who provided the party affiliation information. The data consists of user personal details such as gender, age, political party, income, occupation, religion and other demographics. Apart from the demographics, users also provide their stances on several social, political and economical issues. After the clean up, we collected the stances for the issues that match with those in Table 1. With some preliminary experiments, we found that these issues are useful in party prediction. We first generated User Ideology Matrix, $\mathcal{R}$ described in Section 3 from the stance data. We use $\mathcal{R}$ for our stance prediction experiments. We use users' political party information as gold truth for our party prediction experiments.

Table 3. Statistics of our dataset

Users	1000
Democrats	519
Republicans	481
Issues	death penalty, gay marriage, national health care, flat taxes, gun rights and abortion
User-ideology matrix sparsity rate	22.95% (percentage of missing values)

5.2 Evaluation Criteria

We use standard metrics from information retrieval for evaluating the performance of the models. We compute *Accuracy* (the higher the better) for all the models at all sparsity rates to evaluate the stance prediction performance on collaborating models. We also use *Precision*, *recall* and *F1 score* to evaluate the performance of PMF model on stance prediction. We use *Purity* (the higher the better) and *Entropy* (the lower the better) and *Rand Index* (the higher the better) to evaluate the performance of political affiliation prediction [37]. We also use *Accuracy* by computing the percentage of users that are "classified" correctly after labeling the clusters using external information, $\mathcal{P}$. We also compute *F1 score* for each political party to evaluate the models.

5.3 Ideological Stance Prediction Experiments

Recall that our first step in our solution model is to predict the missing ideological stances of users. We use $\mathcal{R}$ for these experiments. To generate the sparsity on the data, we hide the users stances randomly and predict the hidden data using CF models. We use the same hidden matrix across all models for unbiased evaluation. Through these experiments, we will answer RQ1 and RQ2.

Experimental Settings: We compare User based, Item based, Slope one, SVD and PMF models for evaluation. For matrix sparsity, we hide stances in $\mathcal{R}$ to obtain sparsity rates(percentage of missing stances) of 30%, 40%, 50%, 60% and 70%. We took an average of three random matrices for each sparse matrix rate. We use Accuracy, Precision, Recall and F1 score for comparison.

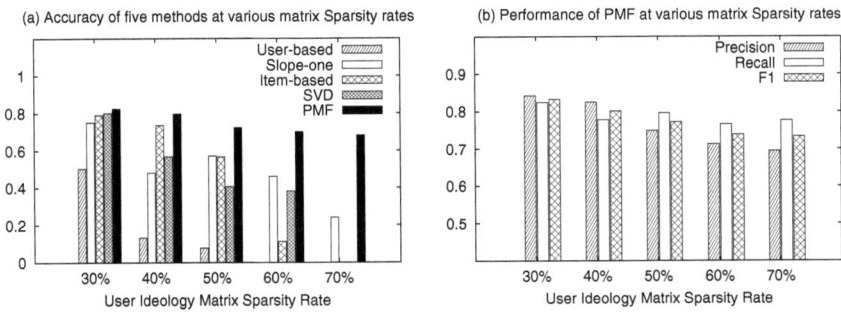

Fig. 1. Ideological stance prediction results at various matrix sparsity rates

Results: The stance prediction accuracy results for all the collaborative model is shown in Figure 1(a). We observe that PMF model outperforms all the other collaborative techniques. At 30% sparsity rate PMF model has an accuracy of 82.5% which is 2.5% higher than second best mode, SVD which has an accuracy of 80%. In case of sparse data, user-based, item-based and SVD completely fail on the model. PMF model still performs the best with an accuracy of 68.2%. Slope-one has an accuracy of only 24% at a sparsity rate of 70%.

We further show precision, recall and F1 scores of PMF model breakdown at various sparsity rates in Figure 1(b). We observe that PMF has a F1 score of 83.3% at 30% sparsity rate. At 70% sparsity rate, the model still performs well with an accuracy of 73.3%. With these observations, we choose PMF for the party prediction experiments which we explain in the next subsection.

Summary: As an answer to RQ1, we show that PMF model outperforms other collaborative models in stance prediction with an accuracy of 82.5%. As an answer to RQ2, our experiments show that PMF has better performance than others on sparse data with an accuracy of 73.3% at a sparsity rate of 70%.

5.4 Party Prediction Experiments

The main goal of our study is to discover the political party of the user. Through this experiment, we would like to study not only the model performance but also the importance of ideological belief in party detection. We will answer RQ3 and RQ4.

Experiment settings: We conduct experiments on various sparsity rates similar to previous experiments. We use the same hidden matrix across all models for unbiased evaluation. The ground truth on the users' political leaning is available from the users' profiles. For baseline model, we use a recent work [38].

Baseline: A direct approach for party prediction can be achieved by measuring the similarity between the user vector and party vector using Hamming distance and assign the user to the party with low Hamming distance.

Discussant Attribute Profile (DAP): [38] proposes to profile discussants by their attribute towards other targets and use standard clustering (K-Means) to cluster discussants, and achieves promising results on a similar task - subgroup detection. We thus incorporate the method on our task by profiling each user by his/her ideologies towards issues stated in Table 1.

Probabilistic Matrix Factorization (PMF): We apply PMF on $\mathcal{R}$ and then cluster users into two clusters. We set the number of latent factors to 10 as we do not observe big difference when vary the latent factor size from 10 to 50. For the other parameters, we select the optimal setting based on the average of 10 runs. λ is chosen from $\{0.1, 0.01\}$.

As discussed in Section 3, the resulting clusters are labeled by using party ideology matrix $\mathcal{P}$. We first calculate the average ideology of each cluster and measure the distance of cluster ideology to party ideology using Hamming distance. The closer the distance to the party, all the users in that cluster are labeled with the corresponding party. We use metrics Purity, Entropy, Accuracy, RandIndex and F1 score for evaluation.

Results: We first present the detailed clustering results in the Table 4. At all sparsity rates PMF outperforms DAP and Baseline on all metrics. We also observed that for Baseline, for some users the Hamming distance to both the parties is equal(when the matrix is sparse) and hence are assigned randomly to one of the major parties. Another observation is that, higher the sparsity rates,

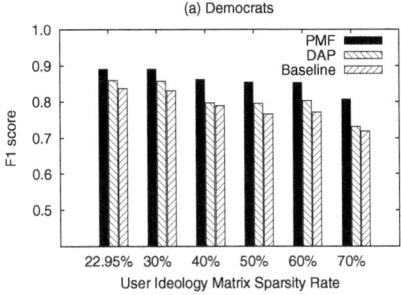
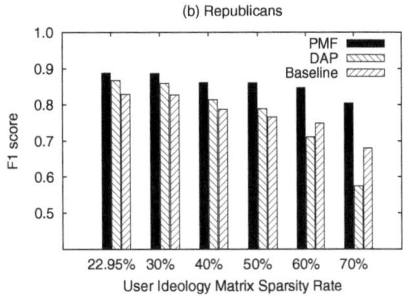

Fig. 2. Party prediction F1 results break down by parties at various matrix sparsity rates. 22.95% is the original sparsity rate of the matrix $\mathcal{R}$.

Table 4. Clustering results for political affiliation detection on all models. 22.95% is the original sparsity rate of the matrix $\mathcal{R}$.

Method	Metric	22.95%	30%	40%	50%	60%	70%
Baseline	P	0.833	0.829	0.788	0.766	0.761	0.701
	E	0.649	0.657	0.740	0.782	0.793	0.880
	A	0.833	0.829	0.788	0.766	0.761	0.701
	R	0.722	0.716	0.666	0.641	0.636	0.580
DAP	P	0.856	0.858	0.806	0.792	0.766	0.670
	E	0.523	0.581	0.694	0.735	0.759	0.904
	A	0.856	0.858	0.806	0.792	0.766	0.670
	R	0.753	0.756	0.687	0.670	0.641	0.557
PMF	P	**0.889**	**0.889**	**0.861**	**0.857**	**0.850**	**0.805**
	E	**0.498**	**0.499**	**0.573**	**0.578**	**0.608**	**0.707**
	A	**0.889**	**0.889**	**0.861**	**0.857**	**0.850**	**0.805**
	R	**0.809**	**0.802**	**0.761**	**0.755**	**0.745**	**0.686**

larger the number of unassigned users. At higher sparsity rates Baseline performs better than DAP as DAP tends to generate unbalanced clusters. On original data all the models have reasonably high accuracies: Baseline has an accuracy of 83.3%, DAP has 85.6% accuracy and PMF has an accuracy of 88.9%. At 30% DAP has an accuracy of 85.8% and PMF has an accuracy of 88.9% which is 4.1% higher. This shows the importance of ideological belief in party prediction task. We observe that the accuracy drops drastically for DAP with the sparsity of the data yielding 67% at 70% sparsity rate, whereas for PMF the accuracy is 80.5% which is 13.5% higher. Even though Baseline also degrades at higher sparsity rates, it perform better than DAP but 10.5% lower than PMF. From our previous experiments, we observe that PMF aids in better prediction of missing ideology stances when compared to other collaborative methods. Such behavior aided in high accuracy by PMF when compared to Baseline and DAP.

We further studied the performance of all the models at the party breakdown. We show F1 scores break down by parties at various matrix sparsity rates for both the models in Fig. 2. We observe that PMF and Baseline has balanced F1 scores for both clusters (Democrats and Republicans) at all sparsity rates, whereas DAP shows high F1 for Democrats than Republicans at 70% sparsity rate. At 30% sparsity rate, Baseline has 83.7% and 82.9%, DAP has 85.7% and 85.9% and PMF has 89% and 88.7% accuracy for Democrats and Republicans respectively. At 70% sparsity rate, Baseline has 71.9% and 68%, DAP has 73.1% and 57.4% and PMF has 80.7% and 80.4% accuracy for Democrats and Republicans respectively. For sparse data, DAP tends to cluster users to the larger cluster, in this case, it is Democrats.

Summary: To answer RQ3, our approach of using collaborative filtering method for prediction task outperforms standard clustering techniques with and accuracy of 88.9%. As an answer to RQ4, users' ideological stances plays vital role in users' party prediction and even standard clustering techniques achieves a high accuracy of 85.5%.

5.5 Discussion

Our experiments demonstrates the benefit of ideological stances of users in predicting their party leaning. While the results are very promising, it is interesting to study the reliability of the issues that we chose for determining the ideological belief of users. For our studies, we used only 6 out of 46 issues[4] for which users provide stances. We collected stances for 46 issues for all users in our corpus and we observed that the sparsity rate of the issue matrix is 52.94%. We then tested the model with all 46 issues and achieved an accuracy of 88.1% which is 0.8% lower than our previous results. Our results with 6 issues are very close to results with 46 issues, which shows that the 6 issues in Figure 1 should suffice for determining the ideology of a user. Furthermore, with high dimensionality where the sparsity rate is high, the model may fail to predict the missing stances [11]. It might be interesting to study which combination of issues (features) have greater impact on the party and we leave it for future studies.

Further, we studied the impact of religion dimension on party prediction task. Some studies observed a correlation between religion and party affiliation such as republicans exhibit greater religiosity compared to democrats [39]. We conducted some experiments where we assign the Christians to Republicans and others to Democrats. We achieved an accuracy of 68.2% for party prediction task, which shows that religion dimension is insufficient for party prediction as religiosity cannot be captured from the religion demographic. It is interesting to study the correlation of other demographics with the party and we leave it for future work. Also another extension can be study of combining demographics with ideological beliefs, exploiting ideological stances with social networks or corpus content such as text, hashtags etc.,

---

[4] http://www.debate.org/big-issues/

6 Conclusion

In this paper, we studied the problem of predicting users political party in social media. In our approach, we exploited users' ideological stances to predict their political party and we proposed a collaborative filtering approach to solve the data sparsity problem of users stances on controversial sociopolitical issues and apply clustering method to group the users with the same party. Our experiment results proves that user's ideological belief is highly correlated with the party affiliation. Evaluation results show that using ideological stances with PMF achieves a high accuracy of 88.9% at 22.9% data sparsity rate and 80.5% at 70% data sparsity rate.

Acknowledgements. This research/project is supported by the Singapore National Research Foundation under its International Research Centre@Singapore Funding Initiative and administered by the IDM Programme Office.

References

1. Benevenuto, F., Rodrigues, T., Cha, M., Almeida, V.: Characterizing user behavior in online social networks. In: Proceedings of the 9th ACM SIGCOMM Conference on Internet Measurement Conference, pp. 49–62 (2009)
2. Yan, X., Yan, L.: Gender classification of weblog authors. In: AAAI Spring Symposium: Computational Approaches to Analyzing Weblogs, pp. 228–230 (2006)
3. Peersman, C., Daelemans, W., Vaerenbergh, L.V.: Predicting age and gender in online social networks. In: SMUC, pp. 37–44 (2011)
4. Conover, M., Gonçalves, B., Ratkiewicz, J., Flammini, A., Menczer, F.: Predicting the political alignment of twitter users. In: Proceedings of 3rd IEEE Conference on Social Computing, SocialCom (2011)
5. Speel, R.W.: The evolution of republican and democratic ideologies. Journal of Policy History 12(7), 413–416 (2000)
6. Saunders, K., Abramowitz, A.: Ideological realignment and active partisans in the american electorate. American Politics Research 32(3), 285–309 (2004)
7. Fiorina, M.P., Abrams, S.J.: Political polarization in the american public. Annual Review of Political Science 11(1), 563–588 (2008)
8. Killian, M., Wilcox, C.: Do abortion attitudes lead to party switching? Political Research Quarterly 61(4), 561–573 (2008)
9. Somasundaran, S., Wiebe, J.: Recognizing stances in ideological on-line debates. In: Proceedings of the NAACL HLT 2010 Workshop on Computational Approaches to Analysis and Generation of Emotion in Text, pp. 116–124 (2010)
10. Walker, M.A., Anand, P., Abbott, R., Tree, J.E.F., Martell, C., King, J.: That is your evidence?: Classifying stance in online political debate. Decis. Support Syst. 53(4), 719–729 (2012)
11. Ma, H., Yang, H., Lyu, M.R., King, I.: Sorec: Social recommendation using probabilistic matrix factorization. In: Proc. of CIKM (2008)
12. Pan, R., Zhou, Y., Cao, B., Liu, N.N., Lukose, R., Scholz, M., Yang, Q.: One-class collaborative filtering. In: ICDM 2008 (2008)
13. Singh, A.P., Gordon, G.J.: Relational learning via collective matrix factorization. In: Proceedings of the 14th KDD, pp. 650–658 (2008)

14. Rao, D., Yarowsky, D., Shreevats, A., Gupta, M.: Classifying latent user attributes in twitter. In: Proceedings of the 2nd International Workshop on Search and Mining User-generated Contents, pp. 37–44 (2010)
15. Mukherjee, A.: 0001, B.L.: Improving gender classification of blog authors. In: EMNLP, pp. 207–217 (2010)
16. Zhou, D.X., Resnick, P., Mei, Q.: Classifying the political leaning of news articles and users from user votes. In: ICWSM (2011)
17. Dahllöf, M.: Automatic prediction of gender, political affiliation, and age in swedish politicians from the wording of their speeches - a comparative study of classifiability. LLC 27(2), 139–153 (2012)
18. Durant, K.T., Smith, M.D.: Predicting the political sentiment of web log posts using supervised machine learning techniques coupled with feature selection. In: Nasraoui, O., Spiliopoulou, M., Srivastava, J., Mobasher, B., Masand, B. (eds.) WebKDD 2006. LNCS (LNAI), vol. 4811, pp. 187–206. Springer, Heidelberg (2007)
19. Durant, K.T., Smith, M.D.: Mining sentiment classification from political web logs. In: Proceedings of Workshop on Web Mining and Web Usage Analysis of the 12th SIGKDD, WebKDD 2006 (2006)
20. Efron, M.: Using cocitation information to estimate political orientation in web documents. Knowl. Inf. Syst. 9(4) (2006)
21. Balasubramanyan, R., Routledge, B.R., Smith, N.A.: From tweets to polls: Linking text sentiment to public opinion time series. In: Proceedings of ICWSM (2010)
22. Abu-Jbara, A., Radev, D.: Subgroup detector: a system for detecting subgroups in online discussions. In: Proc. of the ACL 2012 Demo, pp. 133–138 (2012)
23. Hassan, A., Abu-Jbara, A., Radev, D.: Detecting subgroups in online discussions by modeling positive and negative relations among participants. In: Proceedings of the 2012 Joint Conference on EMNLP and CoNLL (2012)
24. Blondel, V.D., Guillaume, J.L., Lambiotte, R., Lefebvre, E.: Fast unfolding of communities in large networks. Journal of Statistical Mechanics: Theory and Experiment 2008(10) (2008)
25. Traag, V., Bruggeman, J.: Community detection in networks with positive and negative links. Physical Review E 80(3), 036115 (2009)
26. Resnick, P., Iacovou, N., Suchak, M., Bergstrom, P., Riedl, J.: Grouplens: an open architecture for collaborative filtering of netnews. In: CSCW, pp. 175–186 (1994)
27. Salakhutdinov, R., Mnih, A.: Probabilistic matrix factorization. In: Advances in Neural Information Processing Systems (NIPS), vol. 20 (2008)
28. Qiu, M., Yang, L., Jiang, J.: Mining user relations from online discussions using sentiment analysis and probabilistic matrix factorization. In: NAACL, pp. 401–410. Association for Computational Linguistics (2013)
29. Yang, S.H., Long, B., Smola, A., Sadagopan, N., Zheng, Z., Zha, H.: Like like alike: joint friendship and interest propagation in social networks. In: WWW (2011)
30. Su, X., Khoshgoftaar, T.M.: A survey of collaborative filtering techniques. Adv. In: Artif. Intell. 2009, 4:2–4:2 (January 2009)
31. Sarwar, B., Karypis, G., Konstan, J., Riedl, J.: Itembased collaborative filtering recommendation algorithms. In: WWW, pp. 285–295 (2001)
32. Lemire, D., Maclachlan, A.: Slope one predictors for online rating-based collaborative filtering. In: Proceedings of SIAM Data Mining, SDM 2005 (2005)
33. Koren, Y., Bell, R., Volinsky, C.: Matrix factorization techniques for recommender systems. Computer 42(8), 30–37 (2009)
34. Sarwar, B., Karypis, G., Konstan, J., Riedl, J.: Incremental singular value decomposition algorithms for highly scalable recommender systems. In: Fifth International Conference on Computer and Information Science, pp. 27–28 (2002)

35. Srebro, N., Jaakkola, T.: Weighted low rank approximation. In: ICML (2003)
36. Salakhutdinov, R., Mnih, A.: Bayesian probabilistic matrix factorization using markov chain monte carlo. In: ICML (2008)
37. Manning, C.D., Raghavan, P., Schütze, H.: Introduction to Information Retrieval. Cambridge University Press (2008)
38. Abu-Jbara, A., Diab, M., Dasigi, P., Radev, D.: Subgroup detection in ideological discussions. In: Proceedings of the 50th ACL, pp. 399–409 (2012)
39. Glaeser, E.L., Ponzetto, G.A.M., Shapiro, J.M.: Strategic extremism: Why republicans and democrats divide on religious values. The Quarterly Journal of Economics 120(4), 1283–1330 (2005)

A Fast Method for Detecting Communities from Tripartite Networks

Kyohei Ikematsu and Tsuyoshi Murata

Tokyo Institute of Technology, 2-12-1 Ookayama Meguro-ku Tokyo 152-8552, Japan
{ikematsu,murata}@ai.cs.titech.ac.jp
http://www.ai.cs.titech.ac.jp/

Abstract. This paper proposes a fast method for detecting communities from tripartite networks. Our method is based on an optimization of tripartite modularity, and the method combines both edge clustering and Blondel's Fast Unfolding. Experimental results on synthetic tripartite networks show that accurate communities are detected with our method. Furthermore, an experiment on a real tripartite network shows that our method is scalable to tripartite networks of tens of thousands of vertices. To the best of our knowledge, this is the first attempt for analyzing real tripartite networks composed of tens of thousands of vertices.

Keywords: community detection, modularity, tripartite networks.

1 Introduction

Many real social relations can be represented as tripartite networks. For example, users of social bookmarking services, such as Delicious, Digg and Japanese Hatena Bookmark, put tags on web pages. These relationships of users, tags and web pages can be represented as a network composed of three types of vertices (Fig.1(a)). Because of the growth of social media, many large-scale tripartite networks are available online. Analyzing such networks and understanding their structures are thus becoming more and more important.

Community detection is one of the methods for network analysis. A community is a subgraph with many edges inside and relatively few edges outside. It clarifies relationships among vertices and the overall structure of the network. For usual (unipartite) networks composed of only one vertex type, Newman-Girvan modularity is proposed for evaluating network divisions, and it is widely used for the research of community detection. Searching for the divisions with high modularity values, which is called modularity optimization, is also investigated by many researchers.

In order to detect communities from tripartite networks, they are often projected into simpler unipartite networks and then processed with conventional methods. However, projection will lose the information that original tripartite networks have. As a way for solving this problem, Murata [1] and Neubauer [2] respectively propose tripartite modularities that evaluate the qualities of the divisions of tripartite networks. With these tripartite modularities, Murata [3] and

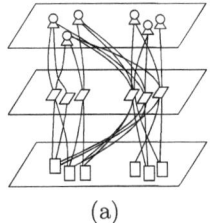 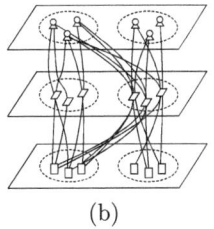

Fig. 1. Examples of a tripartite network (a) and a community structure (b)

Ikematsu [4] detects communities from tripartite networks. But these methods still have some problems such as requiring a parameter, projection and computational cost. In short, previous researches on tripartite networks mainly focus on the proposal of new tripartite modularities for evaluating network divisions, but the optimization of the tripartite modularities are not fully investigated.

This paper proposes a fast method for optimizing tripartite modularity. Our method combines both edge clustering [4] and Blondel's Fast Unfolding [5]. Experimental results for detecting communities from synthetic tripartite networks are presented in order to show the accuracy and computational speed of our method. In addition, a real tripartite network obtained from Delicious, one of the famous social bookmarking services, is also used for our experiments. The results show that our method is scalable to tripartite networks composed of tens of thousands of vertices. To the best of our knowledge, this is the first attempt for analyzing real tripartite networks composed of tens of thousands of vertices.

2 Related Works

2.1 Newman-Girvan Modularity

Newman and Girvan propose the following modularity Q, an objective function for evaluating the qualities of the divisions of unipartite networks [6]. Divisions with denser edges within communities and sparser edges between communities will have high modularity values.

$$Q = \sum_{l}(e_{ll} - a_l^2) \quad (1)$$

e_{lm} and a_l used in the above formulas are defined as follows. M is the number of edges, i and j are vertices, l and m are communities, V is a set of all vertices, V_l and V_m are sets of vertices that constitute communities l and m, and $A(i,j)$ is (i,j)-th element of an adjacency matrix.

$$e_{lm} = \frac{1}{2M} \sum_{i \in V_l} \sum_{j \in V_m} A(i,j) \quad (2)$$

$$a_l = \sum_{m} e_{lm} = \frac{1}{2M} \sum_{i \in V_l} \sum_{j \in V} A(i,j) \quad (3)$$

a_l^2 in (1) is the expected number of edges within community l in the *null model*, which is a network that matches the original network in some of its structural features (such as the number of vertices, number of edges, and degree distributions), but is otherwise taken to be an instance of a random network. The minimum value of Q is -1, and its maximum is close to 1.

2.2 Tripartite Modularity

This section explains the tripartite modularity, which is a quality function for the divisions of tripartite networks. The definitions of tripartite networks and communities in them are also explained.

Tripartite Networks. In general, tripartite networks are the networks whose vertices are composed of three disjoint sets. Tripartite networks are different from multislice networks [7]. Each node in multislice networks appears in every slice of the multislice networks. On the other hand, nodes in tripartite (and n-partite) networks are basically different in each layer. In this paper, we focus on the specific type of tripartite networks, which are called 3-partite 3-uniform hypernetworks (Fig.1(a)). Each edge in 3-partite 3-uniform hypernetworks is a hyperedge that always connects three vertices from each of three disjoint vertex sets. The reason we limit the type of tripartite networks to 3-partite 3-uniform hypernetworks is to simplify the problems.

In the following explanations, 3-partite 3-uniform hypernetworks are simply represented as tripartite networks. Three disjoint vertex types of a tripartite network are called X, Y and Z. M is the number of all edges and V is a set of all vertices in the tripartite network. V^X is a set of vertices that belongs to vertex set X. V_l^X is a set of vertices that belongs to vertex set X and also to communuty l. C^X is a set of communities in vertex set X. An adjacent matrix of a tripartite network is A, and its (i,j,k)-th element is represented as $A(i,j,k)$, where $i \in V^X, j \in V^Y$ and $k \in V^Z$. It takes three arguments because each hyperedge always connects the three vertices from each of $i \in V^X$, $j \in V^Y$ and $k \in V^Z$. $A(i,j,k)$ takes 1 when vertices i, j and k are connected with a hyperedge, and otherwise it takes 0.

Communities in Tripartite Networks. Definition of a community in n-partite networks is controversial. Barber defines it as a set of vertices of arbitrary vertex types [8]. So vertices from V^X, V^Y and V^Z can be mixed together in a community with his definition. We define a community in a different way. In this paper, a community is a set of vertices of only one vertex type (Fig.1(b)). So each community is composed of the same vertex type. The reason we define it this way is that relations of communities of different vertex types will be clarified with this definition.

Murata's Tripartite Modularity. Newman-Girvan modularity is not appropriate for evaluating divisions of tripartite networks. Because it evaluates the density of edges within each community, but in the case of tripartite networks, there is no edge that connects the vertices of the same type. If a tripartite network is projected into unipartite or bipartite networks, Newman-Girvan modularity or other bipartite modularity can be applied [9,10]. But the projection will lose the information that original tripartite network has [1,2].

Murata extended Newman-Girvan modularity for tripartite networks so that no projection is required [1]. Murata's tripartite modularity evaluates the density of connections between communities of different vertex types, not within communities. For example, modularity value Q^X for the communities of vertex type X is computed with (4), where Q_l^X is the partial modularity value of community l in vertex X.

$$Q^X = \sum_l Q_l^X = \sum_l \sum_m \sum_n (e_{lmn} - a_l^X a_m^Y a_n^Z)$$
$$m, n = \underset{(m',n')}{\mathrm{argmax}}(e_{lm'n'}) \tag{4}$$

e_{lmn}, a_l^X, a_m^Y and a_n^Z are defined as follows. $a_l^X a_m^Y a_n^Z$ in (4) is the expected number of edges that connect vertices in communities l, m and n in the *null* model. In (4), argmax is used for specifying communities m and n that have the strongest connection with l.

$$e_{lmn} = \frac{1}{M} \sum_{i \in V_l^X} \sum_{j \in V_m^Y} \sum_{k \in V_n^Z} A(i,j,k) \tag{5}$$

$$a_l^X = \sum_{m \in C^Y} \sum_{n \in C^Z} e_{lmn} = \frac{1}{M} \sum_{i \in V_l^X} \sum_{j \in V^Y} \sum_{k \in V^Z} A(i,j,k) \tag{6}$$

$$a_m^Y = \sum_{l \in C^X} \sum_{n \in C^Z} e_{lmn} = \frac{1}{M} \sum_{i \in V^X} \sum_{j \in V_m^Y} \sum_{k \in V^Z} A(i,j,k) \tag{7}$$

$$a_n^Z = \sum_{l \in C^X} \sum_{m \in C^Y} e_{lmn} = \frac{1}{M} \sum_{i \in V^X} \sum_{j \in V^Y} \sum_{k \in V_n^Z} A(i,j,k) \tag{8}$$

For the communities in vertex types Y and Z, Q^Y and Q^Z are computed in the same way as in (4). The final definition of Murata's tripartite modulrity Q_{Murata} is represented as (9). The fraction $\frac{1}{3}$ is for normalizing its maximum value to 1.

$$Q_{Murata} = \frac{1}{3}(Q^X + Q^Y + Q^Z) \tag{9}$$

Neubauer's Tripartite Modularity. Murata's tripartite modularity does not fully consider the cases when one community corresponds to many communities of different vertex types. His modularity has to choose only one corresponding community because argmax is used in its defintion. If community V_l^X is connected with communities V_{m1}^Y and V_{m2}^Y and the density of both connections are almost the same, for example, Murata's tripartite modularity can be unstable because the correspondences of communities change drastically with an addition of a few noisy edges.

Neubauer overcomes this problem by proposing a new tripartite modularity which takes all correspondences among communities into consideration [2]. Neubauer's tripartite modularity $Q_{Neubauer}$ is computed with (10) using (5)-(8).

$$Q_{Neubauer} = \sum_{l}\sum_{m}\sum_{n} Q_{lmn}$$
$$= \sum_{l}\sum_{m}\sum_{n} \alpha_{lmn}(e_{lmn} - a_l^X a_m^Y a_n^Z) \qquad (10)$$
$$\alpha_{lmn} = \frac{1}{3}(\frac{e_{lmn}}{a_l^X} + \frac{e_{lmn}}{a_m^Y} + \frac{e_{lmn}}{a_n^Z})$$

In (10), a weight α_{lmn} is added to each of the correspondence of community triples (l, m, n). Neubauer's tripartite modularity is the weighted average of partial modularity Q_{lmn} that are for the correspondences of three communities (l, m, n). $Q_{Neubauer}$ takes -1 as its minimum value, and 1 as its maximum.

2.3 Modularity Optimization

As mentioned previously, modularity is a function for evaluating the quality of network divisions. Searching for the divisions with high modularity values is therefore desirable for community detection. This approach is called modularity optimization, and it is one of the most popular approaches for community detection. Finding the maximum modularity value is NP-hard problem, so heuristic method is often used for optimizing modularity. For example, greedy search, spectral partitioning or simulated annealing are often used for community detection from unipartite networks [11]. Newman Fast [12], CNM [13] and Fast Unfolding [5] are widely used for large-scale unipartite networks.

For tripartite networks, Murata [3] employs spectral partitioning for tripartite networks composed of thousands of vertices. However, his method requires the number of communities as a parameter, and the method projects tripartite networks into bipartite networks in the process of community detection. Ikematsu [4] employs edge clustering for the optimization of tripartite modularity. Although high-quality results are obtained with this method, its computational cost is high so it is not easy to apply this method for large-scale tripartite networks.

In the following sections, Fast Unfolding and edge clustering are briefly explained because these are related to our new method proposed in this paper.

Fast Unfolding Method. Fast Unfolding [5] is originally proposed for the optimization of modularity for unipartite networks. This method iterates the following two steps: (1) modularity optimization, and (2) community aggregation. In step (1), a vertex is randomly selected and it is "moved" to each of its adjacent communities in order to check whether the move of the vertex will increase modularity value. Then the vertex is moved to (or stayed in) the community so that the highest modularity value will be obtained. If any move will not contribute to the increase of modularity, step (2) is performed. Step (2) aggregates all vertices in a community into one vertex in order to obtain meta-level network. These steps (1) and (2) are iterated and hierarchical community structure is obtained.

Fast Unfolding is non-deterministic because its results depend on the order of the vertex selection in step (1). However, the method is fast and accurate, and it scales to huge networks that are composed of hundreds of millions of vertices [5]. The reason why the method is fast is that the move of a vertex in step (1) is limited to its adjacent communities, and it is easy to calculate the increase of modularity before and after the move of a vertex.

Edge Clustering Method. Most of modularity optimization methods cluster vertices. But Ikematsu [4] attempts edge clustering for the optimization of tripartite modularity. After edge clusters are obtained, vertex communities are then generated as the results of community detection.

In this method, all edges are set in different clusters as its initial state. Then they are merged in a greedy bottom-up manner so that the biggest increase of modularity values will be obtained for each merge. Each time edge clusters are obtained in the procedures, vertices are assigned to vertex communities based on the edge clusters, and the network division is then obtained. After that, the network division is evaluated based on tripartite modularity, which means edge clusters are indirectly evaluated by tripartite modularity via vertex communities.

Although the method is accurate, it is computationally expensive. The reason for its accurateness is that the edges in n-partite networks contain more information than those of unipartite networks [4]. And the reason for high computational costs is that the method employs naive greedy bottom-up manner for edge clustering, so the number of combination is too large, and the method needs to convert the clusters to vertex communities in order to compute the tripartite modularity [4].

3 Our Method

Our new method is based on edge clustering. In general, community detection based on edge clustering is composed of the following three tasks. The method shown in [4] (explained in the previous section) is one example of the following three tasks: greedy bottom-up clustering is employed in task 1, vertices are assigned to vertex communities (if a vertex connects to edges from more than one edge clusters, bigger edge cluster is selected) in task 2, and tripartite modularity is used for evaluating vertex communities in task 3.

Task 1: Cluster the edges based on some criteria.
Task 2: Generate vertex communities based on the edge clusters.
Task 3: Evaluate the quality of the vertex communities.

There are two relations between these tasks. The first relation is that edges are clustered with some criteria in task 1, and the criteria is evaluated based on the quality of the vertex communities in task 3. The second relation is that network division generated in task 2 is used as an input to task 3. Note that we can choose different strategies from [4] for these tasks if these relations are kept. For example, top-down clustering can be chosen for task 1, although [4] employs greedy bottom-up clustering. In the next section, we explain the concrete procedures of task 1, 2 and 3 that are employed in our new method. We call our new method FUE (Fast Unfolding for Edges) in this paper.

3.1 Fast Unfolding for Edges

Task 1. FUE generates "an adjacent edge network" that represents the adjacency of hyperedges in given tripartite network. We define the adjacency of hyperedges as follows: two hyperedges e_1 and e_2 are adjacent if they share common vertex. Therefore, hyperedges are represented as vertices, and their adjacencies are represented as edges in the adjacent edge network (Fig.2).

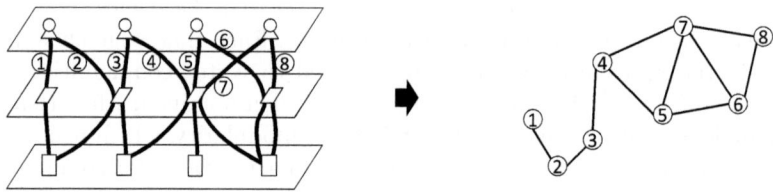

Fig. 2. From a tripartite network (left), an adjacent edge network (right) is generated

After the above procedure, Blondel's Fast Unfolding is applied to the adjacent edge network in order to obtain edge clusters. But Fast Unfolding is originally for unipartite networks and it uses Newman-Girvan modularity to evaluate the moves of a vertex. Since Newman-Girvan modularity is not suitable for the evaluation of edge clusters, task 2 and task 3 are applied and evaluate the moves of a vertex. From edge clusters, vertex communities are generated in task 2 and then they are evaluated based on tripartite modularity in task 3. Therefore, moves of a vertex in task 1 are indirectly evaluated in task 3 based on the final vertex communities in task 2. The procedures of task 1 are shown in Fig.3.

Task 2. In [4], vertices are assigned to vertex communities based on edge clusters. But the procedure in [4] is computationally expensive. So this paper improves this procedure as follows. First, for each vertex v in a tripartite network, find C_{adj}, a set of edge clusters whose elements are the clusters assigned to hyperedges adjcent to the vertex v. After that, the sizes of the edge clusters $c \in C_{adj}$ are computed, and v is assigned to the largest edge cluster $c_{max} \in C_{adj}$. If there is more than one edge clusters of maximum size, v is assigned to a new cluster. Fig.4 explains the process of assigning vertices to vertex communities. Fig.5 shows an example of obtaining vertex communities from edge clusters.

1. An adjacency edge network G_{AE} is generated from given tripartite network.
2. As an initial state of edge clustering, all vertices in G_{AE} are set in different clusters. Task 2 and 3 are applied to an initial state of edge clustering. Suppose its evaluation value by tripartite modularity is $Eval$.
3. Optimization of tripartite modularity
    (a) Vertices in G_{AE} are stored in a queue in a random order.
    (b) A vertex v is removed from the top of the queue. Let C_{adj} be a set of clusters that are adjacent to the vertex v.
    (c) For each $c \in C_{adj}$, the vertex v is tentatively moved to the cluster c and the state of edge clustering is updated. Then the state is processed through task 2 and 3, and evaluated based on tripartite modularity.
    (d) Among all possible $c \in C_{adj}$, the cluster c_{max} with the maximum increase of the modularity is selected. The vertex v is then assigned to the cluster c_{max}. If the tripartite modularity will not increase with any move, v's cluster will not be changed.
    (e) The above steps 3(b), 3(c) and 3(d) are repeated for all vertices in the queue. If the cluster assignment of any vertex has been changed during repetition, go to step 3(a).
4. Task 2 and 3 are applied to the obtained state of edge clustering. Suppose its evaluatation value is $Eval'$. If $Eval' == Eval$, this algorithm is terminated. If $Eval' > Eval$, $Eval$ is updated to the value of $Eval'$ and go to step 5.
5. Aggregation of clusters
    – Each cluster in the state of edge clustering is aggregated as a vertex, and meta-level adjacent edge network G_{AE} is generated. Go to step 3.

Fig. 3. The procedures of task 1

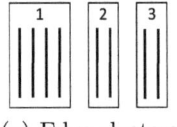
(a) Edge clusters

(b) Assignment of a user to a user community

(c) Assignment of a tag to a tag community

Fig. 4. After obtaining edge clusters (a), users/tags are assigned to user/tag communities. (b) shows that a user connected to two edges whose clusters are 1 and 2 will be assigned to a user community 1 because the former is bigger than the latter. (c) shows that a tag connected to two edges whose clusters are 2 and 3 will be assigned to a new tag community 4 because the size of edge cluster 2 and 3 are equal.

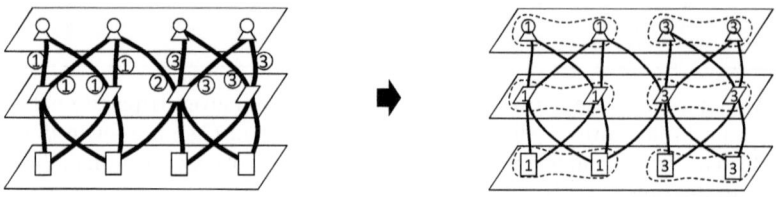

Fig. 5. Based on edge clusters (left), users, web pages and tags are assigned to their communities (right)

Task 3. FUE employs tripartite modularity for the evaluation of vertex communities. Although many tripartite modularities have been proposed, we use Neubauer's tripartite modularity in this paper because its fast implementation is relatively easy.

3.2 Merging Task 2 and Task 3

If task 2 and task 3 are implemented independently, Neubauer's tripartite modularity is computed from scratch for every state of vertex communities at step 3(c) in Fig.3. This part contains redundant computation because most partial modularity of Neubauer's modularity remains the same after the re-assignment of vertex v in G_{AE} to a new edge cluster. Therefore, only the difference of Neubauer's modularity before and after the move of vertex v should be computed at step 3(c) for speedup. Figure 6 explains detailed procedures for avoiding redundant computation at step 3(c).

In Fig.6, P_{old} is the network division before the move of a vertex v at step 3(c) in Fig.3. CO_{old}, $Q_{old}^{partial}$ and Q_{old} are also defined as follows in P_{old}, and Fig.7 shows some examples of these symbols used in Fig.6. CO_{old} is a set of correspondence (l, m, n) among communities $l \in C^X$, $m \in C^Y$ and $n \in C^Z$. $Q_{old}^{partial}$ is a hash function whose keys are correspondences (l, m, n) and whose values are partial modularity Q_{lmn}. Q_{old} is Neubauer's tripartite modularity. P_{new} is the network division after the move of a vertex v and applying task 2, and CO_{new}, $Q_{new}^{partial}$ and Q_{new} are defined similarly as for P_{old}.

4 Experiments on Synthetic Tripartite Networks

In order to show the accuracy and speed of FUE, the following experiments with synthetic and real tripartite networks are performed. First, synthetic tripartite networks with known community structures are used for the experiments. 100 synthetic tripartite networks with several amounts of noises are generated, and our method is applied to detect communities. Synthetic networks in our experiments have scale free properties just like many real networks.

The way to generate the synthetic networks is as follows. Let M and p be the number of edges and the fraction of noises. First, in order to put correct

1. Before the move of a vertex v in G_{AE},
   - Keep P_{old}, CO_{old}, $Q_{old}^{partial}$ and Q_{old} for reference during the following steps.
2. After the move of a vertex v and after applying task 2 (and before task 3),
   (a) Find C_{moved}^{X}, a set of X-part communities whose vertices have moved by applying task 2, in P_{old}. Find C_{moved}^{Y}, C_{moved}^{Z} as well.
   (b) Find $CO_{changed}$, a set of correspondences (l, m, n) such as $l \in C_{moved}^{X}$ or $m \in C_{moved}^{Y}$ or $n \in C_{moved}^{Z}$, in CO_{old}.
   (c) Find CO_{added}, a set of correspondences (l, m, n) which don't exist in CO_{old}, in CO_{new}.
3. Apply the following procedures as task 3
   (a) For each $(l, m, n) \in CO_{changed}$, get partial modularity $Q_{lmn} \in Q_{old}^{partial}$, and let Q' be the sum of all partial modularity Q_{lmn}.
   (b) For each $(l, m, n) \in CO_{changed}$, if $(l, m, n) \in CO_{new}$, calculate partial modularity Q_{lmn} in P_{new}, and let Q'' be the sum of all partial modularity Q_{lmn}.
   (c) For each $(l, m, n) \in CO_{added}$, calculate partial modularity Q_{lmn} in P_{new}, and let Q''' be the sum of all partial modularity Q_{lmn}.
   (d) Calculate Neubauer's tripartite modularity Q_{new} as follows: $Q_{new} = Q_{old} - Q' + Q'' + Q'''$

Fig. 6. Speedup of FUE (improvement of step 3(c) in Fig.3)

The state of a network at step 1 in Fig.6.

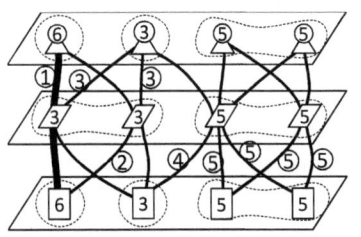

The network division is P_{old}. CO_{old} is $\{(6,3,6),(3,3,3),(3,5,3),(5,5,5)\}$. $CO_{changed}$ is $\{(6,3,6),(3,3,3),(3,5,3)\}$.

The state of a network at step 2 in Fig.6.

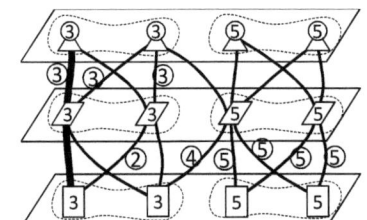

The network division is P_{new}. CO_{new} is $\{(3,3,3),(3,5,3),(5,5,5)\}$. CO_{added} is $\emptyset$.

Fig. 7. Examples of the symbols used in Fig.6. In the above figure, the left side is the state of a network, at step 1 in Fig.6, where there are five edge clusters and some vertex communities, and a bold edge belonging to cluster 1 will move to cluster 3. The right side is also the state of a network, but at step 2 in Fig.6, where task 2 has already been applied after the move of the edge cluster 1, and new vertex communities have been obtained. C_{moved}^{X} is $\{6,3\}$, C_{moved}^{Y} is $\emptyset$ and C_{moved}^{Z} is $\{6,3\}$ at step 2(a) in Fig.6. The other symbols are as described in the above figure.

edges, select a community each from X, Y and Z vertex type so that the three communities have correct relation in the answer community structure. After that, select a vertex each from the three communities, and put an edge between these three vertices. Repeat these procedures until the number of correct edges equals to $M(1-p)$. Note that the selection of the three vertices should be done to have scale free properties. In the case of noise edges, the three communities should be selected to be incorrect relation in the answer community structure, but the way to put edges is the same as the case of the correct edges.

As the baseline method to compare to FUE, naive greedy bottom-up vertex clustering (Fig.8) are also applied. In order to evaluate the qualities of detected communities, NMI (normalized mutual information) [14,15] is used. NMI takes its maximum value 1 when obtained network division is identical to the correct answer, and takes its minimum value 0 when they are not related at all.

1. Initially all vertices are set in different communities. Neubauer's tripart modularity $Q_{Neubauer}$ is computed.
2. For each of X-part, Y-part and Z-part, all possible combinations of community pairs are merged and $Q_{Neubauer}$ is computed for the new state.
3. Among the step 2, the community pair of the maximum increase of $Q_{Neubauer}$ is selected and the communities are merged.
4. Step 2 and 3 are repeated for each of X-part, Y-part and Z-part until all communities are merged into one.
5. The state of maximum $Q_{Neubauer}$ during the above processes is selected as the result of community detection.

Fig. 8. Naive greedy bottom-up approach (baseline method)

Four types of synthetic tripartite networks are used for our experiments. Fig.9 shows the four types of networks and the performances of FUE (solid line) and baseline method (dashed line). The x axis represents the fraction of noises in synthetic networks, and the y axis represents NMI. Of course, higher NMI means better performance. The number of vertices in each community is 10, and the number of edges that connect corresponding communities is 35 in our experiments. So the numbers of vertices and edges are 60 and 70 ($= 35 \times 2$) for the cases of Fig.9(a), and 60 and 105 ($= 35 \times 3$) for Fig.9(b), and 90 and 105 ($= 35 \times 3$) for 9(c), and 90 and 175 ($= 35 \times 5$) for Fig.9(d).

As shown in Fig.9, FUE performs much better than the baseline method. As for speed, for example, baseline method takes 31.39 seconds on average for detecting communities in the case of Fig.9(d), while FUE takes only 2.14 second on average. These experiments are performed using a PC with Core i7-2600 3.4GHz CPU, 16GB RAM and Python 2.7. These results show the accuracy and speed of our proposed method. The reasons for the speedup of FUE are as follows. The first reason is the number of possible combinations. Baseline method tries all possible combinations of community pairs, while FUE tries adjacent edge pairs only. In general, the latter case takes much less time than the former case. The second reason is to avoid redundant computation. FUE without avoiding redundant computation that we mentioned in Sect.3.2 takes 13.21 seconds on average, while FUE takes only 2.14 seconds on average. This means that redundant computation is actually avoided and it successfully makes FUE faster.

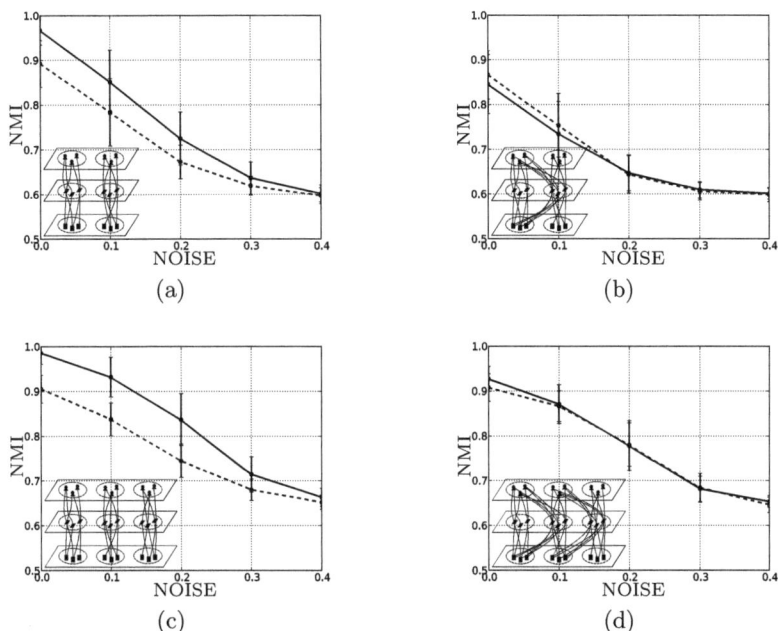

Fig. 9. Results on synthetic networks. FUE is the solid line, baseline method is the dashed line.

5 Experiment on a Real Tripartite Network

In order to show the effectiveness of our method, a real tripartite network from Delicious (a social bookmarking service) [16] is used for the next experiment. The dataset from [17] is the record of tags that users put on web pages with time stamps. We extract the data from January 2010 to February 2010 and generate a real tripartite network of users, tags and web pages. The numbers of users, tags and web pages are 455, 5925, and 4204, respectively. The total numbers of vertices and edges are 10584 and 19194, respectively.

The computational time for community detection from the above network is about 21 hours, and Neubauer's tripartite modularity value is 0.54. The statistics of detected communities are shown in Table 1. The rows represent users, tags and web pages, respectively. The columns represent the number of communities, the average size of communities, and the average numbers of correspondences per community. The values in the last column are much bigger than 1, which means that there are many-to-many correspondences among communities, which cannot be detected if we use projection.

Table 1. Statistics of detected communities

	Number of communities	Average size of communities	Average number of correspondences per community
User	274	1.66	6.39
Tag	399	14.85	4.39
Web page	524	8.02	3.34

Table 2 shows some examples of tag communities. The meanings of these communities are rather easy to understand, while there are some other communities that are difficult to understand. As for web page communities, pages in the same domains often constitute same web page communities.

Table 2. Some examples of detected tag communities

log, logger, messaging, prompt, shell, socket, stl, vim, ...
pop, post-punk, punk, rock, song, sweet-punk, ...
career, hr, recruiter, recruiting, personality_test, ...
baghdad, hands-of-victory, iraq, monument, political, ...
opinion-mining, sentiment, sentiment-analysis, textmining, ...

Previous research on community detection from tripartite networks uses very small networks that are composed of less than 100 vertices. To the best of our knowledge, this is the first attempt for analyzing real tripartite networks composed of tens of thousands of vertices. The above result is the first step for more detailed analysis of real tripartite networks.

6 Conclusions

This paper proposes a new method for detecting communities from tripartite networks using edge clustering [4] and Blondel's Fast Unfolding [5]. Experimental results on synthetic networks show the speed and accuracies of the proposed method. In addition, the result on a real tripartite network from Delicious [16] shows that our method is scalable up to networks of ten thousands of vertices with realistic computational time.

Since our method is not deterministic, detailed analysis of detected communities, such as comparing them or visualizing them, are left for our future work. Another way of implementing tasks that we mention in Sect.3 is also important for further improvement. Further speedup is needed for much larger scale tripartite networks.

Acknowledgement. This work was supported by JSPS KAKENHI Grant Number 22300049.

References

1. Murata, T.: Modularity for heterogeneous networks. In: Proceedings of the 21st ACM Conference on Hypertext and Hypermedia, pp. 129–134 (2010)
2. Neubauer, N., Obermayer, K.: Community detection in tagging-induced hypergraphs. In: Workshop on Information in Networks (2010)
3. Murata, T.: Detecting communities from social tagging networks based on tripartite modularity. In: Workshop on Link Analysis in Heterogeneous Information Networks, pp. 1–4 (2011)
4. Ikematsu, K., Murata, T.: Improvement of a tripartite modularity and its optimization method. Transactions of the Japanese Society for Artificial Intelligence 29(2) (to appear, 2014) (in Japanese)
5. Blondel, V.D., Guillaume, J.L., Lambiotte, R., Lefebvre, E.: Fast unfolding of communities in large networks. Journal of Statistical Mechanics: Theory and Experiment P10008 (2008)
6. Newman, M.E.J., Girvan, M.: Finding and evaluating community structure in networks. Physical Review E 69(026113), 1–15 (2004)
7. Mucha, P.J., Richardson, T., Macon, K., Porter, M.A., Onnela, J.P.: Community structure in time-dependent, multiscale, and multiplex networks. Science 328, 876–878 (2010)
8. Barber, M.J.: Modularity and community detection in bipartite networks. Physical Review E 76(066102), 1–9 (2007)
9. Cattuto, C., Baldassarri, A., Servedio, V.D.P., Loreto, V.: Emergent community structure in social tagging systems. Advances in Complex Systems 11, 597–608 (2008)
10. Neubauer, N., Obermayer, K.: Towards community detection in k-partite k-uniform hypergraphs. In: The NIPS 2009 Workshop on Analyzing Networks and Learning with Graphs (2009)
11. Fortunato, S.: Community detection in graphs. Physics Reports 486, 75–174 (2010)
12. Newman, M.E.J.: Fast algorithm for detecting community structure in networks. Physical Review E 69(066133), 1–5 (2004)
13. Clauset, A., Newman, M.E.J., Moore, C.: Finding community structure in very large networks. Physical Review E 70(066111), 1–6 (2004)
14. Ana, L., Jain, A.: Robust data clustering. In: Proceedings of 2003 IEEE Computer Society Conference on Computer Vision and Pattern Recognition, vol. 2, pp. II–128–II–133 (2003)
15. Danon, L., Duch, J., Diaz-Guilera, A., Arenas, A.: Comparing community structure identification. Journal of Statistical Mechanics P09008, 1–10 (2005)
16. Delicious, http://www.delicious.com
17. Cantador, I., Brusilovsky, P., Kuflik, T.: 2nd workshop on information heterogeneity and fusion in recommender systems (hetrec 2011). In: Proceedings of the 5th ACM Conference on Recommender Systems. RecSys 2011. ACM (2011)

Predicting Social Density in Mass Events to Prevent Crowd Disasters

Bernhard Anzengruber[1], Danilo Pianini[2], Jussi Nieminen[1], and Alois Ferscha[1]

[1] Institute for Pervasive Computing, Johannes Kepler University Linz, Austria
surname@pervasive.jku.at
[2] Department of Computer Science and Engineering - DISI, Alma Mater Studiorum Università di Bologna, Italy
danilo.pianini@unibo.it

Abstract. Human mobility behavior emerging in social events involving huge masses of individuals bears potential hazards for irrational social densities. We study the emergence of such phenomena in the context of very large public sports events, analyzing how individual mobility decision making induces undesirable mass effects. A time series based approach is followed to predict mobility patterns in crowds of spectators, and related to the event agenda over the time it evolves. Evidence is collected from an experiment conducted in one of the biggest international sports events (the Vienna city marathon with 40.000 actives and around 300.000 spectators). A smartphone app has been developed to voluntarily engage people to provide mobility data (1503 high-quality GPS traces and 1092694 Bluetooth relations have been collected), based on which prediction analysis has been performed. Using this data as training set, we compare density estimation approaches and evaluate them based on their forecasting precision. The most promising approach using Support Vector Regression (SMOreg) achieved prediction accuracies below 2 (root-mean-squared deviation) when compared to actual evidenced density distributions for a 12 minute forecasting interval.

1 Human Crowd Density and Safety

This year, more than 60 people died during the "New Year's Eve stampede" in Abidjan [13]. During the Love Parade on July 24, 2010 in Duisburg, 21 people died and more than 500 were injured in a crowd disaster [12]. Similarly at Hillsborough stadium on April 15, 1989, 96 people died and 400 were injured [24]. Many more such sobering statistics exist, explaining why disciplines such as crowd density estimation and human movement flow analysis have received much attention over the years. While human movements generally depend on the environment and the current context – such as the type of event – it was shown that disasters occurred in many different venues at different circumstances and could have been avoided through better management and planning [8]. In [12] the authors published a time line that shows how even 20 minutes prior to the disaster police forces had the opportunity to intervene if crowd density forecasts for that time period had been available. Towards this end, crowd density metrics that indicate the danger for individuals in a crowd have been established. The *(i)* density of people in a defined area, the *(ii)* flow of persons per meter and the *(iii)* pressure exerted on each person have been mentioned [8, 12] with safe levels approximated as *(i)* 2-3 persons per meter and minute, *(ii)* 82 persons per meter and minute and *(iii)* 1500 N respectively. Thus measuring crowd densities and predicting their development can offer valuable information about the current and future overall event safety and help to avoid catastrophes.

Previous work on this topic has been conducted in the field of pedestrian crowd dynamics [1, 5, 10] and similarly in the field of evacuation scenarios [11, 16, 31]. The purpose of many of these works has been to further our understanding of these dangerous phenomena in order to preemptively react to them. Generally the approach of choice has been to use simulations and draw conclusions from their outcome, whereby simulations have been based on models of reality that were constructed by abstracting data from real-world events. Our contribution uses a similar approach, in that we use time series analysis and other approaches to predict the crowd density at future points based on observed crowd densities from the past. Our approach is based on a dataset – comprised of the spatial properties of event spectators – that was created by a collaborative crowd-sourcing/participatory sensing approach using mobile devices of spectators (similar to e. g., [28]; see [4] for more information on participatory sensing). It has been argued that a linear correlation between user density – meaning the observed density of spectator's mobile devices – and crowd density exists [29]. Due to this linear relation a prediction of user density can be assumed to be correlated to actual crowd density. Thus for the purpose of this work we do not distinguish between them. However all presented results should be viewed in this light. We state our research hypotheses as follows:

Research Hypotheses:
(H1): Crowd-sourced data of spatial properties of individuals can be used to develop near real-time prediction mechanisms for crowd densities in urban mass events.
(H2): Such prediction mechanisms can forecast the density of crowds far enough into the future to allow intervention.

Our main contribution is the investigation of these issues, however we further compare multiple approaches to due so thus giving other researchers a head start for doing similar investigations. Previous works to predict crowd density exist, however most use different approaches than the one presented in this paper. [14, 18] use light-based/visual approaches that can only cover areas that were previously equipped with cameras. [22] uses wireless signal strength indicators (RSSI) to estimate the number of devices in an area. However the mentioned approach was only validated in a small-scale experiment. Perhaps most similar to our work is the research conducted in e. g., [27, 29]. The authors also use collaboratively crowd-sourced datasets of users' spatial properties to predict crowd densities at a given time and location. Our work deviates in that it uses a different forecasting approach to predict crowd density.

The rest of this paper is structured as follows: In Section 2 we describe our data collection approach for the mentioned event, the properties of collected data and its potential use in the area of crowd density estimation. In Section 3 we discuss various early and more sophisticated approaches to predict the development of the evidenced crowd densities and compare the results to the actual measured densities. Section 4 discusses future work and finally concludes this paper.

2 Approximating Human Density - A Data Collection Approach

In order to investigate patterns of crowd movement and resulting crowd densities we performed a large–scale experiment during an urban sports event in the city of Vienna. Our goal was to observe and track movements and neighborhoods of spectators in the event area. We developed an event App for Android and iOS that was active on the event day and sent spatial data of willing users to a database. The App was active only on the event day between 7 o'clock and 18 o'clock and was used by 6698 people (android: 5192, iOS: 1506) of which roughly 5230 agreed to participate in our study. We decided to employ GPS sensors for location tracking which are widely distributed in mobile platforms and can be easily used to log the position of devices with fair accuracy. We further enriched the collected traces using proximity-based sensor technologies to capture the user's neighborhoods. Since we did not actually transmit information but simply performed presence detection we decided on using Bluetooth which is very widely distributed. Other technologies have been used – see for example [20] – but all technologies suffer from various drawbacks such as short interaction range, limited availability or sparse scan points. We collected GPS locations in an interval of 1 minute and scanned for Bluetooth devices in an interval of 2 minutes. We filtered the collected dataset until 1503 long, high-quality user traces remained. On average, each trace contains 128 samples with a duration of 265 minutes whereby the median of the interval between samples is 1 minute. The Bluetooth data was collected by 3352 active users that detected 38816 unique ids and generated 1092694 proximity relations overall. We combined the Bluetooth and GPS information using the knowledge that every time a device detected a Bluetooth neighborhood, the detected device was physically close – which we defined as within a circular area of 30 meters based on standard Bluetooth radio transmission ranges – to the source device. We further segmented the event area into squares of 100x100 meters and computed the user densities for all of these segments – i. e., counted the number of data points in each segment. The imprecision that was introduced by the approximation of the devices' locations (about 30m) is somewhat mitigated by the comparably coarser segmentation of the city (100m). Figure 1 visualizes the described segmentation and computed densities for three time intervals in the morning, midday and evening of the event day. In the most crowded area of the event – the city center – we measured peak densities of 397 persons per segment (i. e., 10000 square meters). Based on the number of unique devices (~38000) and actual spectators (~300000) we project roughly 0.31 persons per square meter in the city center during the time intervals in which most of these people were located in the same area.

Many similar datasets for spatial data have been collected in previous works. In respect to GPS-based dataset we acknowledge [15], [17], and [30]. Similarly for Bluetooth-based datasets we acknowledge the work presented in [7], [26], [25] and [23]. For our purposes, the dataset collected during the mentioned urban sports event is more relevant due to the very large number of users that move with common purpose in the same environment at the same time.

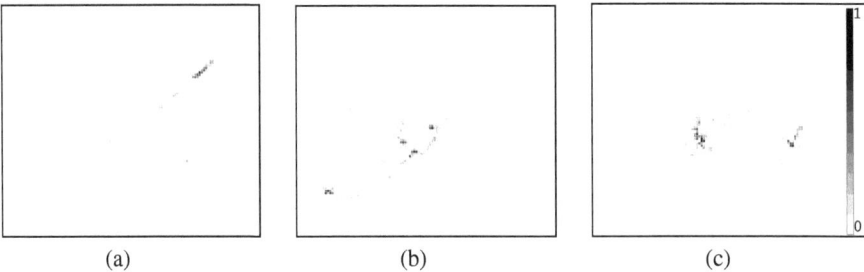

Fig. 1. Segmentation and crowd densities for (a) the morning, (b) midday and (c) the evening of the event day. In this visualization segment densities above 150 spectators per segment were clipped to 150. All values were subsequently divided by 150 and mapped to gray scale.

3 Towards the Prediction of Crowd Densities

3.1 A Human Crowd Density Flow Network

We build the graph of the city by interpreting segments as vertices and connecting them with two unidirectional edges between neighboring segments thus creating a checkboard-like network structure. We abstract the movements of people in such a way that they can only travel along the created edges. In order to determine the flow of people between segments during the event, we computed the difference of the segments' densities between consecutive time steps and used the resulting "δ-densities" as starting values for the traditional minimum-cost flow/transportation problem [21]. Segments with positive δ-densities were hereby treated as flow sources, those with 0 δ-density as transport nodes and segments with negative δ-densities as sinks. All edges between vertices had unlimited capacity and cost 1 associated with them. However due to the nature of the data used to generate this flow network (mobile devices that were turned on or off at random), the resulting flow network cannot be guaranteed to be balanced. As such we created a dummy node that provided/consumed the difference in source/sink flow. The dummy node was connected via network edges to all segments, however to not corrupt the results, all flows over these edges were associated with a cost of 100000 (i.e., so high that such flows are only computed once all normal sources/sinks have been satisfied). We computed these flows using an established algorithm – described in [2] – to solve linear programming challenges such as the minimum-cost flow problem.

As such we created a time series of segment density distributions in 2 minute time intervals – from here on referenced as $d_{0,2}, d_{2,4}, ...$ – and a sequence of flow networks that model how these distributions convey into one another – referenced as $f_{0,2}, f_{2,4}, ...$ –, whereby $f_{0,2}$ describes the flow that runs between $d_{0,2}$ and $d_{2,4}$.

3.2 Prediction Approaches

In order to forecast the development of human densities in the entire tessellated event area we attempted to predict both *(i)* the flow of humans between individual segments based on their δ-densities as well as *(ii)* the segment densities directly. To this end we conducted experiments using various classification methods. Our attempts use the collected and previously presented dataset for training. The created models were used to predict densities for various time intervals. Performance evaluation is performed in relation to the other approaches as well as in relation to the actual measured densities during the event day.

A First Exploratory Approach: To create a baseline comparison using a simple prediction method and in order to investigate the forecasting possibilities using the given dataset, we first conducted experiments using an exploratory forecasting approach. For this purpose we assume that flows between segments are only influenced by the densities of the source and destination segment. Our goal is to compute a prediction function that takes these two densities as input and returns the most likely flow between these segments. To this end we parse the computed values of all arcs of all the generated flow networks and combine them with the associated densities of the source and destination segments for each arc. More precisely, each density pair $(d_{source}, d_{destination})$ taken from $d_{x,x+2}$ is mapped to the respective flow taken from $f_{x,x+2}$. If such a mapping already exists, we add the new reading to the list of previously observed flow values. Further we compute the mean over the entire list of flow values and use it as the predicted flow for the respective density pair in the future. Subsequently we cluster the generated function in order to generate mappings for values that were not part of the training data. We do so by computing the nearest density pair for which we have observed a flow value – using the root mean squared distance metric – and attributing the observed flow to the new density pair. We can subsequently use this static function to forecast the development of an arbitrary density distribution over multiple time steps. The results of such an analysis are given in Section 3.3.

Time Series Forecasting: A deeper analysis on the dataset has been performed by means of time series forecasting. We divided our dataset in multiple smaller datasets, each one considering the data of a single segment and its context. For each single segment, in fact, we can include different contents: its density, the densities of the surrounding segments, the estimated flows as computed in Section 3.1. Moreover, the density of the segments can be measured in two different ways: using the actual absolute density value or using the δ-density value as described in Section 3.1; Consequently, we obtained six different data sets. If we use actual density we have **SA**: single segment; **NA**: segment and surroundings; **FA** segment, surroundings and flows. With δ-density, we similarly have **SD, ND** and **FD**.

This data partitioning is meant to answer the following questions: *(i)* can a forecaster built upon the information of a single segment provide reliable forecasts, or is the information about its surroundings necessary? *(ii)* How useful is the flow estimation performed in Section 3.1? Can it really help in building efficient forecasters? *(iii)* Which measure is better to use when training a forecaster, the actual density value or the δ-density?

In order to perform time series forecasting, we relied on the tools provided by the well-known WEKA software [9]. We performed an evaluation of the forecasters using the training data, namely, once the forecaster has been trained, it is then applied to make a forecast at each point in time by stepping through the data. These predictions are collected by the evaluator provided by WEKA, and are compared to the actual data in order to estimate the error that would have been obtained if such a forecaster had been used. We identified three different learning algorithms suitable for our purpose and already available in the tool: Linear regression, Gaussian processes [19] and Support Vector Machines [6].

Considering the datasets introduced above, we obtain 18 combinations of algorithms and data sets. We will use a single letter to identify the algorithm in such combinations: **-l** for linear regression, **-g** for Gaussian processes and **-s** for support vector machines (e.g. **SA-s** will be the analysis conducted on the single node, with absolute density, using support vector machines).

3.3 Results

In order to investigate the precision and usefulness of the approaches described in Section 3.2 we computed and evaluated their respective prediction errors. For the purpose of this contribution, in which we compare the predictive capabilities of multiple approaches using only a single time series, we use the RMSD (root-mean-squared deviation) metric for comparisons. This gives the inherent benefit that the results are on the same scale as the original time series.

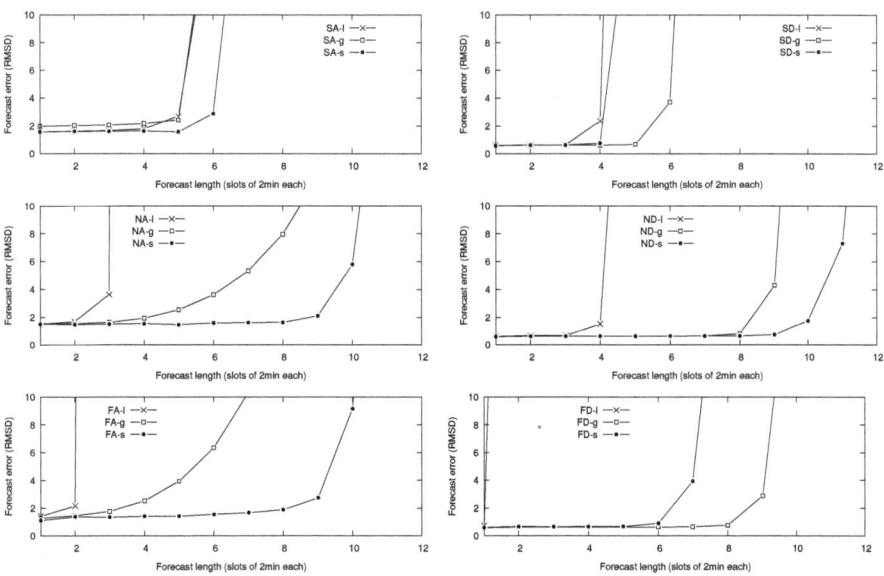

Fig. 2. Performance of the three algorithms with all the available datasets

Initially, we compared all the time series forecasting algorithms. Figure 2 provides a good insight, by showing in six separated charts the error produced by all the eighteen combinations. It is immediately clear that the support vector machines is the most promising approach. In fact, it obtains the best results with all the datasets, except with **SD** and **FD**, where the Gaussian processes algorithm performs better. The last interesting fact is that all those algorithms tend to perform acceptably well for a few time steps, then their performance suddenly diverges making them completely unreliable.

The information about the segment itself is not descriptive enough, and when used alone to train the forecaster the predictions are significant only for the immediate future. The accuracy improves consistently when the the values of the neighborhood are also considered in the prediction. Adding the information about the flows, instead, produced some unexpected results: it does not improve the behavior in case of analysis using the absolute density, while it worsens the forecaster performance when the δ-density is used. This suggests that estimating the people flow by means of solving a minimum cost flow problem could be an oversimplification, and together with the fact that such data increases the number of inputs for the regression function, may lead to errors.

The comparison between absolute density and δ-density shows different behavior depending on the data set: with the single node, the forecaster trained with the absolute values seems to perform slightly worse, but this measure is more stable when the forecast window is wider; with information about estimated flows and neighborhood, the result is curiously exactly the opposite. Without the flow information, the δ-density performs better. Our hypothesis is that δ-density implicitly carries information about the flows when neighboring nodes are considered, and consequently it does not benefit from the additional information carried by the our flow estimates.

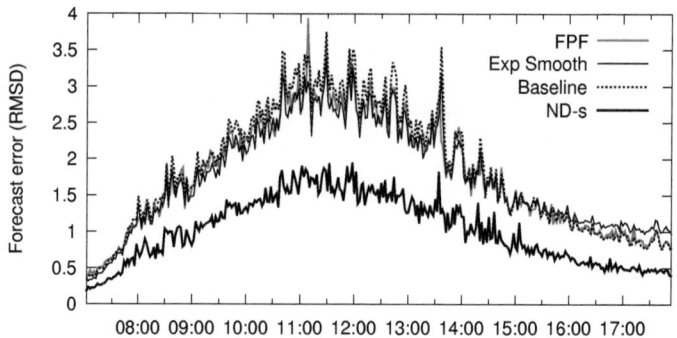

Fig. 3. Accuracy of various approaches for a forecasting depth of a single step

Figure 3 compares the prediction accuracy for a forecasting depth of a single step. The analysis was performed for the entire collected time series, forecasting the next distribution (i.e., the crowd density in 2 minutes) and comparing the result to the actual crowd density at that time. The figure compares our approaches to a simple baseline algorithm that always predicts the previous density distribution and to the performance of Brown's simple exponential smoothing algorithm [3]. Flow prediction function (FPF), exponential smoothing and baseline perform very similarly, while the ND-s forecaster is able to obtain better results, in particular at midday when the people movement is more intense.

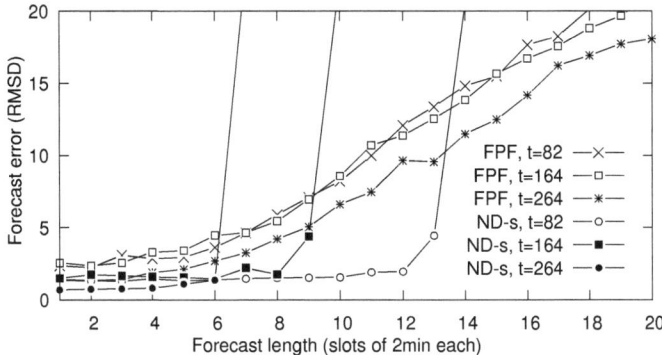

Fig. 4. Comparison of flow prediction function and support vector machines forecaster for various forecast lengths at different time of the day

We run the flow prediction function and the forecaster based on support vector machines at the three different times of the day presented in Figure 1. We chose to predict up to twenty time steps, and results in Figure 4 make clear that the support vector machines algorithm works better for predictions in a short time window, then it completely diverges. On the other hand, the flow prediction function is very stable, but is affected by a bigger error when performing short-term forecast.

Due to the fact that the approaches described in Section 3.2 were trained and evaluated on the same dataset, their results are likely affected by training bias. This is a double edged sword: on the one hand these approaches provides good results for future instances of the given urban sports event and similar occurrences, on the other hand if applied to significantly different case studies they may produce less reliable results.

Algorithms Performance: on our test bed[1] the training time for the **ND-s** was of about two minutes and thirty seconds, thus allowing for a near real time usage. The training phase for datasets in which the flow information was considered was much longer. The flow prediction function is faster and perfectly suitable for real time prediction: the training phase only takes around ten seconds.

4 Future Work and Conclusion

In this paper we have presented data-based approaches to predict crowd-densities of humans during a large-scale urban event. Our results indicate that sophisticated approaches – which were trained specifically for the given event data – are able to predict crowd densities for at least 12 minutes with an overall RMSD smaller than 2 (whereby peak densities for single segments were measured at 397 devices). However, depending on the initial forecasting time, spans of 24 minutes were predicted with similar precision. This gives organizers of the given urban sports event a considerable advantage and reaction time to counter mass disasters in future years. A static flow function-based

---

[1] Intel Core i7 2670QM @ 2.2GHz, 8GB RAM, Sabayon Linux 13.04, Oracle JDK 7u25.

approach predicts crowd densities for up to 20 minutes with an RMSD lower than 7.8 when compared to evidenced densities for the given event. Due to the very general nature of this approach, we expect it to perform comparably for unrelated mass events in different environments. Both approaches are sufficiently performant to be used with live data and deliver results in near real-time. Based on these results we believe that both hypotheses formulated in Section 1 have been satisfied. We plan to further improve the precision of our approaches in future work and to enhance the applicability of the described methods to the prediction of crowd-densities in arbitrary contexts by included data from other real-world mass events.

Acknowledgements. Work supported by the SAPERE project (EU FP7-FET, Contract No. 256873).

References

1. Antonini, G., Bierlaire, M., Weber, M.: Discrete choice models of pedestrian walking behavior. Transportation Research Part B: Methodological 40(8), 667–687 (2006)
2. Berkelaar, M., Eikland, K., Notebaert, P., et al.: lpsolve: Open source (mixed-integer) linear programming system. Eindhoven U. of Technology (2004)
3. Brown, R.: Smoothing, Forecasting and Prediction of Discrete Time Series, Dover Phoenix Editions. Dover Publications (2004)
4. Burke, J., Estrin, D., Hansen, M., Parker, A., Ramanathan, N., Reddy, S., Srivastava, M.B.: Participatory sensing. In: Workshop on World-Sensor-Web (WSW 2006): Mobile Device Centric Sensor Networks and Applications, pp. 117–134 (2006)
5. Burstedde, C., Klauck, K., Schadschneider, A., Zittartz, J.: Simulation of pedestrian dynamics using a two-dimensional cellular automaton. Physica A: Statistical Mechanics and its Applications 295(3-4), 507–525 (2001)
6. Cortes, C., Vapnik, V.: Support-vector networks. Machine Learning 20(3), 273–297 (1995)
7. Eagle, N., Pentland, A.: Social serendipity: mobilizing social software. IEEE Pervasive Computing 4(2), 28–34 (2005)
8. Fruin, J.: The causes and prevention of crowd disasters. In: Smith, R.A., Dickie, J.F. (eds.) Engineering for Crowd Safety, pp. 99–108 (1993)
9. Hall, M., Frank, E., Holmes, G., Pfahringer, B., Reutemann, P., Witten, I.H.: The weka data mining software: an update. SIGKDD Explor. Newsl. 11(1), 10–18 (2009)
10. Helbing, D., Buzna, L., Johansson, A., Werner, T.: Self-organized pedestrian crowd dynamics: Experiments, simulations, and design solutions. Transportation Science 39(1), 1–24 (2005)
11. Helbing, D., Johansson, A.: Pedestrian, Crowd, and Evacuation Dynamics, pp. 6476–6495. Springer, New York (2009)
12. Helbing, D., Mukerji, P.: Crowd disasters as systemic failures: analysis of the love parade disaster. EPJ Data Science 1(1), 7 (2012)
13. Herrmann, S.: Ivory coast mourns after new year stampede in abidjan (January 2013), http://www.bbc.co.uk/news/world-africa-20885814
14. Hsu, W.-L., Lin, K.-F., Tsai, C.-L.: Crowd density estimation based on frequency analysis. In: 2011 Seventh International Conference on Intelligent Information Hiding and Multimedia Signal Processing (IIH-MSP), pp. 348–351 (2011)
15. Jetcheva, J., Hu, Y.-C., PalChaudhuri, S., Saha, A., Johnson, D.: Design and evaluation of a metropolitan area multitier wireless ad hoc network architecture. In: Proceedings of the 5th IEEE Workshop on Mobile Computing Systems and Applications, Monterey, CA, USA, pp. 32–43 (October 2003)

16. Klingsch, W., Rogsch, C., Schadschneider, A.: Pedestrian and Evacuation Dynamics 2008. Springer, Heidelberg (2010)
17. Lee, K., Hong, S., Kim, S.J., Rhee, I., Chong, S.: Slaw: A mobility model for human walks. In: Proceedings of the 28th Annual Joint Conference of the IEEE Computer and Communications Societies (INFOCOM), Rio de Janeiro, Brazil. IEEE (April 2009)
18. Ma, R., Li, L., Huang, W., Tian, Q.: On pixel count based crowd density estimation for visual surveillance. In: 2004 IEEE Conference on Cybernetics and Intelligent Systems, vol. 1, pp. 170–173 (2004)
19. Mackay, D.J.: Introduction to gaussian processes (1998)
20. Mäkelä, K., Belt, S., Greenblatt, D., Häkkilä, J.: Mobile interaction with visual and rfid tags: a field study on user perceptions. In: Proceedings of the SIGCHI Conference on Human Factors in Computing Systems, CHI 2007, pp. 991–994. ACM, New York (2007)
21. Monge, G.: Memoire sur la theorie des delais et des remblais. In: Histoire de lAcademie Royale des Sciences de Paris, avec les Memoires de Mathematique et de Physique pour la meme annee, pp. 666–704 (1781)
22. Nakatsuka, M., Iwatani, H., Katto, J.: A study on passive crowd density estimation using wireless sensors. In: 4th International Conference on Mobile Computing and Ubiquitous Networking (ICMU), Tokyo, Japan (2008)
23. Natarajan, A., Motani, M., Srinivasan, V.: Understanding urban interactions from bluetooth phone contact traces. In: Uhlig, S., Papagiannaki, K., Bonaventure, O. (eds.) PAM 2007. LNCS, vol. 4427, pp. 115–124. Springer, Heidelberg (2007)
24. Nicholson, C., Roebuck, B.: The investigation of the hillsborough disaster by the health and safety executive. Safety Science 18(4), 249–259 (1995)
25. Pietilänen, A.-K., Diot, C.: Dissemination in opportunistic social networks: the role of temporal communities. In: Proceedings of the Thirteenth ACM International Symposium on Mobile Ad Hoc Networking and Computing, MobiHoc 2012, pp. 165–174. ACM, New York (2012)
26. Su, J., Chan, K.K.W., Miklas, A.G., Po, K., Akhavan, A., Saroiu, S., de Lara, E., Goel, A.: A preliminary investigation of worm infections in a bluetooth environment. In: Proceedings of the 4th ACM Workshop on Recurring Malcode, WORM 2006, pp. 9–16. ACM, New York (2006)
27. Weppner, J., Lukowicz, P.: Bluetooth based collaborative crowd density estimation with mobile phones. In: IEEE International Conference on Pervasive Computing and Communications (PerCom), vol. 18, p. 22 (2013)
28. Wirz, M., Franke, T., Roggen, D., Mitleton-Kelly, E., Lukowicz, P., Troster, G.: Inferring crowd conditions from pedestrians' location traces for real-time crowd monitoring during city-scale mass gatherings. In: 2012 IEEE 21st International Workshop on Enabling Technologies: Infrastructure for Collaborative Enterprises, pp. 367–372 (2012)
29. Wirz, M., Franke, T., Roggen, D., Mitleton-Kelly, E., Lukowicz, P., Trster, G.: Probing crowd density through smartphones in city-scale mass gatherings. In: EPJ Data Science, pp. 1–24 (2013)
30. Zheng, Y., Li, Q., Chen, Y., Xie, X., Ma, W.-Y.: Understanding mobility based on gps data. In: Proceedings of ACM Conference on Ubiquitous Computing (UbiComp 2008), pp. 312–321. ACM Press, Seoul (2008)
31. Zia, K., Riener, A., Ferscha, A., Sharpanskykh, A.: Evacuation simulation based on cognitive decision making model in a socio-technical system. In: IEEE/ACM International Symposium on Distributed Simulation and Real Time Applications, pp. 98–107 (2011)

Modeling Social Capital in Bureaucratic Hierarchy for Analyzing Promotion Decisions

Jyi-Shane Liu, Zhuan-Yao Lin, and Ke-Chih Ning

Department of Computer Science
National Chengchi University
64 Sec. 2 Zhi-Nan Rd., Taipei, Taiwan, R.O.C.
jsliu@cs.nccu.edu.tw

Abstract. We report research results in applying social network analysis to develop a data-driven computational approach for social scientists to perform investigative exploration on analyzing bureaucratic promotion. We consider social capital as primary determinants of promotion decisions in bureaucratic hierarchy and propose a hybrid multiplex social network model for representing relational and structural information among entities. The approach develops quantified assessment of social capital and provides objective evaluation of promotion decisions in anterior prediction. Experimental results with actual government officials' career data provide evidence to the effectiveness and the utility of social capital evaluation for bureaucratic promotion decisions.

Keywords: social network analysis, social capital assessment, bureaucratic promotion decision.

1 Introduction

Social informatics is an umbrella term that spans disciplines and research domains sharing common interest in the effects of using information and communication technologies in any sort of social settings. One of the primary themes of social informatics is the design, analyze, and develop information technologies for social applications. Another prominent issue concerns the intriguing phenomena that emerge from the extensive use of information and communication technologies in social behaviors. Common traits of social informatics research include problem-oriented nature, empirical and social theory-based focus, a means of inter-disciplinary linking, and ties to informatics [1].

Among many research topics in social informatics, social network analysis attracts researchers from both disciplines of social science and computer science. With its central focus on modeling relations among entities and investigating component interactions in a broad sense of social contexts, social network analysis provides a middle ground for embedding empirical pheomena with computational perspectives. The interplay between both disciplines stimulates innovative research angles and creates interesting research issues in both modeling and methodology. This has led to a wide range of promising inter-disciplinary research results that are potentially applicable and influential in real-world contexts [2].

Over the years, social network analysis has made great stride in discovering obscure entity relations and understanding complex entity interactions in the context of homogeneous networks. Recent research interest, however, has gradually moved toward modeling and analyzing heterogeneous networks that are closer to real-world settings. Instead of composing of one class of nodes and one type of relations, many real-world problems are more faithfully modeled with multiple classes of nodes with multiple types of relations among nodes. The increased complexity of network modeling and analysis poses considerable challenges but has also seen fruitful research efforts. One type of complex heterogeneous networks involves multiple classes of nodes. As a representative heterogeneous network model, a bipartite network consists of two classes of nodes and relations between nodes of each class [3]. Another type of complexity is derived from the exploration of multiple types of relations between any pair of nodes in one class. Research advances in such heterogeneous networks include modeling and analytic techniques in multi-dimensional networks [4-5], multi-relational networks [6], multi-layered networks [7].

Among the many topics in social science, we are particularly interested in the study of political executives based on cross-disciplinary investigative curiosity and potentially innovative research impact. Past studies of political executives mostly followed social science research methods with hypothesized theories, field survey, investigative sample data, statistical analysis, and qualitative interpretation [8]. For example, a number of empirical studies of political leaders around the world and through time have been conducted by social scientists [9-10]. This form of research methodology mostly stems from constrained access to and limited availability of problem domain data. Recently, some social scientists have reflected on the limitations of traditional methodological repertoires [11].

In this paper, we report research results in applying social network analysis to develop a data-driven computational approach for social scientists to perform investigative exploration on analyzing bureaucratic promotion. In our problem domain, social network analysis provides a flexible analytical platform to organize large quantity of fragmented data such that relations among entities can be processed and exploited. In an attempt to capture the primary determinants in bureaucratic promotion, we consider three aspects of social capital in bureaucratic hierarchy and propose a hybrid multiplex social network model for representing relational and structural information among entities. The approach develops quantified assessment of social capital and provides objective evaluation of promotion decisions in anterior prediction. Experimental results with actual government officials' career data provide evidence to the effectiveness of social capital evaluation for bureaucratic promotion decisions.

2 A Hybrid Mutliplex Network Model of Social Capital in Bureaucratic Career System

Many contemporary goverments in the world are typically characterized with a bureaucratic hierarchy. Government officials are mostly professional bureaucrats with stable and long career in various government branches. The organizational structure of governments supports a hierarchy based career system, where bureaucrats compete for limited promotional opportunities, and climbing up the ladder of a clear and linear career structure [12]. In human resource management, determinants of career success have been linked to social capital, particularly at upper level executive positions [13]. The notion of social capital is concisely explained as the resource available to actors as a function of their location in the structure of their social relations [14]. An actor's position and connections in a social network structure provide access to resources which benefit career success. In other words, individuals with more social capital may receive more favorable considerations when superiors are making promotion decisions for filling vacant post.

Our research interest is focused on incorporating computational approaches to facilitate the analytic process of examining the relations between social capital and career success, especially in the more rigid career systems of government bureaucratic hierarchy. We propose a hybrid multiplex network model to capture social capital of professional bureaucrats in government bureaucratic career systems. In other words, the abstract notion of social capital is approximated and materialized in multiple networks, each representing an aspect of social structures in government bureaucratic hierarchy. The hybrid multiplex network model of social capital in government bureaucratic career systems provides a common reference framework for assessing relative advantages of professional bureaucrats in receiving favorable promotion decisions. Using actual government bureaucratic career data, this bureaucratic social capital reference framework further allows the examination of social capital effects on government executive promotion practices.

As an inital attempt, we approximate bureaucratic social capital with three types of social structures in the government bureaucratic career systems. The first social structure contains the superior-subordinate relationship. Direct work relations tend to accumulate trust, reciprocity, information, and cooperation, which may contribute to career sponsorship. The second social structure concerns with bureaucratic seniority moving up the hierarchy ladder. Senior officials who are closer to the vacant post in the succession ladder may present stronger qualification. The third social structure relates officials to their career paths in the bureaucratic hierarchy. Similarity of career paths to successsful predecessors of the vacant post may indicate greater professional merit. Each social structure is represented in a different social network model and the bureaucratic social capital is derived by a hybrid integration of the three network models.

2.1 A Network Model of Superior-Subordinate Relationship

Direct work relations between superiors and subordinates are modeled in a one-mode network where each node is an official and each directed edge is a command chain from a superior to a subordinate. The directed edges are also weighted for the time length of the work relations. For simplicity, the number of days of a direct command relation is used as edge weight. An official is linked to another official based on both past and current superior-subordinate relations. As a result, entity nodes are linked with multiple paths. A simplified superior-subordinate network that exemplifies the collected social structure of superior-subordinate relationship is shown in Fig. 1. A is the superior of B with a work relation of 200 days. B, in turn, is the superior of C with a work relation of 100 days, of D with a work relation of 500 days, and of E with a work relation of 300 days. A also had past supurior-subordinate work relations with C of 500 days and with J of 400 days. Being a subordinate of D with 600-day work relation, G was also a direct subordinate of B for 200 days in the past. Similarly, J is an ongoing subordinate of E and was also a subordinate of A in the past. The current superior of H is D, while the past superior was E.

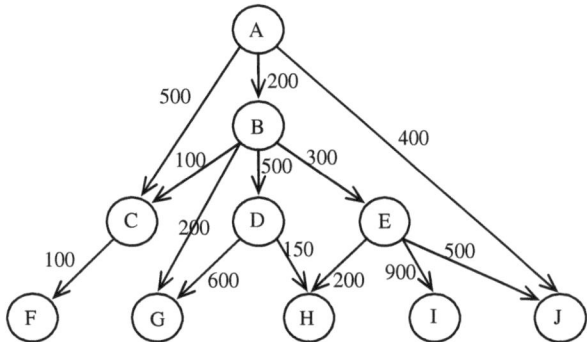

Fig. 1. A simplified superior-subordinate network

The superior-subordinate network provides a social structure to assess career sponsorship based on strength of work relations and relative positions in the bureaucratic hierarchy. Suppose D is leaving his/her post and a total of five officials, including F, G, H, I, and J, are potential candidates for the vacant post. In most cases, promotion decisions are made by the immediate superior of the vacant post, e.g., B. We estimate career sponsorship from a superior to a candidate by accumulating all path weight between them. In case of multiple paths, the weight on a shorter path is further multipled by the path length of the longest path between the two nodes. The rational is to account for the advantageous recognition from the work relation with a higher superior, who may exercise his/her strong influence in making promotion decision. Career sponsorship, CS_{ij}, from a superior, s_i, to a candidate, c_j, is given in equation (1), where p is the set of paths between s_i and c_j, p_k is a path in p, w_e is the weight of an edge, e, in p_k, and l_p is the length of the longest path in p.

$$CS_{ij}(s_i, c_j) = \Sigma_p \Sigma_{pk} \, w_e \times l_p \qquad (1)$$

2.2 A Network Model of Bureaucratic Seniority

Bureaucratic seniority in the hierarchy ladder is modeled in a two-mode network with two sets of actors of different types, officials and bureaucratic posts. The set of officials includes candidates eligible for filling the vacant post. The set of bureaucratic posts includes the vacant post and other posts that have been held by the candidates. Relations are defined between officials and bureaucratic posts and within the set of bureaucratic posts. Note that this network modeling is characterized as a more extensive two-mode network, but not a bipartite network [15]. Relations within the set of bureaucratic posts is defined by the bureaucratic hierarchy of government offices and follow a tree-like structure. In essence, each bureaucratic post is in a linear command chain. While a senior post may command several junior posts, each junior post reports to only one senior post immediately above. Fig. 2 shows an exemplar bureaucratic seniority network that includes a set of officials and a set of bureaucratic posts in a partial bureaucratic hierarchy with four seniority levels. Nodes on each level are given a hierarchy weight, more weight for higher levels.

The bureaucratic career of an official is encoded by multiple relations between the official and a number of bureaucratic posts he/she has been appointed to. Each edge across the two sets of actors connects an official to one of the bureaucratic posts he/she has held. The tenure of appointment is assigned as edge weight in number of days. Exemplar bureaucratic careers are shown in Fig. 2, where an official is linked to multiple bureaucratic posts. A bureaucratic post is also linked by multiple officials who represent a succession of appointees. Sequential information in both bureaucratic career and succession of post appointees is not encoded and is not used for our research purpose.

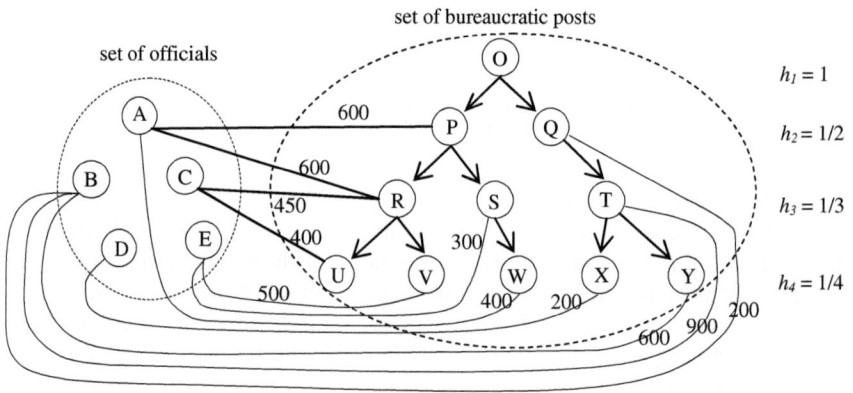

Fig. 2. An exemplar bureaucratic seniority network

A bureaucratic career is usually composed of lateral and upwardmovement in government hierarchy. Bureaucratic promotion mostly follows rigid rank advancement system where current positions and length of services are important determinants. Therefore, we consider bureaucratic seniority as the second aspect of social capital in bureaucratic career systems. Each term of service at an appointed post earns a seniority credit, which is estimated as the multiplication of the length of service and the hierarchy weight of the appointed post. Bureaucratic seniority, BS_i, of an official, o_i, is simply assessed by accumulating all seniority credits in his/her bureaucratic career, and is given in equation (2), where c is the set of appointed posts, t_{ie} is length of service of o_i at an appointed post e, w_h is the weight of an appointed post in the hierarchy.

$$BS_i = \Sigma_c \, t_{ie} \times w_h \qquad (2)$$

2.3 A Network Model of Career Distinction

As in many organizations, bureaucratic posts in governments are typically structured as a pyramid hierarchy. The number of higher level posts are proportionally less than that of lower level posts. Each vacant post at a higher level is highly sought after as a significant step in moving up the hierarchy ladder. While the vertial position of a bureaucratic post in the hierarchy defines its authority level, horizontally equivalent posts do not necessarily carry the same weight in responsibility, mandate, and recognition. In other words, some of the bureaucratic posts are more prestigious than others at the same hierarchy level. Successful tenures at these prestigious posts are expected to receive more favorable consideration for promotion to a higher level vacant post.

We conisder career distinction as the third aspect of social capital in bureaucratic career systems. For a high level post, career paths of a list of preceding holders offer a good reference for distinguishing career quality among potential candidates at lower level posts. We design a two-mode network to model the notion of career distinction with two sets of actors of different types, officials and bureaucratic posts. The set of officials includes a subset of eligible candidates and a subset of preceding holders of the vacant post. The set of bureaucratic posts include those that are in the career paths of the set of officials. Relations are defined between the two sets for occurrences of services, e.g., who held which post. Relations are also defined within the set of bureaucratic posts for their positions in the hierarchy. Fig. 3 shows an exemplar career distinction network, where O is the vacant post with three preceding holders (predecessors), A, B, and C. Eligible candidates include D, E, and F, whose current posts are Q, P, and R, respectively.

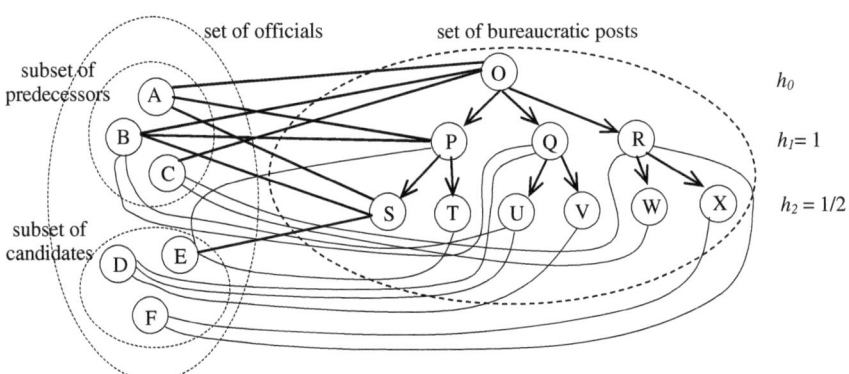

Fig. 3. An exemplar career distinction network

The career distinction of a candidate is estimated by the weighted number of nodes that are common between the career paths of the candidate and the set of predecessors. Each common node is weighted by its hierarchy level, with higher level nodes receiving more weight. Denote the career distinction of a candidate, c_i, by CD_i, the set of nodes in the career path of c_i by cp_i, the set of nodes in the career paths of the set of predecessors by cp_p, and the hierarchy weight of each level by w_j. Career distinction is calculated by Equation (3), where $|(cp_i \cap cp_p) \times w_j|$ is the weighted cardinality of the intersection at hierarchy level j.

$$CD_i = \Sigma_j \mid (cp_i \cap cp_p) \times w_j \mid \qquad (3)$$

2.4 Social Capital Evaluation for Bureaucratic Promotion Decisions

We integrate the career sponsorship network, the bureaucratic seniority network, and the career distinction network into a hybrid mutliplex network for social capital assessment in promotion decisions. An initial approach is a simple weighted accumulation of the three components. A weight is given to a component that represents its relative value in contributing to the social capital for receiving favorable promotion decisions. Equation (4) gives the social capital value, SC_i, of an official, o_i, which is the weighted sum of the normalized values of his/her career sponsorship, CS_i, bureaucratic seniority, BS_i, and career distinction, CD_i.

$$SC_i = w_{cs} \times \text{normalized}(CS_i) + w_{bs} \times \text{normalized}(BS_i) + w_{cd} \times \text{normalized}(CD_i) \qquad (4)$$

The quantified assessment of individual social capital in the bureaucratic systems provides an objective evaluation of promotion decision for a vacant post. A list of potential candidates can be ranked by social capital values or one of the three components. This candidate ranking can be used both in anterior prediction and posterior analysis. Anterior prediction provides leading condidates on expected post vacancy for outside observation. Posterior analysis reveals the significant factors of successful appointee's social capital or a lack of thereof.

Table 1. Assessment of social capital in promotion decision

Candidate	Career Sponsorship	Bureaucratic Seniority	Career Distinction	Social Capital	Rank
A	1.0	1.0	0.89	2.89	1
B	0.94	0.92	0.82	2.68	2
C	0.83	0.81	1.0	2.64	3
D	0.91	0.31	0.94	2.16	4
E	0.78	0.63	0.56	1.97	5
~	~	~	~	~	~
X	0.12	0.17	0.23	0.52	24
Y	0.17	0.15	0.16	0.48	25

Table 1 shows an example of posterior analysis on social capital assessment for a execuitve post at the Ministry of Economic Affairs from actual bureacratic career data of Taiwan government officials. A list of 25 potential candidates were selected for social capital assessment based on a loose eligibility criterion. The official who was promoted to the vacant exective post was C. The posterior analysis indicates that C's successful promotion may be attributed to his/her outstanding career distinction. On the other hand, suppose X was the actual appointee. The unexpected appointment would be noted and characterized by factors other than the appointee's bureaucratic social capital. This illustrates the utility of social capital network modeling and analysis for interdisciplinary practical applications.

3 Experimental Evaluation

Social capital mostly affects high-level executive promotion. Our research proposes a hyrbid multplex network modeling on social capital for bureaucratic executive promotion. In order to evaluate and verify the approach, we perform experiments on social captial effects in promotion decsions for executive posts in government agencies. With access to the actual Taiwanese government post database [16], we extract a set of bureaucratic career data that cover 1,226 officials of 338 government executive posts in a time span from 1988 to 2011. The three government agencies are Ministry of Economic Affairs, Ministry of Finance, and Council of Atomic Energy. Each executive post succession and each bureaucratic career are manually inspected and verified for accuracy and completeness. The experimental data are prepared in sufficient scale for covering representativeness and variability of actual promotion decisions.

The network modeling and analysis of social capital of a group of eligible officials for an event of vacant post result in a ranked list of candidates ordered by social capital values. This result is evaluated by comparing the ranked list of candidates with the actual appointee for the vacant post. The higher order of the actual appointee in the ranked list of candidates the better the result. This indicates that our approach of modeling and analyzing of bureaucratic social capital is feasible and potentially useful for predicting and dissecting actual promotion decisions. To quantify the evaluation results, we define a measure of predictability as a performance index. Given a ranked candidate list, C_{ij}, for a vacant post i at a promotion event j, and the order of the actual appointee in the ranked list, o_{ij}, the predictability of the ranked list, p_{ij}, is calculated by equation (5), where $|C_{ij}|$ is the cardinality of C_{ij}.

$$p_{ij} = \frac{|C_{ij}| + 1 - o_{ij}}{|C_{ij}|} \quad (5)$$

The predictability of the ranked candidate lists reflects how much actuall promotion decisions are determined by our modeling of candidates' bureaucratic social capital. In an ideal situation, a ranked list would include the actual appointee as first in the order and receive a predictability of 1.0. On the other hand, if the actual appointee is ranked last in the list of, say, 20 candidates, the predictability would become 0.05. Assuming bureaucratic social capital is the norm of promotion determinants in bureaucratic career systems, this evaluation also provides a measure of how much an actual appointment adhered to the norm of bureaucratic promotion or how much it was out-of-the-box. Thus, our approach also facilitates posterior analysis on bureaucratic appointment for political observation.

The experiments are designed to evaluate the overall performance of bureaucratic social capital modeling in terms of predictability to actual promotion results; and to determine the effects of component weights in social capital integration. In our model of assessing a candidate's social capital as in equation (4), component weights, e.g., w_{cs}, w_{bs}, w_{cd}, can be adjusted to represent the relative importance of each aspect of social capital. A straightforword approach is to set equal weight to the three components, e.g., $w_{cs} = w_{bs} = w_{cd} = 1$. Another experimental approach is to set one of the components with a weight of one, while the other two receive zero weight, so as to isolate the predictability performance of an individual component of social capital. We can also vary the ratio of component weights to examine the effects of individual social capital component in bureaucratic appointment predictability.

Table 2. Appointment predictability by components of social capital

Component Contribution to Social Capital	Atomic Energy	Finance	Economic Affairs
All equal weights ($w_{cs} = w_{bs} = w_{cd} = 1$)	0.71	0.74	0.78
Only Superior Sponsorship ($w_{cs} = 1$, $w_{bs} = w_{cd} = 0$)	0.69	0.71	0.75
Only Bureaucratic Seniority ($w_{bs} = 1$, $w_{cs} = w_{cd} = 0$)	0.58	0.69	0.74
Only Career Distinction ($w_{cd} = 1$, $w_{cs} = w_{bs} = 0$)	**0.77**	**0.79**	**0.81**

Table 2 summarizes the overall performance of appointment predictability based on inclusive or exclusive component contribution to social capital for post vacancy categorized in government agency. The reported predictability values are averaged over 118, 242, and 188 times of apointments in Ministries of Atomic Energy, Finance, and Economic Affairs, respectively. The preliminary results suggest that career distinction, with the best performance in all four combinations of component contribution, is the most dominant determinant of bureaucratic promotion in all three ministries. The least dependable factor is bureaucratic seniority, which leads to the lowest average predictability in each ministry. Overall, the results show that bureaucratic social capital modeling provides a respectable predictability for actual promotion decisions.

Table 3. Appointment predictability by ingredient strength of social capital

Ingredient Strength of Social Capital	Atomic Energy	Finance	Economic Affairs
All equal weights ($w_{cs} = w_{bs} = w_{cd} = 1$)	0.71	0.74	0.78
MoreCareer Sponsorship I ($w_{cs} = 2$, $w_{bs} = w_{cd} = 1$)	0.69	0.73	0.77
More CareerSponsorship II ($w_{cs} = 3$, $w_{bs} = w_{cd} = 1$)	0.67	0.72	0.76
More CareerSponsorship III ($w_{cs} = 4$, $w_{bs} = w_{cd} = 1$)	0.66	0.71	0.74
More CareerSponsorship IV ($w_{cs} = 5$, $w_{bs} = w_{cd} = 1$)	0.65	0.70	0.73
More Bureaucratic Seniority I ($w_{bs} = 2$, $w_{cs} = w_{cd} = 1$)	0.70	0.73	0.77
More Bureaucratic Seniority II ($w_{bs} = 3$, $w_{cs} = w_{cd} = 1$)	0.68	0.72	0.76
More Bureaucratic Seniority III ($w_{bs} = 4$, $w_{cs} = w_{cd} = 1$)	0.63	0.70	0.75
More Bureaucratic Seniority IV ($w_{bs} = 5$, $w_{cs} = w_{cd} = 1$)	0.62	0.69	0.74
More Career Distinction I ($w_{cd} = 2$, $w_{cs} = w_{bs} = 1$)	0.73	0.75	0.79
More Career Distinction II ($w_{cd} = 3$, $w_{cs} = w_{bs} = 1$)	0.74	0.76	0.80
More Career Distinction III ($w_{cd} = 4$, $w_{cs} = w_{bs} = 1$)	0.75	0.77	0.80
More Career Distinction IV ($w_{cd} = 5$, $w_{cs} = w_{bs} = 1$)	**0.76**	**0.77**	**0.81**

Table 3 reports the results of appoinment predictability with various ingredient strength of social capital. For the components of superior sponsorship and bureaucratic seniority, increased strength suffers performance degradation. On the other hand, increased strength of career distinction leads to performance improvement, albeit in small proportion. This results seem to be consistent with the observation in Table 2 that career distinction is the most dependable social capital for receiving favorable promotion considerations.

4   Conclusions

Social informatics research is closely grounded in real-world social contexts and has gained valuable knowledge in the transformation and creation of new social systems. As researchers in the discipline of computer science, one of our primary research goals has been to apply the computational techniques of social network analysis to address research issues in social science and to work toward interdisciplinary innovations. In this paper, we propose to integrate social network analysis paradigm with the notion of social capital in the study of bureaucratic promotion. Initial experimental evaluation supports the effectiveness of our hybrid multiplex network modeling approach. Our research also extends the applications of social network analysis for evidence-based investigative utility in political problem domain. As a complementary approach to most research that emphasize the design of new algorithmic methods for automatic discovery, our work aims at developing an exploratory tool that integrates problem domain data with social network analytic functions. Such an exploratory tool provides users, e.g., political researchers or observers, with flexible evidence-based analysis and empirical observation by applying domain knowledge for parameter setting and result interpretation. Therefore, this work also has the practical value of providing objective scrutiny on political power transition for the benefit of public interest.

Acknowledgement. This research was partially supported by research grants NSC102-2221-E-004-012, NSC101-2221-E-004-012 from Taiwan's National Science Council and by Top University Project of National Chengchi University.

References

1. Sawyer, S., Eschenfelder, K.R.: Social Informatics: Perspectives, Examples, and Trends. Annual Review of Information Science and Technology 36(1), 427–465 (2002)
2. Scott, J.: Social Network Analysis: Development, Advances, and Prospects. Social Network Analysis and Mining 1(1), 21–26 (2011)
3. Guillaume, J.L., Latapy, M.: Bipartite Graphs as Models of Complex Networks. Physica A: Statistical and Theoretical Physics 371(2), 795–813 (2006)
4. Berlingerio, M., Coscia, M., Giannotti, F., Monreale, A., Pedreschi, D.: Foundations of Multidimensional Network Analysis. In: 2011 International Conference on Advances in Social Network Analysis and Mining, pp. 485–489. IEEE Computer Society (2011)

5. Berlingerio, M., Coscia, M.: Giannotti: Finding and Characterizing Communities in Multi-dimensional Networks. In: 2011 International Conference on Advances in Social Network Analysis and Mining, pp. 490–494. IEEE Computer Society (2011)
6. Davis, D., Lichtenwalter, R., Chawla, N.V.: Multi-Relational Link Prediction in Heterogeneous Information Networks. In: 2011 International Conference on Advances in Social Network Analysis and Mining, pp. 281–288. IEEE Computer Society (2011)
7. Brodka, P., Stawiak, P., Kazienko, P.: Shortest Path Discovery in the Multi-layered Social Network. In: 2011 International Conference on Advances in Social Network Analysis and Mining, pp. 497–501. IEEE Computer Society (2011)
8. Bryman, A.: Social Research Methods. Oxford University Press, Oxford (2001)
9. Eldersveld, S.J.: Political Elites in Modern Societies: Empirical Research and Democratic Theory. University of Michigan Press, Ann Arbor (1989)
10. Inglehart, R.: Modernization and Postmodernization: Cultural, Economic and Political Change in 43 Societies. Princeton University Press, Princeton (1997)
11. Savage, M., Burrows, R.: The Coming Crisis of Empirical Sociology. Sociology 41(5), 885–899 (2007)
12. Baruch, Y.: Managing Careers: Theory and Practice. Prentice Hall/Pearson (2004)
13. Seibert, S.E., Kraimer, M.L., Liden, R.C.: A Social Capital Theory of Career Success. Academy of Management Journal 44, 219–237 (2001)
14. Adler, P.S., Kwon, S.W.: Social Capital: Prospects for a New Concept. The Academy of Management Review 27(1), 17–40 (2002)
15. Wasserman, S., Faust, K.: Social Network Analysis: Methods and Applications. Cambridge University Press (1994)
16. Liu, J.-S., Ning, K.-C.: Applying Link Prediction to Ranking Candidates for High-Level Government Post. In: 2011 International Conference on Advances in Social Network Analysis and Mining, pp. 145–152. IEEE Computer Society (2011)

Information vs Interaction: An Alternative User Ranking Model for Social Networks

Wei Xie, Ai Phuong Hoang, Feida Zhu, and Ee-Peng Lim

Singappre Management University, Singapore
{wei.xie.2012,aphoang.2012,fdzhu,eplim}@smu.edu.sg

Abstract. The recent years have seen an unprecedented boom of social network services, such as Twitter, which boasts over 200 million users. In such big social platforms, the influential users are ideal targets for viral marketing to potentially reach an audience of maximal size. Most proposed algorithms use the linkage structure of the underlying network to measure the information flow and hence evaluate a users influence. Yet that is not the full story for social networks. In this paper, we propose to examine users' influence from a social interaction perspective. We built a ranking model based on the dynamic user interactions taking place on top of these underlying linkage structures. In particular, in the Twitter setting we supposed a principle of balanced retweet reciprocity, and then formulated it to re-evaluate the value of Twitter users. Our experiments on real Twitter data demonstrated that our proposed model presents different yet equally insightful user ranking results.

Keywords: Twitter, user ranking, retweet behaviour.

1 Introduction

The sheer number of indexed webpages online, which is estimated at 3.97 billion[1], has made ranking algorithms indispensable for virtually any practical applications to access individual webpages. Algorithms such as PageRank [11] and HITS [3] have achieved huge success in finding top-ranked authoritative webpages by analyzing the URL linkage structure. Similarly, the recent boom of social network services has posted a need as strong for good algorithms to rank their users for a variety of applications. For example, top-ranked users by social influence are ideal targets for viral marketing to potentially reach an audience of maximal size. Among the social network services, micro-blogging services like Twitter have been the most favorable in terms of marketing due to the fact that information, in the form of tweets, could spread the fastest through the follow links. A number of algorithms have therefore been proposed for the particular setting of Twitter among which TwitterRank [15] has been one of the most noticeable. What TwitterRank and PageRank, including those similar ones they each represent, shared in common is that they both rely on the linkage structure of the

---

[1] http://www.worldwidewebsize.com/

respective underlying network, i.e., the URL linkage network for PageRank and the follow link network for TwitterRank.

A closer examination of these linkage structures shows that they represent primarily how information would flow and tend to be relatively static. For example, the Twitter follow network gives the diffusion of tweets and is relatively static compared to the other user actions such as tweet and retweet. What they fail to capture is the dynamic user interactions constantly taking place on top of these linkage structures, e.g., how users retweet and reply one another. Yet, it is our belief that the dynamic user interactions are also an important part essential to a social network because they reveal more insights into users' social relationships than the underlying linkage structure. For example, it is common that users only interact with a small number of other users with retweet and reply out of the many who follow them and whom they follow, or both. Even among those they indeed interact with, they interact differently, e.g., retweeting with different frequency. Clearly, these user interactions, which are also much more dynamic, shed more interesting insights into their social relationships, e.g., relationship strength, relative social status, etc..

In this paper, we propose an alternative user ranking model based on a user interaction perspective, which could give rather different ranking results compared with the traditional ones based on an information flow perspective. Let's look at a simple illustrative example. In Figure 1, nodes represent Twitter users, directed edges in (a) denote follow links and the weighted directed edges in (b) denote the number of times a user has retweeted the other one. For example, it can be told from the figure that Dave has retweeted Alice three times while Alice has only retweeted Dave once. Now if we run PageRank algorithm on the underlying follow network, the node of Dave would rank the highest as it is the network hub of the information flow. While this makes perfect sense from the information flow perspective, we argue that, if we examine instead how users interact with one another, we could have a different ranking of the nodes. For example, suppose we assume the ratio between the number of retweets between two users corresponds to their relative social relation status in the sense that a user with relatively higher status would be retweeted more times than the other party of relatively lower status. Given this assumption, the node of Alice could now be the highest ranked one from the user interaction perspective since Alice appears superior to Dave. This example illustrates the difference between the rankings from two different perspectives, namely, the information flow one and the user interaction one.

The main contribution of this paper is to re-examine the value of users in social network from the social interaction perspective. In particular, we consider the social interaction in the notion of reciprocity based on the retweet interaction between Twitter users. Reciprocity is a well-established concept in both social science [4] and economics [13]. In our particular Twitter setting, it refers to the mutual adoption of each other's tweets between two users in the form of retweet, the result of which is a boost to both parties' social impact. We formulated the retweet reciprocity, proposed an alternative user ranking model based on retweet

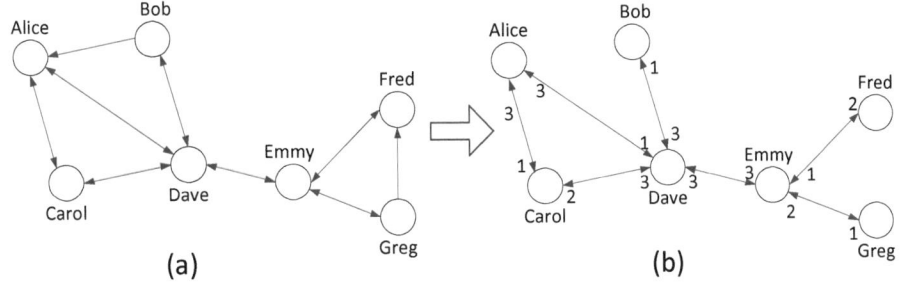

Fig. 1. (a) Twitter follow network. (b) Twitter reciprocal retweet network.

reciprocity and developed an efficient inference solution. Our experiments on real Twitter data demonstrated that our proposed model presents different yet equally insightful user ranking results.

2 Proposed Solution

2.1 Preliminaries

We consider a set of users $U = \{u_i\}_{i=1}^n$, where n is the size of U, from which we want to find influential users. Denote R as all the retweeting behavior performed by U. Here we only consider the number of retweets without considering their content. In other words, R is represented as a set of ordered pairs $\{(u_i, u_j)\}$, where a pair (u_i, u_j) means user u_i retweets a tweet from user u_j. Based on R, we construct a retweet network, which is a weighted directed graph $G = (U, E, W)$, such that the set of users U is the set of vertices of graph G, $E = \{(u_i, u_j)\} \in U \times U$ is the set of retweet behaviour among them, and $W = \{w_{i,j}\}$ where $w_{i,j}$ indicates the number of times u_i retweets u_j. After getting the retweet network, we construct the reciprocal retweet network by simply removing all the one-way edges, as illustrated in Figure 1 (b). At last, denote $V = \{v_i\}_{i=1}^n$ as the values of users. In this paper, we propose to measure such a set of values of V from the retweet interactions among users of U.

2.2 Model

Principle of Balanced Reciprocity
We consider the social interaction in the notion of reciprocity based on the retweet interaction between Twitter users. Particularly, here it refers to the mutual adoption of each other's tweets between two users in the form of retweet. However, we can observe inequality of such retweet reciprocity. For example, as shown in Figure 1, Alice and Dave have retweeted each other. But it is in an unequal way – Dave has retweeted Alice three times while Alice has only retweeted Dave once. We suppose such inequality reflects the inequality of the

users' social status. For example, Alice may be the superior of Dave in their offline real-world workplace, so it is understandable that Dave retweeted Alice's tweets more than Alice retweeted Dave's. The Principle of Balanced Reciprocity, in this setting of retweet, is to state that the relative ratio between the numbers of times two users retweet each other is proportional to the relative ratio between their values as reflected in their social relationship. Accordingly, we measure users' social status as their values of V and we therefore have the Equation 1 as follows.

$$\frac{w_{i,j}}{w_{j,i}} = \frac{v_j}{v_i} \quad (1)$$

We suppose Equation 1 is reflected by the continuous interaction among Twitter users.

Minimizing Error

For the observed data, we expect that the error of Equation 1 should be as small as possible for all pairs of users. For easy optimization, we transform Equation 1 into an equivalent linear formulation as shown in Equation 2 below.

$$w_{i,j} \cdot v_i = w_{j,i} \cdot v_j \quad (2)$$

So Equation 3 below which is the sum of all the square error of all pairs of users should be minimized.

$$e(V) = \sum_{i=1}^{n} \sum_{j=1}^{n} (w_{i,j} \cdot v_i - w_{j,i} \cdot v_j)^2 \quad (3)$$

So we can infer the values of users V by minimizing the error function $e(V)$.

2.3 Inference

In this section, we discuss how to infer the value of each user by minimizing the error function $e(V)$. First, it is quite obvious that $e(V) = 0$, if we set all $v_i = 0$, which makes no sense. So here we introduce a penalty function $p(v)$, and append penalty terms at the end of $e(V)$ as Equation 4 below. So we minimize $e^*(V)$ instead of $e(V)$. Here the penalty function $p(v)$ should have such properties: (I) $p(v)$ is very large at $v = 0$, so that v_i is far from 0 by minimizing $e^*(V)$; (II) $p(v) > 0$ and $\lim_{v \to +\infty} = 0$, so that there is no penalty when v_i is far from 0; and (III) $p(v)$ is a monotonically decreasing function. In this paper, we set $p(v) = M \cdot e^{-kv}$, (where M and k are positive values), but it is not the only formulation for the penalty function $p(v)$.

$$e^*(V) = \sum_{i=1}^{n} \sum_{j=1}^{n} (w_{i,j} \cdot v_i - w_{j,i} \cdot v_j)^2 + \sum_{i=1}^{n} p(v_i) \quad (4)$$

Denote the derivative of $p(v)$ as $p'(v)$. We can get the derivatives of $e^*(V)$ as Equation 5 below.

$$\frac{\partial e^*}{\partial v_i} = 2\sum_{j=1}^{n} w_{i,j}(w_{i,j} \cdot v_i - w_{j,i} \cdot v_j) + \sum_{i=1}^{n} p'(v_i) \quad (5)$$

Based on the the derivatives of $e^*(V)$, the gradient descent method is used to get the optimal value of V.

3 Empirical Evaluation

3.1 Data Set

We use a Twitter data set which contains 3,165,479 users. These users are obtained by a snowball-style crawling starting from a seed set of Singapore local celebrities and active users and tracing their follower/followee links up to two hops. We use the subset of tweets between October 1, 2011 and December 31, 2011, which contains 50,918,021 tweets and 90,205 distinct users and 6,943,189 retweets. Among these 90,205 users, 44,152 users retweet at least one tweet or be retweeted at least once in our data set. We also get the follow links between these 44,152 users, including 653,619 links in total. Using these users and retweets, we constructed the retweet graph described in Section 2.1. The follow graph is also constructed based on the follow links among these users.

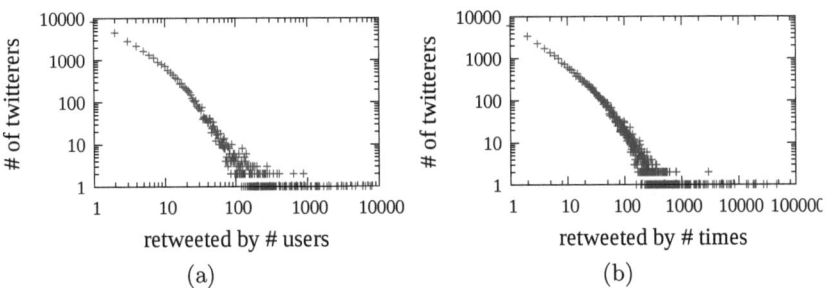

Fig. 2. Power-law phenomenon in the retweet network

First we consider Twitter users' popularity in terms of the number of users by whom they were retweeted, and the number of times they were retweeted. In Figure 2(a), we show the distribution of the number of users by whom Twitter users were retweeted. In Figure 2(b), we show the distribution of the number of times Twitter users were retweeted. As illustrated in [10], the power-law phenomenon was found in the Twitter follow network, similarly, in our data set, from Figure 2(a)(b) we also found that the power-law phenomenon exists in the

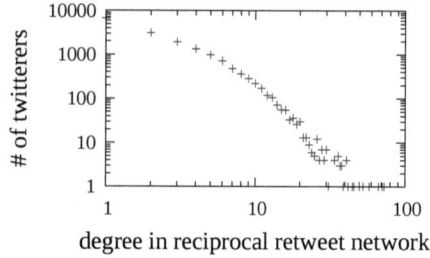

Fig. 3. Power law phenomenon exists in the reciprocal retweet network

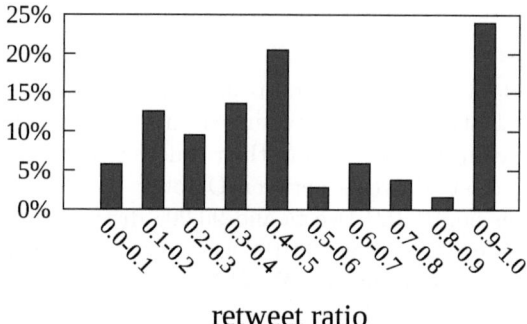

Fig. 4. Retweet ratio

Twitter retweet network. Intuitively, the value of the Twitter users can be measured by their popularity in both terms above. In Section 3.2 we compared our solution with this intuition.

As we try to evaluate the users' values from their reciprocal retweet behaviour, we constructed the retweet reciprocal network based on the retweet network, by simply removing all the one-way edges in retweet network. As shown in Figure 3, the power-law phenomenon also exists in this reciprocal retweet network. For each pair of users who retweeted each other, we studied the ratio of the numbers of tweets retweeted by them (the smaller one divided by the larger one). Figure 4 showed the distribution of these retweet ratios. From Figure 4, we can see that the bars at regions 0.9-1.0 and 0.4-0.5 are higher than others, because a lot of users only retweet others one or two times, which leads to the popularity of "1:1" and "1:2". However, in general, the retweet ratios vary from 0.0 to 1.0, which implies the difference of inherent values of different users.

As the scale is too large to get the optimal values for all the users in our dataset in practical time, here we considered a sub reciprocal retweet network of top 3,000 users who have most numbers of users by whom they were retweeted. Based on this sub network, we calculated the values for all these 3,000 users.

3.2 Comparison with Related Algorithms

In this section, we conducted the comparisons against related algorithms. All the algorithms studied include:

1. **Proposed Method.** Based on the reciprocal retweet network, we calculate the score for each user using our proposed method. In the experiment, the parameters of the penalty function $M = 1e6$, $k = 15$, and the number of iterations is 1000.
2. **PageRank.** Based on the weighted retweet network, (A retweets B means a pointer from A to B), we calculate the PageRank score for each user. In the experiment, the residual probability is set to 0.85, and the epsilon is set to 1e-9.
3. **HITS.** Based on the unweighted retweet network, (A retweets B means a pointer from A to B), we calculate the authority score for each user using HITS algorithm. In the experiment, the epsilon is set to 1e-9.
4. **Users-based.** In this method, we rank the users based on the number of users by whom they were retweeted.
5. **Retweets-based.** In this method, we rank the users based on the number of times they are retweeted by others.
6. **PageRank based on follow network.** In this method, we construct the follow link network for the users. The PaeRank score is calculated based on this graph. In the experiment, the residual probability is set to 0.85, and the epsilon is set to 1e-9.
7. **HITS based on follow network.** In this method, we construct the follow link network for the users. The authority score of HITS is calculated based on this graph. In the experiment, the epsilon is set to 1e-9.
8. **TwitterRank.** In this method, we construct the follow link network for the users, and use LDA [2] to learn topics from all the tweets of these users. Then set the weight of links as mentioned in TwitterRank, which is based on users's topic profile. Combining the ranking lists in different topics, an aggregation of TwitterRank is calculated. The number of topics $T = 50$, Dirichlet hyper-parameters $\alpha = 50/T$, $\beta = 0.1$, and the residual probability is set to 0.85.

For ease of presentation, our proposed method is denoted as **Ours**, and the related algorithms are abbreviated to **PR**, **HS**, **UB**, **RU**, **FN-PR**, **FN-HS** and

Table 1. The top 6 users ranked by different methods

	1	2	3	4	5	6
Ours	SoSingaporean	NaomiNeo_	imwhywhy	speishi	heedyjoee	ShilinKEY
PR	Cursedwithsex	mrbrown	stcom	NaomiNeo_	TommyWee	xavlur
HS	SoSingaporean	fakeMOE	NaomiNeo_	xavlur	stcom	Cursedwithsex
UB	SoSingaporean	stcom	mrbrown	fakeMOE	xavlur	NaomiNeo_
RB	SoSingaporean	stcom	mrbrown	NaomiNeo_	Cursedwithsex	BvsSG
FN-PR	SoSingaporean	Xiaxue	mrbrown	stcom	fakeMOE	JoannePeh
FN-HS	SoSingaporean	Xiaxue	mrbrown	fakeMOE	stcom	BvsSG
TR	stcom	mrbrown	Cursedwithsex	Xiaxue	bongqiuqiu	humsyourlife

TR respectively. Table 1 lists the top users ranked by all the methods above. Due to the limitation of the space, only top 6 users are listed.

Case Studies. We first evaluate the top "valuable" users in our solution (the row starts with "Ours"). The top one is "*SoSingaporean*", who has more than 121,000 followers, and actively shares everything funny, unique and localised in Singapore. "*SoSingaporean*" is so popular that 7,476 users out of 44,152 users in our dataset retweeted him, however, he only retweeted 36 users back. The second one is "*NaomiNeo_*", who is very active online celebrity and tweeted over 34,000 tweets. She also has a lot of followers (more than 64,000) and was retweeted by 3,545 users in our dataset. It is reasonable that such kinds of users are at the top positions in our results.

Then we compare the results of different methods. Although these 8 methods make use of different information (**Ours** based on the reciprocal retweet network of 3,000 users, **PR** and **HS** based on the retweet network of 3,000 users, **UB** and **RU** based on the whole retweet network, **FN-PR** and **FN-HS** based on the whole follow network, **TR** based on the whole follow network, as well as all the tweets), except **PR** and **TR**, all the methods rank "*SoSingaporean*" as the top one, which shows the inherent value of "*SoSingaporean*". Besides, "*NaomiNeo_*" and "*stcom*" are ranked in the top 6 users by most of the methods.

The other case in our experiment is "*fakeMOE*", which is ranked low in our result, but is ranked top by other methods such as PageRank (top 7 in PageRank, not shown in Table 1). "*fakeMOE*" spoofs the official Twitter account of Ministry of Education (MOE). Being followed by more than 22,000 Twitter users and retweeted by 4,671 users in our dataset, "*fakeMOE*" is definitely a hub user, so that it succeed to disguise itself as influential user in the eyes of PageRank and other methods. However, it is actually just a fluffing account, and few real influential Twitter users retweet this account. By examining the retweet interactions between "*fakeMOE*" and other users, our method rank it low.

The other case is one local influential news media "*stcom*" (The Straits Times), which has more than 233,000 followers, and is retweeted by 6,455 users in our dataset, doesn't appear in the top 20 list, and actually it is ranked as the 59th one by our method. By exploring the reciprocal retweet network, we found that "*SoSingaporean*" and "*NaomiNeo_*" behave like a hub in the network, i.e. they interact with some "agents" and these "agents" interact with others. Figure 5 presents the 2-layer eco-network of "*SoSingaporean*", from which we can see the reciprocal retweet network is much sparser than the follow network. However, contrary to "*SoSingaporean*" and "*NaomiNeo_*", "*stcom*" only connected with two nodes (one is an art journalist who and the other is a geek). We can only infer the "value" of "*stcom*" from these two nodes, so that we can not infer the accurate "value" of "*stcom*". The reason maybe the serious news media such as "*stcom*" retweet others very carefully and very rarely, because of the consideration of public influence. For these kind of users, the lack of such interaction behaviour makes us hard to infer the accurate "values" of them.

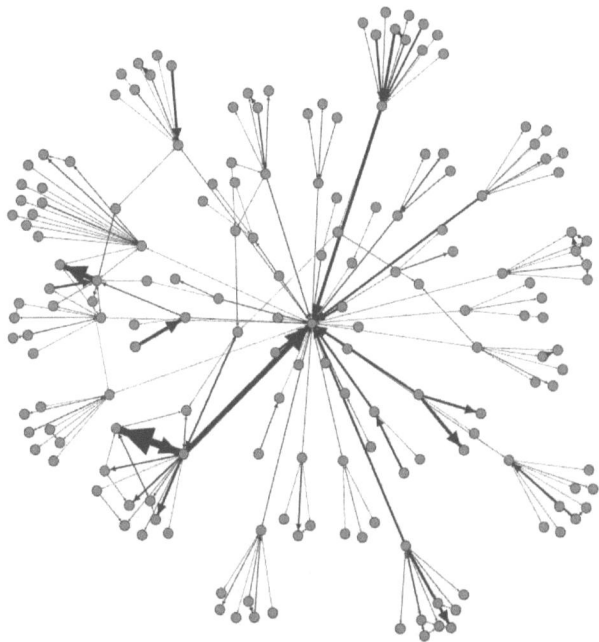

Fig. 5. The eco-network of *"SoSingaporean"*

Correlation. To further study the relationship between these methods, we study the correlation between the rank lists generated by them. The correlation is measured by the Kendall rank correlation coefficient [9], which takes value in the range of $[-1, 1]$. If the two rank lists are exactly the same, the correlation coefficient is 1; if the two rank lists are independent, the correlation coefficient is 0; and it is -1 if two rank lists are opposite to each other. The larger the correlation coefficient is, the stronger the similarity of the two rank lists are.

Table 2 lists the Kendall correlation coefficients between all the rank lists generated by all the methods studied. It is observed that the rank lists generated by **PR**, **HS**, **UB** and **RU**, which all are based on the retweet network, are similar; and that the rank lists generated by **FN-PR**, **FN-HS** and **TR**, which are based on the follow network, are similar. As expected, the rank list generated by our solution does not overlap with results of all other methods, because our model aims to capture totally different values of the users in the network. However, compared with **PR**, **HS**, **UB**, **RU** and **TR**, our result is more similar to **FN-PR** and **FN-HS**. The differences between the networks these method based on and the differences between the modes these methods used lead to the differences between the results of them. And they represent the values of Twitter users in different dimensions.

Table 2. Kendall rank correlation of results between rank lists by different methods

Correlation	Ours	PR	HS	UB	RB	FN-PR	FN-HS	TR
Ours	1.00000	0.02944	-0.05467	0.03128	-0.03412	0.10967	0.11280	-0.06659
PR	0.02944	1.00000	0.52186	0.46060	0.36742	0.24603	0.23552	0.26955
HS	-0.05467	0.52186	1.00000	0.65668	0.50010	0.18476	0.19378	0.34078
UB	0.03128	0.46060	0.65668	1.00000	0.57979	0.25932	0.27937	0.31034
RB	-0.03412	0.36742	0.50010	0.57979	1.00000	0.15418	0.17571	0.29854
FN-PR	0.10967	0.24603	0.18476	0.25932	0.15418	1.00000	0.84048	0.49512
FN-HS	0.11280	0.23552	0.19378	0.27937	0.17571	0.84048	1.00000	0.47802
TR	-0.06659	0.26955	0.34078	0.31034	0.29854	0.49512	0.47802	1.00000

3.3 Retweet Behaviour Prediction

To verify the effectiveness of our model, in the section we conduct a user retweet behaviour prediction test. In this test, an assumption is made that for a pair of Twitter users A and B who retweet each other, if A's value is larger than B's value, then B will retweet A more than A retweet B. Based on this assumption, according to the values of Twitter users, we can predict their retweet behaviour, i.e. predict whether A retweet B more than B retweet A.

First, for all the pairs of Twitter users who retweet each other, we randomly choose 1% pairs. For each pair in them, if for both two Twitter users in this pair, their degrees in the reciprocal retweet network are no less than 3, then we remove all the retweets between them. We do this so that even all retweets between them are removed, they also have at least 2 neighbours in the rest network, which can help us to infer their values. Then based on this new network, the values of Twitter users are recalculated for our method and all other methods, except **FN-PR**, **FN-HS** and **TR**, which are based on the follow network rather than the retweet network. At last, for each pair of users between whom all the retweets are removed, according to the values provided by each method, prediction is made. (We ignore the cases that two users retweet each other equally.) Based on the

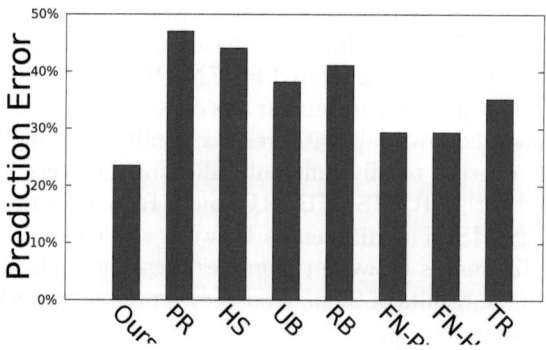

Fig. 6. Prediction error of different methods

ground truth (the retweets between the Twitter users in each removed pair), we verify the prediction of each method in the term of prediction error rate.

Figure 6 shows the prediction error rates of different methods. We can observe three points as below. 1) Our method outperforms other methods to a large extent, which shows that our model has better understanding of the Twitter users' retweet behaviours. 2) Though slightly worse than our method, **FN-PR** and **FN-HS** outperform all other five methods. This result is consistent with the correlation coefficients between different methods shown in Section 3.2, recalling that the rank lists generated by **FN-PR** and **FN-HS** are more similar to our method's. It also implies the inherent relationship between users' state in follow graph and their retweet behaviours. 3) The error rates of all methods presented are less than 50%, which is the expectation of randomly guessing. Under this consideration, our assumption makes sense, that in general higher-value Twitter users retweet lower-value Twitter users less.

4 Related Work

Online social networks such as Facebook and Twitter has been growing rapidly these years. There are several pieces of works to study the influential or valuable users in the online social network for purposes including maximizing the spread of influence [8] and viral marketing [14]. PageRank [11] and HITS [3], which are originally used to rank the web pages in the network which is made up of web pages, are naturally used in this new scenario to rank the users in the network which is made up of users. TwitterRank [15] extends PageRank by introducing a new dimension of the topics of tweets. However, both PageRank and HITS are derived based on their own assumptions. PageRank assumes there is a surfer randomly visiting the web pages. HITS considers a academic scenario in which there are two roles: authority and hub. Due to the limitations of the assumptions, PageRank and HITS don't take into account the interactions of users in social network, which may be the key point to disclose the values of users in social network. Under this consideration, we proposed a quite different model, in which Twitter users' behaviours are treated as reciprocal social behaviours. In this scenario, the inherent values of users are determined by the continuous interaction between them. Especially, we focused on the retweet behaviours of Twitter users, which also attracted the attention of several works such as [17] and [16]. Using factor graph mode, [17] studied the retweet behaviour for each individual user and message. To do the same job, [16] leveraged four different types of features, including social-based, content-based, tweet-based and author-based features. Different from these works, our work treated the retweet behaviours as interactions between pairs of Twitter users, with totally different purpose of ranking Twitter users.

Besides, other works which consider the social features includes [12] and [7]. [12] is an application of HITS in the Twitter setting. It identifies influential users who are able to diffuse information quickly and influent others effectively. It introduces the "passive users" who are reluctant to be influenced. In this model, higher value in the ranking implies that that particular user cans even influent most passive users. [7] models vitality and susceptibility in Twitter. In this model viral information diffusion is due to viral users, viral items and susceptible users. These models provide the other directions to measure the users in the social network.

5 Discussion

In this section, we try to explain the model built in Section 2.2 from an economic perspective. In particular, we try to measure the economic value of the retweet, which leverages the power of word of mouth to help information dramatically spread over the whole Twitter network, and makes some tweets to reach a large number of audiences and to gain huge influential impacts.

In fact, each tweet has its own influential economic value. For example, when a satisfied iPhone user posted a positive tweet about iPhone; that tweet potentially reaches a large number of followers, triggering viral marketing effect for Apple. Eventually, this can help increasing the sales of Apple. In this case, the resulting difference in Apple's sales reflects the economic value of that tweet. Not just only the tweet, retweet also plays an important role in this picture as the power of original tweet is strengthened exponentially by the number of retweets. In term of economic value, it is fair to expect that retweet even has higher value than the tweet itself. The original tweet, most of the time, only expresses anticipatory or evaluative opinion of individual [6]. On the other hand, if someone retweets that original tweet, that action implies that the original opinion is verified, adopted and forwarded to other users.

Now that retweet has big economic value, and there is no such thing as a free lunch, we suggested that there would be an underlying *"virtual retweet market"*, on which Twitter users's retweeting behaviours are based. In this virtual market, for common benefits, Twitter users *"trade"* with each other by exchanging their retweeting behaviours, i.e. retweeting each other. For example, A retweeted B 3 times, and in reciprocation, B retweeted A 2 times. In this case, A makes a deal with B using its 3 retweets for B's 2 retweets. After conducting such kind of trade, their influence is extended by increasing the numbers of their audiences from the followers of others. We further assume that this virtual market is a free price system without external effects [5], in which the prices of good and services are eventually determined by the exchanging behaviour of users [1]. In this system, the interchange of retweeting behaviours determines the prices, which reflect the economic *"values"* of Twitter users. We can mathematically formalise the *"trade"* behaviours between Twitter users as Equation 2 in Section 2.2. It means that each pair of Twitter users make a fair *"trade"* according to their economic *"values"*.

As electronic commerce develops quickly, it is very possible that this virtual retweet market comes true as a real market in the future. In this scenario, our model can be the basis of this market by calculating the prices of Twitter users' retweeting behaviours.

6 Conclusions and Future Work

Finding the valuable users in social network is a quite motivated problem due to the potential commercial interest. Rather than from a perspective of information flow, this paper re-examine the value of users in social network from the social interaction perspective. In particular, we consider the social interaction in the notion of reciprocity based on the retweet interaction between Twitter users. We formulated the retweet reciprocity, proposed an alternative user ranking model based on retweet reciprocity and developed efficient inference solution. Our experiments on real Twitter data demonstrated that our proposed model presents different yet equally insightful ranking results. The conducted prediction test also showed the correctness of our model. Besides, we also discuss the meaning of our proposed model from an economic perspective, and explain Twitter users' retweeting behaviour as economic behaviour.

Our paper is a preliminary study with much room for extension. First, as the experimental results show, there are still some real influential users such as "*stcom*" are not ranked top in our ranking list, which is due to the lack of enough interactions of these users. We plan to incorporate the different kinds of interactions in a social platform, and find influential users by combining all such kinds of interactions. Second, we use gradient descent method to infer the values of users, which is not efficient enough to handle large scale social data. We also plan to improve this by developing approximate efficient algorithms. Third, in near future, social networks will evolve dramatically. Future work of this research will consider the interaction of users in community as well as focus on the interaction between communities. At last, one feasible direction is to add the topic dimension as in TwitterRank [15], and study the interactions between users in different topics.

Acknowledgement. This research is supported by the Singapore National Research Foundation under its International Research Centre @ Singapore Funding Initiative and administered by the IDM Programme Office.

References

1. Altvater, E., Camiller, P.: The future of the market: An essay on the regulation of money and nature after the collapse of'actually existing sociali (1993)
2. Blei, D.M., Ng, A.Y., Jordan, M.I.: Latent dirichlet allocation. The Journal of Machine Learning Research 3, 993–1022 (2003)
3. Chakrabarti, S., Dom, B., Raghavan, P., Rajagopalan, S., Gibson, D., Kleinberg, J.: Automatic resource compilation by analyzing hyperlink structure and associated text. Computer Networks and ISDN Systems 30(1), 65–74 (1998)
4. Easley, D., Kleinberg, J.: Networks, crowds, and markets: Reasoning about a highly connected world (2012)
5. Gregory, P.R., Stuart, R.C.: Comparing economic systems in The Twenty-first century. Houghton Mifflin (2004)

6. Hennig-Thurau, T., Wiertz, C., Feldhaus, F.: Exploring the twitter effect: an investigation of the impact of microblogging word of mouth on consumers early adoption of new products (2012) Available at SSRN 2016548
7. Hoang, T.A., Lim, E.P.: Virality and susceptibility in information diffusions. In: Sixth International AAAI Conference on Weblogs and Social Media (2012)
8. Kempe, D., Kleinberg, J., Tardos, É.: Maximizing the spread of influence through a social network. In: Proceedings of the Ninth ACM SIGKDD International Conference on Knowledge Discovery and Data Mining, pp. 137–146. ACM (2003)
9. Kendall, M.G.: A new measure of rank correlation. Biometrika 30(1-2), 81–93 (1938)
10. Kwak, H., Lee, C., Park, H., Moon, S.: What is twitter, a social network or a news media? In: Proceedings of the 19th International Conference on World Wide Web,, pp. 591–600. ACM (2010)
11. Page, L., Brin, S., Motwani, R., Winograd, T.: The pagerank citation ranking: bringing order to the web (1999)
12. Romero, D., Galuba, W., Asur, S., Huberman, B.: Influence and passivity in social media. In: Machine Learning and Knowledge Discovery in Databases, pp. 18–33 (2011)
13. Sahlins, M.D.: Stone age economics. Aldine de Gruyter (1972)
14. Shakarian, P., Paulo, D.: Large social networks can be targeted for viral marketing with small seed sets. In: 2012 IEEE/ACM International Conference on Advances in Social Networks Analysis and Mining (ASONAM), pp. 1–8 (2012)
15. Weng, J., Lim, E.P., Jiang, J., He, Q.: Twitterrank: finding topic-sensitive influential twitterers. In: Proceedings of the Third ACM International Conference on Web search and Data Mining, pp. 261–270. ACM (2010)
16. Xu, Z., Yang, Q.: Analyzing user retweet behavior on twitter. In: Proceedings of the 2012 International Conference on Advances in Social Networks Analysis and Mining (ASONAM 2012), pp. 46–50. IEEE Computer Society (2012)
17. Yang, Z., Guo, J., Cai, K., Tang, J., Li, J., Zhang, L., Su, Z.: Understanding retweeting behaviors in social networks. In: Proceedings of the 19th ACM International Conference on Information and Knowledge Management, pp. 1633–1636. ACM (2010)

Feature Extraction and Summarization of Recipes Using Flow Graph

Yoko Yamakata[1], Shinji Imahori[2], Yuichi Sugiyama[1], Shinsuke Mori[3], and Katsumi Tanaka[1]

[1] Graduate School of Informatics, Kyoto University
Yoshida-Honmachi, Sakyo-ku, Kyoto 606-8501, Japan
[2] Graduate School of Engineering, Nagoya University
Furo-cho, Chikusa-ku, Nagoya 464-8603, Japan
[3] Academic Center for Computing and Media Studies, Kyoto University
Yoshida-Honmachi, Sakyo-ku, Kyoto 606-8501, Japan

Abstract. These days, there are more than a million recipes on the Web. When you search for a recipe with one query such as *"nikujaga,"* the name of a typical Japanese food, you can find thousands of *"nikujaga"* recipes as the result. Even if you focus on only the top ten results, it is still difficult to find out the characteristic feature of each recipe because a cooking is a work-flow including parallel procedures. According to our survey, people place the most importance on the differences of cooking procedures when they compare the recipes. However, such differences are difficult to be extracted just by comparing the recipe texts as existing methods. Therefore, our system extracts (i) a general way to cook as a summary of cooking procedures and (ii) the characteristic features of each recipe by analyzing the work-flows of the top ten results. In the experiments, our method succeeded in extracting 54% of manually extracted features while the previous research addressed 37% of them.

1 Introduction

Cooking is one of the most fundamental activities of human social life. It is not only connected with the joy of eating but also deeply affects various aspects of human life such as health, dietary, culinary art, entertainment, human communication, and so on. Hene, the number of recipes on the Web has been increasing rapidly in recent years. In Japan, COOKPAD, the biggest recipe portal site, has more than 1.5 million recipes and 12 million users [14]. Rakuten-Recipe has more than 620,000 recipes. In the United States, Food.com has more than 475,000 recipes, while Allrecipe.com and FoodNetWork.com have more than ten million users. Google also offers a service for recipe search.

However, more is not always better. Even if you submit just one query, such as *"nikujaga,"* to COOKPAD, you can find more than 5,600 *"nikujaga"* recipes. Of course, all of the recipes explain how to cook *"nikujaga,"* but they are somewhat different. Some recipes fry meat in advance while others fry meat after an onion. Some recipes add a soy-source during frying while others mix it during stewing.

The title is not useful because it was given by the recipe authors freely and it reflects only his/her subjective evaluation. You try to find your favorite recipe by reading the text parts of several recipes. However, it is very hard because a cooking is a work-flow with parallel procedures and it requires much effort to understand, memorize, and compare these cooking procedures.

Because all of the recipes are obtained using the same query "*nikujaga*," the ways of cooking described by these recipes must be similar. Therefore, we propose a method for finding their general way to cook as the summary by extracting the common structure of the cooking flow-graphs from the top ten search results. Moreover, the system obtains the characteristic features of each recipe by comparing it with the generated general recipe.

As it is introduced in Section 3.1, we are researching a method that converts a recipe text to a flow-graph of cooking procedures. Therefore, we assume that all recipes had been converted to such flow-graphs using this method. Additionally in this paper, a recipe flow graph is assumed to be a tree structure because the most of the recipes' flow-graphs can be represented as tree-type graphs. Hereafter, we refer to it as a "recipe tree." The system conducts node-to-node mapping of all pairs of recipe trees and integrates the most similar pair of recipes. The system repeats this integration and finally obtains one general recipe. The characteristic features of each recipe can be extracted by mapping the recipe tree and the obtained general recipe tree and finding the differences.

2 Feature Types and Their Importances of Recipe

In this section, we analyses which type of feature should be extracted from a recipe in the purpose.

To analyze recipe feature types and their importances, we conducted a survey. We searched on the recipe portal site COOKPAD [14] with the four queries "*nikujaga*," "*carbonara*," "beef stew," and "*nigauri* (a name of an ingredient)," and collected top 10 recipes for each query. Therefore, these four recipe sets of 10 recipes respectively, 40 recipes in total, were obtained as a test data. We asked two annotators, who were undergraduate students, to extract characteristic features of each recipe comparing with the other nine recipes in each set manually. We also asked them to assign a rank to each feature according its importance when they found more than two features for one recipe. Consequently, 197 features with 29 duplication and 168 unique features were obtained from 40 recipes in total. We classified obtained 168 features into seven types as follows.

- **Additional ingredient:** when a recipe has an uncommon ingredient.
- **Reduced ingredient:** when a recipe does not have a common ingredient.
- **Ingredient quantity:** when a quantity of an ingredient is significantly differ from the others.
- **Uncommon action:** when a recipe has an uncommon action.
    ex.) The potato is immersed in water after cutting.
- **Action order:** when an order of actions differs from the other recipes.
    ex.) Soy source is mixed when it fries ingredients in a recipe while soy source is mixed after adding water to the ingredients in other recipes.

- **Tool:** when an uncommon tool is used.
- **Writing type:** when it is different writing type from the others.

The number and the proportions of each type in the manually extracted features are shown in Table 1. As you see in the table, the most common feature is additional ingredient. Though the proposed method is able to extract this feature, even a simple method that compares the member of the ingredient list of the recipe with the others can also extract such as [10]. Reduced ingredient and quantity of ingredient are also able to be extracted in the same way. Meanwhile, the annotators considered the action order is the most important features even though the proportion of it was not very high. Uncommon action is also considered more important than additional ingredient. Since these two features cannot be extracted just by comparing the words of the instructions, it is said that sophisticated analysis is required to find important recipe features.

Table 1. Manually extracted features for each categories

Type	Ave. rank	*Nikujaga*	*Carbonara*	*Nigauri*	Beef stew	Total
Action order	**1.8**	5 (8%)	1 (2%)	1 (2%)	5 (11%)	12 (7%)
Quantity of ingredient	2.1	6 (9%)	0 (0%)	0 (0%)	1 (2%)	7 (4%)
Reduced ingredient	2.2	6 (9%)	1 (2%)	0 (0%)	1 (2%)	8 (5%)
Uncommon action	**2.4**	11 (17%)	2 (5%)	4 (9%)	7 (16%)	24 (14%)
Additional ingredient	2.5	28 (44%)	39 (89%)	3 (7%)	34 (77%)	104 (62%)
Writing style	3.7	1 (2%)	0 (0%)	0 (0%)	2 (5%)	3 (2%)
Cooking tool	3.9	7 (11%)	1 (2%)	0 (0%)	2 (5%)	10 (6%)
Total		64	44	8	52	168

3 Pre-processing for Recipes

3.1 Recipe Tree: The Work-Flow Format of a Recipe

We first convert recipe procedures written in a natural language into a tree-type work-flow graph. Fig. 1 shows an example of a recipe tree. In the recipe tree, each leaf node corresponds to an ingredient of the recipe such as "a potato," "meat," and "sugar." Each intermediate node corresponds to a cooking action of eleven categories, "Mix," "Cut," "Fry," "Roast," "Boil," "Cook in boiling water," "Deep frying," "Heat by instrument," "Steam," "Stop ongoing action" and "Others". The root node corresponds to the completed dish which is ready to serve. The label of each node is a pair of a type, ingredient or cooking action, and a word sequence corresponding to the name. The root node is the only exception and has the dish name. For instance, a sentence "A potato is cut to larger bite-sized pieces, and is immersed in water" corresponds to the sub-tree in the dotted circled in Fig. 1.

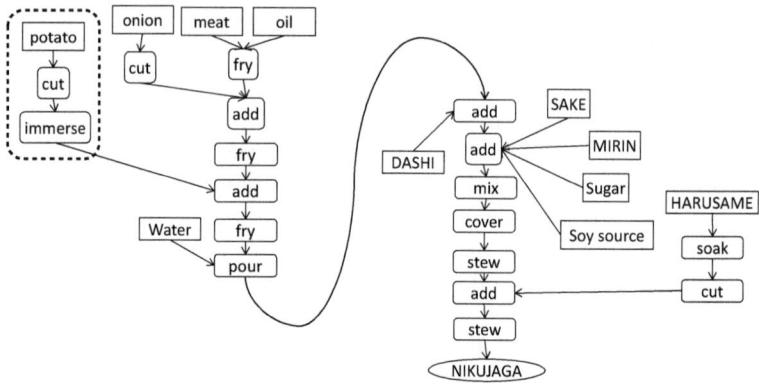

Fig. 1. A recipe tree of one "*nikujaga*"

Table 2. Named entity tags

NE Tag	Meaning	NE Tag	Meaning
F	Food	Ac	Action by the chef
T	Tool	Af	Action by foods
D	Duration	S	State of foods
Q	Quantity		

Such a recipe tree can be automatically converted from a recipe text using natural language processing (NLP) [3–5]. NLP for recipe texts proposed in [5] consists of two important parts based on machine learning methods. The first one is named entity (NE) recognition after word identification which extracts important word sequences shown in Table 2 appearing in the recipe text. The second one is predicate-argument structure (PAS) analysis after syntactic analysis which determine the subject, the direct object, and the indirect object (arguments) for a verb (predicate). In this paper we adopt the same named entity definition but only use F (Food) for ingredient nodes and Ac (Action by the chef) for cooking action nodes. We use PAS of the verbs marked as Ac which corresponds to the arcs in recipe trees.

It is reported that the NE recognition accuracy went up from 53.4% to 67.0% by only 5 hour annotation [5]. The NE recognition accuracy for the general tag set (person name, organization name, place, etc.) is around 80% ~ 90% when enough large training data are available [7]. In addition, there are less variations for food names and cooking action names than for the general tag set. Therefore we can say that it is possible to achieve about 90% accuracy just by preparing practically large training data. Currently the accuracy of PAS analysis is less than NE recognition. In the recipe domain, however, the vocabulary is much more limited than in general domains of NLP such as newspaper articles. Thus a domain adaptation technique for PAS analysis [12] allows us to achieve an enough high accuracy with a practical size of the training data.

As we described above, the NLP community is working on recipe texts, as well as patent disclosures etc., as a domain adaptation example of NE recognition or PAS analysis. Since currently the accuracies are, however, not sufficiently high for the application we propose in this paper, in the experiments we use recipe trees manually converted from recipe texts.

3.2 Tree Mapping Algorithms

Tree is one of the most common and well-studied combinatorial structures in computer science. Comparison of two (or more) trees is a fundamental task in many applications such as computational biology, structured text databases and image analysis. Various measures have been proposed and studied for comparison of two trees. Among such measures, tree edit distance is the most common and well studied. For two labeled trees T and T', the edit distance from T to T' is measured by the minimum cost sequence of edit operations needed to transform T into T'. The edit operations are deletion, insertion, and substitution. For ordered labeled trees, efficient algorithms for computing the edit distance have been proposed in the literature. Tai [9] developed the first polynomial time algorithm for the problem, several improvements followed, and Demaine et al. [1] proposed an optimal algorithm that runs in $O(n^3)$ time for n-node trees.

For unordered labeled trees, including recipe trees, the problem of computing the edit distance between two trees is difficult (more precisely, the problem is known to be NP-hard [13]). Therefore, it is reasonable to try to develop heuristic algorithms for this case. Shasha et al. [8] proposed a simple heuristic algorithm by sorting and iterative improvement algorithms based on metaheuristics. However, they focused on only the number of child for ordering while label matching is required for our purpose. Do and Rahm [2] proposed a system called COMA, which provides an extensible library of simple and hybrid match algorithms, but the editing costs cannot be adjusted flexibly.

For a given set of trees, computing one tree that is similar to all the other trees is a challenging task and has been studied in the literature. Phillips and Warnow [6] showed the hardness of this problem and proposed a heuristic method for computing a tree called the asymmetric median tree. Their method works well for evolutionary trees (in which species label the leaves). However, it is hard to apply this method to our application and different heuristic methods are necessary to compute a general recipe tree.

4 Generation of General Recipe Tree

4.1 Framework

The procedures of the system are as follows:

[**Step 1**] Ten recipes are given as a search result. In advance, every recipe has been converted to a recipe tree, in the form of a rooted, labeled, and unordered tree. Set the weight w of a tree T to one for each recipe tree.

[Step 2] The system calculates an approximate edit distance $d(T, T')$ between every pair of trees (T, T'). For each calculation, two unordered trees are transformed into ordered trees so that the distance of them becomes closer.

[Step 3] The trees of the closest pair T and T' are integrated to one tree. Let T and T' be the two trees to be integrated, and w and w' their weights. The system generates a new tree with the properties that (i) the distance from T is around $d(T, T')w'/(w + w')$, (ii) the distance from T' is around $d(T, T')w/(w + w')$, and (iii) the weight is $w + w'$. After adding the new tree to the current set of trees and removing two trees T and T' from it, our algorithm returns to Step 2, if two or more trees are remaining. Go to Step 4 when the number of trees becomes one. The final integrated tree, the general tree T_{gen}, is the output of our system.

[Step 4] Extract the characteristic features of each recipe tree by mapping the recipe tree with the general tree T_{gen} and finding the differences.

4.2 Transformation to Ordered Tree

As stated in Section 3.2, it is difficult to compute the accurate edit distance between unordered labeled trees. Therefore, the system converts each unordered tree into an ordered tree so that an approximate edit distance between them becomes small.

In our heuristic method, the system decides the order of children for each node from the root node to leaf nodes. At the beginning, the system finds the node that is closest to the root node and that has more than two or more children for each tree. Let u and v be the found nodes of two trees, and $u_1, u_2, \ldots, u_p$ and $v_1, v_2, \ldots, v_q$ be the child nodes of u and v. To decide the orders of these child nodes, the system solves the following problem.

$$\text{Maximize} \sum_{i,j} c(i,j) x(i,j)$$

$$\text{Subject to } y(i) = \sum_j x(i,j) \geq 0 \ (i = 1, 2, \ldots, p)$$

$$z(j) = \sum_i x(i,j) \geq 0 \ (j = 1, 2, \ldots, q)$$

$$x(i,j)(y(i) - 1)(z(j) - 1) = 0 \ (i = 1, \ldots, p, \ j = 1, \ldots, q)$$

$$x(i,j) \in \{0, 1\},$$

where $x(i, j)$ is the decision variable whose value is one if and only if node u_i is mapped to node v_j, and $c(i, j)$ denotes the number of common ingredients which appear in both sub-trees whose root nodes are u_i and v_j. Note that the system does not define node-to-node mapping at this point. One reason is that the number of children p and q may be different, and another reason is that a one-to-many mapping can be suitable for some cases (see Fig. 2 as an example). After one-to-one or one-to-many mappings are obtained by solving this problem at nodes u and v, the procedure goes to their descendants (i.e., solving similar

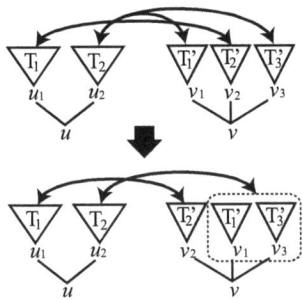

Fig. 2. Example for transformation to ordered trees

problems for each corresponding sub-tree). The procedure stops when it cannot find a descendant.

4.3 Node-to-Node Mapping between Two Trees

For two rooted, labeled and ordered trees T and T', we compute the minimum cost sequence of edit operations needed to transform T into T'. The edit operations are (i) deletion: deleting a node from a tree, (ii) insertion: inserting a node into a tree, and (iii) substitution: changing one node of a tree into another node. Each operation has its cost $c_{del}(u), c_{ins}(u)$, and $c_{sub}(u,v)$, respectively.

Now, let consider such case that "Stir the onion and add the carrot." Note a chef keeps stirring the vegetables when he/she add the carrot. It means that "Mix" including "add", "throw in", "put" and "pour" can be given to the other cooking action. "Other" is also in the same way. Therefore, we set $c_{sub}(u,v)$ depend on combination of u and v' as followings.

When u is a leaf:
$c_{sub}(u,v) = 0$ if the labels of u and v are the same.
$c_{sub}(u,v) = \infty$ otherwise.
When u is a root:
$c_{sub}(u,v) = 0$ if v is a root.
$c_{sub}(u,v) = \infty$ otherwise.
When u is an intermediate node:
$c_{sub}(u,v) = 0$ if the labels of u and v are the same.
$c_{sub}(u,v) = C_{sub1}$ if at least one of the labels of u or v is "Mix" or "Others".
$c_{sub}(u,v) = C_{sub2}$ if the labels of u and v differ.

The cost of $c_{del}(u)$ and $c_{ins}(u)$ are set a constant number C_{del_ins} for all nodes.

The minimum cost sequence of edit operations required to transform T into T' can be computed using the algorithm of Tai [9]. Although this algorithm runs fast enough for our data set, more sophisticated algorithms (e.g., [1]) will be useful for complicated recipes with many ingredients and cooking actions.

4.4 Recipe Tree Integration

To suggest the general way of cooking in a given recipe set, the generated general tree should be almost equally close to each of the recipes. The general tree should have the common characteristics of the given set. That is, ingredients or cooking actions that appear often in the given set must be extracted into the general tree. Moreover, sequences of cooking actions are also important for recipes and should be stored in the general recipe. Therefore, when the system integrates two trees, it counts how many recipes are integrated into each tree and generates a new tree that is affected by each tree in proportion as its integration counts.

For each pair of recipe trees, edit distance is computed using the methods explained in Sections 4.2 and 4.3. The system integrates the closest pair of recipes into one intermediate recipe tree. Let T and T' be the two ordered trees to be integrated, and w and w' their weights. The edit distance $d(T, T')$ and a set of edit operations transforming T into T' are computed. Our system generates a new intermediate tree whose distance from T (resp., T') is around $d(T, T')w'/(w + w')$ (resp., $d(T, T')w/(w + w')$). Concretely, the system adopts n insertion/deletion/substitution operations, where n is calculated as

$$n = d(T, T')w'/(w + w') \times m$$

and m is the number of insertion/deletion/substitution operations. The order of preference in this adoption is as follows:

- The deletion operation is adopted if
  - the deleted node is an ingredient and the number of its occurrences in the ten recipes is fewer than two.
  - the deleted node is a "Mix" or "Other" action.
- The insertion operation is adopted if
  - the inserted node is an ingredient and the number of occurrences in the ten recipes is two or more.
  - the inserted node is not a "Mix", "Other", or "Stop ongoing action" action.

There is no order of preference for substitution operations. After such integration procedures, the generated new tree T'' could have an action node as a leaf, because the connected leaf node of an ingredient was removed. In such cases, the system removes a sub-tree that has no ingredient as its leaf.

Then, T and T' are removed from the current set of trees and T'' with the weight $(w + w')$ is added to the set. The system repeats this integration and finally obtains one general recipe tree T_{gen}.

4.5 Characteristic Feature Extraction

The features of each recipe are extracted by comparing the recipe with the general recipe T_{gen}. Concretely, T is mapped with T_{gen} and deletion/insertion/substitution operations corresponding to the characteristic features of T.

5 Experiments and Results

5.1 Recipe Data Set

The given data set was the top ten results of searching with a query *"Nikujaga"* at COOKPAD [14]. The recipe IDs of obtained recipes were A) 1487670, B) 1485091, C) 1499546, D) 1519874, E) 1521946, F) 1524200, G) 1531094, H) 1531503, I) 1531751, and J) 1531880 (you can find these recipes at COOKPAD by searching with these IDs as a query). Then, we converted them into unordered recipe trees manually, this process is possible to automatized as we stated in section 3.1.

5.2 Examples of Transformation to Ordered Trees

The system calculated the mapping score for all combinations of two of the ten recipe trees. For each pair, the trees were transformed to ordered trees so that these trees could be mapped with lower cost in accordance with the algorithm explained in Section 4.2.

Fig. 3 (a) and (b) show examples of transformation from an unordered tree to an ordered tree when mapping recipe D) to recipe G). As shown in these figures, the trees on the right are closer to each other than the trees on the left. The subtrees indicated by thick lines were reordered in this procedure.

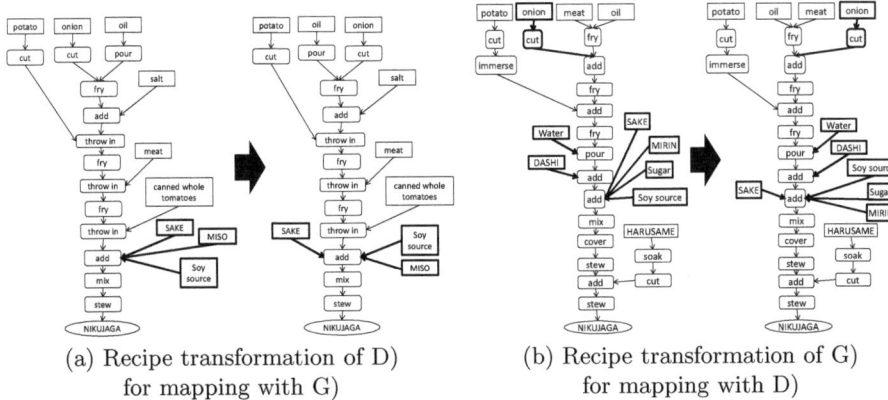

(a) Recipe transformation of D) for mapping with G)

(b) Recipe transformation of G) for mapping with D)

Fig. 3. Results of transforming from unordered to ordered tree

5.3 Node-to-node Mapping

The system calculated the mapping cost of editing distance. In this experiments, we set C_{del_ins}, C_{sub1}, and C_{sub2} as 7, 5, and 10, respectively. The mapping result between recipes D) and G) is shown in Fig. 4. Since it has 12 deletion operations, 5 insertion operations, and 3 substitution operations, the mapping cost was 134.

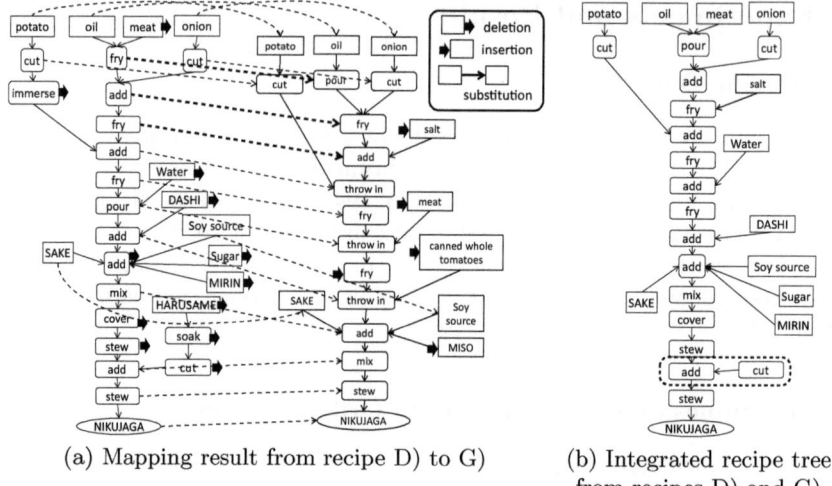

(a) Mapping result from recipe D) to G)   (b) Integrated recipe tree from recipes D) and G)

Fig. 4. Mapping and integration results

Table 3. Edit distances of all combinations of two of ten recipes

	A	B	C	D	E	F	G	H	I	J	Ave.
A	-	148	213	157	166	214	197	211	152	200	184.2
B	148	-	155	150	140	215	152	171	146	183	162.2
C	213	155	-	140	140	230	172	144	155	182	170.1
D	157	150	140	-	140	192	134	162	159	179	157.0
E	166	140	140	140	-	201	143	153	145	182	156.7
F	214	215	230	192	201	-	175	213	234	234	212.0
G	197	152	172	**134**	143	175	-	160	143	204	164.4
H	211	171	144	162	153	213	160	-	204	237	183.9
I	152	146	155	159	145	234	143	204	-	193	170.1
J	200	183	182	179	182	234	204	237	193	-	199.3

The mapping costs of all combinations of pairs of ten recipes are shown in Table 3. The first line on the right of this table shows the average distance from each recipe to the others. In this line, recipe E) gets the closest average distance. This means that recipe E) is the most general of these ten recipes.

5.4 Recipe Tree Integration

Since the pair of lowest cost was the combination of recipes D) and G), the system integrated these recipes and generated one tree, in accordance with the algorithm introduced in Section 4.4. Fig. 4 (b) shows the integrated recipe tree. Because one leaf node is not an ingredient but an action, "cut," the system removed that node and the "add" node for that leaf, as indicated by the dotted circle in Fig. 4 (b). When more than two or more nodes of the same type are

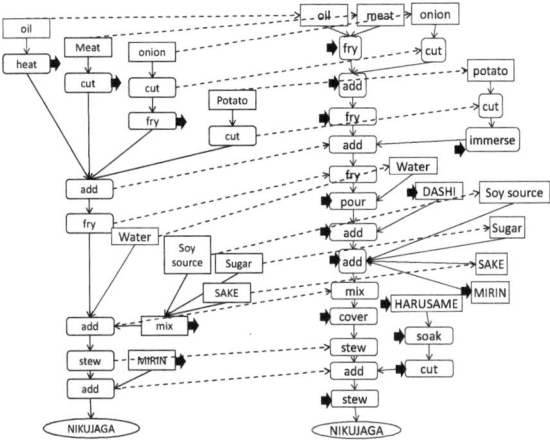

Fig. 5. Mapping result of the general tree (on the left) with recipe D) (on the right)

directly connected and have no other branch, these nodes are combined into one node. In this example, after removing the circled part, two "stew" nodes were directly connected and the system removed one of them.

5.5 General Recipe Tree of Ten Recipes

The system repeated the integration until the ten recipe trees became one. The finally integrated recipe tree is shown on the left side of Fig. 5. The edit distances between the general tree and each recipe tree are shown in Table 4. The average distance of the general tree is 133.7, while the average distance between each recipe with the others is greater than 157, as shown in Table 3. This means that the general recipe is much closer to all of the recipes than any one of them.

Table 4. The edit distances between the general and each recipe tree

	A	B	C	D	E	F	G	H	I	J	Ave.
General	159	63	138	133	87	162	127	131	152	185	**133.7**

5.6 Characteristic Features of Each Recipe

The characteristic features of recipe D) were extracted by mapping it with the general tree. The mapping result is shown in Fig. 5. According to the editing operation of the mapping, the characteristic features of recipe D) can be extracted as follows: (i) "MIRIN is added at the same timing with other seasoners," (ii) "place a small lid directly on the food," (iii) "It uses HARUSAME and DASHI," (iv) "meat is fried without cutting," and (v) "Potato is immersed in water after cutting." (ii), (iii), and (v) were matched with manually extracted features. However such manually extracted features as "use mince," "use sesame oil," and "use a small amount of sugar" could not be extracted because the recipe tree did not include such data. On the other hand, (iv) was extracted incorrectly because the action of cutting meat might just be omitted in the instruction.

We adopted the feature extraction method to all of the ten recipes respectively and obtained 85 features in total. Then we compared them with the manually extracted features that were introduced in Section 2. The number of consistences between the manually extracted features and automatically extracted features, and precisions and recalls of our method are shown in Table 5.

As shown in the table, the features of "Action order", that was evaluated the most important features by the annotators and is difficult to be extracted by the previous researches, were extracted with 12% precision and 60% accuracy. Since 84% of the incorrectly extracted features could also be agreed as "Action order" feature when we recheck the recipe text, one of the reasons for these false positive results could be that the annotators could not find these features. If so, it means that the proposed methods has higher performance than human for extracting such features. The features of "Action type", which is also difficult to be extracted by previous researches, could be extracted with 45% precision and 45% recall. In total, our method succeeded in extracting 54% of manually extracted features while the previous researches address only 37% of them.

We also conducted the experiment on the recipes of "Carbonara" and it achieved precision of 47% with recall of 60%. Speaking about "Action order" and "Uncommon action", two of three features were successfully extracted.

Table 5. Feature type of recipe

Type	Ave. ranking	Manually extracted	Automatically extracted	# of consistence	Precision	Recall
Action order	**1.8**	5 (8%)	25 (29%)	3	12%	60%
Quantity of ingredient	2.1	6 (9%)	0 (0%)	0	-%	0%
Reduced ingredient	2.1	6 (9%)	14 (16%)	6	43%	100%
Action type	**2.4**	11 (17%)	11 (13%)	5	45%	45%
Additional ingredient	2.5	28 (44%)	29 (34%)	17	59%	61%
Writing style	3.7	1 (2%)	0 (0%)	0	-	0%
Cooking tool	3.9	7 (11%)	6 (7%)	3	50%	43%
Total		63	85	34	40%	54%

6 Discussions

Mapping Cost. In the proposed method, a cost of node-to-node mapping is set according to their types so that ingredients or actions of the same broad categories are treated as the same. However, such a difference is sometimes very meaningful for finding characteristic features of a recipe. Moreover, not all actions are equally important. For example, a washing action on a potato is abbreviated very often, because it goes without saying that a potato should be washed. As future work, we will make the mapping cost of insertion, deletion, and substitution operations depend on the name of the ingredient/action, so that the mapping costs between similar types of ingredients or actions will be lower than between dissimilar types.

Integration Weight. The system integrated two trees to one according to their weight. The weight of a tree means the number of trees that are integrated into the tree. However, not all parts of the tree are overlapped in all previous integrations; some parts can be joined at the last integration. As future work, we will give a weight not to the whole of a tree but to each node. The weight of a node should be counted according to how many times the node is overlapped in the previous integrations.

Applications. In this paper, we described a method for generating a general recipe and for extracting characteristic features of a recipe. In a previous study, we used a recipe tree as a scenario of a chef's behavior for recognizing the chef's cooking in a cooking video [11]. Recipe-tree mapping can be used also for recipe rewriting. For example, a simple recipe can be transformed into a detailed recipe, because the system can obtain pairs of sentences in simple and detailed descriptions by tree mapping.

Though we generated a general recipe from the top ten results in this paper, it is possible to address more than ten results, if there are sufficient calculation time, memory, and processors.

7 Conclusions

In this paper, we proposed a method for obtaining a general way to cook from a set of multiple recipes and extracting characteristic features of each recipe. All recipes were converted in advance to recipe trees. This process will be automatized as we stated in section 3.1. The system calculated the edit distance of all combinations of pairs of recipes and integrated the recipes of the closest pair into one tree. In these processes, the system took into account the differences in importance between the action types. As the result, our method succeeded in extracting 54% of manually extracted features while the previous research addressed 37% of them.

Acknowledgments. This work was supported by JSPS KAKENHI Grant Numbers 23700144, 23500177 and 24240030.

References

1. Demaine, E.D., Mozes, S., Rossman, B., Weimann, O.: An optimal decomposition algorithm for tree edit distance. ACM Trans. on Algorithms 6, Article No. 2 (2009)
2. Do, H.-H., Rahm, E.: COMA – a system for flexible combination of schema matching approaches. In: Proc. of the 28th International Conference on Very Large Data Bases, pp. 610–621 (2002)

3. Hamada, R., Ide, I., Sakai, S., Tanaka, H.: Structural analysis of cooking preparation steps. IEICE Trans. J85-D-II, 79–89 (2002) (in Japanese)
4. Karikome, S., Fujii, A.: Improving structural analysis of cooking recipe text. IEICE Technical Report 112(75), DE2012-8, 43–48 (2012) (in Japanese)
5. Mori, S., Sasada, T., Yamakata, Y., Yoshino, K.: A machine learning approach to recipe text processing. In: Cooking with Computers Workshop, pp. 1–6 (2012)
6. Phillips, C., Warnow, T.J.: The asymmetric median tree – a new model for building consensus trees. Discrete Applied Mathematics 71, 311–335 (1996)
7. Sang, E.F.T.K., Meulder, F.D.: Introduction to the CoNLL-2003 shared task: language-independent named entity recognition. In: Proc. of CoNLL 2003, pp. 142–147 (2003)
8. Shasha, D., Wang, J.T.-L., Zhang, K., Shih, F.Y.: Exact and approximate algorithms for unordered tree matching. IEEE Trans. on Systems, Man, and Cybernetics 24, 668–678 (1994)
9. Tai, K.-C.: The tree-to-tree correction problem. Journal of the ACM 26, 422–433 (1979)
10. Tsukuda, K., Nakamura, S., Yamamoto, T., Tanaka, K.: Recommendation of addition and deletion ingredients based on the recipe structure and its stability for exploration of recipes. IEICE Trans. J94-A, 476–487 (2011) (in Japanese)
11. Yamakata, Y., Kakusho, K., Minoh, M.: A method of recipe to cooking video mapping for automated cooking content construction. IEICE Trans. Inf. & Syst. J90-D, 2817–2829 (2007) (in Japanese)
12. Yoshino, K., Mori, S., Kawahara, T.: Predicate argument structure analysis using partially annotated corpora. In: Proc. of the Sixth International Joint Conference on Natural Language Processing (to appear)
13. Zhang, K., Jiang, T.: Some MAX SNP-hard results concerning unordered labeled trees. Information Processing Letters 49, 249–254 (1994)
14. COOKPAD (May 29, 2013), http://cookpad.com/ (in Japanese)

Unsupervised Opinion Targets Expansion and Modification Relation Identification for Microblog Sentiment Analysis

Jenq-Haur Wang and Ting-Wei Ye

National Taipei University of Technology, Taiwan
jhwang@csie.ntut.edu.tw, bad00124@gmail.com

Abstract. Microblog brings challenges to existing researches on sentiment analysis. First, microblog short messages might contain fewer content features. Second, it's difficult to know what users want to express without suitable contexts. On the other hand, people tend to express their opinions in microblog messages, which could be helpful to sentiment analysis. In this paper, we propose a sentiment analysis approach based on opinion target finding and modification relations identification in microblog. First, user comments on specific topics are collected from microblog and preprocessed to reduce noises. Then, opinion targets are expanded by discovering the most frequently co-occurring terms, named entities, and synonyms of the topic. Finally, according to modification relations among part-of-speech (POS) tags, we extract entities or aspects of the entities about which an opinion has been expressed and calculate the overall score of sentiment orientation. In our experiment on 1,000 reviews of 50 movies collected from Twitter, the proposed method can achieve an average accuracy of 84.4% and an average precision of 87.1%, which is better than content similarity with SVM and Naive Bayes. This validates the higher precision in sentiment orientation identification for the proposed approach.

1 Introduction

With the rapid growth of social networking communities, huge amount of user-generated contents are being published everyday. One of the most important characteristics of social Web is that it's very common for users to express their comments and opinions on products, events, or any entities that might interest their friends or fans. Thus, social Web becomes one of the new media from which user opinions can be collected and analyzed. However, there are some challenges for opinion analysis in social Web. First, user postings on social networking sites might be very short texts, especially in microblogs. Content features might be very limited comparing to conventional text documents. Second, given the casual nature of social media, much noise could be generated along with user comments. It's a great challenge to distinguish user opinions from noises in microblogs. Existing methods for opinion analysis in social Web focus on establishing linguistic resources for extracting opinion words with their orientation and strength. Then, short text classification methods are utilized for sentiment orientation identification. To estimate the sentiment orientation and strength of user comments, opinion words are identified by either opinion lexicon or machine learning methods. However, without suitable contexts of opinion targets, it's difficult to distinguish what users really want to express their opinions on. Thus, opinion target extraction has become one of the major tasks in sentiment analysis.

However, there are some problems in existing methods. First, short texts in microblogs might contain few content features. Conventional text classification might not be accurate. Second, simply counting the opinion words and aggregating their orientations might not be enough. Different opinion words might modify different aspects of the entities that might be related or unrelated to the targets that we are interested in. In this paper, we propose a sentiment analysis approach based on opinion targets expansion and modification relation identification in microblogs. First, opinion targets including related entities and aspects are expanded by co-occurrence analysis, named entity recognition, and synonym finding from online reviews. Second, given a short text, opinion words are extracted by matching with opinion lexicons. Then, modification relations between opinion words and expanded targets are identified by Part-of-Speech (POS) tag relations in linguistic rules. Finally, opinion scores for the given topic are accumulated. In our experiments on movie reviews in Twitter, the proposed method can accurately identify opinion targets and modification relations which give an average accuracy of 84.4% and an average precision of 87.1%. This validates the higher precision for estimating opinion orientation in microblog comments. Further study is needed to apply the method in more topic domains.

2   Related Work

Subjectivity detection and sentiment orientation identification are among the most important tasks in the research field of opinion mining and sentiment analysis [11]. For ordinary documents, opinionated sentences are identified and opinion words are usually extracted using sentiment lexicons such as SentiWordNet, or machine learning methods such as Support Vector Machine (SVM) and Naïve Bayes (NB) classifiers with word n-gram features to classify the sentiment orientation of product reviews [11]. Then, given linguistic resources, it's common to estimate the sentiment orientation from the frequency of positive and negative opinion words. In the tasks of opinion summarization or recommender systems, more sophisticated methods try to identify the opinion holders and opinion targets [4]. Thus, aspect-based opinion mining has gained more and more attention [8, 11].

In general, opinion target extraction can be done by supervised [5, 7, 15] or unsupervised [1, 2, 12] methods. In the domain of product reviews, product features are opinion targets or aspects, which are identified by calculating frequent nouns and noun phrases [1]. For class-specific opinion mining on product reviews, Point-wise Mutual Information (PMI) is used to calculate the frequently co-occurring terms [12]. Jakob and Gurevych [2] models opinion target extraction as an unsupervised word sequence labeling problem. To improve the precision of opinion target extraction, syntactic structures are often used to understand the grammatical modification relation between opinion words and their targets. For example, Zhuang et al. [15] utilizes supervised opinion target extraction to obtain dependency relation templates for opinion summary generation. In Kobayashi et al. [5], syntactic patterns are learned to extract aspect-evaluation pairs using various linguistic resources such as dependency trees. Qiu et al.

[13] uses syntactic relations from dependency trees. Sentence parsers are used to parse the text document and find out the syntactic structure. Li et al. [7] presents a semi-supervised shallow semantic parsing approach to extracting opinion targets. They address the problem at parse tree level. Although supervised learning usually achieves better performance [7], domain portability could be the major problem [2].

In the case of social media, sentiment analysis has gained more popularity especially in microblogs since user generated content usually contains rich opinions and sentiment expressions. However, given the short texts in microblogs, content features are lacking and sentences might be more casual without formal linguistic structure. Conventional opinion analysis techniques on documents cannot be directly applied. First, short text classification might not be accurate. Second, it might not be easy to parse the syntactic structure for casual postings in short texts. Simple counting or aggregation [9] of positive and negative opinion terms might be useful in certain topic domains such as political discussions on forums. However, it's not accurate in our experiments on tweets movie reviews. Thus, instead of utilizing the unreliable syntactic structure in microblogs, we focus on the part-of-speech (POS) tags of each term and their relations in a sentence.

3   The Proposed Approach

In our proposed approach, there are four major components: topic-specific target expansion, opinion extraction, modification relation identification, and opinion score estimation, as shown in Fig.1.

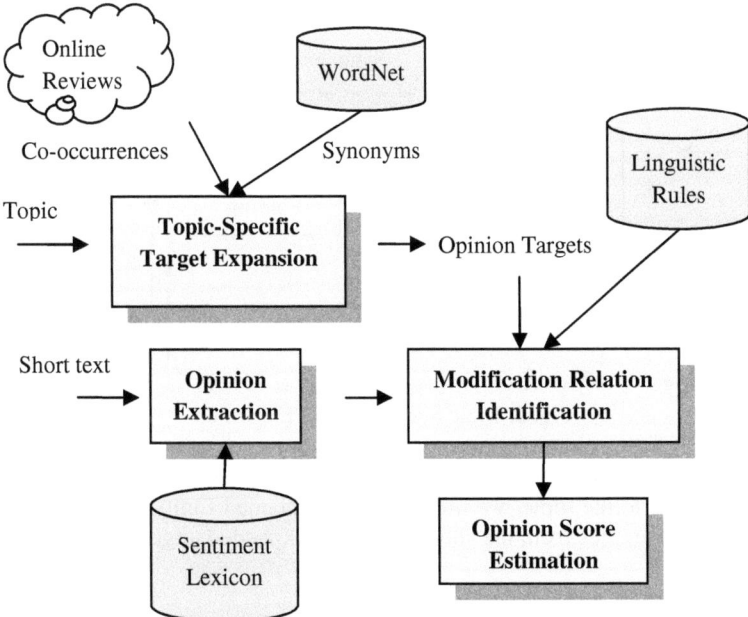

Fig. 1. System architecture of the proposed approach

Given a specific topic, we first utilize online reviews and lexicons for topic-specific target expansion. In particular, frequently co-occurring terms in online review datasets are analyzed using co-occurrence statistics such as mutual information, and their synonyms are extracted with lexicons such as WordNet[1]. These terms form the opinion target candidates, which will be used for further processing.

Next, given a short text review in microblogs, we first perform preprocessing tasks including spell checking by Google Spell Check[2] and stemming by Porter's Stemmer. These can be useful to correct typos in microblogs since users tend to write in a casual way. Then, opinion words are extracted from the short text by matching with sentiment lexicons such as SentiWordNet[3] or NTUSD [6]. To further identify the modification relations among opinion words and candidate targets, we design a set of linguistic rules based on their POS tags. Finally, the overall opinion score of a short text is estimated by a weighted sum of individual scores for each modification relation.

3.1 Topic-Specific Target Expansion

Users might not directly comment on the specific topic in reviews. For example, in the comment: "I watched "Battleship" last night, Rihanna's acting is amazing." The idea of target expansion is to find related aspects or entities that are related to the specific topic in user reviews, as shown in Fig.2.

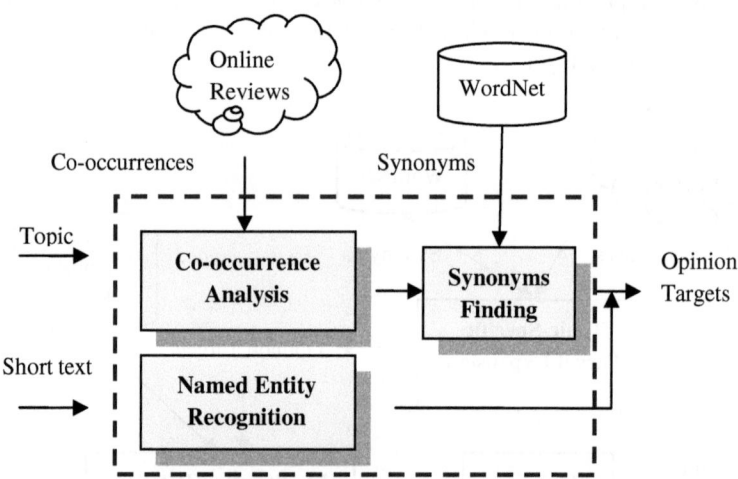

Fig. 2. Topic-Specific Target Expansion

First, since the opinion targets are likely to contain people, events, time, location, or objects related to the topic, we utilize Stanford Named Entity Recognizer[4] (NER) to extract named entities from the short text as the expanded targets.

---

[1] http://wordnet.princeton.edu/
[2] https://code.google.com/p/google-api-spelling-java/
[3] http://sentiwordnet.isti.cnr.it/
[4] http://nlp.stanford.edu/software/CRF-NER.shtml

Second, since different domains might utilize different terms to comment on the various aspects of the target, we include a domain-specific resource when discovering the terms. In the case of sentiment analysis on movies, review datasets such as Movie Review Data[5] is used. By calculating co-occurrence statistics such as Pointwise Mutual Information (PMI), we can obtain a list of terms co-occurring with movies.

$$\text{PMI}(w_1, w_2) = \log_2 \frac{P(w_1, w_2)}{P(w_1)P(w_2)} \quad (1)$$

Finally, since the same opinion on aspects or entities can be expressed in different ways by different people, we further utilize lexical resources such as WordNet to supplement various expressions of the possible expanded targets.

3.2 Modification Relation Identification

After extracting the opinion targets, the next step is to find out the opinions which have modification relations with the opinion targets. In our approach, opinion words are extracted by matching against opinion lexicons such as SentiWordNet for English, and NTUSD [6] for Chinese.

Although microblogs do not have strict linguistic structure, they are still composed of sentences written by common users. Instead of fully understanding each sentence by natural language processing techniques, we only focus on the modification relations between opinion words and the expanded targets in a sentence. Also, comparative sentences and negations are taken into consideration.

POS Tag Relations. First, we adopt common punctuation marks as the delimiter of sentences. Then, in each sentence, we utilize Stanford POS Tagger[6] to find out the POS tags of each word, as defined in Penn Treebank [10]. Next, we design a set of rules $\{r_k\}$ of modification relations between opinion words op_i and any of the expanded targets t_j from the previous step. According to previous study, opinion words are usually verbs and adjectives [4]. Thus, we design the following rules of modification relations in two sets: opinion words as verbs and adjectives.

(1) Verbs: VB/ VBD/ VBG/ VBN/ VBP/ VBZ
$op_i + t_j$: for example, "I love Battleship."
$t_j + op_i$: for example, "The film bored me to death."

(2) Adjectives: JJ
$t_j + V + op_i$: for example, "This movie is worth seeing."
$op_i + t_j$: for example, "It's my favorite movie."

For any pair $p_{ij} = <op_i, t_j>$ matched with the corresponding modification relation r_k, we assign an opinion score $os(p_{ij})$ for the pair p_{ij} with the opinion score of op_i. That is, we assume that the sentiment orientation and score of the target are propagated from its modifying opinion word through the modification relation.

---
[5] http://www.cs.cornell.edu/people/pabo/movie-review-data/
[6] http://nlp.stanford.edu/software/tagger.shtml

When there are more than one rules matched, the rule with the nearest match between opinion words and targets is chosen.

Comparative Sentences. In additional to simple statements, users might also compare the target with other entities using comparative sentences. Since we only focus on sentiment orientation identification, we can determine the sentiment orientation of entities being compared as similar or different according to the following types of comparative sentences [3]:

(1) Non-equal adjectives and adverbs: JJR, RBR

The patterns "$t_j + op_i$" and "$op_i +$ than $+ t_j$" are two possible ways to express comparisons between opinion targets and other entities. The two entities are assumed to have different orientations.

(2) Equal comparisons can be useful to propagate the sentiment orientation. For example, the pattern "as + JJ + as" might be used to assign similar orientations to the entities being compared.

Negations. According to Tottie [14], there are three types of negations: not-negation, no-negation, and affixal negations. Since we utilize SentiWordNet as our opinion lexicon, affixal negations can be simply identified. When we encounter terms in not-negation and no-negation, such as "not", "no", "nothing", "never", to name a few, we simply reverse the sentiment orientation of the opinion word. For example, "I don't like this move, the plot is so boring." The orientation of "like" is reversed to negative by the term "not".

3.3 Opinion Score Estimation

After identifying modification relations, we can obtain one or more pairs of opinion word-target modification relations in a sentence s. The opinion score of sentence s can then be calculated as a weighted sum of the opinion scores for each pair of opinion word-target, that is:

$$\text{score}(s) = \sum_{p_{ij} \in s} \frac{os(p_{ij})}{d(p_{ij})} \qquad (2)$$

where p_{ij} is the pair $< op_i, t_j >$ of opinion words and targets matched within sentence s, and $os(p_{ij})$ is the corresponding opinion score for the pair p_{ij}, and $d(p_{ij})$ is the word distance between op_i and t_j. The idea is that: the longer the word distance between opinion word and target, the lower the corresponding weight in calculating the overall opinion score. The overall score of the short text is then aggregated for all sentences.

4 Experiments and Discussions

In the experiments, we used Twitter as our major source of microblog data. During Feb. – Mar. 2013, we randomly selected 50 movies and collected the corresponding review tweets with Twitter API[7]. To avoid over concentration of user reviews in similar movies, we sampled 200 tweets per day, at an interval of every five days from Mar. 20 – Apr. 10, 2013. Thus, a total of 1,000 tweets were collected as the test data. Then, each tweet was manually labeled by five individuals as positive (+1), negative (-1), or objective (0). In total, there are 468 positive tweets, 231 negative tweets, and 301 objective tweets. In addition to the preprocessing using Google Spell Check and Porter's Stemmer, we remove replies, links, and retweets in a tweet to reduce possible noises.

In the following experiments, we evaluated the performance of subjectivity classification and sentiment orientation identification with the standard metrics: precision, recall, F-measure, and accuracy. For sentiment orientation identification, we reported the macro-averaging precision, recall, and F-measure since the distribution of instances in each class is relatively balanced. We took the baseline as the setting before applying opinion target expansion. Then, the performance of our proposed method is compared with content classification using Naïve Bayes and SVM classifiers.

First, the baseline performances of subjectivity classification and sentiment orientation identification are shown in Table 1.

Table 1. The baseline performances of subjectivity classification and sentiment orientation identification

	Subjectivity Classification Baseline	Sentiment Orientation Identification Baseline
Average Precision	0.892	0.899
Average Recall	0.382	0.398
Average F-measure	0.535	0.552
Accuracy	0.574	0.588

As shown in Table 1, the precisions of subjectivity classification and sentiment orientation identification using our proposed method are high. This is due to the modification rule-based matching. However, the recalls are quite low without opinion target expansion. Thus, we investigate the effects of different components in our design in the following sections.

4.1 The Effects of Named Entity Recognition

First, we compared the performance of subjectivity classification before and after using Stanford Named Entity Recognizer (NER) as shown in Table 2.

---

[7] https://dev.twitter.com/docs/api

Table 2. The performance of subjectivity classification before and after Named Entity Recognition

Subjectivity Classification	Baseline	with NER	Improvement (%)
Precision	0.892	0.907	1.6 %
Recall	0.382	0.403	5.6 %
F-measure	0.535	0.558	4.4 %
Accuracy	0.574	0.596	3.8 %

As shown in Table 2, all metrics were improved with NER. After inspecting the results, we found that the names of directors and actors are extracted as well as other irrelevant named entities. Thus, the improvement in precision is not as prominent as recall.

Table 3. The performance of sentiment orientation identification before and after Named Entity Recognition

Sentiment Orientation Identification	Baseline	With NER	Improvement (%)
Average Precision	0.899	0.912	1.4 %
Average Recall	0.398	0.418	4.8 %
Average F-measure	0.552	0.573	3.7 %
Accuracy	0.588	0.610	3.6 %

Similarly, as shown in Table 3, we found more improvement in recall than precision. This shows the effectiveness of NER in opinion target expansion.

4.2 The Effects of Co-occurrence Analysis

In order to find out the common terms for movie reviews, we utilized the Movie Review Data[8] as our topic-specific resource for co-occurrence analysis. There are 2,000 reviews, including 1,000 positive and 1,000 negative reviews. We checked the effects of co-occurrence analysis using PMI with different parameters of k, the number of highly co-occurring terms to be included.

---

[8] http://www.cs.cornell.edu/people/pabo/movie-review-data/

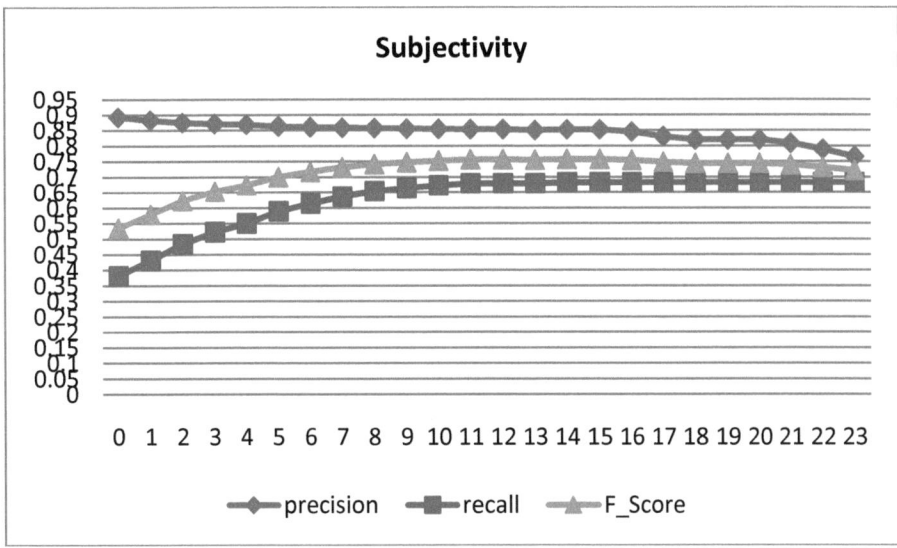

Fig. 3. The effects of *k* on the performance of subjectivity classification

As shown in Fig.3, we can see an increase in recall but decrease in precision when k increases from 1 to around 15. The F-measure reaches the highest when *k*=15.

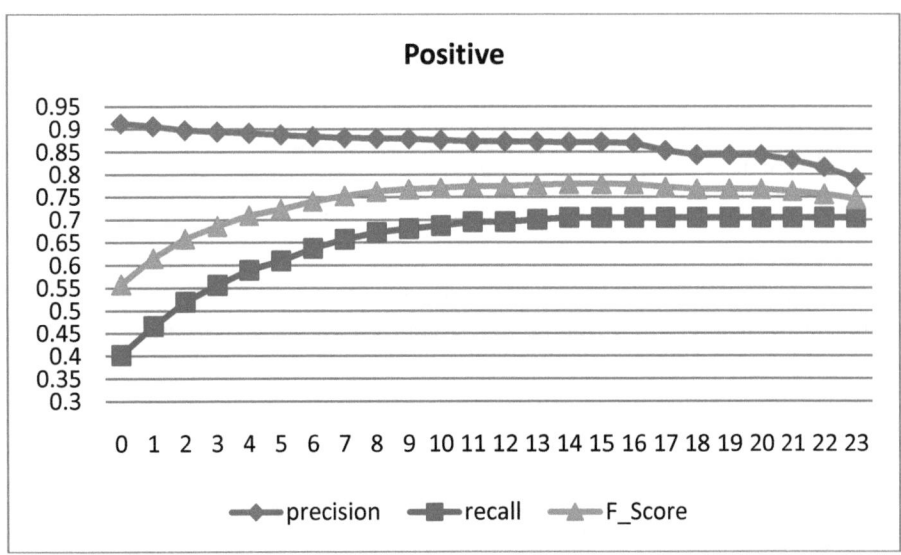

Fig. 4. The effects of *k* on the performance of sentiment orientation identification

As shown in Fig.4, we can see a similar trend in the performance of sentiment orientation identification. The F-measure reaches the highest when *k*=15, after which the performance started to decline.

Table 4. The performance of subjectivity classification by including co-occurrence analysis with $k=15$

Subjectivity Classification	Baseline	$k = 15$	Improvement (%)
Precision	0.892	0.852	-4.4 %
Recall	0.382	0.682	78.6 %
F-measure	0.535	0.758	41.7 %
Accuracy	0.574	0.793	38.1 %

As shown in Table 4, the recall was greatly improved, while the precision was slightly reduced when we include co-occurrence analysis using PMI with $k=15$. The reason for this reduction in precision when we increase k is that: as the number of expanded targets increase, more unrelated terms might be included which happen to co-occur frequently with the topic.

Table 5. The performance of sentiment orientation identification by including co-occurrence analysis with $k=15$

Sentiment Orientation Identification	Baseline	$k = 15$	Improvement (%)
Average Precision	0.899	0.860	-4.3 %
Average Recall	0.398	0.692	73.7 %
Average F-measure	0.552	0.767	38.9 %
Accuracy	0.588	0.827	40.7 %

As shown in Table 5, there's a similar trend in all metrics for the performance of sentiment orientation identification. The overall F-measure and accuracy were both improved by around 40%.

4.3 The Effects of Synonym Finding

Since people might express the same aspect or entity in different terms, we further include WordNet as our major source of synonyms. After including the synonyms as our expanded targets, the performance of subjectivity classification is shown in Table 6.

Table 6. The performance of subjectivity classification by including synonyms of expanded taregts

Subjectivity Classification	PMI (k = 15)	PMI (k = 15) + Synonyms	Improvement (%)
Precision	0.852	0.870	2.0 %
Recall	0.682	0.697	2.2 %
F-measure	0.758	0.774	2.1 %
Accuracy	0.793	0.810	2.0 %

As shown in Table 6, all metrics are slightly improved by including synonyms of our expanded targets.

Table 7. The performance of sentiment orientation identification by including synonyms of expanded targets

Sentiment Orientation Identification	PMI (k = 15)	PMI (k = 15) + Synonyms	Improvement (%)
Average Precision	0.860	0.870	1.2 %
Average Recall	0.692	0.703	1.5 %
Average F-measure	0.767	0.778	1.4 %
Accuracy	0.827	0.839	1.4 %

As shown in Table 7, similar trends can be observed for the performance of sentiment orientation identification when including synonyms of expanded targets. This validates the usefulness of opinion targets expansion using synonyms.

4.4 Comparing with Content Classification

Finally, we put together all components of our proposed approach, and compared its performance with content classifiers such as Naive Bayes (NB) and SVM. From our observation, we can obtain better performance of NB and SVM classifiers when the features of unigrams and bigrams are used.

Table 8. Comparing the performance of subjectivity classification with SVM and NB classifiers

Subjectivity Classification	The Proposed Approach	LibSVM	Naive Bayes
Precision	**0.873**	0.694	0.640
Recall	0.713	0.882	0.908
F-measure	**0.785**	0.776	0.750
Accuracy	**0.823**	0.817	0.789

As shown in Table 8, the proposed approach has better precision than both SVM and NB classifiers, but the recall is worse. The overall F-measure and accuracy of the proposed approach are both slightly better than SVM.

Table 9. Comparing the performance of sentiment orientation identification with SVM and Naive Bayes classifiers

Sentiment Orientation Identification	The Proposed Approach	LibSVM	Naive Bayes Classifier
Average Precision	**0.871**	0.7137	0.659
Average Recall	0.713	0.882	0.849
Average F-measure	0.784	0.788	0.741
Accuracy	**0.844**	0.834	0.812

As shown in Table 9, we can obtain better precision but worse recall than NB and SVM classifiers. The overall accuracy of the proposed approach is better then NB and SVM classifiers. Thus, our unsupervised appraoch can obtain comparable performance to supervised ones with much higher efficiency since it's based on lingusitics resources and modification rules. This verifies the effectiveness of the proposed approach in sentiment orientation identification.

5 Conclusion

In this paper, we have proposed a simple but effective approach to subjectivity classification and sentiment orientation identification for microblogs. By opinion targets expansion, we can effectively include more potential candidate opinion targets for sentiment analysis. The modification relations between opinion targets and opinion words are identified by POS tag relations in linguistic rules. Our experimental results on movie reviews showed an average accuracy of 84.4% and an average precision of

87.1%. This validates the effectiveness of our proposed approach for microblog sentiment analysis. Further investigation is needed to apply the proposed approach in different topic domains.

References

1. Hu, M., Liu, B.: Mining and Summarizing Customer Reviews. In: Proceedings of SIGKDD 2004, pp. 168–177 (2004)
2. Jakob, N., Gurevych, I.: Extracting Opinion Targets in a Single and Cross Domain Setting with Conditional Random Fields. In: Proceedings of EMNLP 2010, pp. 1035–1045 (2010)
3. Jindal, N., Liu, B.: Identifying Comparative Sentences in Text Documents. In: Proceedings of SIGIR 2006 (2006)
4. Kim, S., Hovy, E.: Extracting Opinions, Opinion Holders, and Topics Expressed in Online News Media Text. In: Proceedingsof ACL/COLING Workshop on Sentiment and Subjectivity in Text (2006)
5. Kobayashi, N., Inui, K., Matsumoto, Y.: ExtractingAspect-evaluation and Aspect-of Relationsin Opinion Mining. In: Proceedings of EMNLP 2007, pp. 1065–1074 (2007)
6. Ku, L.W., Chen, H.H.: Mining Opinions from the Web: Beyond Relevance Retrieval. Journal of American Society for Information Science and Technology 58(12), 1838–1850 (2007), Dictionary available at
   http://nlg18.csie.ntu.edu.tw:8080/opinion/index.html
7. Li, S., Wang, R., Zhou, G.: Opinion Target Extraction Using a Shallow Semantic Parsing Framework. In: Proceedings of AAAI 2012, pp. 1671–1677 (2012)
8. Liu, B.: Sentiment Analysis and Opinion Mining. Morgan & Claypool Publishers (2012)
9. Liu, H.C., Wang, J.H.: Aggregating Opinions on Hot Topics from Microblog Responses. In: Hou, Y., Nie, J.-Y., Sun, L., Wang, B., Zhang, P. (eds.) AIRS 2012. LNCS, vol. 7675, pp. 447–456. Springer, Heidelberg (2012)
10. Marcus, M., Santorini, B., Marcinkiewicz, M.A.: Building a Large Annotated Corpus of English: The Penn Treebank. Computational Linguistics 19(2), 313–330 (1993)
11. Pang, B., Lee, L.: Opinion Mining and Sentiment Analysis. Foundations and Trends in Information Retrieval 2(1-2), 1–135 (2008)
12. Popescu, A.M., Etzioni, O.: Extracting Product Features and Opinions from Reviews. In: Proceedings of the Conference on Human Language Technology and Empirical Methods in Natural Language Processing (EMNLP 2005), pp. 339–346 (2005)
13. Qiu, G., Liu, B., Bu, J., Chen, C.: Opinion Word Expansion and Target Extraction through Double Propagation. Computational Linguistics 37(1), 9–27 (2011)
14. Tottie, G.: Negation in English Speech and Writing: A Study in Variation. Language 69(3), 590–593 (1993)
15. Zhuang, L., Jing, F., Zhu, X.Y.: Movie Review Mining and Summarization. In: Proceedings of the 15th ACM International Conference on Information and Knowledge Management, CIKM 2006 (2006)

Pilot Study toward Realizing Social Effect in O2O Commerce Services

Tse-Ming Tsai, Ping-Che Yang, and Wen-Nan Wang

Institute for Information Industry (III)
{eric,maciaclark,wennen}@iii.org.tw

Abstract. Social media has become the most convenient space to retrieve the tremendous consumers' experience, opinion and preference—toward each brands, products or even specific features. The real-time and big amount characteristics of Social media provide great opportunity for producer to know their customers (and potential ones) well. This paper proposes an Online to Offline (O2O) Commerce Service Model and takes the social relationship dashboard as an pilot study, which can help retailers or brands to understand their customers via social network existing data (especially Facebook for this case) by which we can adapt the current social commerce marketing strategy more quickly and responsively.

Keywords: Social Effect, Social Relation Management, O2O.

1 Introduction

Social media and mobile device not only change the way people communicate with friends, but also change the way providers communicate with consumers. We keep posting and reading all the time and place, no matter we are online surfing/blogging or offline walking/shopping. The ubiquity of people getting connected dramatically changes the landscape of post Electronic Commerce.

1.1 Offline to Online

With the growth of local commerce on the Web, the linkage between online and physical commerce are becoming stronger. Alex Rampell, explored the forces behind what he called "Online2Offline" commerce. O2O means finding consumers online and bringing them into real-world stores [1]. Users can visit real store and also get virtual service online. For example, users can purchase products online, and get the products or service at the nearest real store. John Doerr created another buzz word, "SoLoMo", which stands for Social, Location and Mobile. As the mobile platform hits critical mass in these years, social networking accelerates the growth of mobile device. Real-time social features accelerating mobile usage growth from sharing, likes, tweets, friending and so on [2]. Retailer can collect social word-of-mouth by creating social events, as to explore valuable users and promote products or services to social customers, through referrals and guiding users to real-world store. Location based service is accessible with mobile devices through internet, which enables the users to find useful context based on the geographical position of the mobile device. For example, users can use their mobile device to find coupons or discounts on online stores or social networks, and find the nearest store based on their current location. No matter offline to online or online to offline, the core value of O2O is to provide a precise consuming experience.

1.2 Social Commerce

We further focus from O2O to social commerce, which is a prevailing commerce type nowadays. In [3] explains how social influence can be used by E-commerce websites to aid the user decision-making process, which indicates the importance of social commerce, and in [4] it proposed an a three-stage system architecture to visually display opinions from social networks for customers' decision-making. However, when business operators push information onto social networks to sell items, they first encounter marketing issue, and in [5] emphasizes the importance of using Facebook fan page for marketing. Being such a case in [6], to fully understand how and when to post contents in order to attract Facebook users becomes a new research topic. In [7], and [8], based on [9] we have developed a social networking-based service platform to real-time monitor social networks events, and in this paper we integrate some of its components into our purposed system, aiming at helping business operators to have full comprehension on their and competitors' fan pages, including posting time, user preference and user activity tracking.

2 System Architecture

This paper proposes the O2O commerce service model representing in Figure 1. The top area represents the real-world marketing service model. Manufactures produce and sent their products to channels or retailers. According the location based service and proximity commerce marketing strategy, users can use their mobile devices to interactive with OOH (out-of-home) digital signage or kiosk to get online coupon then shopping in the nearby real world store.

The bottom area represents the online marketing service model. In tradition, manufacture is used to apply E-commerce marketing strategy. With explosion of the social network users, we find out users would like to survey others' opinions online before they make purchasing decision. Whatever users like the brand, they are very willing to participate the events founded by brands on Social Network, such as Facebook fans group. They will click likes, shares, check-in, and comments in any social events.

There are more and more brands and retailers have created their own fans groups. But not everyone knows how to create topics, interact with fans and attract them to the real-world store. So it becomes a hot topic to create online to offline service model through social networks. In Figure 1, manufacture and brands can apply social network technologies, such as semantic analysis, user preferences analysis, social relationships, social events spreading, referral strategies, Return on Investment and competitors' performance monitoring.

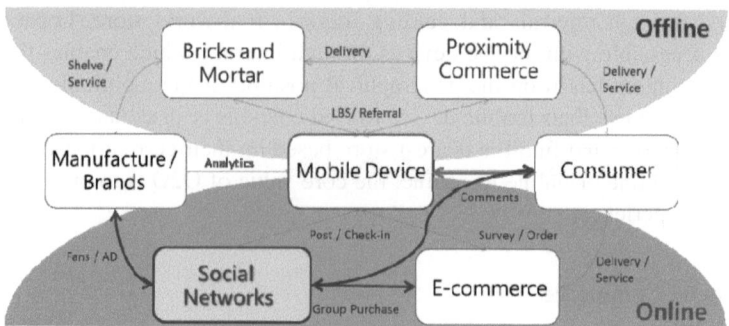

Fig. 1. O2O Commerce Service Model

This pilot study focuses on realizing the social effect such as (1) when the competitors create social events, (2) how the social events are spread and (3) what the fans are interested in. In Figure 2, we construct a social relationship dashboard. The bottom layer is the social crawler and scalable distributed DB allocated in Amazon EC2. So that whenever a social event is created on Facebook fans group, the social crawlers will collect any user interaction through Facebook SDK with licensed access token and store in the scalable distributed DB.

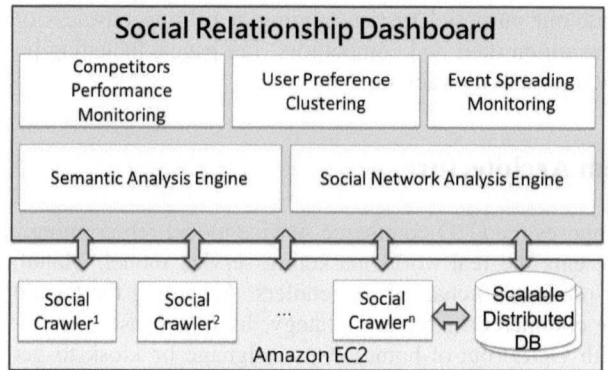

Fig. 2. System Architecture

The middle layer of Figure2 is Semantic Analysis Engine and Social Network Analysis Engine. The Semantic Analysis Engine can detect the positive and negative opinions comments of users' posts. The Social Network Analysis Engine can analyze the social event by date and hours and provide social relationship information.

The top layer of Figure 2 is Competitors Performance Monitoring, User Preference Clustering and Event Spreading Monitoring. Competitors Performance Monitoring model can help fans group manager to know the interactions between posts and fans replies. The fans group manager can know when the best time to post on Facebook wall is. And what kind of posts will possible be popular posts and get most users' replies. User Preference Clustering can cluster the users' preferences into the interesting domains defined in Facebook. Event Spreading Monitoring can monitor the spread of each target social event by feedbacks (replies) and time and create a spread dashboard so that the fans group manager can know who are the level 1 spreaders, level 2 spreaders and the important spreaders.

3 Pilot Demonstration

3.1 Competitors Performance Monitoring

We take two major retail companies, Company A and Company B in Taiwan as example. These two retail companies are leading companies in the convenient store business, and they are competitors to each other, and they are struggling for popularity on social networks, especially on Facebook fan page. Currently, Company B is second place to Company A on social networks. To assist Company B with Social Relationship Dashboard, searching administrator operation behavior difference on the fan page of social networks is the first thing that we focus on. Figure 3 shows the sum of posted numbers of each time period (24 hours) by Company A and Company B. In April, Company B submitted 3 posts and Company A submitted 14 posts from 11:00 to 12:00 within a month. Company B submitted 6 posts and Company A submitted 18 posts from 18:00 to 19:00 within a month. On the contrary, Company B paid more attention from 8:00 to 9:00 and from 12:00 to 13:00, whereas Company A didn't.

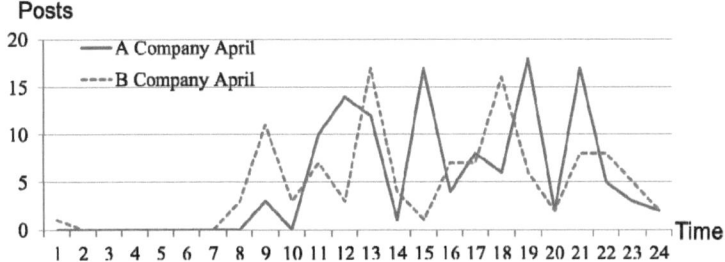

Fig. 3. The sum of posted numbers of each time period (24 hours) by Company A and Company B in April

Next, the second step is to observe the result of such posting operations, namely, the comparison of fans' replies. In Figure 4, Company B received 87 fans' replies, from 11:00 to 12:00, while Company A received 1140 posts; Company B received 102 fans' replies, from 18:00 to 19:00, and Company A received 579 posts. However, the time period that Company B paid much attention on did not receive the expected higher replies but a comparatively fewer replies. It is obvious that different the posting time period and numbers of the day may result in different replies on fan page.

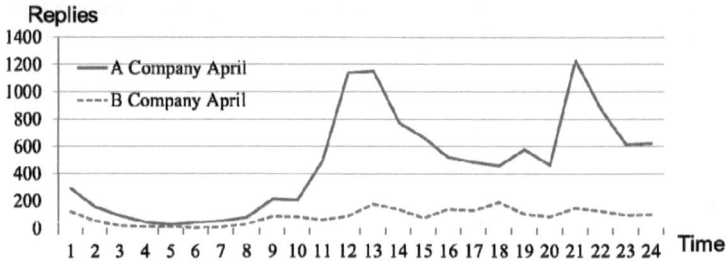

Fig. 4. The sum of reply numbers of each time period (24 hours) by Company A and Company B in April

As the result shown above, we suppose a deduction that much rapid raising of fans' replies would be presented if the post were posted at the right time period. Company B made the wrong posting time strategies. The administrator focus on posting at the period from 8:00 to 9:00, which is the commuting time for most Facebook users, most users are office workers according to market analysis in Taiwan. Company B made another posting peak from 12:00 to 13:00, lunch time for most of workers, which results in fewer replies on fan page. Company B focused posting time period on from 19:00 to 21:00, but Company A focused posting time period on from 11:00 to 13:00 and from 20:00 to 23:00, for majority of Facebook users, the former time period is the time before lunch, and the latter one is the time users finish their work, which are leisure time for using social networks. From the analysis, the system helped Company B changes their posting time period next month, and empirically raise fans replies better, as shown in Figure 5.

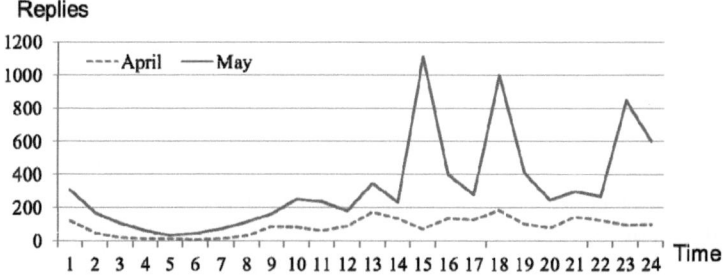

Fig. 5. The sum of reply numbers of each time period (24 hours) by Company B in April

4 Conclusion

Social media has become the most convenient space to retrieve the tremendous consumers' experience, opinion and preference—toward each brands, products or even specific features. The real-time and big amount characteristics of Social media provide great opportunity for producer to know their customers (and potential ones) well. Entering the O2O era we are looking forward every chance and occasion to get in touch the consumers, no matter they are online or offline. This work demonstrates to understand customers via social network existing data (especially Facebook for this case) by which we can adapt the current marketing strategy more quickly and responsively. Next, we will continue to explore the possibility that how social media influence the other commerce activity, such as mobile, cooperative and location-based commerce transaction.

Acknowledgement. This study is conducted under the "Digital Convergence Service Open Platform" of the Institute for Information Industry which is subsidized by the Ministry of Economy Affairs of the Republic of China.

References

1. TechCrunch, Why Online2Offline Commerce Is A Trillion Dollar Opportunity, http://techcrunch.com/2010/08/07/why-online2offline-commerce-is-a-trillion-dollar-opportunity/
2. Doerr, J.: Top Mobile Internet Trends (2011), http://www.slideshare.net/kleinerperkins/kpcb-top-10-mobile-trends-feb-2011
3. Kim, Y., Jaideep, S.: Impact of Social Influence in E-commerce Decision Making. In: Proceedings of the Ninth International Conference on Electronic Commerce. ACM (2007)
4. Chen, L., Qi, L.: Social Opinion Mining for Supporting Buyers' Complex Decision Making: Exploratory User Study and Algorithm Comparison. J. SNAM. 1(4), 301–320 (2011)
5. Lin, K.Y., Lu, H.P.: Intention to Continue using Facebook Fan Pages from the Perspective of Social Capital Theory. Cyberpsychology, Behavior, and Social Networking 14(10), 565–570 (2011)
6. Bushelow, E.E.: Facebook Pages and Benefits to Brands. Elon Journal of Undergraduate Research in Communications 3(2), 5–17 (2012)
7. Hsieh, W.T., et al.: Social Event Radar: A Bilingual Context Mining and Sentiment Analysis Summarization System. In: Proceedings of the ACL 2012 System Demonstrations. Association for Computational Linguistics (2012)
8. Lin, Y.-C., et al.: Technology Trend Analysis Tool using Twitter as a Source. In: Proceedings of International Conference on Information Technology, E-Government and Application (2012)
9. Chou, S.C., et al.: Semantic web technologies for context-aware museum tour guide applications. In: 19th International Conference on Advanced Information Networking and Applications, AINA 2005, vol. 2. IEEE (2005)

The Estimation of aNobii Users' Reading Diversity Using Book Co-ownership Data: A Social Analytical Approach

Muh-Chyun Tang, Yi-Ling Ke, and Yi-Jin Sie

Department of Library and Information Science, National Taiwan University, Taipei, Taiwan
{mctang,r98126021,r98126003}@ntu.edu.tw

Abstract. Usage data available through social media provides a great many opportunities to capture users' preference. Using books saved in users' online bookshelves, the study set out to explore social network analytical methods to capture the diversity of a reader's reading interests. "Reading diversity" denotes how widely scattered one's reading interests are. Drawing data from aNobii, a social networking site for booklovers, users' reading diversity was defined by the number of components created by the book co-ownership network of the books in their bookshelves. Five book-book similarity measures were proposed and their clustering results were tested against users' self-assessed reading diversity in order to identify the best suited similarity measure and threshold for such a task. One of the proposed similar measures produce a clustering results that is significantly correlated with users' self-assessed diversity. Furthermore, a multiple regression analysis showed that the proposed measure was able to provide explanatory power for reading diversity over and above mere counting the number of books in the bookshelf.

Keywords: social network analysis, preference structure, book co-ownership network.

1 Introduction

1.1 Personalization and Users' Preference Structure

With the growing visibility of recommender systems in e-commerce sites, researchers have started looking into individual traits that might influence consumer response to personalized recommendations [1]. For example, it was found that individuals of different cultural orientations responded differently to different recommendation methods. Individuals with higher individualistic or independent tendencies respond more favorably to personalized recommendations, compared to targeted recommendations[2]. Customers' cognitive differences have been shown to influence recommender success[3]. Individuals with higher adaptive-innovation style and involvement level toward recommendation agents were found to give more deliberation to the recommendation agent's advice. In Liang, Lai, and Ku[4], an experimental personalized news service was shown to perform better than a non-personalized one in terms of both prediction accuracy and user satisfaction. More noticeably, their study went beyond simple comparison of system performance to include several contextual variables such as degree of user participation and motivations for information access that might affect the effectiveness of the system. It was also found that the effects of personalized news service on user satisfaction were moderated by different motivations for information access.

We suspected that readers with diverse reading interests might be more open to non-obvious recommendations. One of the consistent issues in the design and evaluation of recommender systems is the balance between accuracy and "non-obviousness" of recommendations[5]. While the recommendation of accurate, yet previously known items might foster user confidence, their presence represents less novelty and serendipity. The proper balance between the two might very much lie in how willing an individual is to try novel recommendations. We therefore wish to use the concept of preference diversity to represent how willing a reader is to venture into previously unfamiliar reading interests. For example, more adventurous readers might prefer novelty and serendipity over accuracy and vice versa.

A novel concept of users' "preference diversity" was therefore proposed in this study that aimed to represent how narrow/wide or diverse/convergent one's reading interests are. By "preference diversity" we meant to represent the range of reading interests of an individual, which, as shown below, would be represent by the number of clusters in the book co-ownership network composed of the books saved in an individual's bookshelf. Individuals with heterogeneous preference are those who have diverse reading interests and less confined to a certain genre or types of readings. Individuals with highly homogenous preference, on the other hand, are those who, when choosing books to read, do not divert much from their favorite genres or author. One can imagine how one's preference diversity might influence his/her reactions to recommendations. We speculated that people with diverse reading interests are more willing to venture out of their familiar genres and favorite authors and therefore more receptive to novelty in recommendation.

1.2 Representation of Users' Reading Diversity with Online Bookshelf Data

With users' previous usage or purchase data, marketers can now analyze their preference for better personalized recommendation. Data available on social network sites for readers such as aNobii provides a great opportunity to study different aspects of users' reading preference. This study set out to explore different social network analytical approaches to representing individual aNobii user's reading diversity. A user's bookshelf can be represented as a book-book similarity network based on book co-ownership frequency data. The number of components or clusters in such a book-book network can then be used as an indicator of the user's reading diversity. We assumed that the more components or meaningful clusters in a bookshelf, the more diverse the user's reading interest is.

Furthermore, to determine the number of meaningful groupings in one's bookshelf, a proper threshold is needed to dichotomize the originally numeric similarity measures. As each bookshelf includes only a small portion of the whole book-book network, we believe that the proper threshold should be determined by the global network, instead of each individual bookshelf, that includes all the books surveyed. The process involves two empirical issues: one is to choose the most appropriate similarity measures for the global network; the other is to determine the proper threshold for each of the corresponding similarity measure. In the following section, we will explain the five similarity measures tested and how the proper threshold was determined.

2 Methodology

2.1 Research Procedures

A total of 50 aNobii users were recruited, all met the criteria of having at least 5 friends and 80 books, which was done as to ensure sufficient data for analysis. With their consent, their bookshelf data were downloaded for the analysis and representation of their reading preference. They were also asked to fill out a questionnaire aimed at capturing three dimensions of their reading preference: preference insight, preference diversity, and reading involvement. Our analysis in this study focused mainly on the diversity dimension. The user self-perceived reading diversity would then be used as the benchmark against which different similarity/threshold combination could be tested.

2.2 Book Coownership Network

Data of the books from the 50 bookshelves were collected, including title, author, as well as the ID of the list of aNobii users who also owned each of the books. With the book co-ownership data, a variety of book similarity measures were then generated based on which book-book networks could be created. One of the research questions of this study was to identify proper similarity measures and determine their proper critical threshold of this global network that included all books collected. The purpose of finding the threshold was to dichotomizing the originally weighted similarity data. The threshold was then applied to each individual bookshelf for the purpose of determining its number of components, which was then used as an indicator of its diversity, with the assumption that the more components or clusters in a users' bookshelf, the more diverse his/her reading interest is.

2.3 Similarity Measures and Their Cut-Off Thresholds

Two groups of similarity measures were tested: one used the owners as the feature vector to represent each books so book pair similarity can be calculated by cosine and correlation measures, the other group of measures defines the similarity of two books by the normalized frequency of their co-occurring in the same bookshelves. Three book co-occurrence based measure were tested in the study:

1. Jaccard coefficient

$$\frac{A \cap B}{A \cup B} \quad (1)$$

Where A, B denote the set of owners of a certain book.

2. Normalized intersection: The second measure involves normalizing the intersection by the smaller of the two sets:

$$\frac{A \cap B}{Min(A,B)} \quad (2)$$

3. "Intersection minus 1": The third measure is a modification of the second measure by taking account of the situation where a book is owned by very few aNobii users.

$$\frac{A \cap B - 1}{Min(A,B) - 1} \quad (3)$$

Normalized intersection by the smaller of the two sets allows us to rescale the similarity values into the range between 0 and 1. However, it also caused a complication where two very different books might end up having a perfect score of 1 simply by their co-occurring in a single bookshelf. To rectify the randomness induced anomaly, a variation was created where both the denominator and numerator in normalized intersection (Intersection) was subtracted by 1. See Fig. 1 for a graphic representation of the procedures.

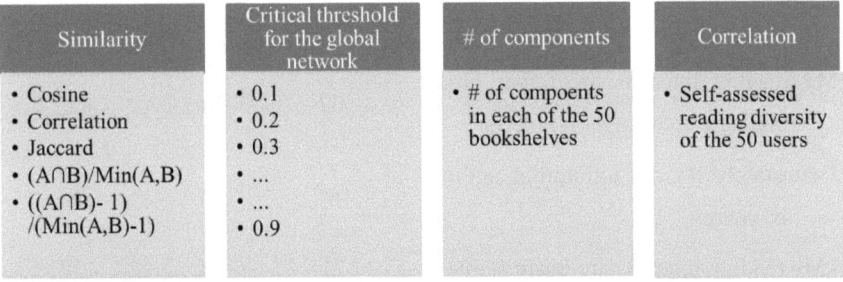

Fig. 1. The process of determining the proper similarity measure

3 Results

3.1 Descriptive Statistics

The number of books in the 50 bookshelves ranged from 3780 to 80, with a medium of 336.5. The data of a total of 21199 books was collected, 970 of which, or 4%, were owned by only one user. The most popular book was owned by 8776 users and the medium number of owners per book is 895.

Table 1 shows the text of the question items in the preference questionnaire and their factor analysis of the factor loading along the three preference dimensions (for a detailed explanation of the creation of the questionnaire, see Tang[6]).

Table 1. Factor analysis of user reading preference structure

Question	Preference Factors		
	Insight	Diversity	Involvement
I have little difficulty judging whether I would enjoy a previously unread work	.90	-.09	-.00
I have a fairly good idea about what I want to read	.87	-.00	.17
I know where to find books that might be of interest to me	.63	.23	.45
I have trusted book alerting sources, which I follow faithfully	.59	-.21	.14
I trust my own judgment of books and am not easily swayed by others	.44	-.23	.29
*I rarely venture out of the authors or genres that I have enjoyed	-.19	.81	-.00
My reading interests are rather broad and hard to be pigeon-holed	-.09	.81	.12
I constantly try out unfamiliar authors or genres	.16	.75	.28
*My reading interests are fairly stable	-.21	.68	-.27
I keep a habit of reading, even when I am busy	.17	.18	.82
I keep monitoring new publications for interesting things to read	.13	-.24	.79
How important would you say reading is to you?	.17	.16	.74

*the scores were reversed in accordance to the semantics of the factor

Fig. 2 shows the distribution of the 50 participants along the dimensions of preference diversity and involvement.

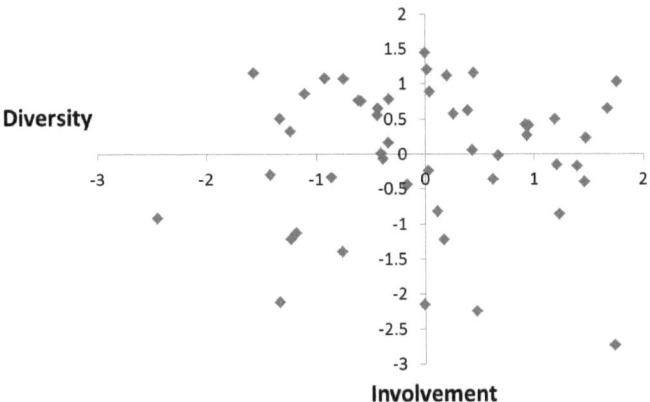

Fig. 2. Dimensions of preference diversity and involvement

3.2 Determine the Proper Threshold for the Similarity Measures

Our approach to capture a user's reading diversity from bookshelf data involves identifying meaningful groupings in an individual's bookshelf, the number of which might be used as an indicator of his/her reading diversity. The question is, then, whether there is a threshold or criterion appropriate for each similarity measure for delineating groups of books. In his study of co-citation clusters, Small[7] demonstrated the existence of percolation transition marked by a sudden expansion of relative cluster size and the emergence of a giant component with a small decrease in similarity. In Small[7], the critical threshold was defined as the value of the similarity (in his case, cosine similarity) just above the level at which the largest percentage increase in size is observed, comparing the size of the cluster at the lower similarity value to its size at the next higher level. We followed Small's approach, but in a reverse manner. In Small, [7] the task is to identify the point at which a specific scientific are dissolve into the broader community of science as a whole (i.e. the giant component), our task, on the other hand, was to start with a giant component where all the nodes are connected, then, by gradually raising the threshold, to weed out the spurious linkages. Therefore we observed the relative decrease, instead of increase as in Small's case[7], in linkages as the threshold was gradually raised. The critical threshold was defined at the level above the greatest relative decrease of linkages occurred. See Table 2 for the relative decrease of links in cosine similarity network when the threshold gradually increased, in which case, 0.7 was identified as the critical threshold as the decrease of threshold from 0.8 to 0.8 caused the largest relative decrease of linkages. The same critical threshold finding procedure was applied to all the five similarity measures. Fig. 3 shows how the size of the giant components and the linkages decreased with different levels of threshold in all five similarity measures.

Table 2. The composition of the book-book cosine similarity network at different thresholds

Threshold	# node Giant Component	# total link	ave_link	Density	# Giant Component decrease	% Giant Component decrease
0	21202	224751801	21201.00	1		
0.1	21122	7299165	688.54	0.032477	80	0.38%
0.2	16901	2797888	263.93	0.012449	4221	19.98%
0.3	10862	1402086	132.26	0.006238	6039	35.73%
0.4	7351	854556	80.61	0.003802	3511	32.32%
0.5	4851	560315	52.85	0.002493	2500	34.01%
0.6	2206	266740	25.16	0.001187	2645	54.52%
0.7	1761	256696	24.21	0.001142	445	20.17%
0.8	369	102958	9.71	0.000458	1392	79.05%
0.9	369	95265	8.99	0.000424	0	0.00%
1	369	89512	8.44	0.000398	0	0.00%

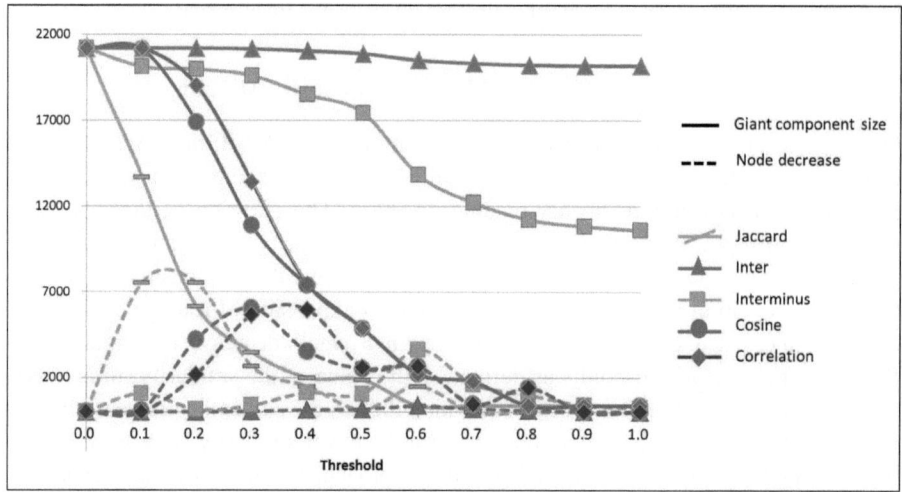

Fig. 3. The size of the giant component and the linkages with different levels of threshold

3.3 Correlation between Number of Components and Self-assessed Reading Diversity

The aforementioned critical thresholds were then applied to the five respective similarity based individual book-book networks so that the numbers of components in the 50 individual bookshelves could be obtained. The numbers of components in the bookshelves were then correlated with the self-assessed diversity measure. Table 3 shows the results of the correlation analyses. Only the components produced by InterMinus had a minor significant correlation with the users' self-assessed diversity. Notice that almost all other measures had a high correlation with the number of the books in the bookshelves, which seems to suggest the failure of these measures in weeding out spurious linkages.

Table 3. Correlations between diversity and the clustering results from the similarity measures

	Jaccard	Inter	InterMinus	Cosine	Correlation
Diversity	.037	-.190	.286*	.052	.051
# of books	.978**	-.132	.559**	.984**	.985**

*p<.05; **p<.01

A correlation analysis was also conducted to test whether the results from the InterMinus measure had any significant correlation with all the three preference dimensions.

Table 4. Correlation between results from InterMinus measure and the preference dimensions

	Preference Insight	Preference Diversity	Involvement
InterMinus	-.032	.286*	.035

*p<.05; **, p<.01

A multiple regression analysis was conducted with users' self-assessed reading diversity as dependent variable, with the number of book and the number of components generated by the clustering results from the InterMinus measure as predictors. The results showed that, even though the number of books has a stronger correlation (.328, p<.05) with reading diversity than the number of components (.286, p<.05), using stepwise procedure, the number of components was chosen as the sole significant predictor (β = .286, t = 2.07, p<.05) over the number of books. The result indicted that the number of components was a better predictor of users' reading diversity than mere counting the number of books in a bookshelf.

4 Discussion

In this study we attempted to explore the idea of using the users' book-book co-ownership network as the profile of their reading interests. Specifically, it was proposed to use the number of meaningful groupings in one's bookshelf to represent his/her reading diversity. Following Small[7], our method involves finding the critical threshold for a variety of similarity measures at the global level, which was then applied at the individual bookshelf level. The only similarity that showed promise of such a task was InterMinsu, which seeks to rectify the anomaly of unduly high similarity simply because of their rareness. Our results showed that the InerMinus measure, combined with a proper critical level, was able to produced meaningful grouping in the individual bookshelves somewhat correlated with the user self-assessed reading diversity. Furthermore, the resulting number of components was able to provide more predictive power for reading diversity than the number of books in the bookshelf.

It was also found that Jaccard, cosine and correlation coefficients all produced the same number of nodes in the giant components of 369 even when the critical threshold was set at 1, which indicated a total agreement of co-owner patterns of these books. It would be interesting to look into these books to examine to what degree those high similarity values were caused by their rarity in the dataset, which might help create proper ways of adjusting book rarity for these measures. Of the five measures, the normalized interaction produced the poorest results (see Fig. 3) as the size of the giant components remained constantly large even at the critical threshold of 1, which showed that the randomness in the book co-ownership, i.e. books co-occurred because of user idiosyncrasy rather than similarity, created some sort of weak ties that is akin to those in the social network that decreased the number of the steps needed to link seemingly unrelated books. The results indicate the challenge to our efforts lies mostly in how to filter out linkages created out of randomness or user idiosyncrasy, which calls for further research for the proper clustering of book co-ownership network.

5 Conclusion

With the growing prevalence of recommender systems for the access of cultural products, researchers have started looking into cognitive traits that might influence users' response to recommended items. We were particularly interested in the notion of reading diversity as its potential influence on users' response to personalized recommendations. In this study we set out to explore social network analytical methods that might capture users' reading diversity. While network theoretic methods have been widely used in peer-to-peer network to identify similar peers for recommendation purpose [8], [9] , to the best of our knowledge, they have not been applied to items within an individual user's bookshelf for the purpose of representing his/her preference structure. In this study five different similarity measures and a simple clustering method, components identification, were used, which is admittedly a rather basic network clustering method. More studies can be done to address the issue of randomness of co-ownership of books and also to explore more sophisticate clustering method for such a task. Admittedly the value of our research of users' reading diversity depends very much on the assumption that it would influence one's response to recommendations. Individuals with more concentrated reading preference, on the other hand, might appreciate familiar genres or authors, though recommendations might be less effective for those readers as they are more likely to be knowledgeable about those genres in the first place and thus less likely to be impressed by what recommender systems have to offer. Future studies should also been done, to test our assumption that reading diversity might influence the user's response to different recommendation strategies.

References

1. Simonson, I.: Determinants of customers' responses to customized offers: Conceptual framework and presearch propositions. J. Mark., 32–45 (2005)
2. Kramer, T., Spolter-Weisfeld, S., Thakkar, M.: The Effect of Cultural Orientation on Consumer Responses to Personalization. Mark. Sci. 26(2), 246–258 (2007)

3. Wang, H.-C., Doong, H.-S.: Online customers' cognitive differences and their impact on the success of recommendation agents. Inf. Manage. 47(2), 109–114 (2010)
4. Liang, T.-P., Lai, H.-J., Ku, Y.-C.: Personalized Content Recommendation and User Satisfaction: Theoretical Synthesis and Empirical Findings. J. Manage. Inf. Syst. 23(3), 45–70 (2007)
5. Herlocker, J., Konstan, J., Terveen, L., Riedl, J.: Evaluating collaborative filtering recommender systems. ACM Trans. Inf. Syst. 22(1), 5–53 (2004)
6. Tang, M.-C., Sie, Y.-J., Ting, P.-H.: Exploring the mediating effect of reader preference on the effectiveness of social navigational tools on social media: A case study of aNobii. Inf. Process. Manag. (in press)
7. Small, H.: Critical thresholds for co-citation clusters and emergence of the giant component. J. Inf. 3(4), 332–340 (2009)
8. Anglade, A., Tiemann, M., Vignoli, F.: Complex-network theoretic clustering for identifying groups of similar listeners in p2p systems. In: Proceedings of the 2007 ACM Conference on Recommender Systems, New York, NY, USA, pp. 41–48 (2007)
9. Perugini, S., Gonçalves, M.A., Fox, E.A.: Recommender Systems Research: A Connection-Centric Survey. J. Intell. Inf. Syst. 23(2), 107–143 (2004)

An Ontology-Based Technique for Online Profile Resolution[*]

Keith Cortis, Simon Scerri, Ismael Rivera, and Siegfried Handschuh

Digital Enterprise Research Institute, National University of Ireland, Galway, Ireland
firstname.surname@deri.org

Abstract. Instance matching targets the extraction, integration and matching of instances referring to the same real-world entity. In this paper we present a weighted ontology-based user profile resolution technique which targets the discovery of multiple online profiles that refer to the same person identity. The elaborate technique takes into account profile similarities at both the syntactic and semantic levels, employing text analytics on top of open data knowledge to improve its performance. A two-staged evaluation of the technique performs various experiments to determine the best out of alternative approaches. These results are then considered in an improved algorithm, which is evaluated by real users, based on their real social network data. Here, a profile matching precision rate of 0.816 is obtained. The presented Social Semantic Web technique has a number of useful applications, such as detection of untrusted known persons behind anonymous profiles, and information sharing management across multiple social networks.

Keywords: profile resolution, person identity, named entity recognition, semantic relatedness, social networks, linked open data, social semantic web.

1 Introduction

Instance matching is the process of finding whether two or more instances refer to the same real-world entity (e.g. persons, products, organisations, events) within a particular domain. It targets the problem of extracting, integrating and matching such instances to a unique representation. For example, two profiles pointing to the same person, may have different profile attributes, such as addresses, name variants, etc. These profiles consist of structured and semi-structured/unstructured data, acquired from heterogeneous sources, such as social networks, and are modelled on different schemas depending on the source. Several profile matching techniques, such as image recognition, syntactic, semantic, statistical, linguistic and network-based measures exist, all of which are built for solving specific problems according to their target domain.

The main motivation behind this work arises from the di.me[1] EU project, which amongst others targets the integration of duplicated personal information found across multiple local and remote sources. Such a profile matcher can be universally applied for a variety of applications within any target domain, such as:

---

[*] This work is supported by the European Commission under the Seventh Framework Program FP7/2007-2013 (*digital.me* – ICT-257787)
[1] http://www.dime-project.eu/

1. better management of one's personal data in order to control its sharing both on and across multiple social networks [16];
2. the detection of fully or partly-anonymous contacts to minimise privacy threats [4];
3. suggest the establishment of new contacts that might be of direct interest to the user, based on several factors, such as common interests, locations and activities.

Other applications target identity resolution (as in our case), anonymous identity detection, fraud detection, and terrorist screening, to mention a few.

In this paper we extend an earlier technique that we proposed in [6], which targets the discovery of duplicate person profiles on LinkedIn, Twitter and Facebook. The choice of these social networks was based on a survey[2] we conducted. Results outlined that the most popular social networks amongst the people were Facebook, LinkedIn and Twitter, with 83.1% having at least one profile with the first two sites, whilst 64.6% have at least one profile with the third. We present an improved profile matching system, extended with text analytics, linked open data (knowledge) and semantic matching. Its nature provides additional benefit since we use an ontology as a background schema. Therefore, it can easily be extended to support additional social networks, since we map all the profile attributes as retrieved from heterogeneous social networks to it, in order to have a standard schema that enables cross-network interoperability.

The main objectives of this paper are:

1. The extension of a profile matching algorithm with Text Analytics on semi-structured/unstructured profile information and the use of Linked Open Data (LOD) to improve the Named Entity Recognition (NER) process;
2. A Semantic-based matching extension to find any possible semantic relations between incomplete and unstructured profile attributes;
3. Various experiments to evaluate the profile resolution technique and its usability.

In the first objective we extend a state-of-the-art information extraction system for performing NER on some profile attributes that are semi-structured/unstructured. We also make use of open public knowledge in the LOD cloud[3] to enhance the NER process, since it is made up of several sources from multiple domains, and contains rich knowledge regarding all entities [18]. The second contribution consists of a semantic-based matching extension to the profile matching technique, where we try to find any semantic relations between profile attributes that are not of same type but are closely-related (e.g. city vs. country). To demonstrate such concept, we apply this technique on the address/location information, since this can be retrieved from all targeted social network APIs. Both local and open public knowledge is used for the matching process. The NER process will also impact highly on this process, since any address sub-entities discovered will increase the probability of finding any known semantic relations.

The last and major contribution is the presentation of the setup and results of a two-staged evaluation. In the first stage, we evaluate the matching technique against a controlled dataset to establish the best attribute similarity score alternative out of the ones we propose. We also find out whether the enhanced hybrid profile matching technique improves the F-measure results and determine its computational performance. An additional technique evaluation over a set of real online profiles is performed to establish a

---
[2] http://smile.deri.ie/online-profile-management
[3] http://lod-cloud.net/

threshold for determining if two online profiles represent the same person or not, with an acceptable degree of confidence. The second stage is a usability evaluation where users were invited to run the profile matcher over their social networks.

The rest of the paper is organised as follows. In Section 2 we discuss and compare related work. In-depth details of our Profile Resolution technique is provided within Section 3, whereas in Section 4 we present and discuss the results of our Technique and Usability Evaluations. Concluding remarks and future work are outlined in Section 5.

2 Related Work

Instance matching (also known as entity resolution, record linkage, data de-duplication in other domains) is all about calculating the degree of similarity between pairs of instances within heterogeneous knowledge sources, and finding whether they refer to the same real world entity, despite having different descriptions [5]. This process is required for several applications like identity recognition, data integration [11] and data cleaning [10]. Several matching techniques exist, such as learning, similarity, rule and context-based. We use a hybrid similarity-based approach made up of syntactic and semantic methods, and enhanced through NER. The advantage of using this method is that it is capable of comparing and matching poorly-/un-structured data e.g., plain text. Instances that do not conform to the same background schema can have different property values, structures and formats [3], [11], [10], [2]. It is important that the data matching process is able to analyse the data structure and value, since both can be heterogeneous [10]. In our context, heterogeneous social networks have different schemas resulting in different user profile structures [16], [10]. We overcome structural heterogeneity by standardising all the profile information across a comprehensive conceptualisation, such that different persons will have the same representation structure, and their attributes the same data value formats. This makes our system interoperable, since any social network can be integrated, given that their schema is mapped to our background ontology. Evaluation of instance matching algorithms is calculated through parameters, such as precision, recall, F-measure and execution time [11], [1]. We perform a complete technique evaluation, such that multiple parameters are calculated as obtained by our system. One shortcoming of such algorithms is also addressed; that is, the identification of an adequate threshold which distinguishes between matching and non-matching instances [5], [10], [15]. This is overcome in our application through the identification of a threshold with an acceptable degree of confidence, for suggesting possible duplicate contacts to the user, who will then make the final decision of whether they match or not.

Semantic relatedness is based on the type of relationship between concepts e.g. 'bicycle' and 'pedal' are more closely related than 'bicycle' and 'car', even though 'bicycle' and 'car' are more similar. Therefore, the *semantic similarity* is a special case of relatedness [8]. There is not a lot of previous work on the computation of semantic relatedness between named entities, which task is really important for discovering the semantic associations between documents [18]. In [14], entity linking is defined as the linking of textual entities possibly extracted through NER from unstructured text, with their equivalent real world entity in a knowledge base (e.g. Wikipedia, WordNet). This process is beneficial for several information extraction and linguistic analysis applications [14], [18]. The approach adopted in LODDO [18] uses LOD for overcoming several limitations of existing knowledge bases attributed to entity coverage. In our NER process, LOD is used to improve our entity linking (profile resolution) process. Moreover, our semantic-based matching approach is different in nature than [17] and [18]. Even though we make use of LOD, we try to find any existing semantic relations between different address sub-entities through a dataset that is part of the LOD cloud.

The profile matching technique in [15], considered only a few profile attributes. If several attributes are considered, the chance of wrong user identification between two people is lowered, even if some personal data is missing or incomplete [16]. The similarity approach used was based on string matching only, however Veldman outlines that this is not enough to determine a match between certain complex attributes, such as an address. The authors in [16], also use string matching functions for attribute comparisons. In contrast to [15], a recursive vector approach is applied for calculating the similarity for attributes containing complex types. Similar to our work, [2] and [15] merge multiple occurrences of the same person into one representation for a more comprehensive view of all the data. The techniques in [15], [16] and [2], were evaluated over a controlled dataset. However, no usability evaluation was conducted in either work. The user profile matching technique in [13], computes the semantic relatedness between semantic-based attributes (e.g. topic/user interest), which cannot always be matched syntactically. The approach is different than ours, since they aim to find the relationships between same type entities, through the Explicit Semantic Analysis method.

3 Ontology-Driven Profile Resolution Technique

The profile resolution technique presented in this section extends earlier work that we proposed in [6]. We will provide some formal notations along the way to help with understanding the technique better.

Our profile matching pipeline is made up of six successive steps (1-6), as outlined by Figure 1. A practical example of the matching input, process and results is also included. In the first (1) step, profiles for the users themselves, and for each of their contacts, is extracted from each targeted social network (LinkedIn/Twitter/Facebook) through its respective API. Semantic Lifting (step 2) is then performed to lift profile information from the remote schema and transform it into instances of the Contact Ontology (NCO)[4]. This ontology represents generic contact/profile information, and is used by our technique as a standard format that enables the profiles to be matched. Therefore, the retrieved profile information from each service API is declaratively leveraged through XSPARQL[5], for transforming it to a Resource Description Framework (RDF)[6] representation. Hence, a person's online profile p has a number of personal attributes a, such that:

$$p = \{a_1, a_2, \ldots, a_m\} \qquad (1)$$

where $m \in \mathbb{N}$, m is a number of defining profile attributes derived from the NCO ontology. The attributes are derived from NCO either directly (e.g., email address from

---
[4] http://www.semanticdesktop.org/ontologies/nco/
[5] http://xsparql.deri.org/
[6] http://www.w3.org/RDF/

nco:hasEmailAddress, phone number from *nco:hasPhoneNumber*), or from a corresponding set of attributes (e.g., current address from *nco:streetAddress*, *nco:locality*, *nco:country*, etc.). Once the transformation is complete, Text Analytics (step 3) is performed on some profile attributes that have complex unstructured information, e.g., entire postal addresses. The method will be explained in Section 3.1. Once the NCO profile generation is complete, a retrieved profile is then matched (step 4) against all previously-retrieved profiles. The matching is performed on the profile attributes at both syntactic (step 4-a) and semantic (step 4-b) levels. Therefore, a generic similarity function between two profiles p_i and p_j can be defined as follows:

$$sim(p_i, p_j) = \frac{\sum_{n=1 \in \mathbb{N}}^{o} sim(a_i^n, a_j^n)}{o} \quad (2)$$

where a_i^n and a_j^n are attributes of the same type e.g. first name, for p_i and p_j respectively, a is a member of the common attributes between p_i and p_j and $o \in \mathbb{N}$, $o \leq m$. That is, the similarity of two profiles p_i and p_j is computed by averaging the similarity of o attributes that are common to both profiles.

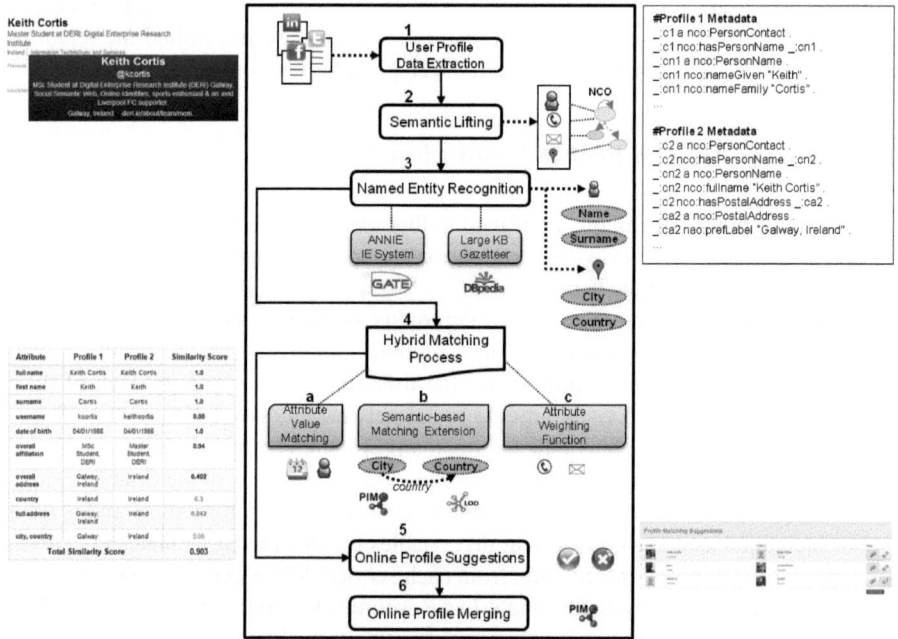

Fig. 1. Profile Resolution Technique Pipeline

Attribute value (syntactic) matching (step 4-a) is applied directly between each attribute of the new profile, and the corresponding attribute in all known profiles. Value matching varies based on the attribute type. A direct value comparison is conducted on the ones that have a strict predefined structure (e.g. date of birth or gender), whereas string matching is performed on attributes having a string value (e.g. first name). In this case, we apply the recursive field matching algorithm proposed by Monge and Elkan

[12]. In addition, rather than applying a blanket algorithm to all attribute types, in [7] we explain how various experiments were carried out to obtain an optimal Monge-Elkan variant for each type[7]. The novel semantic matching extension (step 4-b), considers the semantic relations in the underlying ontologies to discover indirect similarities (e.g. between a city of residence in one profile, and a residential region in the other, even when the country of residence is missing in both), as explained in Section 3.2. The final step in the profile resolution technique (step 4-c) refers to the application of a suitable attribute weighting function, in order to skew the results of the similarity function towards attributes which are considered to be more unique in determining a person's identity. The chosen weighting strategy is explained in Section 3.3.

Results of the profile matching technique yield the similarity score, $sim(p_i, p_j)$, in step 5. Potential systems adopting our technique are not expected to automatically merge profiles with a high score. Instead, they intend to display those profiles matched with a suitable degree of confidence to the system's user, as suggestions. For this purpose, we have established a threshold such that only profiles with a similarity score beyond it would be shown (refer to Section 4.1). The number of profile comparisons performed by the technique for each user is $a \times b$, with a and b being the number of contact profiles obtained from two respective social networks. Some personal information about the suggested matches (e.g. photo, name, affiliation) is displayed alongside them, in order to help the user make a decision. If two profiles are merged (step 6), the link is stored at the Personal Information Model (PIM) level, which serves as the user's personal repository. This will enable the applications introduced beforehand, such as easier management and sharing of personal data across multiple social networks.

3.1 Text Analytics

Named Entity Recognition (NER) is performed on two specific profile attributes; full person names and addresses, in order to decompose them further, where possible. This technique is here useful to cater for limitations of some social network APIs (in terms of retrievable profile data) and enable a more exact comparison. For this purpose, we chose the open-source General Architecture for Text Engineering (GATE) platform as the Language Engineering tool, as it could be extended for our task's requirements. The Nearly-New IE System (ANNIE) [9], a built-in Information Extraction (IE) component set within GATE, was used as the NER technique to perform sub-entity discovery and resolution. This component is made up of several main processing resources for common Text Analytics tasks, such as a tokeniser, sentence splitter, Part-of-speech (POS) tagger, gazetteer, finite state named entity transducer (semantic tagger), orthomatcher and coreference resolver. Specifically, attribute decomposition and sub-attribute resolution are enabled by applying the following two GATE processing resources (PRs): i) the *ANNIE semantic tagger*, which is responsible for processing the outputs of any extracted entities (i.e. name and address). For the purpose, this PR was extended through Java Annotation Patterns Engine (JAPE) rules[8], which are based on

---

[7] Results can be found at:
http://smile.deri.ie/syntactic-matching-experiments
[8] http://gate.ac.uk/sale/tao/splitch8.html#x12-2060008

regular expressions.; and ii) the *Large Knowledge Base (LKB) Gazetteer*[9], which is able to retrieve local and public knowledge from open repositories, based on RDF.

The extended JAPE rules in ANNIE target the decomposition of a person's full name into five possible sub-entities: a prefix and a suffix, and first name, middle name and surname; and full addresses into: a street, city, country, region, postcode, extended address and P.O. box. Since the default ANNIE gazetteers are quite limited for the address sub-entity resolution task, we extend them through the LKB Gazetteer, by extracting additional city and country values from the open public knowledge in DBPedia[10]. This remote dataset forms part of the LOD cloud, and contains over 573,000 locations, 387,000 of which are populated places. DBPedia was chosen since it is automatically updated based on Wikipedia, i.e., it is based on global community knowledge that is quite reliable and covers a lot of domains. Due to the nature of the retrieved sub-entities, the LKB Gazetteer extension has a particular positive impact on the semantic-based matching extension, described in Section 3.2 below.

The full address attribute is composite, since it comprises more than two sub-attributes (similar to the full person name) as mentioned above. Given that we are employing NER to decompose it into further sub-entities, each atomic sub-attribute is assigned a predefined weight according to the importance of the address sub-entities. Predefined weights were assigned as follows: city, country: *0.30*; region/province: *0.20*; full address, street, postcode: *0.05*; extended address, p.o. box: *0.025*. The importance of the address sub-entities was based on the entities that are most commonly specified across all targeted social networks. Hence, we extend Equality 2 as follows:

$$sim(addr_i, addr_j) = \sum_{n=1 \in \mathbb{N}}^{q} sim(s_i^n, s_j^n) * stype_w(s_i^n) \qquad (3)$$

where $addr_i$ and $addr_j$ are address attributes for p_i and p_j respectively, $q \in \mathbb{N}$ is the number of sub-attribute comparisons; s_i^n and s_j^n are same type address sub-entities (s) e.g. country, for $addr_i$ and $addr_j$ respectively, such that $\forall s \in a$; and $stype_w(s_i^n)$ is a predefined weight of the sub-entity type as listed above. The total of all sub-attribute predefined weights in one comparison add up to 1, where the similarity score of each sub-attribute pair obtained from the string matching metric or semantic-based matching extension is represented as a proportion of the respective sub-entity weight.

3.2 Semantic-Based Matching Extension

The semantic-based matching extends the attribute value matching process, so that profile attributes that do not have a 1-1 relationship but are related regardless, will also be taken into consideration. The said relationships are derived from semantic relations defined within standard ontologies. Hence, this extension is one of the main advantages for using ontologies, rather than flat database schemas, to represent person profiles, since entities can be linked at a semantic level, rather than just a syntactic or logical level. In our profile matching task, this extension is mostly relevant to the location-related profile attributes, whereby two profiles that have a different address attribute can still be

---
[9] http://gate.ac.uk/sale/tao/splitch13.html#x18-34600013.9
[10] http://dbpedia.org/

compared, with the help of the ANNIE extensions and LKB gazetteer described in the previous section. For example, rather than achieving a low string-only based similarity score for the address attribute 'Dublin' in profile p_i against 'Ireland' in profile p_j, the resolution of sub-entity 'Dublin' as a DBPedia city and 'Ireland' as a DBPedia country, and a discovery of semantic relationship in between (e.g., dbpedia-owl:country), will achieve a higher similarity score for the address. Therefore, the address attribute similarity between the two profiles in Equality 3 is in this case extended as follows:

$$sim(addr_i, addr_j) = \left(\frac{\sum_{n=1 \in \mathbb{N}}^{q} sim(s_i^n, k_j^n)}{q}\right) * stype_w(s_i^n)$$

such that:

$$sim(s_i^n, k_j^n) = \frac{rel(s_i^n, k_j^n) + rel(k_j^n, s_i^n)}{2} \quad (4)$$

where s_i^n and k_j^n are address sub-entities of a different type such that $s_i^n \neq k_j^n$, $q \in \mathbb{N}$ is the number of distinguishable semantic matching comparisons; $stype_w(s_i^n)$ is a predefined weight for the sub-entity type (s_i^n) being matched (specified in Section 3.1); and $rel()$ is a function that computes the semantic similarity between two sub-entities.

In the current implementation, the city, region and country[11] address sub-entities are being semantically compared. As opposed to string-based matching, this extension attempts to find the similarity based on the semantic relation between the sub-attributes in question in a bi-directional manner, according to the ontology relationships listed in Table 1. The Personal Information Model Ontology (PIMO)[12] and DBPedia Ontology[13] define several location relationships. In all, three PIMO and five DBPedia properties were chosen, being the ones mostly effective thus are bound to yield a semantic relation.

Table 1 also shows the arbitrary values of these semantic relationships, based on observations of their use in both closed and open data repositories. Thus, for example, if we want to find any semantic relationship that may exist between the city *Dublin* (s_i) and the country *Ireland* (k_j) entities, each of which is separately used as the address in the two profiles p_i and p_j, the semantic-based extension will look for any relations in a bi-directional manner. If all the listed DBPedia (*country, isPartOf, largestCity, capital*) and PIMO (*locatedWithin, containsLocation, isLocationOf*) relationships exist between the mentioned entities, the semantic similarity $sim(s_i, k_j)$ will total 0.30. Hence, the address semantic similarity $sim(addr_i, addr_j)$ equals 0.09. The number of available relations between two entities, will determine the strength of their semantic relationship.

The ontology relationships shown in Table 1 are not relevant to all combinations of address sub-entities. For example, given the definition of its RDF domain and range, it is never expected for a country to be defined as a capital of a city, but the other-way round. Similarly, the capital-of relationship is neither expected to be used to relate a region to a country. Therefore, to optimise the semantic search extension, we identified which properties to look for when matching a sub-entity of a certain type to a sub-entity of another type. Table 2, which must be read using a row(domain)-column(range) manner,

---

[11] These are the ones that are most commonly specified across all targeted social networks.
[12] http://www.semanticdesktop.org/ontologies/2007/11/01/pimo/
[13] http://dbpedia.org/Ontology

Table 1. Ontology Properties for Semantic Matching

#	DBPedia	Weight	#	PIMO	Weight
1.	largestCity*	0.10	6.	locatedWithin*	0.10
2.	capital*	0.10	7.	containsLocation*	0.10
3.	country*	0.10	8.	isLocationOf	0.05
4.	isPartOf	0.05			
5.	region	0.05			

Table 2. Semantic relations between the location sub-entities

Domain	Range		
	City	Region/Province	Country
City		4, 5, 6	3, 4, 6
Region/Province	1, 2, 3, 4, 7, 8		3, 6
Country	1, 2, 7, 8	7, 8	

lists the relevant relationships that we check for (e.g. the possible semantic relations between a city and a region/province, are: *isPartOf*, *region* and *locatedWithin*).

The semantic-based matching process was implemented as follows. A SPARQL[14] query is executed against the user's PIM and the public DBPedia SPARQL endpoint, in order to find out if any relations between two entities, as listed above, exists. In cases when for some reason (e.g. maintenance) the DBPedia endpoint to which we submit this query is unavailable, this process is executed against the user's local knowledge base (in our case the PIM) alone given that some information is available, otherwise the profile matching technique reverts to the attribute value-based technique only.

3.3 Ontology-Based Attribute Weighting

The attribute weighting function is enhanced by the cardinality and Inverse Functional Properties (IFPs) attribute constraints, as defined in the NCO ontology. Hence, different predefined weights are assigned to the profile attributes according to the attribute constraints as specified in Table 3. For obvious reasons, IFPs are the attributes assigned the highest additional weights (0.5). Attributes that have a maximum or exact cardinality of one also carry a higher additional weight (0.25), whereas all other attribute-types are assigned no additional weight at all. Thus, we extend Equality 2 as follows:

$$sim(p_i, p_j) = \frac{\sum_{n=1 \in \mathbb{N}}^{o} \left(\frac{sim(a_i^n, a_j^n) + type_w(a_i^n)}{1 + type_w(a_i^n)} \right)}{o} \quad (5)$$

where $type_w(a_i^n)$ is the additional attribute-type weight, as shown in Table 3.

Table 3. List of Profile Attributes

Constraint	Type	Additional Weight
IFP	email, phone number, instant messaging ID	0.5
Cardinality One	gender, first name, surname, full name, birth date, full address, street, extended address, city, country, region, postcode, po box, organisation, role start date, role end date	0.25
Other	username, additional name, name prefix, name suffix, url, profile url, overall address, overall affiliation, role	0

After computing the above weights, two different approaches can be used to calculate the final similarity score of an attribute pair, $sim(a_i^n, a_j^n)$. The first approach (Approach 1) applies the obtained similarity score directly as shown in Equality 5. In

---
[14] http://www.w3.org/TR/rdf-sparql-query/

contrast, the second approach (Approach 2) applies a normalised similarity score based on a threshold obtained for each attribute type, as determined from several experiments carried out in our earlier work [7]. In this case, the threshold acts as a function over all $sim(a_i^n, a_j^n)$ scores in Equality 5, such that:

$$\text{if } sim(a_i^n, a_j^n) \geq threshold(a_i^n), \text{ then } sim(a_i^n, a_j^n) = 1,$$
$$\text{else, } sim(a_i^n, a_j^n) = 0. \qquad (6)$$

where $threshold(a_i^n)$ refers to the established attribute-type thresholds. Thus, in the second approach, if the obtained attribute similarity score is above the corresponding attribute-type threshold, it is rounded up to 1, elsewhere it is reduced to 0. Both approaches will be considered in the first stage of our evaluation, as discussed below.

4 Experiments and Evaluation

In this Section we discuss several experiments conducted within a two-staged evaluation that sought to evaluate the implemented technique and its usability both quantitatively and qualitatively. In particular, we determine: 1) which attribute similarity approach out of the two proposed (Section 3.3) fares best; 2) if the text analytics and semantic-based matching extension improve the overall technique; 3) the computational performance of the hybrid technique against the syntactic-based one; 4) a final threshold that determines profile equivalence within a satisfactory degree of confidence; and 5) the precision level for the profile matching as derived from the usability evaluation.

4.1 Technique Evaluation

In evaluating the technique[15], we performed several experiments based on two datasets: a controlled dataset of public profiles obtained from the Web, and private personal and contact-list profiles obtained from five consenting participants. The controlled dataset was purposely selected in order to make the profile matching task more difficult, with the intent of exposing its limitations. This dataset was used to tackle points (1-3) above. The real 5-person dataset was used to determine the threshold in point (4) above.

Controlled Dataset Experiments. The controlled dataset was selected manually by means of an online profile extractor. Out of the resulting 182 online profiles, 70 referred to 35 real-world people (being well-known sports journalists having public accounts on both LinkedIn and Twitter). The other 112 were used as 'control' for this test and were purposely ambiguous; referring to different real-world persons that coincidentally had a number of common attributes. The collection of these ambiguous profiles was based on an info-graphic report provided by LinkedIn[16], which listed the most common eight member first names (male and female) obtained from four different countries. Therefore, the profile extractor was implemented to retrieve profiles of people that have the same first name and surname, and lived in the same country. In addition, the extractor also gave precedence to people that resided in the same location area, and had a

---

[15] Details and full results at: http://smile.deri.ie/technique-evaluation
[16] http://www.linkedin.com/200million?trk=200fb

similar current job (role and/or affiliation). The retrieved ambiguous profiles were then manually filtered to confirm that they indeed referred to different persons.

This dataset was first used to observe which of the two attribute similarity approaches performs best. We executed the profile resolution technique with both alternatives in two separate tests, to find out the F-measure for five thresholds: from 0.7 till 0.9, with an interval of 0.05. Results for each approach are shown in Fig. 2. As clearly visible from these results, Approach 1 outperforms Approach 2. The F-measures themselves, which never exceed a high of 0.479 (Approach 1), are not considered to be indicative of the profile matching technique per se, due to the purposely-selected ambiguous profiles. Instead, the point of this exercise was purely to determine which of the two similarity approaches to employ in the rest of the evaluation, as well as in the final implementation. For this experiment, we performed 8631 online profile pair comparisons.

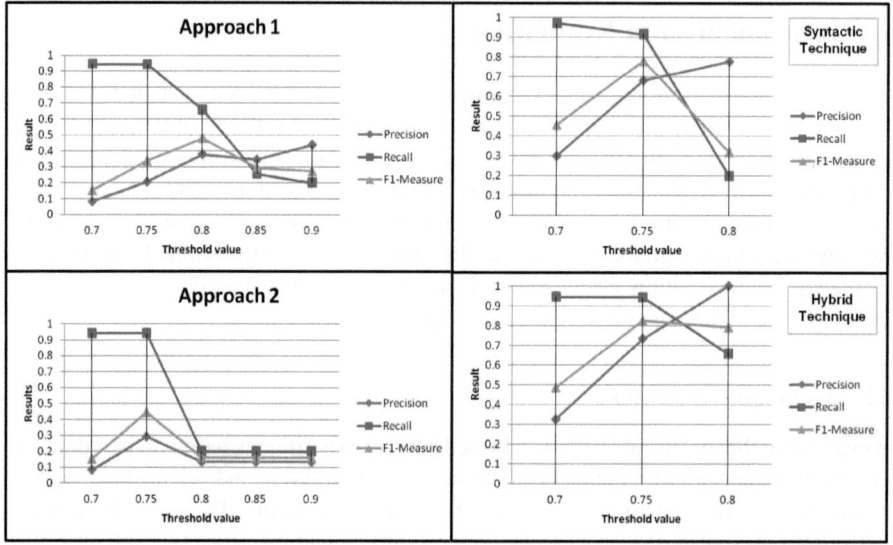

Fig. 2. Attribute Similarity Approaches   **Fig. 3.** Syntactic vs. Hybrid Technique

In a second experiment, we compared the syntactic-based technique introduced in [7] versus the hybrid technique presented in this paper, to identify whether the text analytics and semantic-based extensions improve the similarity scores. Here, three thresholds were considered: 0.7 till 0.8 (interval of 0.05). These were narrowed down from the five in the previous experiment, since the best F-measures (for the selected Approach 1) were obtained at a threshold of 0.8. Results (Fig. 3) confirm that the hybrid technique improves the results considerably over the syntactic only-based one. Particularly, they show how the F-measure remains more or less stable for thresholds of 0.75 and 0.8.

A benchmark was performed for the third experiment[17] to determine the computational performance of the above mentioned techniques, and how both perform over the sparsity of each matched profile pair (with 1 to 15 common attributes). For this test we

---

[17] Details of setup and discussion of results at:
http://smile.deri.ie/performance-evaluation

selected i) several profiles having a number of common attributes and ii) at least 1 attribute candidate for semantic matching. Results in Fig. 4 indicate that both techniques perform within a reasonable time frame. On average the hybrid technique takes ≈15ms more when compared to the syntactic-only based one. The significant leap in time at 15 common attributes is due to the profiles compared having two address sub-attributes to which semantic matching applies i.e. country vs. city and country vs. region.

5-Person Dataset Experiments. The sole objective behind this experiment is to find a deterministic threshold that can be used to generally state whether two compared profiles constitute a person match, or otherwise, with a high degree of confidence. The profile matching algorithm was first executed over each of the five user's own, and their contact's profiles, as retrieved from two social networks in Facebook, LinkedIn and Twitter, based on their highest activity levels in that particular network. They were then required to manually mark all those contacts that should match, to enable the identification of true / false positives and false negatives, and consequently the calculation of the F-measure. This was calculated for 9 different thresholds: from 0.8 (based on the best F-measure in the previous experiments) till 0.96, with an interval of 0.02. The results in Fig. 5, suggest that the optimal threshold is 0.9, yielding an F-measure of 0.693.

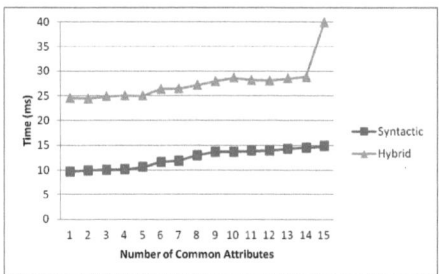

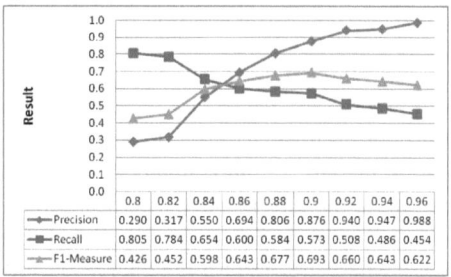

Fig. 4. Technique Performance Results   **Fig. 5.** Threshold Experiment Results

4.2 Usability Evaluation

A user evaluation was conducted to test the performance of our final profile matching technique, and gauge its acceptance by potential users. Interested participants were invited to run the profile matcher, against the two social networks that they are most active in[18]. The participants, totalling 16, were later presented with a profile merging suggestion page, where they could either link (correct match) or dismiss (incorrect match) the results. They were also asked to answer a short survey about their user experience.

Overall statistics and results for this experiment outline that 8415 distinct profiles were retrieved from the two preferred social networks of each of the 16 participants (an average of 262 profiles per social network per participant). In total, 1,041,279 profile comparisons (an average of 65,080 per participant) were conducted, where our matching technique identified 1,195 as positive matches. Out of these, the participants marked 975 as correct, and 220 as incorrect. This yields a satisfactory precision rate of *0.816*.

---

[18] Details of complete setup and results at:
http://smile.deri.ie/usability-evaluation

The user-survey sought to qualitatively gauge the user's level of satisfaction with the technique through the results. A first question asked users how satisfied they were with the result[19], with 50% reporting being *extremely* satisfied and 43.8% being satisfied *quite a bit*. We also asked participants whether they ever interacted with any profile matching technique. Here, 62.5% claim to have used the Android matched contact suggestion feature. When asked to compare the latter to our feature, the participants commented that the usage experience is similar or better, with no-one saying it fares worse. In addition, we asked participants to share their views about three selected applications of our technique: i) whether they agree that it can enable better control of personal data sharing across multiple social networks (Application 1); ii) whether it would make them feel more secure about displaying public personal information to fully or partly-anonymous contacts, knowing that it would warn them if it detects that they are in fact known (and possibly untrusted) persons (Application 2); and iii) whether it could be useful to suggest the establishment of new contacts, based on the identification of shared locations, activities, or interests (Application 3). The encouraging results for each question are discussed as follows. For Application 1, 31.3% *extremely*, 50% *quite a bit*, and 18.8% *moderately* agree that, a reliable contact matcher would enable such functionality. Secondly, 37.5% *extremely*, 37.5% *quite a bit*, and 18.8% *moderately* agree on Application 2. Third and last, 31.3% *extremely*, 50% *quite a bit* and 6.3% *moderately* agree that a reliable contact matcher could also be used for Application 3.

Some limitations that we intend to address were also outlined in the evaluation. The first is related to a person's gender, whereby, even though we eliminate matches found between profiles stating a different gender, this information is only provided by the Facebook API. We will attempt to overcome this issue through the NER process, which can quite reliably identify a person profile's gender based on the first/last name. We have also observed that weights of the full name sub-entities are too high, and there are cases where they impact the final result too strongly. Therefore, more experiments will be performed to fine-tune these weights, which were initially based on a heuristic.

5 Conclusion and Future Work

In this paper we discuss an ontology-driven profile resolution technique that targets the discovery of multiple online profiles referring to the same person, as retrieved from widely-used applications and social networks. This Social Semantic Web technique employs text analytics on top of open data knowledge, to improve the structure of the retrieved profiles against the NCO standard. The semantically-lifted profiles are then compared at both syntactic and semantic levels, to identify profile resolution. Our profile matcher can be used for various real life applications, such as privacy-sensitive detectors of anonymous profiles and contact recommendations based on common locations, activities, friends and interests. From the technique evaluation performed we determined the best attribute similarity approach of the technique, the precision, recall, F-measure and performance results of the improved technique, and an acceptable threshold. The profile matcher obtained a precision rate of 0.816, which result was determined from the usability evaluation. Due to its use of ontologies and RDF as standard

---

[19] A 5-point rating scale was used: Extremely, Quite a bit, Moderately, A little, Not at all.

knowledge representation formats, the designed matching algorithm can be generalised and with some effort adapted to other instance matching tasks in other domains.

As part of our future work directions we will consider all the limitations that were outlined in the usability evaluation. Another possible future enhancement related to the name issue, is to consider the frequency that each attribute value occurs [5]. Our aim is also to apply a method where the weights in Tables 1 and 3 are automatically learned from the retrieved user profile data. In the future, we will consider the identification of higher degrees of semantic relatedness as part the semantic matching extension, e.g., although cities 'Galway' and 'Cork' are not directly related, they are both related to country 'Ireland' by the 'country' relationship (2nd degree distance similarity). Other remote datasets part of the LOD cloud will also be targeted for this extension. This will enhance the open public knowledge, and enrich the technique further for other profile attributes, such as affiliations and user interests. More social networks are being targeted in the future, so as to provide more options to the users of our system.

References

1. Araújo, S., Hidders, J., Schwabe, D., de Vries, A.P.: Serimi - resource description similarity, rdf instance matching and interlinking. CoRR, abs/1107.1104 (2011)
2. Bartunov, S., Korshunov, A., Park, S.-T., Ryu, W., Lee, H.: Joint link-attribute user identity resolution in online social networks. In: SNA-KDD 2012 at KDD 2012, Beijing, China (2012)
3. Bilenko, M., Mooney, R., Cohen, W., Ravikumar, P., Fienberg, S.: Adaptive name matching in information integration. IEEE Intelligent Systems 18(5), 16–23 (2003)
4. Bourimi, M., Scerri, S., Cortis, K., Rivera, I., Heupel, M., Thiel, S.: Integrating multi-source user data to enhance privacy in social interaction. In: INTERACCION 2012, pp. 51:1–51:7 (2012)
5. Castano, S., Ferrara, A., Montanelli, S., Varese, G.: Ontology and instance matching. In: Paliouras, G., Spyropoulos, C.D., Tsatsaronis, G. (eds.) Multimedia Information Extraction. LNCS, vol. 6050, pp. 167–195. Springer, Heidelberg (2011)
6. Cortis, K., Scerri, S., Rivera, I., Handschuh, S.: Discovering semantic equivalence of people behind online profiles. In: RED 2012 at ESWC 2012, Heraklion, Greece, pp. 104–118 (2012)
7. Cortis, K., Scerri, S., Rivera, I., Handschuh, S.: Techniques for the identification of semantically-equivalent online identities. In: Lacroix, Z., Ruckhaus, E., Vidal, M.-E. (eds.) RED 2012. LNCS, vol. 8194, pp. 1–22. Springer, Heidelberg (2013)
8. Cross, V.: Fuzzy semantic distance measures between ontological concepts. In: NAFIPS 2004, vol. 2, pp. 635–640 (June 2004)
9. Cunningham, H., Maynard, D., Bontcheva, K., Tablan, V.: GATE: A Framework and Graphical Development Environment for Robust NLP Tools and Applications. In: ACL 2002 (2002)
10. Dorneles, C.F., Gonçalves, R., dos Santos Mello, R.: Approximate data instance matching: a survey. Knowl. Inf. Syst. 27(1), 1–21 (2011)
11. Ferrara, A., Lorusso, D., Montanelli, S., Varese, G.: Towards a benchmark for instance matching. In: OM. CEUR Workshop Proceedings, vol. 431. CEUR-WS.org (2008)
12. Monge, A., Elkan, C.: The field matching problem: Algorithms and applications. In: KDD 1996, pp. 267–270 (1996)
13. Raad, E., Chbeir, R., Dipanda, A.: User profile matching in social networks. In: NBiS, Takayama, Gifu, Japan, 2010, pp. 297–304. IEEE Computer Society (2010)

14. Shen, W., Wang, J., Luo, P., Wang, M.: Linden: linking named entities with knowledge base via semantic knowledge. In: WWW 2012, pp. 449–458. ACM, New York (2012)
15. Veldman, I.: Matching profiles from social network sites: Similarity calculations with social network support. In: Master Thesis, University of Twente (2009)
16. Vosecky, D., Hong, J., Shen, V.Y.: User identification across multiple social networks. In: Proc. of the Int. Conference on Networked Digital Technologies (NDT 2009), pp. 360–365 (2009)
17. You, G., Park, J., Hwang, S., Nie, Z., Wen, J.: Socialsearch+: enriching social network with web evidences. World Wide Web (May 2012)
18. Zhou, W., Wang, H., Chao, J., Zhang, W., Yu, Y.: LODDO: Using linked open data description overlap to measure semantic relatedness between named entities. In: Pan, J.Z., Chen, H., Kim, H.-G., Li, J., Wu, Z., Horrocks, I., Mizoguchi, R., Wu, Z. (eds.) JIST 2011. LNCS, vol. 7185, pp. 268–283. Springer, Heidelberg (2012)

Aspects of Rumor Spreading on a Microblog Network

Sejeong Kwon[1], Meeyoung Cha[1], Kyomin Jung[2], Wei Chen[3], and Yajun Wang[3]

[1] Korea Advanced Institue of Science and Technology, Republic of Korea
{gsj1029,meeyoungcha}@kaist.ac.kr
[2] Seoul National University, Republic of Korea
kjung@snu.ac.kr
[3] Microsoft Research Asia, China
{weic,yajunw}@microsoft.com

Abstract. Rumors have been studied for several decades in social and psychological fields, where most studies were theory-driven and relied on surveys due to difficulties in gathering data. Rumor research is now gaining new perspectives, because online social media enable researchers to examine closely various kinds of information dissemination on the Internet. In this paper, we review social psychology literature on rumors and try to identify the key differences in the dissemination of rumors and non-rumors. The insights from this study can shed light on improving automatic classification of rumors and better comprehending rumor theories in online social media.

Keywords: Rumor, Social Media, Diffusion Structure, Linguistic Properties.

1 Introduction

A rumor is defined as an unverified explanation of an event at the time of circulation [16]. Nwokocha et al. says that the essence of rumors is in their ambiguity [13], where ambiguity of evidence makes rumors spread more widely. Another study about rumors says that a cognitive mechanism exists in the way people tend to modify a message they heard in the past [8]. Definitions of rumors vary in research [14]. A piece of information can be considered either *verified* or *unverified*, based on the judgments made at the time of circulation. The latter, a piece of information that cannot be verified at the time of circulation (i.e., unverified), is commonly considered to be a rumor in social psychology fields. In this paper, we rigorously divide the latter further into three types: *true*, *false*, and *unknown*, based on the judgments made after the time of circulation. The first type, true, describes when a piece of information that was unverified during circulation is officially confirmed as true after some time. This could be interpreted as information leakage, marketing, or prediction with enough reliable evidence. The other two types, false and unknown, which later in time are confirmed as false or remain unverified respectively, are what we define as rumors. Based on this definition, we built a rigorous set of ground truth data on rumors by recruiting four coders to manually annotate a large amount of social media data and identify rumors.

We test numerous theories and beliefs about rumor propagation in a social net- work. For instance, Alison says that people spread rumors to feel superior, to feel like part of the group, to get attention, or out of anger, boredom, envy, or unhappiness [17]. Others hypothesize that rumors are dominated by certain sentiments and polarities [18, 20]. These studies, which are based on surveys, bring interesting insights into the characteristics of rumor spreading.

The growth of online social media has made propagation of informative and creative content, as well as rumors, spam, and misinformation more prevalent. In order to handle the spread of potentially harmful information, researchers have investigated the problem of detecting unusual behaviors such as misbehaving users [6] and spammers [9]. Similarly, our main goal is to identify the patterns of spreading that are unique to rumors. In doing so, we also try to explain how the findings from social media research are related to well-known social and psychological theories on rumors.

We bridge theory and practice in this work and characterize the key properties of rumor spreading based on human-annotated data. We use near-complete data from Twitter and examine real rumor spreading cases in this network. We start by reviewing the social psychology literature on the theories and ideas related to rumors which we will then test one by one.

2 Theories on Rumor Spreading

Examining how a rumor spreads has been challenging, because the researcher had to be at the right place at the right time. Since this was nearly impossible prior to the use of social media data, previous studies on social and psychological aspects of rumors have mainly been theory-driven and have relied on a small amount of manually collected anecdotal evidence. We summarize four main hypotheses from the literature for an in-depth investigation in this paper.

Rumor Spreaders and the Direction of Information Flow
Besides its ambiguity, another essential characteristic of a rumor is its influence [13]. A rumor has the power to arouse people's interest; therefore, people gossip or spread rumors to get attention. This means that rumors are one of the ways that people gain influence over friends. However, highly influential individuals, who do not want to put their reputations at risk, will not likely initiate conversations on rumors because rumors have low information credibility [4, 20]. As a reasonable proxy of measuring a user's influence, we consider the time the user has been on Twitter (i.e., registration) and the user's number of followers (i.e., degree) in this paper. The first hypothesis we test is,

H1: Rumor spreaders are likely new based on registration time and has fewer followers; thus, rumors more likely disseminate from low-degree users to high-degree users.

Skeptics and Participation
Psychological theories describe how people react to a given rumor. When a person hears about a rumor, he will first doubt the meaning and rely on his knowledge [8]. He will then check with factual sources to verify the rumor [3]. This process of doubt ends when he gathers enough evidence, at which points he either accepts the rumor and propagates it further or disapproves it and expresses negating comments. Solove [19] says that reputation gives people a strong incentive to conform to social norms. Because

rumors spread without strong evidence, rumor receivers may simply neglect the message, incurring low infection rate and often terminating the propagation process. The low credibility of rumors and the doubts incurred by the rumor's audience will result in a different writing style in rumor conversations compared to non-rumors.

H2: Rumors contain more words related to skepticism and doubts such as negation and speculation and are less successful as conversation topics.

Sentimental Difference
Now we examine what kinds of rumors have been studied in social psychology. A classical study was done by Knapp [12], where he gathered a large collection of World War II rumors printed in the Boston Heralds Rumor Clinic column and categorized them into several types: pipe-dream (or wish-fulfillment), bogie (or fear), and wedge-driving (or aggression). The same approach was adopted in a study of 966 rumors from the Iraq War [11], giving insights into the societal attitudes and motivations of rumor spreaders. Wish-fulfillment rumors are fantasies about the world in which all desires are fulfilled [1]. Such rumors contain positive emotions like satisfaction and happiness. On the other hand, there is a general lay belief that rumors are dominated by negative sentiment and polarity [20].

H3: Rumors contain several characteristic sentiments (e.g., anger) compared to other types of information.

Social Relationships and Communication
While unverified information like rumors are often neglected and have low infection rates, this does not mean all rumors are short-lived. In contrast, certain rumors have been reported to be alive for a long period of time. What are the dissemination channels for those successful rumors? Could portals and prominent websites play a role (as they often do for other viral content)? We could not confirm this since the popularity of even the most famous rumor websites like snopes.com and networkworld.com was far lower than mass media websites and portals according to Alexa.com. This means that the primary channel of rumor dissemination is not through websites but through other means. The word-of-mouth of individual users can be one alternative mean, in which case rumor spreaders will attribute their source to social relations like friend, mate and family. Based on this assumption, we hypothesize that a large portion of rumors spread from person to person. Knapp's theory also supports this [12].

H4: Rumors will more likely contain words related to social relationships (e.g., family, mate) and actions like hearing.

3 Methods

We use data crawled from Twitter as explained in previous work [5]. The dataset contains profile information for 54 million users, 1.9 billion follow links between them, and the 1.7 billion public tweets posted from March 2006, when Twitter was launched, through August 2009. The link information is based on a snapshot of the network in August 2009. The complete set of users, links, and tweets provides us a unique opportunity to study user behaviors surrounding real information diffusion.

Collecting Events and Annotation

Given a data set of tweets, we need to collect real rumor cases that circulated on Twitter. We rigorously define a rumor as follows: (i) a statement that was unverified at the time of circulation and (ii) either remains unverified or is verified to be false after some time (i.e., at the time of this study).

Table 1. Representative rumor and non-rumor cases and their tweet data summary

Topic	Spreaders (Audience) Example tweet	Tweets (Mentions)	Description (Regular Expression)
		Rumor	
Bigfoot	462 (1731926)	1006 (40)	The dead body of bigfoot is found (bigfoot & (corpse | (dead body))
	"Bigfoot Trackers Say They've Got a Body, I Say They Don't"		
AdCall	325 (780300)	719 (151)	Call a specific number to avoid advertisement (888-382-1222)
	"Tired of telemarketers? call 888-382-1222 from the phone you want registered"		
ObamaAnti	119 (780300)	135 (19)	Obama is muslim and antichrist (obama & (muslim | antichrist)) "
	"Obama may reach out to world's Muslims on first international trip as president."		
Swineflu	21896 (5300366)	26290 (7710)	Don't eat pork killed by swine flu (swine flu & pork)
	"swine flu...don't eat pork it's disgusting"		
		Non-rumor	
Dell	1581 (1814798)	1909 (389)	Dell enters into smartphone market (dell & smartphone & market)
	"Would you buy a Dell smartphone? Seems you'll soon have the chance."		
Iphone3G	16056 (433215)	31003 (4454)	iphone3G is launched and its review (iphone3g)
	"got Iphone 3G and it is amazing"		
Havard	219 (603911)	448 (111)	A black Harvard professor is arrested at his house ((harvard & arrest) | (henry louis & arrest))
	"Arrest of Harvard prof H.L. last week in his own home by cops "		
Summize	2054 (4367672)	969 (285)	Twitter buys an IT company (twitter & buy & summize) (twitter & buy & summize)
	"Twitter buying summize is BRILLIANT. I bet it powers the home screen."		

In order to understand the diffusion characteristics of rumors, we first had to identify real rumor cases from the Twitter data. For this, we searched lists of popular events from three websites: snopes.com, urbanlegends.about.com, and networkworld.com. Once target rumors were identified, we further identified a set of keywords describing each target rumor by consulting these websites and informed individuals in order to extract relevant tweets. We focused on a period of 90 days starting from a key date; this either corresponds to the date when the event occurred or the date when the event was widely reported in the traditional mass media (e.g., TV and newspapers). These rumors span political, health, urban legend, and celebrity topics. For a control group, we also searched a list of popular events from various media and websites. These non-rumor events are about political controversies, IT product launches, and movie releases.

We first identified 125 topics of interest, out of which 68 were rumors and 57 were non-rumors. To ensure that all rumors and non-rumors are valid, we recruited four well-trained human coders and asked them to classify each topic as either rumor or non-rumor. For each topic, we provided four randomly chosen tweets and a list of URLs on the topic to the annotators. We tested the annotators' agreement level and found an intraclass correlation coefficient (ICC) of 0.992. This indicates that the human coders' annotations were highly reliable. Table 1 lists examples of rumors and non-rumors, respectively. In this study, we further limited our data to only those topics that contained at least 60 tweets and as a result retained 102 topics (47 rumors and 55 non-rumors).

Variables

In Section 2, variables related to the hypotheses can be divided into three categories: personal, topological and linguistic. In case of personal characteristics, we define Age and $Follower$. Both are proxies of user influence. For each topic, Age is defined as the average time between user registration and the key date of the topic as described above. $Follower$ is an average number of followers.

For topological characteristics, we first define *friendship network* and *diffusion set*. *Friendship network* is defined as a subgraph of the original follower-followee graph induced by those users who posted at least one related tweet and follow links among them. From the friendship network, we define diffusion set as a set of ordered pairs, $D = \{e_1, e_2, \dots\}$, where each element in D represents a type of information flow from one user to another. We say information flows from user A (source) to user B (target), if and only if (1) B follows A on Twitter and (2) B posts about a given topic only after A did so. Then, we represent this information flow as an ordered pair, (A, B). If a target has multiple potential sources (e.g., $(s_1, t), (s_2, t) \dots, (s_n, t)$), we pick only the source of the most recent tweets the ordered set. Thus, a target cannot have multiple sources in this work.

Next, we introduce two measures from the diffusion set; $Flow$ and $Singleton$. $Flow$, the proportion of information flow from low-degree user to high-degree user, is defined as follows where $t(e)$, $s(e)$, and ind represent target, source of a given e and number of followers of a given node in the Twitter network, respectively.

$$Flow = \frac{|\{e \in D | ind(t(e)) > ind(s(e))\}|}{|D|}$$

$Singleton$ represents the proportion of users who posted about the topic without influencing others, i.e., having none of their followers reply or talk about the topic. If rumors are not successful conversation topics, $Singleton$ will be higher for rumors than non-rumors. We formulate $Singleton$ as follows where s_i, t_i and V are source and target of a given element, e_i, in D and set of nodes (i.e., users) in the friendship network, respectively.

$$Singleton = \frac{|V \setminus \bigcup_{\forall e_i \in D}\{s_i, t_i \in e_i\}|}{|V|}$$

In addition to topological aspects, we investigate linguistic characteristics of rumor spreading by utilizing a widely used sentiment analysis tool. LIWC (Linguistic Inquiry and Word Count) has been used for text analysis of psychological and behavioral dimensions [15]. Empirical results demonstrate that it can detect meanings in a wide variety of experimental settings, including attention focus, emotionality, social relationships, thinking styles, and individual differences [21].[1] Since the tool requires some minimum amount of text as input (e.g., 50 words), we group all the tweets belonging to a single topic as an input and collectively measured the score of sentiment (e.g., anger, sad) and linguistic (e.g., negate) categories.

Table 2. Variables related to hypotheses. In the "Expectation" column, we list whether rumors or non-rumors are expected to have a higher value. In case of the Linguistic features, "Definition" column lists words related to a given symbol.

Characteristic	Symbol	Definition	Expectation
H1: Rumor spreaders and the direction of information flow			
Personal	Age	Average of registration age	Non-rumor
Personal	$Follower$	Average number of followers	Non-rumor
Topological	$Flow$	Fraction of information flow from low to high degree users	Rumor
H2: Skeptics and participation			
Topological	$Singleton$	Fraction of users whose content is ignored	Rumor
Linguistic	$negate$	no, not never	Rumor
Linguistic	$cogmech$	cause, know, ought	Rumor
Linguistic	$exclusive$	but, without, exclude	Rumor
Linguistic	$insight$	think, know, consider	Rumor
Linguistic	$tentative$	may be, perhaps, guess	Rumor
H3: Sentimental difference			
Linguistic	$affect$	happy, cried, abandon	Non-rumor
Linguistic	$negemo$	hurt, ugly, nasty	Rumor
Linguistic	$anxiety$	worried, fearful, nervous	Rumor
Linguistic	$anger$	hate, kill, annoyed	Rumor
Linguistic	sad	crying, grief, sad	Rumor
Linguistic	$posemo$	love, nice, sweet	Non-rumor
H4: Social relationship and communication			
Linguistic	$social$	mate, talk, they, child	Rumor
Linguistic	$hear$	listen, hearing	Rumor

Table 2 lists variables related to the hypotheses we will test. In the, 'Characteristic' column, "Topological" and "Linguistic" mean the corresponding variables are estimated from *diffusion set* and LIWC, respectively.

4 Result

In this section, we test the significance of the variables described in Table 3 between rumors and non-rumors. Table 3 shows the result of the comparisons for each variable.

The first hypothesis, $H1$, considers three variables: Age, $Follower$ and $Flow$. These variables describe who the rumor spreaders are and how information flows. Table 3

---
[1] Full list available at http://www.liwc.net/descriptiontable1.php

Table 3. Extracted features and their p-values in t-test. In the "Type" column, "Non-rumor' and "Rumor" mean the feature had a higher value for non- rumors and rumors, respectively. In the "Expectation" column, we list whether rumors or non-rumors are expected to have a higher value.

Hypothesis	Symbol	Type	Expectation	p-value
$H1$	Age	None	Non-rumor	0.73
	$Follower$	None	Non-rumor	0.68
	$Flow$	Rumor	Rumor	**
$H2$	$Singleton$	Rumor	Rumor	***
	$negate$	Rumor	Rumor	***
	$cogmech$	Rumor	Rumor	*
	$exclusive$	Rumor	Rumor	***
	$insight$	None	Rumor	0.15
	$tentative$	Rumor	Rumor	**
$H3$	$affect$	Non-rumor	Non-rumor	*
	$negemo$	None	Rumor	0.92
	$anxiety$	None	Rumor	0.61
	$anger$	None	Rumor	0.85
	sad	None	Rumor	0.11
	$posemo$	Non-rumor	Non-rumor	*
$H4$	$social$	Rumor	Rumor	**
	$hear$	Rumor	Rumor	**

*P<0.05; **P<0.01; ***P<0.001

demonstrates that a rumor flows from low-degree users to high-degree users with a statistically significantly high probability. This is in stark contrast to typical information propagation, which mostly involves flow from high-degree to low-degree users (i.e., two-step-flow of information [10]). However, we could not confirm the hypothesis that non-popular users utilize rumors to increase their influence over their friends. Rumors and non-rumors had similar registration age and number of followers.

The second hypothesis, $H2$, deals with the existence of different participation rates and writing styles in rumors. In Table 3 and Figure 1, we can see that most of the related variables show statistically significant distinct ranges of values between rumors and non-rumors, as predicted by the social and psychological theories. The high value of $Singleton$ indicates that rumor rarely initiate a conversation (i.e., no one talks about a rumor after seeing it). Other variables in $H2$ are about the speculative words that indicate doubt about the content of rumors. Our results support that process of doubt properly works to users for rumors and it induces different writing styles. The presence of a lot of negation in rumors can be attributed to people exhibiting uncertainty in their tweets. Our statistical test confirms that rumors have a clearly different writing style, providing empirical confirmation of the social and psychological theories about rumor spreading.

In third hypothesis, $H3$, our purpose is to test which sentiments are more dominant in rumors compared to non-rumors. In Table 3, we can see that rumors do not necessarily contain different sentiments than non-rumors. In fact, negative sentiments like anger, sadness, and anxiety may depend on the topic rather than on information credibility. Figure 2 shows the 95% confidence intervals of sentiment variables. For instance, news about a crime and an accident show higher negative sentiments. Lower *affect* score of

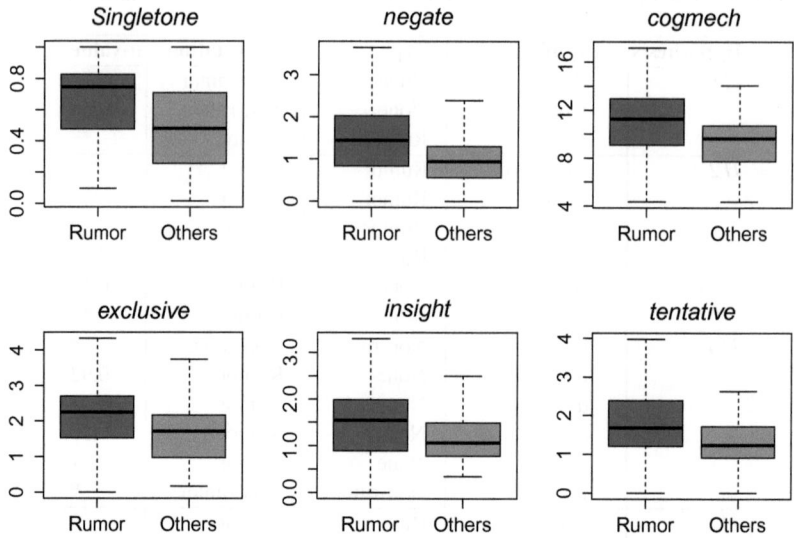

Fig. 1. 95% confidence intervals of variables on participation and writing style

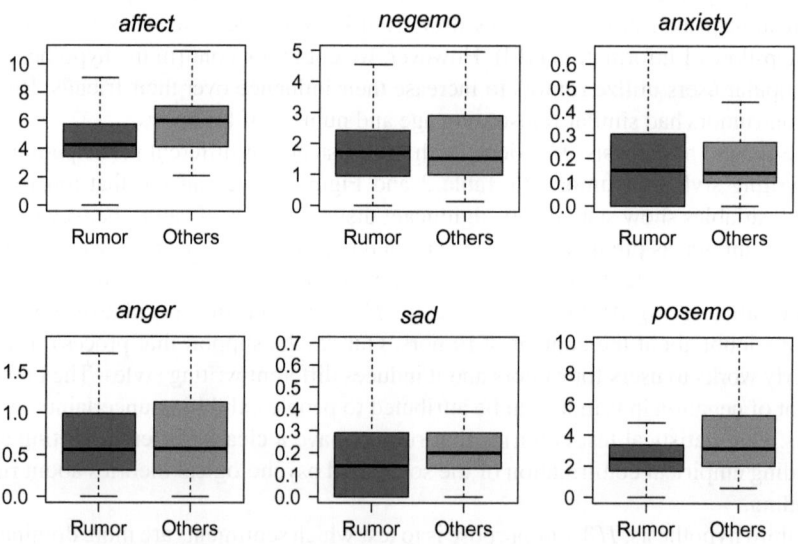

Fig. 2. 95% confidence intervals of variables on sentiments

rumor indicates that less appearance of words like happy, cried, and abandon. From this, we can infer that rumors are less likely influences human emotion. Combining these results, we conclude that unlike what a wide range of theories suggest, sentiment in rumors depends on the topic and is not statistically distinct from sentiment in non-rumors.

In Table 3, we see that rumors have a significantly higher fraction of words related to *social* and *hear*. That is, words related to social relation like 'friend', 'buddy', and 'neighborhood' are more showed up in the rumor tweets. This would be an indicator of main propagation mechanism that rumors are more likely to be disseminated through social relation. This is quite different from other information, which originates from mass media. Thus, we conclude that there is enough evidence to support H4.

Table 4. Variable importance by Random Forest. MDA is Mean Decrease Accuracy value. Higher MDA means higher discriminative power.

Rank	Variable	Characteristic	MDA	Rank	Variable	Characteristic	MDA
1	negate	linguistic	18.46	10	social	linguistic	5.52
2	affect	linguistic	17.67	11	sad	linguistic	4.73
3	Flow	topological	16.16	12	insight	linguistic	2.84
4	Singleton	topological	11.63	13	anxiety	linguistic	2.66
5	hear	linguistic	10.94	14	Follower	personal	1.45
6	tentat	linguistic	10.20	15	negemo	linguistic	1.34
7	excl	linguistic	10.14	16	Age	personal	1.20
8	posemo	linguistic	9.29	17	anger	linguistic	0.30
9	cogmech	linguistic	9.10				

In addition, we estimated the discriminative power of the features. To prevent the problem of over fitting [7], we used *Random Forest*, a modified algorithm of bagging, which utilizes a large collection of de-correlated trees to measure variable importance [2]. Analysis of the discriminative power of the 17 features yields three insights. First, the personal characteristics, *Age* and *Follower* (ranked 16th and 14th respectively), do not have much discriminative power. Second, sentimental differences (e.g., anger, sad) inferred from our literature reviews have no discriminative power. Third, writing style, the fraction of words related to speculation, and topological variables have the highest predictive power.

5 Discussion

Studies on rumors always have a data problem, because one must be at the right place at the right time. Thus, existing social psychology studies have been conducted only on very small-scale data (containing up to tens of users) and are mostly theory-driven. On the other hand, recent studies on rumors in online social media investigate many characteristics of rumors using large-scale data, but those are not related to theories in social and psychological studies. Our study on Twitter rumors (containing up to thousands and tens of thousands of users) serves as useful large-scale empirical data. Using the Twitter data, we examine the actual, complete diffusion instances, testing the hypotheses generated by the social and psychological literature. Hence, our tested features are more intuitive to understand a mechanism of rumor spread than others introduced in recent research and are directly applicable to real online networks for the task of classifying data as rumor or non-rumor.

Acknowledgement. This research was supported by Basic Science Research Program through the National Research Foundation of Korea (2011-0012988).

References

1. Allport, G., Postman, L.: The psychology of rumor. Rinehart & Winston (1947)
2. Breiman, L.: Random forests. Machine Learning 45(1), 5–32 (2001)
3. Cantril, H.: The invasion from Mars: A study in the psychology of panic. Transaction Pub. (1940)
4. Castillo, C., Mendoza, M., Poblete, B.: Information Credibility on Twitter. In: Proc. of the International World Wide Web Conference (2011)
5. Cha, M., Haddadi, H., Benevenuto, F., Gummadi, K.: Measuring User Influence in Twitter: The Million Follower Fallacy. In: Proc. of the International AAAI Conference on Weblogs and Social Media (2010)
6. Cheng, H., Liang, Y.-L., Xing, X., Liu, X., Han, R., Lv, Q., Mishra, S.: Efficient Misbehaving User Detection in Online Video Chat Services. In: Proc. of the ACM International Conference on Web Search and Data Mining (2012)
7. Dietterich, T.: Overfitting and undercomputing in machine learning. ACM Computing Surveys (CSUR) 27(3), 326–327 (1995)
8. Festinger, L., Cartwright, D., Barber, K., Fleischl, J., Gottsdanker, J., Keysen, A., Leavitt, G.: A study of a rumor: its origin and spread. Human Relations (1948)
9. Kant, R., Sengamedu, S., Kumar, K.: Comment Spam Detection by Sequence Mining. In: Proc. of the ACM International Conference on Web Search and Data Mining (2012)
10. Katz, E.: The two-step flow of communication: An up-to-date report on an hypothesis. Public Opinion Quarterly 21(1), 61–78 (1957)
11. Kelley, S.: Rumors in Iraq: A guide to winning hearts and minds. Master's Thesis, Monterey, CA: Naval Postgraduate School (2004)
12. Knapp, R.: A psychology of rumor. Public Opinions Quarterly 8, 22–37 (1944)
13. Nkpa, N.K.: Rumor mongering in war time. The Journal of Social Psychology 96(1), 27–35 (1975)
14. Pendleton, S.C.: Rumor research revisited and expanded. Language & Communication 18 (1998)
15. Pennebaker, J.W., Mehl, M.R., Niederhoffer, K.G.: Psychological Aspects of Natural Language Use: Our Words, Ourselves. Annual Review of Psychology 54, 547–577 (2003)
16. Peterson, W., Gist, N.: Rumor and public opinion. The American Journal of Sociology 57 (1951)
17. Poulsen, A.: Why people gossip and how to avoid it (2013), http://www.sowhatireallymeant.com/articles/personality-traits/gossip/
18. Rosnow, R.: Inside rumor: A personal journey. American Psychologist 46, 484–496 (1991)
19. Solove, D.J.: The future of reputation: Gossip, rumor, and privacy on the Internet. Yale Univ. Pr. (2007)
20. Sunstein, C.R.: On Rumours: How Falsehoods Spread, Why We Believe Them, What Can Be Done. Penguin (2011)
21. Tausczik, Y.R., Pennebaker, J.W.: The Psychological Meaning of Words: LIWC and Computerized Text Analysis Methods. Journal of Language and Social Psychology 29(1), 24–54 (2010)

Traffic Condition Is More Than Colored Lines on a Map: Characterization of Waze Alerts

Thiago H. Silva[1,3], Pedro O.S. Vaz de Melo[1] Aline Carneiro Viana[3], Jussara M. Almeida[1], Juliana Salles[2], and Antonio A.F. Loureiro[1]

[1] Federal University of Minas Gerais, Belo Horizonte, Brazil
[2] Microsoft Research, Redmond, USA
[3] INRIA, France

Abstract. Participatory sensor network (PSN) enables the understanding of city dynamics and the urban behavioral patterns of their inhabitants. In this work, we focus our analysis on a specific PSN, derived from Waze, for sensing traffic conditions. Our objective is to characterize the properties of this PSN, its broad and global spatial coverage as well as its limitations. We also bring discussions on different opportunities for application design using this network. We claim that the PSN derived from Waze has the potential to help us in the better understanding of traffic problem reasons. Besides that, it could be useful for improving algorithms used in navigation services: (1) by exploiting the provided real-time traffic information or (2) by helping in the identification of valuable pieces of information that are hard to detect with traditional sensors, such as car accidents and potholes.

Keywords: Urban social behavior, city dynamics, participatory sensing, mobile social networks, social big data.

1 Introduction

Participatory Sensing Systems (PSSs) [1,2] are revolutionizing the way we see cities, societies and the interactions among people. PSSs provide a mobile interface that allows people carrying smartphones to share data about the environment (or context) they are inserted in at any time and place. These systems certainly have the power to contribute in the process of making ubiquitous computing a reality. Consider the large variety of PSSs already deployed and functioning at global scale, such as Foursquare[1], Instagram[2], Weddar[3], and Waze[4]. Each of these systems can provide valuable information about an aspect of a given city or society in almost real-time, such as its traffic and weather conditions, local parties and festivals, riots, among others. More importantly, the cost for obtaining this data is almost negligible, since it is distributed among all the people who are sharing it.

From participatory sensing systems we can derive participatory sensor networks (PSNs), where each node in the network consists of a user equipped with a mobile device, sending data to web services. In this direction, we can view PSNs as sensing layers

---
[1] http://www.foursquare.com
[2] http://www.instagram.com
[3] http://www.weddar.com
[4] http://www.waze.com

of a global scale sensor network that uses humans in the sensing process. For example, from Waze we can obtain a layer about the traffic, from Instagram a layer containing pictures of places, from Foursquare, we can obtain a layer about the category of locations, and from Weddar, a layer about weather conditions. Each layer is responsible for sensing data related to a certain aspect, for instance traffic or weather conditions, of a specific area in the globe, such as countries, cities, or neighborhoods. In this work, we focus our analysis on a specific sensing layer, the one responsible for sensing traffic conditions. Data collected from this layer, as well as from others as mentioned above, have the potential to transform society. They enable the understanding of city dynamics and the urban behavioral patterns of their inhabitants, supporting smarter decision making. In fact, real-time traffic maps could inform more than the traffic flow's conditions (usually represented by colored lines in the map), for example, it could provide routes that cause less pollution to the city, dangerous areas to avoid, among others.

In order to evaluate the potential of the traffic sensing layer, we here analyze participatory data coming from Waze – the most popular traffic report application. Waze was created in 2008 and recently, had approximately 50 million users [3]. Waze periodically collects sensor data from mobile phones, and uses it to compute the speed of their devices to infer traffic conditions. The system also offers to its users predefined alerts stating incidents such as traffic jams and police traps, extending the information about traffic conditions. One of Waze's main features is the user engagement to contribute to the common good, i.e., Waze is not just crowdsourcing, but personal participation [3].

The objective of this work is to characterize the properties of the PSN derived from Waze, its broad and global spatial coverage as well as its limitations. Moreover, we discuss different opportunities for application design using data collected from Waze. For example, such data could be exploited to drive improvements in algorithms for navigation services and to support quicker identification of information about car accidents, potholes, and slippery roads, which are valuable information that are hard to detect with traditional sensors.

The rest of this paper is organized as follows. In Section 2 we present the related work. In Section 3 we discuss the participation of human in the process of sensing. In Section 4 we present the characterization of a PSN derived from Waze. In Section 5 we illustrate some of the possible applications based on the data shared in Waze. Finally, in Section 6 we present the conclusions and future work.

2 Related Work

Data obtained from participatory sensor networks (PSNs) may be very complex and, therefore, a fundamental step in any investigation is to analyze the collected data to understand its characteristics and usefulness. There are several proposals devoted to the study of specific characteristics of PSNs. For example, in location sharing services like Foursquare, Cranshaw et al. [4] presented a model to extract distinct regions of a city that reflect current collective activity patterns. In a previous work [1], we have characterized data collected from three distinct PSNs derived from location sharing services, such as Foursquare. Among the results, we showed the planetary scale of those networks, as well as the highly unequal frequency of data sharing, both spatially and temporally, which is highly correlated with the typical routine of people. In another previous work [5], we performed the first characterization of Instagram using photos shared by users, analyzing them from a sensor network point of view. We showed that photo-sharing systems, particularly the Instagram, can also be used to map the characteristics of urban locations at a low cost.

Quercia et al. [6] studied how social media communities resemble real-life ones. They tested whether established sociological theories of real-life social networks hold in Twitter. They found, for example, that social brokers in Twitter are opinion leaders who take the risk of tweeting about different topics. Frias-Martinez et al. [7] proposed a technique to determine the most common activities in a city by studying tweeting patterns. Sakaki et al. [8] studied the real-time interaction of events (e.g., earthquakes) in Twitter and proposed an algorithm to monitor tweets to detect a target event.

To the best of our knowledge, Fire et al. [9] is the only prior work to analyze Waze. They showed that it might be possible to identify areas where accidents are more likely to occur by analyzing user accident reports in Waze.

Our work differs from all previous studies as is the first characterization of Waze from a crowdsensing point of view. Moreover, continuing our recent studies [1,5], we show that traffic alert sharing systems, particularly Waze, can also be exploited for mapping the characteristics of urban locations at a low cost, providing complementary data in relation to those obtained from other types of systems, such as location or photo sharing system. As previously mentioned, we believe that the personal involvement of users in such system can allow inferring much richer conclusions about traffic conditions than the usual colored information about traffic jam provided by on-line traffic websites. This work also discuss possible ways towards this goal.

3 Social Media as a Source Sensing

Social media websites based on location (also known as Location-based Social Networks – LBSNs), such as Foursquare, Instagram and Waze, build new virtual environments that integrate user interactions. Such systems have been extensively used for different applications. The ubiquity of smartphones, associated with the adoption of social media websites, enable unprecedented opportunities to study city dynamics and urban social behavior by mining the social big data shared by users of these applications.

Social media websites based on location allow people sharing useful data about the city area where they are located at any given moment, acting as a source of sensing and, thus, leading to the so-called participatory sensing systems [1,2]. Participatory sensing networks (PSNs) can be derived from such systems, where nodes represent users equipped with mobile devices sending data to web services. For example, in a Waze derived PSN, the sensed data consists of reports on traffic conditions of a specific road/street. Data from PSNs can be usually collected through the API of the specific system (e.g., Foursquare, Instagram or Waze APIs). More details about PSNs can be found in [5,1], in [1] is also discussed in more details the challenges faced by this emerging type of network. Different systems, e.g. Foursquare, Waze, and Instagram, lead to different PSNs. Each PSN may enables the access to data related to a certain city aspect, being considered a sensing layer.

4 Characterization of Waze

This section investigates the participatory sensor network (PSN) derived from Waze.

4.1 Overview of the Dataset

Waze is a popular navigation system that uses crowdsensing to offer near real-time traffic information and routing. The system was created in 2008 and registered approximately 50 million users in 2013. Waze periodically collects data from the built-in GPS typically found in smart phones, and uses it to compute the speed of the device. With that, Waze can provide useful information about traffic conditions in different areas. The system also offers to its users predefined alerts stating incidents such as traffic jams and police traps, which extends the information about traffic conditions. It is also possible to use subcategories of incidents to better specify them, for example, "heavy traffic jam" instead of just "traffic jam".

Here, we are interested in characterizing user participation in the dissemination of alerts about traffic. To that end, we collected a dataset of Waze alerts directly from Twitter, since Waze traffic information is not publicly accessible by an API. Our dataset covers the period from December 21st, 2012 to June 28th, 2013, and consists of 212,814 tweets containing alerts about traffic shared by Waze users, each one providing the user id, type of incident (e.g., traffic jam), and the address of the incident. In order to obtain the latitude and longitude of the provided address, we performed a geocoding process using the Bing Maps API[5], which provides the confidence of the result's quality: low, medium, and high. We excluded all results classified as low. After this filtering process, we extracted 162,212 tweets containing alerts, shared by 21,852 users.

In Figure 1, we provide an overview of types of alerts reported by users of our dataset, using word clouds to represent the relative frequency[6]. Alerts were translated into English using a manually created dictionary of translation. As we can see, the most common type of reported alert is traffic jam[7], though police and hazard are also very popular.

Fig. 1. Overview of reported alerts

Fig. 2. Spatial coverage of Waze in Rio de Janeiro

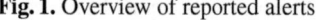

[5] http://www.microsoft.com/maps/developers/web.aspx
[6] The size of the word indicates its popularity.
[7] Alerts containing a subcategory of an incident were unified to its main category, for example, "heavy traffic jam" was associated to the word "traffic jam".

4.2 Sensing Layer Coverage

In this section, we discuss the spatial coverage of the PSN derived from Waze. Towards a global view of this coverage, we first built a heatmap with all alerts shared by users in our dataset[8], and then we selected the most popular cities for further analysis.

A popular city from our dataset is shown in Figure 2. In this figure we show the number of alerts in different regions of Rio de Janeiro by a heat map, where the scale varies from yellow to red (more intense activity)[9]. The spatial coverage is not as proliferated as the one observed in location and photo sharing systems [1,5]. A factor that might help to explain it is the user population of our dataset, which is smaller than those reported in the mentioned studies. Another factor is that users might have fewer opportunities to share traffic alerts, compared to opportunities to share photos or check-ins.

In order to evaluate user participation across different regions at a finer granularity, we propose to divide the geographical area of each city into smaller rectangular spaces (or quadrants), as in a grid, and analyze the distribution of the number of alerts across quadrants. We here consider a quadrant delimited by steps of $0.0001°$ in both latitude and longitude. This scheme produces quadrants of different spatial areas, depending on the geographic location of the city, but this does not affect our analysis. For instance, it represents an area of $\approx 8 \times 11$ meters in New York and 10×11 meters in Rio de Janeiro.

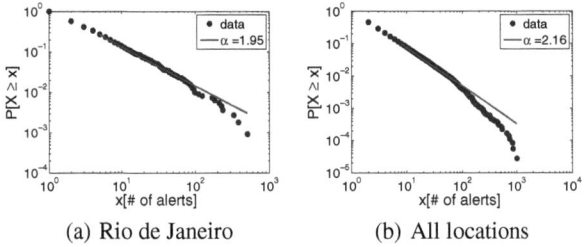

(a) Rio de Janeiro (b) All locations

Fig. 3. Distribution of the number of alerts

The complementary cumulative distribution functions (CCDF) of the number of alerts shared in a quadrant of the city of Rio de Janeiro, as well as across all locations in our dataset are presented in Figures 3a and 3b, respectively. Note that a power law describes well this distribution in both cases. This means that few areas have hundreds of shared alerts, while most of the quadrants have just a small number. This finding is consistent with previous results about user participation in location sharing services [1,10] and photo sharing services [5]. As in those other services, it is likely that some areas, such as large avenues in downtown, have more activity of traffic alerts. Note that the number of vehicles circulating on each region greatly impacts the local coverage of a traffic alert sharing system such as Waze, as shared alerts often refer to traffic jams and hazards, or even police traps (see Figure 1), which tend to occur more often in locations with heavier car flow. This is in contrast to location and photo sharing services, where places often visited by a large number of people are not necessarily covered by a large

---
[8] Figure omitted due to space limitations. We note, however, that user participation in Waze is low is certain regions, particularly Asia.
[9] The darkest red represents a region with 508 alerts.

amount of shared data, because the motivation of users to share data in such systems is different from Waze. For example, a large supermarket may be visited by a large number of people on a daily basis, but it is not likely that those people will share many check-ins or photos at it.

4.3 Time Intervals between Traffic Alerts

We now analyze the frequency in which users share alerts in Waze. This is important because the success of a PSN depends on the continuous participation of the users, since nodes are autonomous and responsible for their own operation and functioning.

The histogram of the inter-sharing time Δ_t between consecutive alerts (performed not necessarily by the same user), in a popular quadrant is shown in Figure 4a. Note that a log-logistic distribution[10] ($\mu = 2.931$, $\sigma = 1.065$) fits well the data, reflecting the fact that there are times when many alerts are shared within a few minutes and there are times when there is no sharing for hours. As also observed for photo [5] and location [1] sharing services, this result may indicate that the majority of alert sharing occurs at specific intervals. For instance, alerts are more likely to be common in urban areas during rush hours.

In Figure 4b, we show the odds ratio function (OR)[11] of inter-sharing time Δ_t. This function highlights the behavior of the distribution at both the head and in the tail. As also observed in previous analyses of phone SMS usage and photo sharing [5,11], the OR function of the inter-sharing time between alerts also presents a power law behavior with slope $\rho \approx 1$. This suggests that the mechanisms behind human activities can be simpler than those proposed in the literature, which depend on a larger number of parameters [12].

The CDF of all observed inter-sharing times performed by any user in the same quadrant is shown in Figure 4c. As we can observe, a considerable portion of users perform consecutive alert sharing in a short time interval. In the present case, the portion is around 25%. This is expected to happen because, for example, when an accident happens many users tend to share it in a short interval.

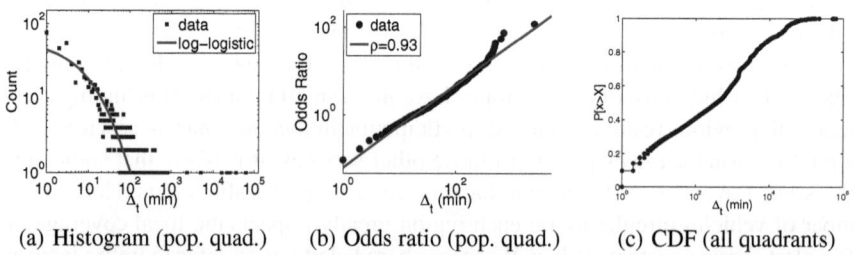

(a) Histogram (pop. quad.)   (b) Odds ratio (pop. quad.)   (c) CDF (all quadrants)

Fig. 4. Time intervals between consecutive alerts, not necessarily done by the same user

---

[10] Probability Density Function: $f(x|\mu,\sigma) = \frac{1}{\sigma}\frac{1}{x}\frac{e^z}{(1+e^z)^2}$; $x \geq 0$, where $z = \frac{log(x)-\mu}{\sigma}$.

[11] $OR(x) = \frac{CDF(x)}{1-CDF(x)}$, where $CDF(x)$ is the cumulative density function, in this case, of the inter-sharing time Δ_t distribution.

4.4 User Activity

We now analyze the contribution of individual users in the PSN derived from Waze. In Figure 5, we show that the distribution of the number of alerts shared by each user of our dataset has a heavy tail, as observed for photo sharing [5]. This implies in a great variability of user participation. For instance, 35% of the users contributed with only one alert during approximately the six-month period covered by our dataset, while 16% and 0.006% of users contributed with more than 10 and 100 alerts, respectively, in the same period. These proportions are similar to those observed in photo sharing [5].

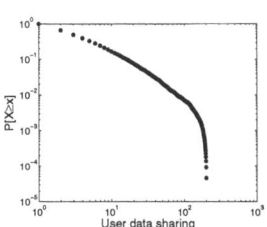
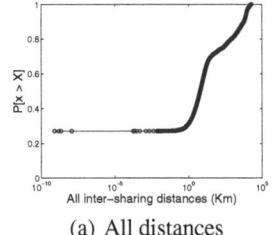
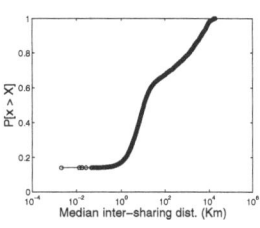

(a) All distances   (b) Median distance per user

Fig. 5. CCDF of the number of shared alerts (same user)

Fig. 6. Distribution of the geographical distance between consecutive data of the same person

We now analyze the spatial distance between consecutive alerts by the same user, by taking the distance [13] between the geographic coordinates associated with both alerts. In Figure 6a, we show the CDF of the distances between consecutive alerts shared by each user, for all users. Note that a large portion of the distances are very short: for instance, around 30% are below 1 meter. Such large fraction of small distances between consecutive sharings were also observed in photo sharing [5] and, to a lesser extent, location sharing services [10]. In the latter, Noulas et al. [10] observed that 20% of the consecutive sharings by the same user were in locations that were apart from each other by up to 1 km. For photos and alerts, this fraction raises to approximately 45% and 80%, respectively. This suggests that users tend to share multiple alerts in the same location.

In Figure 6b, we show similar results for the distribution of the median distance between consecutive sharings for each user. That is, even aggregating results for each user, we still observe that alerts are shared at very short distances: around 15% of users share alerts 1 meter apart from each other.

4.5 Influence of User Routines

In this section, we study how user routines affect the temporal frequency of alert sharing. In Figure 7a, we show the temporal variations[12] of the number of alerts shared throughout the week (Monday to Sunday), for all locations of our dataset. As expected, user participation presents a diurnal pattern, and the activity during late night hours and dawn is much lower than previously observed in location and photo sharing patterns [1,5]. During that period, traffic problems are typically rare, whereas users have

---

[12] The time of sharing was normalized according to the timezone where the alert was shared.

more opportunities to share data in location and photo sharing systems (e.g., in a night club or in a concert).

Intense user activity during the weekends, as observed in location sharing services [1,14] and photo sharing services [5], is not observed here. This might indicate that the reasons motivating users to contribute alerts are distinct from the ones to perform check-ins. In Figure 7b, we show the average number of data sharings throughout the day, separately for weekdays (Monday to Friday) and weekends (Saturday and Sunday). Note the two clear peaks of activity, one around 7 to 8 AM and the other around 6 PM, coinciding with typical rush hours in urban areas. This result is different from the three clear peaks previously observed in location sharing services [1,14], around breakfast, lunch and dinner times, as well as from the two peaks during lunch and dinner times in photo sharing [5].

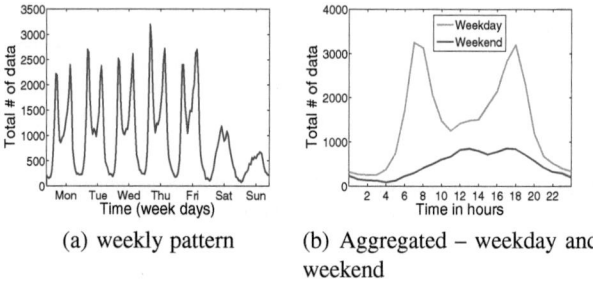

(a) weekly pattern

(b) Aggregated – weekday and weekend

Fig. 7. General temporal sharing pattern (all locations)

We now analyze the hourly variations of alert sharings in six large cities: Chicago and New York (Figure 8a[13]); Belo Horizonte and Sao Paulo in Brazil (Figure 8b); and London, and Paris (Figure 8c). Note that the curve of each city follows the general trend observed for all locations (Figure 7b).

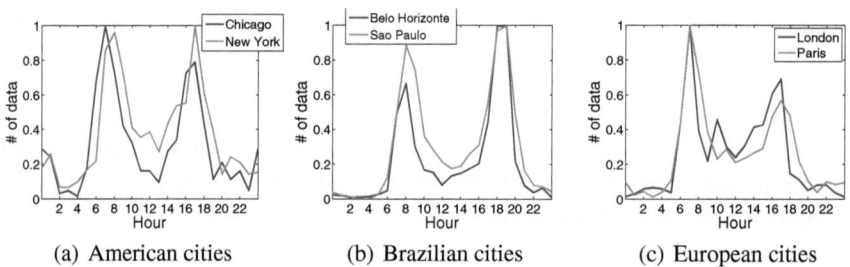

(a) American cities

(b) Brazilian cities

(c) European cities

Fig. 8. Alerts sharing throughout the day in different cities around the world

We can also observe that the peaks reflect distinct rush times that are related to the common working hours of different cities. In Chicago (Figure 8a) the morning peak is around 7 AM, as in the two European cities (Figure 8c). In contrast, in New York and in the Brazilian cities (Figure 8b), the morning peak is usually one hour later, suggesting that people tend to leave later to work in those cities. The second most expressive peak in

---

[13] Each curve is normalized by the maximum number of alerts shared in the city in question.

both American cities is around 5 PM, which is similar to the European cities. However, this is distinct from the Brazilian cities, which have a peak of activity around 6 PM.

To complement this analysis, we performed, from July 16th to July 18th, an hourly collection of traffic conditions of Paris, using Google Maps. We note that the time of the observed peaks reflects relatively well intense traffic conditions reported by Google Maps, whereas the reduced activity prior and after the peaks also reflects better traffic conditions. This suggests that this information could be used to assure the quality and improve traffic condition information services, such as those offered by Google Maps.

4.6 Discussion

We showed the planetary scale of the studied PSN, derived from Waze. We also showed the highly unequal frequency of data sharing, both spatially and temporally, which is highly correlated with the typical routine of people. Our characterization provided a deep understanding of the properties of this particular PSN, revealing its potential to drive various studies on city dynamics and urban social behavior, as discussed next.

5 Opportunities for New Services and Applications

This section discusses some possible situations where a PSN derived from Waze can be exploited to build new services and applications. As discussed in Section 4.1, the most often reported problem by Waze users is traffic jam. Since this is a common cause of complaints and many other problems may end up resulting in traffic jam, a natural question that arises is: What are the causes of traffic jam? This is not an easy question to answer, and it may vary from place to place. However, the shared alerts in Waze might help us to understand the causes.

More specifically, we note that the analysis of the traffic alerts can lead to a more detailed investigation of traffic conditions. For instance, the real-time identification of locations with potholes or animals in the road, whose detection is hard with traditional sensors, becomes more feasible when users participate in the sensing process. This is useful to discover not obvious reasons for a frequent traffic jam. Besides that, such detection opens opportunities for various services, such as, help smart cars in the correct identification of problems on the road.

In the same direction, the identification of problematic roads might also be possible by looking at the number of alerts reported on a road. For example, if we take the top five reported locations in Belo Horizonte (Cristiano Machado Av; Raja Gabaglia Av; Contorno Av; Beltline Rd; Amazonas Av) and Paris (A15; N104; A6 - E15; A13 - E05; N118), we observe that all of them are roads that typically present traffic problems, especially on rush hours. This shows that it is possible to identify problematic roads using traffic alerts. However, the main advantages of using a PSN of traffic alerts do not lie on discovering common problematic roads, but on detecting unusual ones. This is possible thanks to the capability of Waze alerts in describing real-time incidents, what can help to discover particular patterns not generally known.

This information could be used to improve algorithms for navigation services. Besides that, traffic information services, such as Bing Maps, could also benefit from this information to assure the quality about the reported traffic condition, as we observed in the Section 4.5. Moreover, an urban planner could use this information to assess the effectiveness of previous roadworks. For instance, it could be verified if roadworks in the Cristiano Machado Avenue, a very problematic road in Belo Horizonte, were effective to reduce the number of problems reported in that road.

6 Conclusion and Future Work

The better understanding of dynamics of cities and the social behavior that happens on them can be achieved with the help of participatory sensor networks (PSNs). This understanding allows system designers to offer smart services that meet people's needs. In this paper, we studied a PSN derived from Waze, the most popular traffic report application, and characterized the properties of this network. Among other results, we showed that the routines have a considerable impact on the data sharing, for example, the peaks of activity reflect distinct rush times that are related to the common working hours of the analyzed cities. A future direction we intend to pursue is the design of new applications that explore traffic alerts shared in Waze, following the opportunities presented in Section 5.

References

1. Silva, T.H., Vaz de Melo, P.O.S., Almeida, J.M., Loureiro, A.A.F.: Challenges and opportunities on the large scale study of city dynamics using participatory sensing. In: Proc. of ISCC 2013, Split, Croatia (July 2013)
2. Burke, J., Estrin, D., Hansen, M., Parker, A., Ramanathan, N., Reddy, S., Srivastava, M.B.: Participatory sensing. In: Workshop on World-Sensor-Web, WSW 2006 (2006)
3. Goel, V.: Maps That Live and Breathe With Data. The New York Times (June 2013)
4. Cranshaw, J., Schwartz, R., Hong, J.I., Sadeh, N.: The Livehoods Project: Utilizing Social Media to Understand the Dynamics of a City. In: Proc. of ICWSM 2012, Dublin, Ireland (2012)
5. Silva, T.H., Vaz de Melo, P.O.S., Almeida, J.M., Salles, J., Loureiro, A.A.F.: A picture of Instagram is worth more than a thousand words: Workload characterization and application. In: Proc. of DCOSS 2013, Cambridge, USA, pp. 123–132 (May 2013)
6. Quercia, D., Capra, L., Crowcroft, J.: The social world of twitter: Topics, geography, and emotions. In: Proc. of ICWSM 2012, Dublin, Ireland (June 2012)
7. Frias-Martinez, V., Soto, V., Hohwald, H., Frias-Martinez, E.: Characterizing urban landscapes using geolocated tweets. In: Proc. of SocialCom 2012, Washington, USA (2012)
8. Sakaki, T., Okazaki, M., Matsuo, Y.: Earthquake shakes twitter users: real-time event detection by social sensors. In: Proc. of WWW 2010, Raleigh, USA, 851–860. ACM (2010)
9. Fire, M., Kagan, D., Puzis, R., Rokach, L., Elovici, Y.: Data mining opportunities in geosocial networks for improving road safety. In: Proc. of IEEEI 2012, pp. 1–4 (2012)
10. Noulas, A., Scellato, S., Mascolo, C., Pontil, M.: An Empirical Study of Geographic User Activity Patterns in Foursquare. In: Proc. of ICWSM 2011, Barcelona, Spain (2011)
11. Vaz de Melo, P.O.S., Faloutsos, C., Loureiro, A.A.: Human dynamics in large communication networks. In: Proc. of SDM 2011, Mesa, USA (2011)
12. Malmgren, R.D., Stouffer, D.B., Motter, A.E., Amaral, L.A.N.: A poissonian explanation for heavy tails in e-mail communication. PNAS 105(47), 18153–18158 (2008)
13. Sinnott, R.W.: Virtues of the Haversine. Sky and Telescope 68(2), 159+ (1984)
14. Cheng, Z., Caverlee, J., Lee, K., Sui, D.Z.: Exploring Millions of Footprints in Location Sharing Services. In: Proc. of ICWSM 2011, Barcelona, Spain (2011)

The Three Dimensions of Social Prominence

Diego Pennacchioli[1], Giulio Rossetti[1] Luca Pappalardo[1], Dino Pedreschi[1],
Fosca Giannotti[1], and Michele Coscia[2]

[1] KDDLab ISTI-CNR, via G. Moruzzi 1, Pisa, Italy
name.surname@isti.cnr.it
[2] CID Harvard University, 79 JFK St, Cambridge, MA, US
michele_coscia@hks.harvard.edu

Abstract. One classic problem definition in social network analysis is the study of diffusion in networks, which enables us to tackle problems like favoring the adoption of positive technologies. Most of the attention has been turned to how to maximize the number of influenced nodes, but this approach misses the fact that different scenarios imply different diffusion dynamics, only slightly related to maximizing the number of nodes involved. In this paper we measure three different dimensions of social prominence: the Width, i.e. the ratio of neighbors influenced by a node; the Depth, i.e. the degrees of separation from a node to the nodes perceiving its prominence; and the Strength, i.e. the intensity of the prominence of a node. By defining a procedure to extract prominent users in complex networks, we detect associations between the three dimensions of social prominence and classical network statistics. We validate our results on a social network extracted from the Last.Fm music platform.

1 Introduction

One classic problem in social network analysis is understanding diffusion effects in networks. Modeling diffusion processes on complex networks enables us to tackle problems like preventing epidemic outbreaks [6] or favoring the adoption of new technologies or behaviors by designing an effective word-of-mouth communication strategy. In our paper, we are focused on the social prominence aspect of the diffusion problem in networks.

In the setting of favoring social influence, most of the attention of researchers has been put on how to maximize the number of nodes subject to the spreading process. This is done by choosing appropriate seeds in critical parts of the network, such that their likelihood of being prominent users, i.e. nodes that are active on an innovation before all the other nodes, is maximum, to possibly achieve larger cascades. While larger cascades are obviously part of the overall aim, we argue that it is not the unique dimension of this problem. Three other dimensions are relevant: the *width*, the *depth* and the *strength* of the social prominence of any given node in a network. The width of a node is being prominent for its immediate neighbors; the depth is its ability to be the root of long cascades; the strength is being the root of an intense activity.

Real-world scenarios focus on specific diffusion patterns requiring a multidimensional understanding of the prominence mechanics at play, along the three mentioned dimensions. Some examples are: (i) an analyst needs information from the personal acquaintances of a subject, the important aspect is that many subject's direct connections respond, ignoring people two steps away or more; (ii) a person wants to find another person with a given object, the important aspect is that some people are able to pass her message through a chain pointing to the target; (iii) an artist wants to influence people in a social network to her art, the important aspect is that some people are influenced above the threshold that will make them aware of the art. In (i) we want a broad diffusion in the first degree of separation. In (ii) we require a targeted diffusion similar to a Depth First Search. In (iii) there is the need of a high-intensity diffusion. Different scenarios may require any combination of the three.

In this paper, we make use of three measures to capture the characteristics of these three scenarios: the Width, Depth and Strength of social prominence. The Width measures the ratio of the neighbors of a node that follows the node's actions. The Depth measures how many degrees of separation there are between a node and the other nodes that followed its actions. The Strength measures the intensity of the action performed by some nodes after the leader.

We study what the relations are between these three measures to understand if we are capturing three orthogonal dimensions of social prominence. We also study the relations between the Width, Depth and Strength measures and different node properties, with the aim of predicting the diffusion patterns of different events, given the characteristics of the nodes that lead their diffusion.

To validate our concepts, we constructed a social network from the music platform Last.Fm[1], along with the data about how many times and when each user listens to a song performed by a given artist. We detect who are the prominent users for each artist, i.e. the users who start listening to an artist before any of their neighbors. We calculate for each prominent user its Width, Depth and Strength, along with its network statistics such as the degree and the betweenness centrality, looking for associations between them. We then create a case study to understand what are the different dynamics in the spread of artists belonging to different music genres, by using the artists' tags.

To sum up, the contributions of our paper are: (i) a proof that social diffusion indeed follows at least these three dimensions, which are uncorrelated or anticorrelated; (ii) the discovery of some significant associations between the three dimensions of social prominence and some traditional network measures; (iii) the ability to predict the patterns of diffusion of particular events by looking at the characteristics of the leaders spreading them.

2 Related Work

In the last decade, there has been growing interest in the studies of diffusion processes. Two phenomena are tightly linked to the concept of diffusion: the spread

---

[1] http://www.last.fm/

of biological [6] or computer [17] viruses, and the spread of ideas and innovation through social networks, the so-called "social contagion" [2], [8]. In both cases, the patterns through which the spreading takes place are determined not just by the properties of the pathogen/idea, but also by the network structures of the population it is affecting.

Some models have been defined to understand the contagion dynamics: the SIR [11], SIS and SIRS [16] models. The idea behind them is that each individual transits between some stages in the life cycle of a disease: from Susceptible (S) to Infected (I), and from Infected to either Recovered (R) or again Susceptible. The availability of Big Data conveying information about human interactions and movements encouraged the production of more accurate data-driven epidemic models. For example, [6] takes into account the spatio-temporal dimension. In [17], authors study the spreading patterns of a mobile virus outbreak.

Christakis and Fowler studied the role of social prominence in the spread of obesity [4], smoking [5] and happiness [9]. Their results suggest that these health conditions may exhibit some amount of "contagion" in a social sense: although the dynamics of diffusion are different from the biological virus case, they nonetheless can spread through the social network.

3 Leader Detection

Each diffusion process has its starting points. Any idea, disease or trend is firstly adopted by particular kinds of actors. Such actors are not like every other actor: they have an increased sensibility and they are the first to perform an action in a given social context. We call such actors prominent users, or *leaders*, because they are able to anticipate how other actors will behave. Given a graph, several interesting problems arise regarding how information spreads over its topology: can we identify the *leaders*? Can we characterize them? What kind of knowledge should we expect to extract from their analysis?

Our approach aims to detect *leaders* through the analysis of two correlated entities: the topology of the social graph and the set of actions performed by the actors (nodes). When discussing the roles of those entities, we refer respectively to the following definitions:

Definition 1 (Social Graph). *A social graph $\mathcal{G}$ is composed by a set of actors (nodes) V connected by their social relationships (edges) E. Each edge $e \in E$ is defined as a couple (u,v) with $u,v \in V$ and, where not otherwise specified, has to be considered undirected. With $\Gamma(u)$ we identify the neighbor set of a node u.*

Definition 2 (Action). *An action $a_{u,\psi} = (w,t)$ defines the adoption by an actor $u \in V$, at a certain time t, of a specific object ψ with a weight $w \in \mathcal{R}$. The set of all the actions of nodes belonging to a social graph $\mathcal{G}$ will be identified by $\mathcal{A}$, while the object set will be called Ψ.*

We identify with $\mathcal{G}_\psi = (V_\psi, E_\psi)$, where $V_\psi \subset V$ and $E_\psi \subset E$, the induced subgraph on $\mathcal{G}$ representing respectively the set of all the actors that have performed an action on ψ, and the edges connecting them. We depict an example

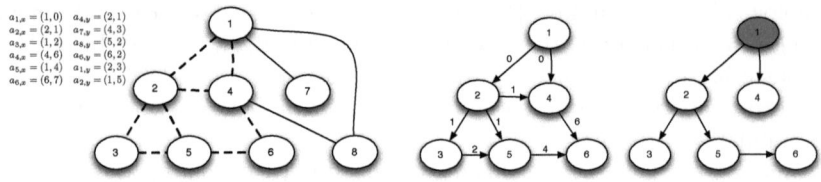

Fig. 1. Toy Example. On the *left* the social graph $\mathcal{G}$ and action set $\mathcal{A}$, where $x, y \in \Psi$ are the objects of the actions; in the *center* the induced subgraph for the action x; on the *right* the diffusion tree for x. In red we highlighted the leader (root) for the given tree.

of the social graph and the set of actions in Figure 1 *(left)*, where the induced subgraph for the object x is highlighted with a dashed line. In the Figure, $a_{1,x}$ refers to the user 1 performing the action x; and $a_{1,x} = (1, 0)$ means that user 1 performed x one time, starting at the timestep 0.

Given the nature of a diffusion process, we would expect that each *leader* will be prominent among its neighbors, being the root of a cascade event that follows some rigid temporal constraints. Our constraint is that a node u precedes a neighbor v iff given $t_{u,\psi} \in a_{u,\psi}$ and $t_{v,\psi} \in a_{v,\psi}$ is verified that $t_{v,\psi} > t_{u,\psi}$ and $t_{v,\psi} - t_{u,\psi} \leq \delta$. Here, δ is a temporal resolution parameter that limits the cascade effect: if $t_{v,\psi} - t_{u,\psi} > \delta$, we say that v executed action $a_{v,\psi}$ independently from u, as u's prominence interval is over.

We transform each undirected subgraph $\mathcal{G}_\psi$ in a directed one imposing that the source node of an edge must have performed its action before the target node. After that, each edge (u, v) will be labeled with $min(t_{u,\psi}, t_{v,\psi})$ to identify when the diffusion started going from one node to the other. The directed version of $\mathcal{G}_\psi$ represent all the possible diffusion paths that connect leaders with their "tribes" (Figure 1 *(center)* an example for the object $x \in \Psi$).

From now on, for a given object ψ, we will refer to the corresponding leader set as $\mathcal{L}_\psi$: when no action is specified the set $\mathcal{L}$ will be used to describe the union of all the $\mathcal{L}_\psi$ for the graph $\mathcal{G}$. To be defined a *leader* an actor should not have any incoming edges in $\mathcal{G}_\psi$. This is because a prominent user cannot act after another user (they are, in their surroundings, innovators), and is a direct consequence to the adoption of a directed graph to express diffusion patterns. Given this definition, for each directed connected component $\mathcal{C}_\psi \subset \mathcal{G}_\psi$ multiple nodes can belong to $\mathcal{L}_\psi$.

Realistically, a leader may be influenced by exogenous events. This is not a problem as we are not measuring a node's influence, but a node's prominence, i.e. its propensity to act faster than others to any kind of exogenous and/or endogenous influence. To study the path of diffusion given an action a and a leader l we use a minimum diffusion tree:

Definition 3 (Leader's Minimum Diffusion Tree). *Given an action a_ψ, a directed connected component $\mathcal{C}_\psi$ and a leader $l \in \mathcal{L}_\psi$, the minimum diffusion tree $T_{l,\psi} \subset \mathcal{C}_\psi$ is the Minimum spanning tree (MST) having its root in l and built minimizing the temporal label assigned at the edges.*

An example of minimum diffusion tree for the node 1 and object x is shown in Figure 1 *(right)*. For each object, the diffusion process on a given network is independent. Moreover, given temporal dependencies on its adoption (expressed through actions $a_{*,\psi} \in \mathcal{A}$), it is possible to identify the origin points of the diffusion. The identified *leaders* will show different topological characteristic and will be prominent in their surroundings in different ways: our aim is to classify diffusion *leaders* characterizing some of their common traits.

4 Measures

To capture the three dimensions of social prominence we need three network measures. We call these measures Width, the ratio of neighbors mirroring an action after a node; Depth, how many degrees of separation are in between a node and the most distant of the nodes mirroring its actions; and Strength, how strongly nodes are mirroring a node's action.

Given a leader, the Width aims to capture the direct impact of her actions on her neighbors, i.e. the degree of importance that a leader has over her friends.

Definition 4 (Width). *Let G be a social graph, $\psi \in \Psi$ an object and $l \in \mathcal{L}_\psi \subset V$ a leader: the function width $: \mathcal{L}_\psi \to [0,1]$ is defined as:*

$$width(l, \psi) = \frac{|\{u | u \in \Gamma(l) \wedge \exists a_{u,\psi} \in \mathcal{A}\}|}{|\Gamma(l)|} \tag{1}$$

The value returned is the ratio of all the neighbors that, after the action of the leader, have adopted the same object.

The Depth measure evaluates how much a leader can be prominent among other prominent leaders, which can be prominent on other leaders and so on.

Definition 5 (Depth). *Let $T_{l,\psi}$ be a minimum diffusion tree for a leader $l \in \mathcal{L}_\psi$ and a given object $\psi \in \Psi$: the function depth $: T_{l,\psi} \to \mathbb{N}$ computes the length of the maximal path from l to a node $u \in T_{l,\psi}$. The function $depth_{avg} : T_{l,\psi} \to \mathbb{R}$ computes the average length of paths from l to any leaf of the tree.*

The last proposed measure, the Strength, tries to capture quantitatively the total weight of the adoption of an object after the leader's action. A leader is strongly prominent if the nodes among which she is prominent are very engaged in adopting what she adopted. Direct prominence diminishes as new adopters become more distant, in the network sense, from the original innovator. Therefore, we decided to introduce a distance damping factor.

Definition 6 (Strength). *Let $T_{l,\psi}$ be a minimum diffusion tree for a leader $l \in \mathcal{L}_\psi$ and an object $\psi \in \Psi$; $0 < \beta < 1$ a damping factor: the function strength $: T_{l,\psi} \times (0,1) \to \mathbb{R}$ is defined as:*

$$strength(T_{l,\psi}, \beta) = \sum_{i \in [0, depth(l)]} \beta^i L(T_{l,\psi}, i) \tag{2}$$

where $L : T_{l,\psi} \times \mathbb{N} \to \mathbb{R}$ is defined as:

$$L(T_{l,\psi}, i) = \sum_{\{u | u \in T_{l,\psi} \wedge distance(l,u)=i\}} \frac{w_{u,\psi}}{w_u} \quad (3)$$

and represents the sum, over all the nodes u at distance i from l, of the ratio between the weight of action ψ and the total weight of all the actions taken.

Given the example in Figure 1, what are the Width, Depth and Strength values for the red node leader and the action x?

Width: from Figure 1 *(left)* we see that $\Gamma(1) = \{2, 4, 7, 8\}$, i.e. 4 nodes. Given that $\Gamma_x(1) = \{u | u \in \Gamma(1) \wedge \exists a_{u,x}\} = \{2, 4\}$, we have $width(1, x) = \frac{|\Gamma_x(1)|}{|\Gamma(1)|} = 0.5$.

Depth: the leaves in Figure 1 *(rigth)* are nodes 3, 4 and 6. Node 4 is a direct neighbor of 1, while node 3 is two edges away. The longest chain is $1 \to 2 \to 5 \to 6$, therefore $depth(T_{1,x}) = 3$. We can also calculate $depth_{avg}(T_{1,x})$, that is the average path length in the tree from node 1 to all the leaves: $\frac{1+2+3}{3} = 2$.

Strength: we need to use the number of times each node performed action x. We also set our damping faction $\beta = 0.5$. At the first degree we have nodes 2 and 4, that performed action x 2 and 4 times respectively; they also performed action y 1 and 2 times respectively: their contribution is then $\beta^0 \times (\frac{2}{2+1} + \frac{4}{4+2})$. Nodes 2 and 5 are at the second degree of separation as they never performed action y, therefore they add: $\beta^1 \times (1+1)$. Finally, at the third degree of separation, node 6 adds $\beta^2 \times \frac{6}{6+6}$. Wrapping up, $strength(T_{1,x}, 0.5) = 2.458\bar{3}$.

5 Experiments

In this section we present our data extracted from the music social media Last.Fm. We use the data to characterize the Width, Depth and Strength measures, by searching for associations with network topology measures. Finally, we analyze the prominence of different users for different musical genres.

5.1 Data

Last.Fm is an online social network platform, where people can share their own music tastes and discover new artists and genres basing on what they, or their friends, like. Users send data about their own listenings. For each song, a user can express her preferences and add tags (e.g. genre of the song). Lastly, a user can add friends (undirected connections, the friendship request must be confirmed) and search her neighbors w.r.t. musical tastes. A user can see, in her homepage, her friends' activities. The co-presence of these characteristics makes Last.Fm the ideal platform on which test our method, as it contains everything we need: social connections that can convey social prominence, a measure of intensity proportional to the number of listening of an artist, rich metadata attached to each song/artist and an intrinsic temporal dimension of users' actions.

Using Last.Fm APIs[2], we obtained a sample of the UK user graph, exploring the network with a breadth-first approach, up until the fifth degree of separation from our seeds. For each user, we retrieved: (a) her connections, and (b) for each week in the time window from Jan-10 to Dec-11, the number of single listenings of a given artist (e.g. in the week between April 11,2010 and April 18,2010 the user 1234 has listened 66 songs from the artist Metallica).

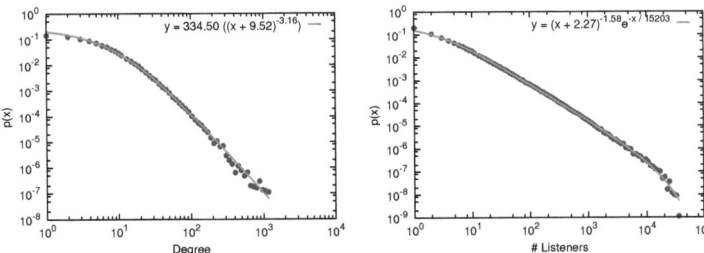

Fig. 2. Log-binned distribution of the nodes' degree. Log-binned distribution of number of listeners per artist.

For each artist we have a list of tags, weighted with the number of users that assigned the tag to the artist (e.g. Metallica has 4 tags: "metal" with counter 50670, "hard rock" with 23405, "punk" with 10500 and "adrenaline" with 670). We split tags, associating the counter to each single word (in the last example: (metal, 50670), (punk, 10500), (hard, 23405), (rock, 23405), (adrenaline, 670)), then we filtered the words referring to a musical genre ((metal, 50670), (punk, 10500), (rock, 23405)). Finally, we assigned a musical genre to an artist iff the survived tag with the greater counter had the relative rate ≥ 0.5 (in the example: $r_{metal}(Metallica) = \frac{50670}{50670+10500+23405} \simeq 0.6$, so Metallica are definitely metal).

After the crawl and cleaning stages, we built our social graph $\mathcal{G}$. In $\mathcal{G}$ each node is a user and each edge is generated using the user's friends in the social media platform. The total amount of nodes is 75,969, with 389,639 edges connecting them. In Figure 2 (left) we depicted the log-binned degree distribution of $\mathcal{G}$, along with the best fit. Each action in the data is one user listening to an artist w times in week t. In Figure 2 (right) we depicted the log-binned distribution of the number of listeners per artist, along with the best fit.

Since we are interested in leaders, we need to focus only on new artists that were previously not existent. If an artist was in activity before our observation time window, there is no way to know if a user has listened to it before, therefore nullifying our leader detection strategy. For this reason, we focus only on artists whose first listening is recorded six months after the beginning of our observation period. Each artist belongs to a music genre (coded in its tag) and we want to use this information in Section 5.3. We decided to focus on music genres with sufficient popularity, namely: dance, electronic, folk, jazz, metal, pop, punk, rap and rock. A genre's popularity is determined by having at least 10 artists with at least 100 listeners. To sum up, we focus on the artists who appear for the first

---

[2] http://www.last.fm/api/

time after six months in our observation period, with at least 100 listeners and belonging to one of the mentioned nine tags. The cardinality of our action set $\mathcal{A}$ is 168,216 actions, while the object set Ψ contains a total of 402 artists.

In our experimental settings, we set our damping factor $\beta = 0.5$ for the calculation of the Strength measure. We also set $\delta = 3$, meaning that if a user listened to a particular artist three weeks or more after its neighbor then we do not consider her neighbor to be prominent for her for that action.[3]

5.2 Characterization of the Measures

For each leader, besides Width, Depth and Strength, we calculated also the Degree (number of edges connected to the node), the Clustering coefficient (ratio of triangles over the possible triads centered on the node), the Neighbor Degree (average degree of the neighbors of the node), the Betweenness (share of the shortest paths that pass through the node) and Closeness Centrality (inverse average distance between the node and all the other nodes of the network).

Table 1. Pearson correlation coefficient ρ between Width, Depth, Strength and other network statistics for our leaders

	Width	Strength	Degree	Clustering	Neigh Deg	Bet Centr	Clo Centr
AVG Depth	-0.03	**-0.23**	-0.08	0.05	-0.08	-0.02	**-0.13**
Width	-	0.01	**-0.31**	**0.13**	0.05	-0.07	**-0.59**
Strength	-	-	0.02	-0.02	0.03	0.00	0.04
Degree	-	-	-	**-0.16**	-0.02	**0.77**	**0.56**
Clustering	-	-	-	-	-0.05	-0.06	**-0.32**
Neigh Deg	-	-	-	-	-	-0.00	**0.39**
Bet Centr	-	-	-	-	-	-	**0.22**

In Table 1 we report the Pearson correlation coefficient ρ between the network measures. We highlighted the correlations whose p-value was significant or whose absolute value was strong enough to draw some conclusions. For the significance of p-values, the traditional choice is to set the threshold at $p < 0.01$. However, given our number of observations, we decided to be more restrictive, setting our threshold at $p < 0.0005$. We also consider a ρ value significant if $|\rho| > 0.1$.

The Depth measure is associated with low Closeness Centrality. This means that a deep prominence is associated to nodes at the margin of the network. It is expected that nodes with high Closeness Centrality have also low Depth: being central, they cannot generate long chains of diffusion. The eccentricity of all the nodes of the network ranges from 6 to 10, meaning that some leaders cannot have a Depth larger than 5. To make a fair comparison, we recalculate the Depth value capping it at 5, meaning that any Depth value larger than 5 is manually reduced to 5. Then, we recalculate the correlation ρ between the Depth capped to 5 and the Closeness Centrality obtaining as result $\rho = -0.1366$, with $p < 0.0005$. We can conclude that central nodes are not associated with deep spread of their prominence in a social network.

---

[3] To assure experiment repeatability, we made our cleaned dataset and our code available at the page http://www.michelecoscia.com/?page_id=606

For the Width measure, the anti-correlation with the Degree is not meaningful, as the Degree is in the denominator of Definition 4. However, we observe a positive association with Clustering, i.e. nodes could be prominent in a tightly connected community; and a negative association with Closeness Centrality, i.e. central nodes could not spread a wide influence. Both associations could be explained with the negative correlation with Degree. Therefore, for both measures we run a partial correlation, controlling for the Degree. In practice, we calculate the correlation between Width and Clustering (or Closeness Centrality) by keeping the Degree constant. Results are in Table 2: even if significant according to the p-value, the relationship between Width and Clustering is very weak and deserves further investigation. On the other hand, it is confirmed that central nodes are also associated with low Width, regardless their degree.

Table 2. Partial correlation and p-value of Clustering and Closeness Centrality with Width, controlling for Degree values

	Clustering	Clo Centr
Partial ρ	0.087216	-0.536861
p-value	1.57×10^{-14}	0

From Table 1, we see that the Strength measure is not correlated with traditional network statistics. As a consequence, hubs associated with low Depth and low Width, do not have necessarily high Strength, making their prominence in a network questionable. Moreover, Strength appears to be negatively associated with Depth, suggesting a trade-off between how deeply a node can be prominent in a network and how strong this prominence is on the involved nodes.

The anti-correlation between the Strength and the Depth may be due to β: from Definition 6 β decreases nodes' contributions at each degree of separation (i.e. at increasing Depths). As a consequence, nodes farther from the leader contribute less to its Strength, i.e. the highest the Depth the smallest are the contributions to the Strength. We recalculated the Strength values by setting $\beta = 1$, therefore ignoring any damping factor and nullifying this effect. We obtained as result $\rho = -0.4168$ and a significant p-value, therefore concluding that β is not causing the anti-correlation between Depth and Strength.

To sum up, we summarize the associations as follows: (i) central nodes are not necessarily prominent in a social network (low Width and Depth), a result that confirms [3] and [1]; (ii) longer cascades (higher Depths) are associated with a lower degree of engagement (lower Strengths), a phenomenon possibly related to the role played by "weak ties"; (iii) be prominent among neighbors is probably easier if the node is in a tightly connected community, but more evidences have to be brought to reject the role played by the node's degree.

5.3 Case Study

Here, we present a case study based on Last.Fm data. Our aim is to use our Leader extraction technique and the proposed Width, Depth and Strength measures to characterize the spread of musical genres among the users of the service.

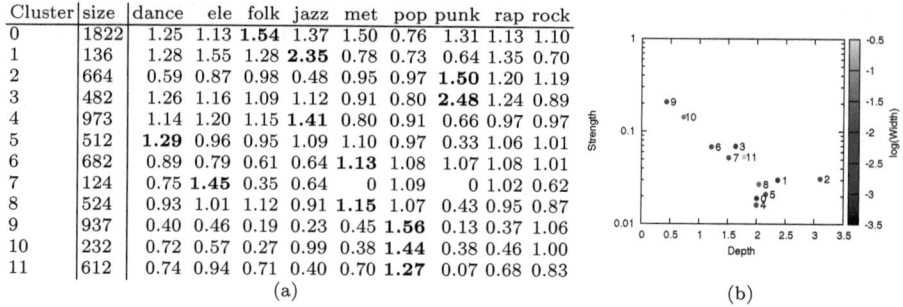

Cluster	size	dance	ele	folk	jazz	met	pop	punk	rap	rock
0	1822	1.25	1.13	**1.54**	1.37	1.50	0.76	1.31	1.13	1.10
1	136	1.28	1.55	1.28	**2.35**	0.78	0.73	0.64	1.35	0.70
2	664	0.59	0.87	0.98	0.48	0.95	0.97	**1.50**	1.20	1.19
3	482	1.26	1.16	1.09	1.12	0.91	0.80	**2.48**	1.24	0.89
4	973	1.14	1.20	1.15	**1.41**	0.80	0.91	0.66	0.97	0.97
5	512	**1.29**	0.96	0.95	1.09	1.10	0.97	0.33	1.06	1.01
6	682	0.89	0.79	0.61	0.64	**1.13**	1.08	1.07	1.08	1.01
7	124	0.75	**1.45**	0.35	0.64	0	1.09	0	1.02	0.62
8	524	0.93	1.01	1.12	0.91	**1.15**	1.07	0.43	0.95	0.87
9	937	0.40	0.46	0.19	0.23	0.45	**1.56**	0.13	0.37	1.06
10	232	0.72	0.57	0.27	0.99	0.38	**1.44**	0.38	0.46	1.00
11	612	0.74	0.94	0.71	0.40	0.70	**1.27**	0.07	0.68	0.83

(a)　　　　　　　　　　　　　　　　　　(b)

Fig. 3. (a) The RCA scores of the presence of each tag in each cluster; (b) The centroids of our clusters

We recall that, as described in Section 5.1, the object set Ψ is composed by 402 artists, each one having a tag corresponding to her main music genre.

For each couple leader l and object ψ, we calculate Depth, Width and Strength values; we compute the size of the Leader's Minimum Diffusion Tree ($|T_{l,\psi}|$); and we group together the objects with the same tag. To characterize the typical values of Width, Depth and Strength for each tag we cannot use the average or the median. This is because Strength and Width values are skewed, and it is the combination of the three measures that really characterizes the leaders. We cluster leaders using as features their Width, Depth and Strength values. We used the Self-Organizing Map (SOM) method [13] because: (i) SOM does not require to set the number of clusters k; (ii) k-means outperforms SOM only if the number of resulting clusters is very small (less than 7) [14], but our study of the best k to be used in k-means with the Sum of Squared Errors (SSE) methodology resulted in a optimal number of clusters falling in a range between 9 and 13 (in fact, SOM returned 12 clusters); and (iii) SOM performs better if the data points are contained in a warped space [12], which is our case.

In Table 3(a), we report a presence score for each tag in each cluster. There are larger and smaller clusters and some tags attract more listeners than others. To report just the share of leaders with a given tag in a given cluster is not meaningful. We correct the ratio with the expected number of leaders with the given tag in the cluster, a measure known as Revealed Comparative Advantage: $RCA(i,j) = \frac{freq_{i,j}}{freq_{i,*}} \Big/ \frac{freq_{*,j}}{freq_{*,*}}$, where i is a tag, j is a cluster, $freq_{i,j}$ is the number of leaders who spread an artist tagged with tag i that is present in cluster j. For each cluster we highlighted the tag with the highest unexpected presence.

The centroids of the SOM are depicted in Figure 3(b): Depth on the x-axis, Strength on the y-axis and the Width as the color (Strength and Width are in log scale). We can identify the clusters characterized by the highest and lowest Strength (9 and 4 respectively); by the highest and lowest Depth (2 and 9 respectively); and by the highest and lowest Width (11 and 1 respectively). There are also clusters with relatively high combinations of two measures: cluster 10 with high Strength and Width or cluster 5 with high Depth and Width.

From Table 3(a) we obtain a description of what values of Width, Depth and Strength are generally associated with each tag. For space constraints, we report only a handful of them for the clusters with extreme values. Jazz dominates clusters 1 (with the lowest Width) and 4 (with the lowest Strength): this fact suggests that jazz is a genre for which it is not easy to be prominent.

Cluster 9, with the lowest Depth but the highest Strength, is dominated by pop (that dominates also clusters 10 and 11, both with high Strength but low Depth). As a result, we can conclude that prominent leaders for pop artists are embedded in groups of users very engaged with the new artist. On the other hand, it is unlikely that these users will be prominent among their friends too.

Finally, cluster 2 with the highest density has a large majority of punk leaders. From this evidence, we can conclude that punk is a genre that can achieve long cascades, exactly the opposite of the pop genre.

We move on to the topological characteristics of the leaders per tag. A caveat: a leader is not bounded to be leader just for one object ψ, but she is free to be prominent in many ψ. Thus, one leader can be counted in more than one tag. To help understand the magnitude of the issue, we depicted in Figure 4 the number of leaders influencing their neighbors for a given amount of actions (left) and for a given amount of tags (right). The y axis is logarithmic. The typical leader influences one neighbor for one artist. However, some leaders express their leadership for 8 objects and 4 tags.

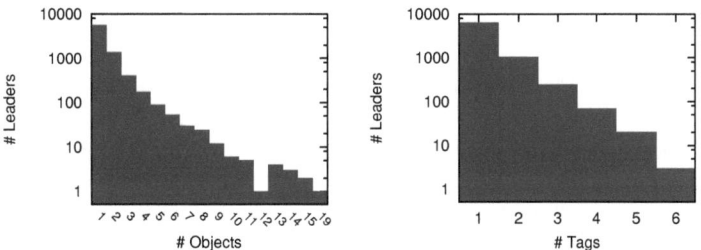

Fig. 4. Distribution of number of objects (left) and of tags (right) per leader

In Figure 5 we depict the log-binned distributions, for the leaders of each tag, of four of the topological measures studied in Section 5.2: Degree, Closeness Centrality, Clustering and Neighbor Degree. We omit Betweenness Centrality for its very high correlation with Degree. Overall, there is no significant distinction between the tags in the distributions of the topological features.

The most noticeable information is carried by the Degree distributions (Figure 5, top left). Each distribution appears very different from the overall degree distribution (Figure 2 (left)). There are fewer leaders with low Degree than expected, therefore it appears that a high Degree increases the probability of being a leader. On the other hand, we know that central hubs have on average lower Depth and Width. As a consequence, it appears that the best leader candidates are the nodes with an average degree, and from Figure 5 (top left) we see that each tag has many leaders with a Degree between 10 and 100.

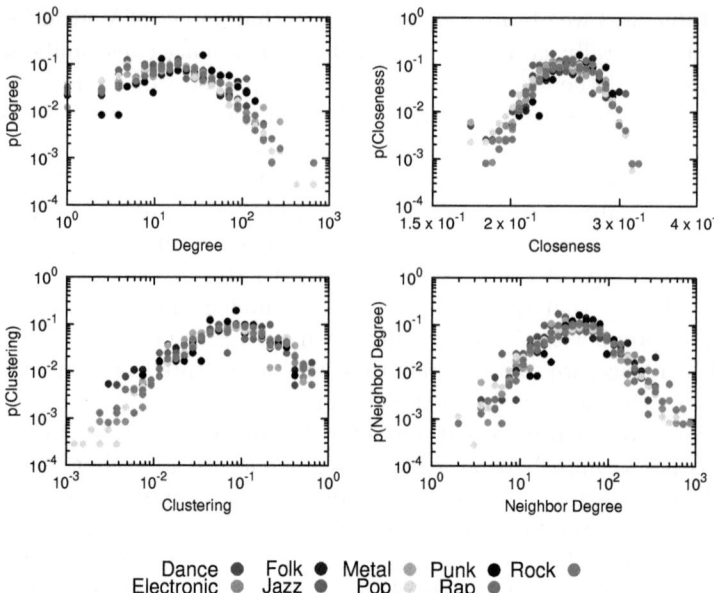

Fig. 5. Distribution of leaders' Degree (top left), Closeness Centrality (top right), Clustering (bottom left) and Neighbor Degree (bottom right) per tag

Using our leaders' Minimum Diffusion Trees, we extract some patterns that help us obtaining a complementary point of view over the leader prominence for different music genres. We mine a graph dataset composed by all diffusion trees $T_{l,\psi}$ with the VF2 algorithm [7]. Suppose we are interested in counting how frequent is the following star pattern: a leader influences three of its neighbors in the diffusion trees of pop artists. In our data, we have 5,043 diffusion trees for pop artists, and 581 have at least four nodes. Since the VF2 algorithm found the star pattern in 186 of these graphs, we say that it appears in 3.69% of the trees, or in 32.01% of the trees that have enough nodes to contain it.

In Table 3 we report the results of mining three patterns of four nodes: i) the star-like pattern described above; ii) a chain where each node is prominent for (at least) one neighbor; iii) a split where the leader is prominent for a node, which itself is prominent for two other neighbors. Two values are associated to each pattern and tag pair: the relative overall frequency, and the relative frequency considering only the trees with at least four nodes (in parentheses).

There is no necessary relation between the patterns and Width, Depth and Strength measures: a low Depth does not imply the absence of the chain pattern, nor does a high Width imply a high presence of the star pattern. However, the combination of the two measures may provide some insights. For instance, we saw in Table 3(a) that jazz leaders are concentrated in the lowest Width cluster. However, many jazz leaders who affect at least three nodes tend to be prominent in their neighbors, much more than in any other genre (7.25% of all leaders, 62.5% of leaders who are prominent for at least three other nodes). Therefore,

Table 3. Presence of different diffusion patterns per tag

Pattern	dance	electronic	folk	jazz	metal	pop	punk	rap	rock
(Y-shape)	3.62% (35.42%)	3.04% (22.50%)	3.94% (30.30%)	7.25% (62.50%)	4.14% (23.08%)	3.69% (32.01%)	6.56% (27.59%)	4.01% (27.97%)	4.22% (30.43%)
(chain)	2.55% (25.00%)	3.92% (29.00%)	3.15% (24.24%)	4.35% (37.50%)	4.83% (26.92%)	3.61% (31.29%)	10.66% (44.83%)	5.60% (38.98%)	4.12% (29.71%)
(split)	3.40% (33.33%)	3.79% (28.00%)	3.94% (30.30%)	4.35% (37.50%)	6.90% (38.46%)	4.73% (41.01%)	12.30% (51.72%)	4.99% (34.75%)	4.52% (32.61%)

jazz leaders have low prominence among their friends, however they are likely to have at least three neighbors for which they are prominent.

The chain pattern is more commonly found in pop leaders than in folk ones, even though the clusters of their leaders described in Table 3(a) would suggest the opposite. It seems that pop leaders are not likely to be prominent for nodes any further than the third degree of separation, while folk leaders tend to generate longer cascade chains. Also in this case, punk leaders are commonly found in correspondence with chain patterns, just as Table 3(a) suggested.

Although pop leaders show a much greater Strength value than metal ones (by confronting in Table 3(a) their presence in high Strength clusters like 9 or 10 and low Strength clusters like 8 and 0), the split pattern tends to be more frequent in the metal genre (6.90% against 4.73% of the trees). This phenomenon suggests us that metal leaders tend to be prominent for nodes strongly devoted to metal, inducing them to spread the music to their neighbors. Pop leaders, on the other hand, affect more neighbors with higher Width and Strength, presumably flooding their ego networks with the songs they like.

6 Conclusion

In this paper, we presented a study of the propagation of behaviors in a social network. Instead of just studying cascade effects and the maximization of influence by a given starting seed, we decided to analyze three different dimensions: the prominence of a leader on how many neighbors, on how distant nodes and on how engaged nodes. We characterized each of these concepts with a different measure: Width, Depth and Strength. We applied our leader detection algorithm to a real world network. Our results show that: (i) central hubs are usually incapable of having a strong effect in influencing the behavior of the entire network; (ii) there is a trade-off between how long the cascade chains are and how engaged each element of the chain is; (iii) to achieve maximum engagement it is better to target leaders in tightly connected communities, although for this last point we do not have conclusive evidence. We also included a case study in which we show how artists in different musical genres are spread through the network.

Many future developments are possible. The limited prominence that central hubs have on the overall network may be studied in conjunction with the problem of network controllability [15]. Alternative leader detection techniques, such as the ones presented in [10], can be confronted with our proposed algorithm. Finally, a deeper analysis of the properties of the Width, Depth and Strength measures can be performed, using additional techniques and exploiting data from other social media services like Twitter and Facebook.

Acknowledgments. The authors want to thank Prof. Otto Koppius for his presentation about the prominence network measures and useful discussions. This work has been partially supported by the European Commission under the FET-Open Project n. FP7-ICT-270833, DATA SIM.

References

1. Berlingerio, M., Coscia, M., Giannotti, F.: Mining the temporal dimension of the information propagation. In: Adams, N.M., Robardet, C., Siebes, A., Boulicaut, J.-F. (eds.) IDA 2009. LNCS, vol. 5772, pp. 237–248. Springer, Heidelberg (2009)
2. Burt, R.S.: Social contagion and innovation: Cohesion versus structural equivalence. American Journal of Sociology 92(6), 1287–1335 (1987)
3. Cha, M., Haddadi, H., Benevenuto, F., Gummadi, K.P.: Measuring user influence in twitter: The million follower fallacy. In: ICWSM (2010)
4. Christakis, N.A., Fowler, J.H.: The spread of obesity in a large social network over 32 years. New England Journal of Medicine 357(4), 370–379 (2007)
5. Christakis, N.A., Fowler, J.H.: The collective dynamics of smoking in a large social network. New England Jou. of Medicine 358(21), 2249–2258 (2008)
6. Colizza, V., Barrat, A., Barthelemy, M., Valleron, A.-J., Vespignani, A.: Modeling the worldwide spread of pandemic influenza: Baseline case and containment interventions. PLoS Medicine 4(1), e13 (2007)
7. Cordella, L.P., Foggia, P., Sansone, C., Vento, M.: A (sub)graph isomorphism algorithm for matching large graphs. IEEE Transactions on Pattern Analysis and Machine Intelligence 26(10), 1367–1372 (2004)
8. Coscia, M.: Competition and success in the meme pool: a case study on quickmeme.com. In: ICWSM (2013)
9. Fowler, J.H., Christakis, N.A.: Dynamic spread of happiness in a large social network: longitudinal analysis over 20 years in the framingham heart study. Bmj Clinical Research Ed. 337(2), a2338–a2338 (2008)
10. Goyal, A., Bonchi, F., Lakshmanan, L.V.S.: Discovering leaders from community actions. In: CIKM, pp. 499–508 (2008)
11. Kermack, W.O., McKendrick, A.G.: A contribution to the mathematical theory of epidemics. The Royal Society of London Series A 115(772), 700–721 (1927)
12. Kiang, M.Y., Kumar, A.: A comparative analysis of an extended som network and k-means analysis. Int. J. Know.-Based Intell. Eng. Syst. 8(1), 9–15 (2004)
13. Kohonen, T.: The self-organizing map. IEEE 78, 1464–1480 (1990)
14. Kumar, U.A., Dhamija, Y.: A comparative analysis of som neural network with k-means clustering algorithm. In: Proceedings of IEEE International Conference on Management of Innovation and Technology, pp. 55–59 (2004)
15. Liu, Y.-Y., Slotine, J.-J., Barabasi, A.-L.: Controllability of complex networks. Nature 473(7346), 167–173 (2011)
16. Pastor-Satorras, R., Vespignani, A.: Epidemic spreading in scale-free networks. Physical Review Letters 86(14), 3200–3203 (2001)
17. Wang, P., González, M.C., Hidalgo, C.A., Barabási, A.-L.: Understanding the spreading patterns of mobile phone viruses. Science 324(5930), 1071–1076 (2009)

Automatic Thematic Content Analysis: Finding Frames in News

Daan Odijk[1], Björn Burscher[2], Rens Vliegenthart[2], and Maarten de Rijke[1]

[1] Intelligent System Labs Amsterdam (ISLA), University of Amsterdam
{d.odijk,derijke}@uva.nl
[2] Department of Communication Science, University of Amsterdam
and Amsterdam School of Communication Research (ASCoR)
{b.burscher,r.vliegenthart}@uva.nl

Abstract. Framing in news is the way in which journalists depict an issue in terms of a 'central organizing idea.' Frames can be a perspective on an issue. We explore the automatic classification of four generic news frames: conflict, human interest, economic consequences, and morality. Complex characteristics of messages such as frames have been studied using thematic content analysis. Indicator questions are formulated, which are then manually coded by humans after reading a text and combined into a characterization of the message. We operationalize this as a classification task and, inspired by the way-of-working of media analysts, we propose a two-stage approach, where we first rate a news article using indicator questions for a frame and then use the outcomes to predict whether a frame is present. We approach human accuracy on almost all indicator questions and frames.

1 Introduction

There is a growing trend of applying computational thinking and linguistic approaches to social science research. In particular, language technology is proving to be a useful but underutilized approach that may be able to make significant contributions to research in a wide range of social science domains [2]. One particular domain in which this is happening is the study of news and its impact. Early examples focus mostly on analyzing factual aspects in news, such as [15], who analyzed the impact of news on corporate reputation by measuring the amount of news about specific issues. Increasingly, however, we are also seeing the use of language technology to analyze more subjective aspects of news for the purposes of social science research [12]. In this paper, we report on work aimed at analyzing the use of framing in news.

Framing in news is the way in which journalists depict an issue in terms of a 'central organizing idea' [6]. Frames can be regarded as a perspective on an issue. In the social sciences, mass communication (e.g., news) is often studied through a methodology called content analysis: "Who says what, to whom, why, to what extent and with what effect?" [11]. The aim of content analysis is to systematically quantify specified characteristics of messages. When these characteristics are complex, thematic content analysis can be applied: first, texts are annotated for indicator questions (e.g., "Does the item refer to winners and losers?") and the answers to such questions are subsequently aggregated to support a more complex judgment about the text (e.g., the presence of a conflict frame). Content analysis is a laborious process, and there is a clear need for a computational approach. This approach can improve the consistency, efficiency, reliability and replicability of the analyses, as larger volumes of news can be studied in a replicable manner, allowing the study of long-term trends.

For studying framing, some first computational approaches have been attempted, using dictionary-based methods. We approach the problem of frame detection in news as a two-stage classification task. We start by predicting the outcomes to indicator questions associated with a frame and then use the predicted outcomes to decide about the presence of the frame in a given text.

Our contribution in this paper consists in a two-stage approach to finding frames that allows us to answer the following research questions:

RQ1. Can we approach human performance on predicting answers to indicator questions?

RQ2. Can we approach human performance on predicting the presence of a frame?

The remainder of this article is organized as follows: in Section 2 we discuss media frame analysis and related work; Section 3 describes our proposed methods and Section 4 describes the experimental setup. We present and discuss our results in Section 5, after which we conclude in Section 6.

2 Media Frame Analysis

News coverage can be approached as an accumulation of "interpretative packages" in which journalists depict an issue in terms of a *frame*, i.e., a central organizing idea [6]. Frames are the dependent variable when studying the process of how frames emerge (*frame building*) and the independent variable when studying effects of frames on predispositions of the public (*frame setting*) [22]. When studying the adoption of frames in the news, content analysis of news media is the most dominant research technique.

Using questions as indicators of news frames in manual content analysis is the most widely used approach to manually detecting frames in text. Indicator questions are added to a codebook and answered by human coders while reading the text unit to be analyzed [26]. Each question is designed such that it captures the semantics of a given frame. Generally, several questions are combined as indicators for the same frame. This way of making inferences from texts is also referred to as thematic content analysis [20].

Automatic or semi-automatic frame detection is rare. The approaches that do exist follow a dictionary-based or rule-based approach. E.g., Ruigrok and Van Atteveldt [21] define search strings for the automatic extraction of a priori defined concepts in newspaper articles, and then apply a probabilistic measure to indicate associations between such concepts. Similarly, Shah et al. [25] first define "idea categories," then specify words that reveal those categories, and finally, program rules that combine the idea categories in order to give a more complex meaning as a frame.

In this paper we focus on four commonly used frames [24]. For convenience they are listed in Section 4, together with their indicator questions. The *conflict frame* highlights conflict between individuals, groups or institutions. Prior research has shown that the depiction of conflict is common in political news coverage [16], and that it has inherent news value [4, 27].

By emphasizing individual examples in the illustration of issues, the *human interest frame* adds a human face to news coverage. According to Iyengar [7], news coverage can be framed in a thematic manner, taking a macro perspective, or in an episodic manner, focusing on the role of the individual concerned by an issue. Such use of exemplars in news coverage is observed by several scholars [16, 24, 28] and connects to research on personalization of political news [7].

The *economic consequence frame* approaches an event in terms of its economic impact on individuals, groups, countries or institutions. Covering an event with respect to its consequences is argued to possess high news value and to increase the pertinence of the event among the audience [5].

The *morality frame* puts moral prescriptions or moral tenets central when discussing an issue or event. Morality as a news frame has been studied in various academic publications and is found to be applied in the context of various issues as, for example, gay rights [17] and biotechnology [1].

Over the past decade, language technology has witnessed a rapid broadening, along two dimensions. First, moving beyond an almost exclusive focus on working with news corpora, different genres of text are now being subjected to, e.g., semantic analysis [13, 14]. Second, from a strong focus on analyzing facts the field is broadening to also include more subjective aspects of language, such as opinions and sentiment [19], human values [3], argumentation [18] and user experiences from online forums [8].

In this paper, we contribute over and above the related work discussed, by presenting and evaluating an ensemble-based classification approach for frame detection in news. To the best of our knowledge, this is the first work in which statistical classification methods are applied to this central issue in studying media. Furthermore, we investigate whether explicitly modeling the thematic content analysis approach improves performance.

3 Frame Classification

We approach the task of frame detection in news as a classification task. The assumption underlying thematic content analysis is that frames manifest themselves in a news article in a manner that is measured using indicator questions. We follow this assumption and analyze the wording in a news article in order to make a decision about the presence of frames.

Given a collection of documents D and a set of frames U for which a set of indicator questions V have been defined, we estimate the probability $P(u_m|d)$

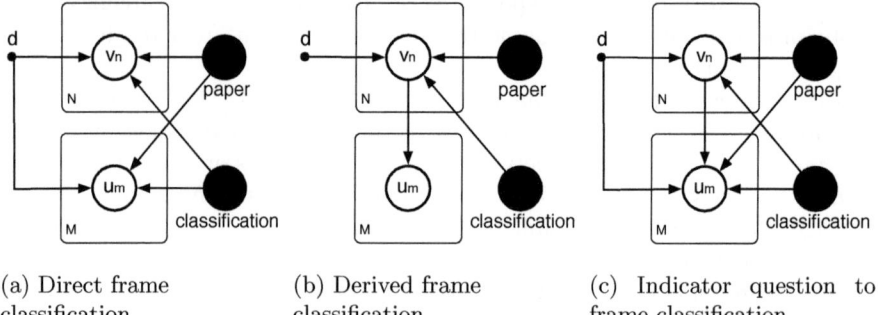

(a) Direct frame classification

(b) Derived frame classification

(c) Indicator question to frame classification

Fig. 1. Graphical models of the three classification approaches. The circles represent random variables, where the filled are observable. The rectangular plates indicate multiple of these variables.

that a frame $u_m \in U$ is present in document $d \in D$. In thematic content analysis this probability is deconstructed into $P(u_m|v_1,\ldots,v_N)$ and a set of probabilities for each $v_n \in V$: $P(v_n|d)$. This formal definition of the task can be used for the automatic classification and for manual content analysis. In the latter case, the probability $P(v_n|d)$ is estimated using manually coding by humans after reading the document.

Document Representation. We represent documents as a bag-of-words with TF.IDF scores for each word. We apply sublinear term frequency scaling, i.e., replace TF with $1 + \log(TF)$, use l2 normalization and smooth IDF weights by adding one to document frequencies. We have evaluated other representation (e.g., n-grams and topic models), but these did not improve classification performance and will not be reported here.

Besides the words represented in the document, we extend the document representation with information on the source of the document and with a classification for each document (i.e., a topic, such as finance, infrastructure, etc.). The extended bag-of-words document representation serves as the features for classification.

Frame and Indicator Question Classification. We propose two baselines and three approaches for automatic indicator question and frame classification. These methods differ in how the coherence between indicator question and frame is modeled. The approaches are depicted in graphical models in Figure 1 and will be described below.

Stratified Random Classification Baseline. Our first baseline approach is very naive and intended as a lower bound. It randomly choses the answer to a indicator question or whether a frame is present or not, taking into account only the prevalence in training set. This naive baseline randomly assigns a classification, without considering the document and its representation, with a probability based on the class distributions. This naive baseline will be more likely to randomly pick the majority class than the minority class.

Direct Classification Baseline. Our second baseline approach is to classify answers to indicator questions and the presence of each frame directly. More formally, we train a classifier to estimate $P(u_m|d)$ for each frame $u_m \in U$. This approach is the simplest approach and is depicted in Figure 1a. Note that for frames, we completely ignore the indicator questions in this baseline approach.

For classification we use Logistic Regression to optimize logistic loss using Pegasos-style regularization. For training we alternate between pairwise ROC-optimization and standard stochastic gradient steps on single examples [23]. This baseline approach aims to be flexible in dealing with issues such as class imbalance.

Ensemble-Based Direct Classification. Our first approach is to improve binary classification decisions for indicator questions and for the presence of a frame by using an ensemble of binary-class linear classifiers (also depicted in Figure 1a). The predictions of all these classifiers are the features for a final classifier. The ensemble includes different linear support vector machines (SVMs), linear rank-based SVMs [9, 23], and Perceptron-based algorithms [10]. This ensemble-based approach aims to be flexible in dealing with the different complex characteristic of each of the classifications. We combine the classifiers in the ensemble using the same classifier as described above for the baseline approach.

Derived Frame Classification. Our second approach is to derive the presence or absence of a frame based on the classification for indicator questions. More formally, we train an ensemble-based classifier to estimate $P(\hat{v}_n|d)$ for each indicator question $v_n \in V$. We then derive the probability $P(u_m|d)$ for each frame $u_m \in U$ from $P(\hat{u}_m|\hat{v}_1, \ldots, \hat{v}_N)$ for all indicator questions $v_m \in V$. This approach is depicted in Figute 1b and closely resembles the manual approach, where human coders make binary decisions for $P(v_n|d)$ for each $v_n \in V$ and $d \in D$.

Indicator Question to Frame Classification. Our third approach is a cascade approach, where we first classify for the indicator question and then use the outcomes to classify the frames. More formally, we train an ensemble-based classifier to estimate $P(\hat{v}_n|d)$ for each indicator question $v_n \in V$. We then train an ensemble-based classifier to estimate the probability $P(u_m|d, \hat{v}_1, \ldots, \hat{v}_N)$ for each frame $u_m \in U$. This approach is depicted in Figure 1c. Practically, we implement this by adding ensemble-based predictions for indicator questions as features for the frames classifiers.

4 Experimental Setup

To evaluate our methods we run a number of experiments. We describe the document collection used, outline how the four frames have been coded in the manual content analysis that we use as training and test data, and explain how we evaluate the performance of our classification models.

Document Collection. Our document collection consists of digital versions of front page news articles of three Dutch national daily newspapers (*De Volkskrant*,

NRC Handelsblad and De Telegraaf) for the period between 1995 and 2011. These articles come from the Dutch Lexis-Nexis newspaper archive, and each article has a topical classification (based, e.g., on the location in the newspaper). The used sample is a stratified sample of 13% for each year.

Indicator Questions Annotations. For each year covered in our collection, a random sample of news articles was taken. This sample was filtered (based on manually assigned labels) to only contain articles that were political in nature. The resulting 5,875 documents have been manually coded for the presence of four generic news frames (described in Section 2). Indicator questions were used to code the news frames.

A total of thirteen yes-or-no-questions were used as indicators of the news frames. In previous research, these questions have been shown to be reliable indicators of the four frames [24]. The indicator questions for each frame are:

C *Conflict frame*:
   C1 Does the item reflect disagreement between parties, individuals, groups or countries?
   C2 Does the item refer to winners and losers?
   C3 Does the item refer to two sides or more than two sides of the problem?
E *Economic consequence frame*:
   E1 Is there a reference to the financial costs/degree of expense involved, or to financial losses or gains, now or in the future?
   E2 Is there a reference to the non-financial costs/degree of expense involved, or to non-financial losses or gains, now or in the future?
   E3 Is there a reference to economic consequences of pursuing or not pursuing a course of action?
H *Human interest frame*:
   H1 Does the item provide a human example or human face on the issue?
   H2 Does the item employ adjectives or personal vignettes that generate feelings of outrage, empathy caring?
   H3 Does the item mention how individuals and groups are affected by the issue or problem?
   H4 Does the item go into the private or personal lives of the actors?
M *Morality frame*:
   M1 Does the item contain any moral message?
   M2 Does the item make reference to morality, God or other religious tenets?
   M3 Does the item offer specific social prescriptions about how to behave?

Manual coding was conducted by a total of 30 trained coders. All coders were communication science students and native speakers of the Dutch language. In order to assess inter-coder reliability, a random subset of 159 articles was coded by multiple coders. Measures of the percentage of inter-coder agreement range from 70% to 94%. The inter-coder reliability is included in the results in Table 1 and Table 3, with the label 'Human.'

Frame Annotations. Based on the annotations for indicator questions, a second annotation round gave rise to the construction of frame annotations, following the methodology described in [24]. To establish the coherence of the indicator questions and their relation to the frames a factor analysis is performed. We find a four factor solution for the answers to the indicator questions. In this solution each indicator question has a loading onto each factor (i.e., a weight).

In these factor loadings, we can identify the four frames, i.e., for each frame there is a factor with high loads for the corresponding indicator questions and low loadings for the others. For two indicator questions (C2 and E2) the factor load is below 0.5, and hence these were considered unreliable indicators (in line with [24]). This means that the remaining indicator questions can be considered reliable indicators of the four frames: a frame is considered present in a news document whenever any of the indicator questions corresponding to the frame is answered positively.

Evaluation Metrics. We perform ten-fold cross-validation and compare the agreement between human annotators and our automatic approach in terms of agreement. Where possible, we evaluate both the answers to indicator questions and the frame annotations. Furthermore, we compare the approaches in receiver operating characteristics (ROC) space. We compare the ability to distinguish true positive classifications from false positives for different operating characteristics that will produce increasingly more positive results. In this ROC space, we can compute the area under the curve (AUC). The AUC metric for a classifier expresses the probability that the classifier will rank a positive document above a negative document.

5 Results and Discussion

Table 1 and Table 3 describe the agreement between our approaches and the human annotations for each of the eleven indicator questions and the four frames. For comparison, these tables also include the inter-annotator agreement for human coders. Table 2 and Table 4 describe the area under the curve (AUC) metric for our approaches.

Indicator Questions Classification Results. We can observe in Table 1 that our baseline single classifier direct approach ("Direct") performs well on some of the indicator questions, but worse on others. The direct baseline is unable to consistently improve over the naive stratified random baseline ("Random"). Our ensemble-based approach ("Ensemble") substantially improves over these baselines and achieves accuracy scores ranging from 65% accuracy upwards. While we observe that the accuracy varies among the four frames and the corresponding indicator questions, our ensemble-based approach is able to capture the complex characteristics for all questions and frames. The conflict indicator questions (C1 and C3) and human interest question H3 perform below average in the baselines, but perform substantially better in the ensemble-based approach.

Human interest question H4 and the morality questions (M1, M2 and M3) show high baseline performance, but do not show substantially improvements for the direct approaches, despite our pairwise optimization approach. This suggests

Table 1. Agreement between automatic classification predictions and human annotations for each of the eleven indicator questions and the three approaches (two baselines and ensemble)

	C1	C3	E1	E3	H1	H2	H3	H4	M1	M2	M3
Random	.5214	.5980	.7093	.8419	.7963	.8346	.5144	.9122	.9348	.9397	.9535
Direct	.5709	.6140	.7093	.8419	.7963	.8346	.5750	.9122	.9348	.9397	.9535
Ensemble	.7064	.6945	.8511	.8650	.8007	.8393	.6489	.9137	.9345	.9460	.9535

Coder biased ensemble run is included below for analysis.

	C1	C3	E1	E3	H1	H2	H3	H4	M1	M2	M3
Biased	.7200	.7413	.8553	.8819	.8213	.8494	.7045	.9185	.9346	.9501	.9525

Human inter-coder agreement is included below for comparison. Note that this is evaluated on a small dataset.

	C1	C3	E1	E3	H1	H2	H3	H4	M1	M2	M3
Human	.7239	.6994	.8282	.8466	.7546	.7055	.6748	.8405	.9080	.9041	.9202

Table 2. Area under the curve (AUC) for ROC of automatic classification predictions compared to human annotations for each of the eleven indicator questions and the two direct approaches (baseline and ensemble)

	C1	C3	E1	E3	H1	H2	H3	H4	M1	M2	M3
Direct	.6235	.6601	.6973	.6885	.6283	.5802	.6027	.5960	.5572	.6591	.4903
Ensemble	.7744	.7672	.8966	.8432	.7483	.7419	.7051	.7990	.6917	.8884	.6509

that these questions are underrepresented and possibly less well represented using a bag-of-words approach than the other questions.

Looking at the AUC metric results in Table 2, we see the same substantial improvements of the ensemble-based approach over the direct classification baseline. We can also observe a substantial improvement for the aforementioned indicator questions H4, M1, M2 and M3. This suggest that while we are not better in terms of accuracy for these questions, we are indeed better at estimating the probability of a document belonging to a class.

Frame Classification Results. We can observe in Table 3 that accuracy scores on frames follow the same pattern as the indicator questions. The conflict and human interest frame prediction again performs worse than the others. Interestingly, we can observe a substantial improvement for the morality frame over the stratified random baseline. The ensemble-based approach is able to obtain substantial improvements over the baselines approaches. We can also observe that deriving the scores from the indicator questions does not perform well, directly predicting scores for frames using the ensemble-based approach performs substantially better. Interestingly, the two-stage indicator question to frame classification approach does not perform better than the direct approach. The additional information we add by first classifying the indicator questions does not help in classifying the frames. The results for the AUC metric (described in Table 4) show a qualitatively similar pattern as the agreement.

Table 3. Agreement between automatic classification predictions and human annotations for each of the four frames and the five frame classification approaches

	C	E	H	M
Random	.6403	.5755	.6231	.8679
Direct	.6654	.8134	.7779	.9668
Ensemble	.7241	.8506	.7949	.9668
Derived	.5709	.7093	.6158	.9348
IQ → F	.7202	.8489	.8014	.9677
Coder biased ensemble run is included below for analysis.				
Biased	.7501	.8642	.8141	.9685
Human agreement on small dataset included for comparison.				
Human	.7730	.8160	.6442	.8528

Table 4. Area under the curve (AUC) for ROC of automatic classification predictions compared to human annotations for each of the four frames and four frame classification approaches

	C	E	H	M
Direct	.6379	.6956	.6008	.5909
Ensemble	.7802	.8496	.7580	.7597
Derived	.5575	.5000	.5897	.5000
IQ → F	.7677	.8436	.7748	.8025

Furthermore, we can observe from Table 1 and Table 3 that the morality frame and the corresponding questions perform strikingly well in all approaches in terms of agreement. A plausible explanation for this is that this frame is a lot less prevalent than the other three (present in 13% of the documents, compared to 64% for conflict, 58% for economic consequence and 62% for human interest). The AUC results in Table 2 and Table 4 provide some evidence that these classifiers still perform up to par.

To validate this, and to obtain more insight into the operation characteristics of the classifiers we take a more detailed look at the ROC curves. Figure 2 shows these ROC curves for the direct ensemble-based approach for the frames. We can observe a similar curve for each of the frames. From these graphs and the AUC results, we can conclude that while we can not perfectly classify the frame annotations, we are able to obtain a good rate of true positives if we allow some false positives.

Human Inter-coder Agreement. Compared to human inter-coder agreement, nearly all accuracy scores for the ensemble-based and two-stage approaches are at or above that level. Note, however, that human agreement is evaluated on a much smaller dataset. We observe a lower performance compared to human

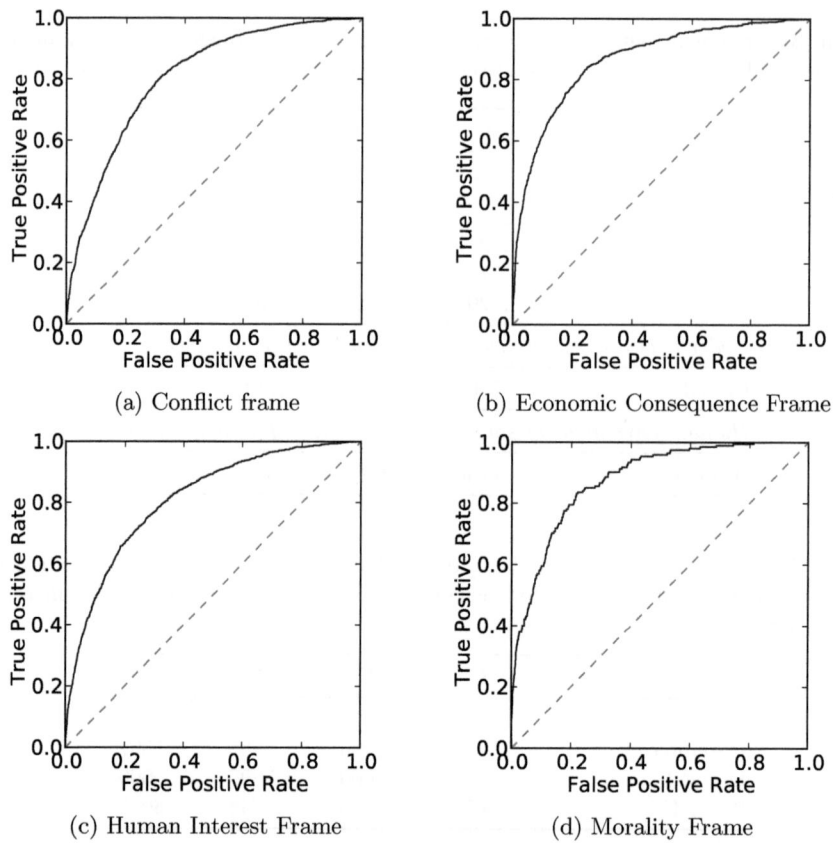

Fig. 2. ROC Curves for the ensemble-based direct approach for the four frames

agreement for question H3, the conflict frame and corresponding questions C1 and C3. For the morality frame and the human interest and morality questions the human inter-coder agreement is even below the stratified random baseline.

To investigate the difficulty of each question and the quality of the human annotations, we look at whether the annotations for questions are stable across coders. We measure this by evaluating a new ensemble-based model where the document representation is extended with variables representing the coder. This creates the unrealistic but insightful scenario where we predict the answer of a specific coder to a specific question. This model allows us to compensate for a bias a coder might have, possibly resulting in higher performance compared to the regular ensemble-based approach.

Agreement for the biased model is included in Table 1 and Table 3. We observe increased performance for most questions, with C3, E3 and H3 standing out. For frames the performance is increased for the human interest frame, economic consequence frame and most substantially for the conflict frame.

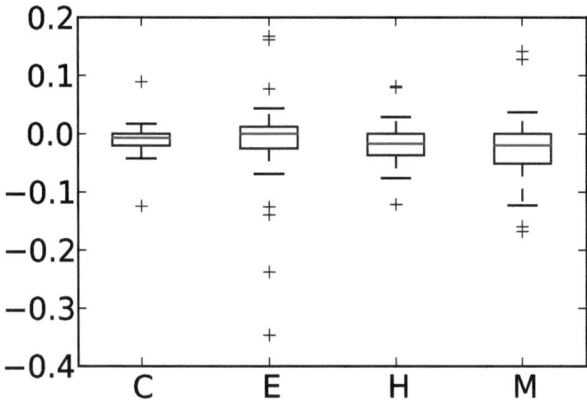

Fig. 3. Box plot of the weights on the binary coder variables for the four frames in one of the ensemble SVM classifiers

To further investigate this, we look at the weights that the coder features get assigned in the biased model. If all coders would answer the indicator questions exactly the same, the coder features will have a weight very close to zero. A weight that differs from zero suggests a consistent difference in answers from one coder compared to the other coders. Figure 3 shows these weights for each frame in one of the classifiers in the direct frame ensemble classifier. We see that the weights do indeed deviate from zero, with a different range per frame. The economic consequence frame has the highest range, with a maximum of 0.5 bias per coder on a scale of −1 to 1. These weights suggest consistently different interpretations of the indicator questions across coders.

6 Conclusion

We have proposed algorithmic approaches to finding frames in news that follow the manual thematic content analysis approach. Our results provide strong evidence that we are able to approach human performance on predicting both the answers to indicator questions as well as the presence of a frame.

Our ensemble-based approach to directly predicting the presence of a frame is the most effective and improves substantially over the baseline approach. The derived approach, which directly follows the manual approach, was the least effective. Surprisingly, the more informed indicator question to frame classification approach did not perform better than the ensemble-based direct classification approach. This suggests that for the task of frame classification, explicitly modeling the manual thematic content analysis does not improve performance. Our ensemble-based direct classification approach is sufficient to capture the complex characteristics of frames that the indicator questions are aimed to represent.

The results of an analysis using a model that explicitly models coder bias and the relatively low inter-coder agreement suggest that coders have different interpretations of the indicator questions for the frames. Like the indicator questions that represent different aspects of complex characteristics of messages, it seems that human coders represent different views on these aspects and characteristics.

Finally, for the task of frame detection in news, we have shown that using an ensemble-based classification approach we are able to approach human performance in terms of accuracy on this task. A combined approach of human and automated frame detection seems to be the logical way forward.

Acknowledgments. This research was supported by the European Community's Seventh Framework Programme (FP7/2007-2013) under grant agreements nr 258191 (PROMISE Network of Excellence) and 288024 (LiMoSINe project), the Netherlands Organisation for Scientific Research (NWO) under project nrs 640.004.802, 727.011.005, 612.001.116, HOR-11-10, 451-09-011, the Center for Creation, Content and Technology (CCCT), the QuaMerdes project funded by the CLARIN-nl program, the TROVe project funded by the CLARIAH program, the Dutch national program COMMIT, the ESF Research Network Program ELIAS, the Elite Network Shifts project funded by the Royal Dutch Academy of Sciences (KNAW), the Netherlands eScience Center under project number 027.012.105 and the Yahoo! Faculty Research and Engagement Program.

References

[1] Brewer, P.R.: Framing, value words, and citizens' explanations of their issue opinions. Political Communication 19(3), 303–316 (2002)

[2] Cheng, A.-S., Fleischmann, K.R., Wang, P., Oard, D.W.: Advancing social science research by applying computational linguistics. In: Proceedings of the Annual Conference of the American Society for Information Science and Technology (2008)

[3] Fleischmann, K.R., Oard, D.W., Cheng, A.-S., Wang, P., Ishita, E.: Automatic classification of human values: Applying computational thinking to information ethics. Proceedings of the American Society for Information Science and Technology 46(1), 1–4 (2009)

[4] Galtung, J., Ruge, M.H.: The structure of foreign news. Journal of Peace Research 2(1), 64–90 (1965)

[5] Gamson, W.: Talking Politics. Cambridge University Press (1992)

[6] Gamson, W., Modigliani, A.: Media discourse and public opinion on nuclear power: A constructionist approach. American Journal of Sociology, 1–37 (1989)

[7] Iyengar, S.: Is anyone responsible?: How television frames political issues. University of Chicago Press (1991)

[8] Jijkoun, V., de Rijke, M., Weerkamp, W., Ackermans, P., Geleijnse, G.: Mining user experiences from online forums: an exploration. In: Proceedings of the NAACL HLT 2010 Workshop on Computational Linguistics in a World of Social Media, pp. 17–18. Association for Computational Linguistics (2010)

[9] Joachims, T.: Optimizing search engines using clickthrough data. In: SIGKDD 2002, pp. 133–142. ACM (2002)

[10] Krauth, W., Mézard, M.: Learning algorithms with optimal stability in neural networks. Journal of Physics A: Mathematical and General 20(11), L745 (1999)

[11] Lasswell, H.D.: The structure and function of communication in society. The Communication of Ideas, 37 (1948)

12. Lazer, D., Pentland, A.S., Adamic, L., Aral, S., Barabasi, A.L., Brewer, D., Christakis, N., Contractor, N., Fowler, J., Gutmann, M., et al.: Life in the network: the coming age of computational social science. Science 323(5915), 721 (2009)
13. Meij, E., Bron, M., Hollink, L., Huurnink, B., de Rijke, M.: Learning semantic query suggestions. In: Bernstein, A., Karger, D.R., Heath, T., Feigenbaum, L., Maynard, D., Motta, E., Thirunarayan, K. (eds.) ISWC 2009. LNCS, vol. 5823, pp. 424–440. Springer, Heidelberg (2009)
14. Meij, E., Weerkamp, W., de Rijke, M.: Adding semantics to microblog posts. In: WSDM 2012: Fifth ACM International Conference on Web Search and Data Mining (February 2012)
15. Meijer, M., Kleinnijenhuis, J.: ssue news and corporate reputation: Applying the theories of agenda setting and issue ownership in the field of business communication. Journal of Communication 56(3), 543–559 (2006)
16. Neuman, W.R., Just, M.R., Crigler, A.N.: Common knowledge: News and the construction of political meaning. University of Chicago Press (1992)
17. Nisbet, M.C., Huge, M.: Attention cycles and frames in the plant biotechnology debate managing power and participation through the press/policy connection. The Harvard International Journal of Press/Politics 11(2), 3–40 (2006)
18. Palau, R.M., Moens, M.-F.: Argumentation mining: the detection, classification and structure of arguments in text. In: Proceedings of the 12th International Conference on Artificial Intelligence and Law, pp. 98–107. ACM (2009)
19. Pang, B., Lee, L.: Opinion mining and sentiment analysis. Foundations and Trends in Information Retrieval 2(1-2), 1–135 (2008)
20. Roberts, C.: Text analysis for the social sciences: Methods for drawing statistical inferences from texts and transcripts. Lawrence Erlbaum, New York (1997)
21. Ruigrok, N., Van Atteveldt, W.: Global angling with a local angle: How US, British, and Dutch newspapers frame global and local terrorist attacks. The Harvard International Journal of Press/Politics 12(1), 68–90 (2007)
22. Scheufele, D.A.: Framing as a theory of media effects. Journal of Communication 49(1), 103–122 (1999)
23. Sculley, D.: Combined regression and ranking. In: SIGKDD 2010, pp. 979–988. ACM (2010)
24. Semetko, H.A., Valkenburg, P.M.: Framing european politics: A content analysis of press and television news. Journal of Communication 50(2), 93–109 (2000)
25. Shah, D.V., Watts, M.D., Domke, D., Fan, D.P.: News framing and cueing of issue regimes: Explaining clinton's public approval in spite of scandal. Public Opinion Quarterly 66(3), 339–370 (2002)
26. Simon, A., Xenos, M.: Media framing and effective public deliberation. Political Communication 17(4), 363–376 (2000)
27. Vliegenthart, R., Boomgaarden, H.G., Boumans, J.W.: Changes in political news coverage. Palgrave Macmillan (2011)
28. Zillmann, D., Brosius, H.B.: Exemplification in communication. Hogrefe and Huber (2000)

Optimal Scales in Weighted Networks

Diego Garlaschelli[1], Sebastian E. Ahnert[2], Thomas M.A. Fink[3], and Guido Caldarelli[4,3,5]

[1] Lorentz Institute of Theoretical Physics, University of Leiden,
Niels Bohrweg 2, 2333 CA Leiden, The Netherlands
[2] Cavendish Laboratory, University of Cambridge,
JJ Thomson Avenue, CB3 0HE Cambridge, United Kingdom
[3] London Institute for Mathematical Sciences,
22 South Audley St., W1K 2NY London, United Kingdom
[4] IMT Alti Studi Lucca, Piazza S. Ponziano 6, 55100 Lucca, Italy
[5] ISC-CNR, Dipartimento di Fisica, Università La Sapienza,
P.le A. Moro 2, 00185 Roma, Italy

Abstract. The analysis of networks characterized by links with heterogeneous intensity or weight suffers from two long-standing problems of arbitrariness. On one hand, the definitions of topological properties introduced for binary graphs can be generalized in non-unique ways to weighted networks. On the other hand, even when a definition is given, there is no natural choice of the (optimal) scale of link intensities (e.g. the money unit in economic networks). Here we show that these two seemingly independent problems can be regarded as intimately related, and propose a common solution to both. Using a formalism that we recently proposed in order to map a weighted network to an ensemble of binary graphs, we introduce an information-theoretic approach leading to the least biased generalization of binary properties to weighted networks, and at the same time fixing the optimal scale of link intensities. We illustrate our method on various social and economic networks.

Keywords: Weighted Networks, Maximum Entropy Principle, Graph Theory, Network Science.

1 Introduction

A large number of social, economic, biological and information systems can be conveniently described as networks (or graphs) where N nodes (or vertices) are connected by L links (or edges). Over the last fifteen years, Network Science has emerged as a fast-growing discipline crossing the boundaries of many research fields [1]. The aim of Network Science is that of characterizing and modelling the structure and dynamics of real-world networks, as opposed to abstract mathematical specifications such as those studied by Graph Theory.

One of the challenges in Network Science is that of extending the relatively well-developed tools available for *binary networks* (where links are either present or absent, with no possible variation in their intensity) to the more general case of *weigthed networks* (where links can have heterogeneous weights) [2–6]. For instance, in binary social networks a link may represent the existence of a friendship relation between two people, irrespective of the strength of such a relation, while in weighted social networks a link may be attached a value indicating the amount of shared time or the degree of intimacy between two friends. While the analysis of social networks has traditionally focused on binary graphs, the recent availability of detailed data about the magnitude of interactions in large-scale social systems offers a new potential for the study of social networks as weighted graphs. However, in the transition from binary graphs to weighted networks two main problems of arbitrariness are encountered, and are still largely unsolved.

First, while several definitions of basic topological quantities have been introduced for binary graphs, the corresponding generalizations to weighted networks are non-unique. An important example is that of the *clustering coefficient* c_i, defined in binary undirected graphs as the fraction of neighbours of node i that are also neighbours of each other, or equivalently the fraction of triangles in which node i participates [1]. In weighted networks, the clustering coefficient can be generalized in many ways [2–6], and there is no natural criterion indicating the optimal definition. Another example is the *reciprocity*, defined in binary directed graphs as the ratio of reciprocated to total links [7]. In weighted networks, there are many possible generalizations requiring sophisticated comparisons and calculations [8]. In general, on one hand the heterogeneity of link intensity observed in weighted networks provides important additional information that one would like to exploit in order to define generalized quantities that reduce to the ordinary and well-studied ones in the particular case of binary graphs, but on the other hand a large degree of arbitrariness makes the problem not well defined.

Second, even when a definition of a weighted quantity is given, one is left with the problem of the arbitrary scale of link intensities. The simplest example is perhaps the *total weight* W of a network, defined as the sum of all link weights in the graph. If the links of the network represent e.g. flows of money between the units of an economic system, or the time spent by two friends in their phone calls, the quantity W clearly depends on the units chosen (e.g. Euros or thousands of Euros, minutes or seconds, etc.). Similarly, any other quantity depending on the edge weights will suffer from the same arbitrariness. This problem can be circumvented by defining adimensional weights that are invariant under rescaling, e.g. dividing each edge weight by the average weight over all pairs of vertices. However, this still does not solve the problem entirely. Consider for instance, as one of the simplest properties of binary graphs, the *link density* defined as the ratio of the number of observed links to the total number of pairs of vertices. This quantity ranges between zero (empty graph) and one (fully connected network). The corresponding weighted quantity, if defined as the ratio of the total weight W to the number of pairs of vertices, ranges between zero and infinity and thus loses the properties of a density. This problem persists irrespective of the preliminary rescaling of the edge weights. Similar considerations apply to the *global* clustering coefficient defined as the fraction of realized triangles: in weighted networks, the weighted counterpart of such a 'fraction' can actually range from zero to infinity.

In this paper, we show that the two seemingly unrelated problems discussed above can actually be rephrased as two sides of the same coin. In sec. 2 we first briefly recall a general method that we proposed in order to generalize the definition of any topological property valid for binary graphs to one valid for weighted networks [6]. While powerful, this approach still does not uniquely fix the scale of edge weights and the functional form of the mapping from binary to weighted properties. For these reasons, in sec. 3 we show that this approach can be rephrased within a statistical physics formalism fixing the functional form of the mapping [9]. Then, in sec. 4 we apply the Maximum Entropy principle to further fix the scale of edge weights in such a way that the weighted topological properties induced by the binary ones are defined in the least biased way. As a result, we obtain an information-theoretic method that fixes the optimal scale of edge weights in the original network and at the same time induces unique and least biased definitions of weighted properties from the well-known binary ones. In sec.5 we finally illustrate our method on various real-world social and economic networks.

2 Weighted Networks as Ensembles of Binary Graphs

Mathematically, a binary directed network with N vertices is uniquely specified by a $N \times N$ *adjacency matrix* $\mathbf{A}$ with entries $a_{ij} = 1$ if a directed link from vertex i to vertex j is present, and $a_{ij} = 0$ otherwise. For binary undirected networks, where links have no orientation, the matrix $\mathbf{A}$ is symmetric. Weighted directed networks are instead characterized by a $N \times N$ *weight matrix* $\mathbf{W}$ where the (non-negative, for the purposes of this article) entry w_{ij} represents the intensity of the directed link connecting vertex i to vertex j (including $w_{ij} = 0$ if the link is absent). Again, in weighted undirected networks the matrix $\mathbf{W}$ is symmetric. In this paper, we will consider both directed and undirected networks, where it is intended that undirected networks can be obtained as the special situation where $\mathbf{A}$, $\mathbf{W}$ and other similar quantities are symmetric.

Quite recently [6], we proposed a method to extend any definition of topological property valid for binary graphs, i.e. any function $f^{(b)}(\mathbf{A})$ of the binary adjacency matrix $\mathbf{A}$, to a corresponding function $f^{(w)}(\mathbf{W})$ of the weight matrix $\mathbf{W}$. Our method is based on the idea that the matrix $\mathbf{W}$ specifying the original weighted network can be mapped to an ensemble of binary graphs defined by a conditional probability $P(\mathbf{A}|\mathbf{W})$. The latter represents the occurrence probability, given $\mathbf{W}$, of a possible graph $\mathbf{A}$ in the ensemble. This mapping from $\mathbf{W}$ to $P(\mathbf{A}|\mathbf{W})$ allows one to define the weighted counterpart $f^{(w)}(\mathbf{W})$ of any binary property $f^{(b)}(\mathbf{A})$ as the expected value of the latter over the ensemble of binary graphs, i.e.

$$f^{(w)}(\mathbf{W}) \equiv \langle f^{(b)}(\mathbf{A}) \rangle_{\mathbf{W}} = \sum_{\mathbf{A}} P(\mathbf{A}|\mathbf{W}) f^{(b)}(\mathbf{A}). \tag{1}$$

If we require that each edge weight w_{ij} only determines the probability $p_{ij} = p(w_{ij})$ of existence of a binary link from vertex i to vertex j (while having no effect on a different pair of vertices), then $P(\mathbf{A}|\mathbf{W})$ simply factorizes over pairs of vertices, i.e.

$$P(\mathbf{A}|\mathbf{W}) = \prod_{i,j} [p(w_{ij})]^{a_{ij}} [1 - p(w_{ij})]^{1-a_{ij}} \quad (2)$$

where $i < j$ for undirected networks and $i \neq j$ for directed networks with no self-loops (if self-loops are allowed, then we should set $i \leq j$ for undirected networks and impose no constraint for directed networks). The problem then reduces to specifying the functional form of the (monotonic) edge-specific probabilities $p_{ij} = p(w_{ij})$ [6]. If these probabilities are regarded as entries of a $N \times N$ matrix $\mathbf{P}(\mathbf{W})$, the factorized form (2) allows to considerably simplify the definition of any weighted properties given in eq.(1). For instance, for any quantity $f^{(b)}(\mathbf{A})$ that is polynomial or multilinear in the entries a_{ij} of the adjacency matrix, the corresponding weighted property reduces to [6]

$$f^{(w)}(\mathbf{W}) \equiv \langle f^{(b)}(\mathbf{A}) \rangle_\mathbf{W} = f^{(b)}[\mathbf{P}(\mathbf{W})] . \quad (3)$$

In our first approaches to the problem [6, 10], we chose the linear mapping

$$p(w_{ij}) \equiv \frac{w_{ij} - w_{min}}{w_{max} - w_{min}} \quad (4)$$

where w_{min} and w_{max} represent the minimum and maximum observed weight in the network, respectively. The above choice ensures that $p(w_{ij})$, as required in order to be a probability, ranges between 0 and 1. We showed that this approach can effectively exploit the additional topological information encoded in the weights, in particular for fully connected networks [6, 10]. However, eq.(4) violates two desirable properties of $p(w_{ij})$, namely $p(0) = 0$ and $p(+\infty) = 1$, i.e. the fact that (only) missing links in the original network are associated with zero connection probability in the binary ensemble, and that (only) infinite connection intensities in the original network are associated with unit connection probability.

In general, the choice of the functional form of $p(w_{ij})$ remains somewhat arbitrary, and eq.(4) can be viewed as the mathematically simplest possibility. This translates the arbitrariness of the initial problem, i.e. the non-uniqueness of the generalization of a binary topological property to a weighted counterpart, to the arbitrariness of the choice of $p(w_{ij})$. This also implies that the second problem of non-uniqueness, i.e. the fact that any weighted topological property $f^{(w)}(\mathbf{W})$ has in general an undesired dependence on the choice of the units of $\mathbf{W}$ in the orginal network, is still unsolved. While the linear choice in eq.(4) is invariant under changes of units (i.e. it is scale-invariant), this will not be the case for more general non-linear choices of $p(w_{ij})$.

3 Statistical Physics of Network Ensembles

We now show that the above approach can be rephrased within a statistical physics formalism in such a way that the first arbitrariness, i.e. the choice of the functional form of $p(w_{ij})$, can be fixed.

Very recently [9], we introduced a general ensemble of binary graphs that, as in statistical physics, is defined (here in slightly simplified form) by the occurrence probability

$$P(\mathbf{A}) = \frac{1}{\mathcal{Z}} \exp\left[\frac{-E(\mathbf{A})}{T}\right]. \qquad (5)$$

In the above equation, $E(\mathbf{A})$ is the *energy* of the particular graph $\mathbf{A}$ (a function of one or more topological properties of $\mathbf{A}$, representing the 'cost' of realizing that graph), T is the *temperature* (representing the degree of topological optimization, with lower T corresponding to a probability concentrated on energetically 'cheaper' configurations) and

$$\mathcal{Z} \equiv \sum_{\mathbf{A}} \exp\left[\frac{-E(\mathbf{A})}{T}\right] \qquad (6)$$

is the normalizing constant, or *grand partition function* of the ensemble. Graph ensembles like the one defined above are extensively used in the statistical physics literature [8, 9, 11–13] as well as in social science [14, 15], where they are known as p^* models or *Exponential Random Graphs*.

Since $E(\mathbf{A})$ represents the cost of realizing the particular graph $\mathbf{A}$, we can regard the ensemble of binary graphs discussed in sec. 2 and defined by the probability $P(\mathbf{A}|\mathbf{W})$ as a particular case of the ensemble defined by eq.(5) where the energy is a function $E(\mathbf{A}, \mathbf{W})$ of the weight matrix $\mathbf{W}$ [9]. In particular, the requirements for $P(\mathbf{A}|\mathbf{W})$ leading to the factorized form (2) translate into the requirement of the additivity of $E(\mathbf{A}, \mathbf{W})$, i.e.

$$E(\mathbf{A}, \mathbf{W}) \equiv \sum_{i,j} \epsilon_{ij} a_{ij} = \sum_{i,j} \epsilon(w_{ij}) a_{ij} \qquad (7)$$

where, again, $i \neq j$ for directed networks and $i < j$ for undirected networks. In the above expression, $\epsilon_{ij} = \epsilon(w_{ij})$ must be interpreted as an edge-specific energy, i.e. the energetic cost contributed by the existence of a link from vertex i to vertex j ($a_{ij} = 1$). In this way, the choice of the functional form of $p(w_{ij})$ translates to the choice of the functional form of $\epsilon(w_{ij})$. Indeed, it is easy to show that inserting eq.(7) into eq.(5) leads precisely to eq.(2) where

$$p(w_{ij}) = \frac{e^{-\epsilon(w_{ij})/T}}{1 + e^{-\epsilon(w_{ij})/T}}. \qquad (8)$$

The above expression is particularly useful in order to select the appropriate form of $\epsilon(w_{ij})$. Specifically, we see that a linear dependence of the type $\epsilon(w_{ij}) \propto w_{ij}$ is not suitable, since it would assign a probability $p(0) = 1/2$ (rather than $p(0) = 0$) to the pairs of vertices connected by no link ($w_{ij} = 0$) in the original weighted network. We also see that the linear choice (4) is not natural, since it would correspond to a very complicated, and difficult to justify, form of $\epsilon(w_{ij})$. On the other hand, as we recently noted [9], the simplest satisfactory choice involves a proportionality between $e^{-\epsilon(w_{ij})/T}$ and w_{ij}, i.e. $e^{-\epsilon(w_{ij})/T} = zw_{ij}$ or in other words

$$p(w_{ij}, z) \equiv \frac{zw_{ij}}{1 + zw_{ij}}. \qquad (9)$$

This means that the dependence of the binary link energy on the observed edge weight is given by
$$\epsilon(w_{ij}, z) = -T\ln(zw_{ij}), \qquad (10)$$
i.e. w_{ij} has a logarithmic effect on $\epsilon(w_{ij}, z)$. In real networks with a power-law weight distribution of the form $\rho(w) \propto w^{-\alpha}$, the above relation can be used to measure the empirical temperature as $T = \alpha - 1$ [9]. Typical observed values are $0.5 \lesssim T \lesssim 2.5$.

Equation (9) fixes the functional form of $p(w_{ij}, z)$ in a very reasonable manner. With such a choice, we recover, for all values of z, the desired properties $p(0, z) = 0$ and $p(+\infty, z) = 1$. Note that if $z = [w_{max} - w_{min}]^{-1}$ and $zw_{ij} \ll 1$ then we have $p(w_{ij}, z) \approx w_{ij}/[w_{max} - w_{min}]$, which is approximately equivalent to the choice in eq.(4). This corresponds to a 'sparse graph' limit for the binary ensemble induced by the weighted network. However, in general the value of z in eq.(9) is arbitrary. This leads us to the main point of this paper, which is discussed in the next section.

4 Maximum-Entropy Scale of Edge Weights

We can regard the arbitrariness of z in eq.(9) as equivalent to the arbitrariness of the unit of edge weights in the original network. Indeed, changing the scale of w_{ij} to λw_{ij}, where λ is any positive constant, is mathematically equivalent to changing z to λz. In particular, from eqs.(9) and (10) it is clear that
$$p(\lambda w_{ij}, z) = p(w_{ij}, \lambda z) \quad \text{and} \quad \epsilon(\lambda w_{ij}, z) = \epsilon(w_{ij}, \lambda z). \qquad (11)$$
This shows that the scale λ can be completely reabsorbed in a redefinition of the parameter z, i.e. $z \to \lambda z$. Therefore, without loss of generality, we can regard z in eq.(9) as the parameter specifying the scale of weights. If we introduce a unique way to fix z, we have automatically eliminated the second and last source of arbitrariness discussed in the Introduction, i.e. the units of edge weights in the original network.

In what follows, we propose the Maximum Entropy principle as a rigorous criterion to fix the value of z, and further show that this value is unique. Our main idea is that, in line with other uses of the Maximum Entropy principle [11, 16], the least biased choice of a quantity should correspond, in absence of any other indication, to the one that maximizes Shannon's entropy (given the available information). Given a real-world weighted network specified by the matrix $\mathbf{W}$ and the corresponding binary ensemble specified by the conditional probability $P(\mathbf{A}|\mathbf{W})$ as given by eqs.(2) and (9), Shannon's entropy reads
$$S(z) \equiv -K \sum_{\mathbf{A}} P(\mathbf{A}|\mathbf{W}) \ln P(\mathbf{A}|\mathbf{W}) \qquad (12)$$
where K is an arbitrary constant, that we fix later for convenience. Now, due to the factorization of $P(\mathbf{A}|\mathbf{W})$ as in eq.(2), and since the entropy of a factorized process is additive, we can simply write

$$S(z) = K \sum_{i,j} s_{ij}(z) \qquad (13)$$

(with the usual convention on i, j for directed and undirected graphs) where $s_{ij}(z)$ is the edge-specific entropy

$$s_{ij}(z) = -p(w_{ij}, z) \ln p(w_{ij}, z) - [1 - p(w_{ij}, z)] \ln[1 - p(w_{ij}, z)] \qquad (14)$$

Note that both missing links ($w_{ij} = 0$) and very large weights ($w_{ij} \to +\infty$) generate a zero entropy $s_{ij}(z) = 0$, and therefore have no effect on the choice of the optimal scale. This is consistent with the fact that both zero and infinite weights are independent of any chosen scale λ. Inserting eq.(14) into eq.(13), we find that the entropy of the ensemble is

$$S(z) = -K \ln \prod_{i,j} [p(w_{ij}, z)]^{p(w_{ij}, z)} [1 - p(w_{ij}, z)]^{1-p(w_{ij}, z)} \qquad (15)$$

If we want $S(z)$ to be normalized between 0 and 1 (although this has no effect on the following results), we can set

$$K \equiv \frac{1}{M \ln 2} \qquad (16)$$

where M is the number of possible pairs of vertices, i.e. $M = N(N-1)$ for a directed network with no self-loops and $M = N(N-1)/2$ for an undirected network with no self-loops. If self-loops are allowed, then the above values of M must be increased by N.

We can now look for the value of z that maximizes $S(z)$ as given by eq.(15). To this end, we write the first derivative of $S(z)$ as

$$\begin{aligned} S'(z) &= K \sum_{i,j} \frac{\partial p(w_{ij}, z)}{\partial z} \ln \frac{1 - p(w_{ij}, z)}{p(w_{ij}, z)} \\ &= K \sum_{i,j} \frac{w_{ij}}{(1 + zw_{ij})^2} \ln \frac{1}{zw_{ij}} \end{aligned} \qquad (17)$$

and the second derivative as

$$S''(z) = K \sum_{i,j} \frac{w_{ij}}{(1 + zw_{ij})^2} \left[-\frac{2w_{ij}}{1 + zw_{ij}} \ln \frac{1}{zw_{ij}} - \frac{1}{z} \right] \qquad (18)$$

Now let w_{max} denote the maximum weight and w_{min} the minimum *non-zero* weight in the original network. As z increases from 0 to $+\infty$, we find that there are five regimes, listed below.

4.1 $z = 0$

This gives a deterministic ensemble with $p(w_{ij}, 0) = 0 \; \forall i, j$. Therefore the entropy has the minimum value $S(0) = 0$, and we are sure that this is not the maximum we are looking for.

4.2 $0 < z < 1/w_{max}$

Consider first the case $z \ll 1/w_{max}$. In this regime, $zw_{ij} \ll 1 \; \forall i,j$, therefore $p(w_{ij}, z) \approx zw_{ij}$. So $S(z)$ increases as z increases and no (local) maxima or minima are encountered. Also in the less strict situation $z < 1/w_{max}$, we have $z < 1/w_{ij} \; \forall i,j$ which implies $\ln(1/zw_{ij}) > 0 \; \forall i,j$. Looking at eq.(17), this means that $S'(z) > 0$, so $S(z)$ increases in the entire range $0 < z < 1/w_{max}$.

4.3 $1/w_{max} < z < 1/w_{min}$

This is the nontrivial range. It can be shown that if a maximum of $S(z)$ exists, it must be within this range. As we showed above, when $z < 1/w_{max}$ we have $S'(z) > 0$. Similarly, below we will show that when $z > 1/w_{min}$ one has $S'(z) < 0$. Taken together, these results imply that, since $S'(z)$ is a continuous function, there must exist a value z^* in the range $1/w_{max} < z^* < 1/w_{min}$ such that $S'(z^*) = 0$. As we show later, this corresponds to a maximum of the entropy.

4.4 $z > 1/w_{min}$

When $z > 1/w_{min}$, we have $z > 1/w_{ij} \; \forall i,j$ which implies $\ln(1/zw_{ij}) < 0 \; \forall i,j$. Looking at eq.(17), this means that $S'(z) < 0$, so $S(z)$ decreases in the entire range $z > 1/w_{min}$. Note that in the extreme case $z \gg 1/w_{min}$ we have $zw_{ij} \gg 1$ $\forall i,j$ and $p(w_{ij}, z) \approx 1 - 1/zw_{ij}$.

4.5 $z \to +\infty$

Now $p(w_{ij}, +\infty) = \Theta(w_{ij})$, and the entropy tends again to the minimum value $S(+\infty) = 0$. Interestingly, this limit corresponds to the situation when the original weighted network is regarded as a binary graph by simply setting each non-zero weight to one, and leaving the other values equal to zero. Within our formalism, we find that this oversimplification corresponds to the minimum entropy, i.e. it is maximally biased.

5 Real-World Social and Economic Networks

We finally illustrate an application of our method to various real-world social and economic networks. We consider snapshots of the World Trade Web (WTW), the network of world countries connected by import/export relationships [17, 18], the RyanAir (RA) airport network [19], the European Union (EU) aviation network [20] and the Cond-Mat (CM) scientific collaboration network [21]. The WTW is a directed network with no self-loops (hence the number of pairs of vertices is $M = N(N-1)$), the RA and the CM are undirected networks with no self-loops ($M = N(N-1)/2$), and finally the EU is a directed network with self-loops ($M = N(N-1) + N$).

For each of these networks, we consider the weight matrix **W** as given in the original dataset, and use it to calculate the ensemble probabilities defined in

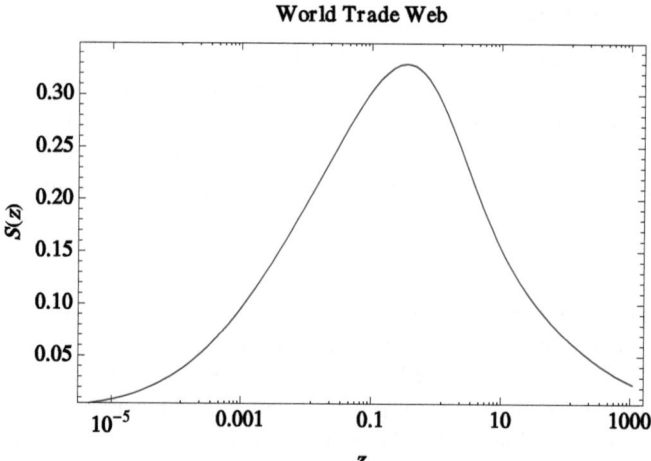

Fig. 1. Log-linear plot of the ensemble entropy $S(z)$ of the World Trade Web (year 2000) as a function of z in the range $1/w_{max} < z < 1/w_{min}$. The number of nodes is 187. The maximum is placed at $z^*_{WTW} = 0.34$.

eq.(9) and consequently the entropy $S(z)$ as defined in eq.(15). So the weight w_{ij} is expressed in the (necessarily arbitrary) units used in the original dataset. We then look for the optimal value z^* that maximizes $S(z)$. Clearly, z^* corresponds to the optimal weight scale $w^* \equiv 1/z^*$, so that the quantity $z^* w_{ij}$ appearing in eqs.(9) and (10) can be rewritten as

$$x_{ij} = z^* w_{ij} = \frac{w_{ij}}{w^*} \qquad (19)$$

The above expression gives us the optimally rescaled weights x_{ij} of the network, i.e. the weights expressed in terms of the non-arbitrary unit w^*. Note that the rescaled weights x_{ij} are independent of the units used in the data, and hence of the original scale of w_{ij}.

The curves of $S(z)$, plotted in the nontrivial range $1/w_{max} < z < 1/w_{min}$ where the entropy has a maximum, are shown in fig.1 for the WTW, in fig.2 for the RA network, in fig.3 for the EU network, and in fig.4 for the CM dataset. As expected, all curves displays a clear maximum for the value z^* such that $S'(z^*) = 0$. The values of z^* are:

$$z^*_{WTW} = 0.34$$
$$z^*_{RA} = 0.47$$
$$z^*_{EU} = 6.69 \cdot 10^{-6}$$
$$z^*_{CM} = 3.034$$

The above values give the following optimal units w^* required in order to rescale the original arbitrary matrix **W** for each network:

$$w^*_{WTW} = 2.92$$

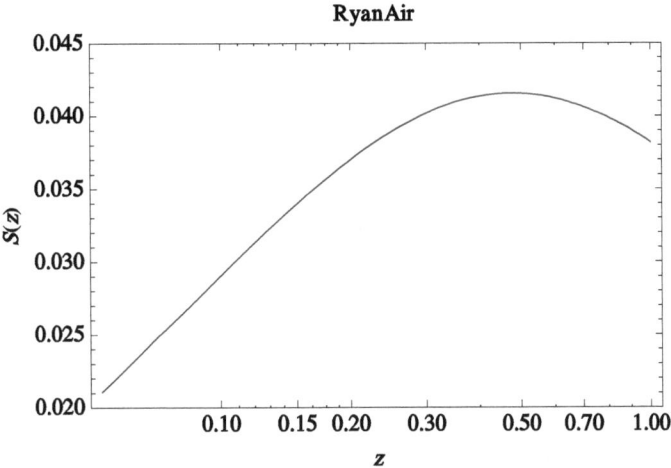

Fig. 2. Log-linear plot of the ensemble entropy $S(z)$ of the RyanAir network (year 2005) as a function of z in the range $1/w_{max} < z < 1/w_{min}$. The number of nodes is 109. The maximum is placed at $z^*_{RA} = 0.47$.

$$w^*_{RA} = 2.09$$
$$w^*_{EU} = 149355$$
$$w^*_{CM} = 0.33$$

Although an analysis of the topological properties of the four networks considered is beyond the scope of this paper, we briefly note that our procedure yields a unique final weight matrix $\mathbf{X}$ expressed in non-arbitrary units, and a corresponding probability matrix $\mathbf{P}(\mathbf{X})$ with entries given by

$$p_{ij} = p(w_{ij}, z^*) = p(x_{ij}, 1) = \frac{x_{ij}}{1+x_{ij}}. \quad (20)$$

Using eq.(1) or (3), $\mathbf{P}(\mathbf{X})$ can be finally used in order to compute the least biased weighted generalization $f^{(w)}(\mathbf{X})$ of any binary property $f^{(b)}(\mathbf{A})$. For instance, for polynomial or multilinear properties

$$f^{(w)}(\mathbf{X}) = f^{(b)}[\mathbf{P}(\mathbf{X})]. \quad (21)$$

The above formula can be used to compute the otherwise problematic weighted counterparts of many topological properties, e.g. the weighted density and the weighted clustering coefficient mentioned in the Introduction. For instance, let us consider the ordinary definition of the density $d^{(b)}(\mathbf{A})$ of a binary network $\mathbf{A}$:

$$d^{(b)}(\mathbf{A}) \equiv \frac{L(\mathbf{A})}{M} = M^{-1} \sum_{i,j} a_{ij} \quad (22)$$

where $L(\mathbf{A}) = \sum_{i,j} a_{ij}$ is the total number of links in $\mathbf{A}$ (our usual conventions for i,j in the sum and for the number M of pairs of nodes hold). Using eq.(21),

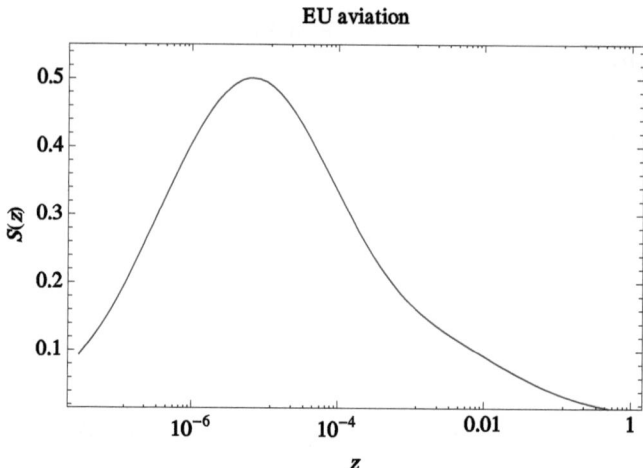

Fig. 3. Log-linear plot of the ensemble entropy $S(z)$ of the EU aviation network (year 2005) as a function of z in the range $1/w_{max} < z < 1/w_{min}$. The number of nodes is 28. The maximum is placed at $z^*_{EU} = 6.69 \cdot 10^{-6}$.

the weighted density of a network with (optimally rescaled) weights $\mathbf{X}$ can be defined as

$$d^{(w)}(\mathbf{X}) = d^{(b)}[\mathbf{P}(\mathbf{X})] = M^{-1}\sum_{i,j} p_{ij} = M^{-1}\sum_{i,j} \frac{x_{ij}}{1+x_{ij}}. \tag{23}$$

By construction, the above definition takes values between 0 and 1, as any proper density measure. This desirable property nicely overcomes the limitations of other naive generalizations of the binary density, illustrating the usefulness of the above approach.

For the four networks in our analysis, the values of the weighted density are:

$$d^{(w)}_{WTW} = 0.31$$
$$d^{(w)}_{RA} = 0.022$$
$$d^{(w)}_{EU} = 0.39$$
$$d^{(w)}_{CM} = 1.75 \cdot 10^{-4}$$

We stress again that the above values are independent of any (necessarily arbitrary) choice of the unit of weight in the original data. It is interesting to compare the above values of the weighted density $d^{(w)}$ with the corresponding values of the ordinary binary density $d^{(b)}$, as measured on the adjacency matrix characterizing the bare topology of the original network:

$$d^{(b)}_{WTW} = 0.58$$
$$d^{(b)}_{RA} = 0.044$$

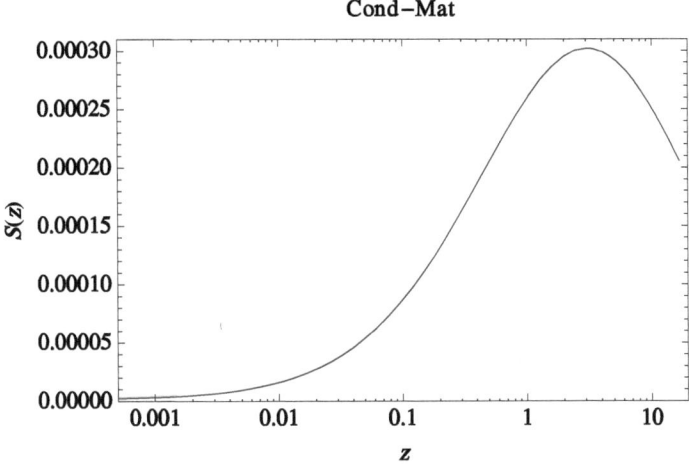

Fig. 4. Log-linear plot of the ensemble entropy $S(z)$ of the Cond-Mat collaboration network (year 2001) as a function of z in the range $1/w_{max} < z < 1/w_{min}$. The number of nodes is 16726. The maximum is placed at $z^*_{CM} = 3.034$.

$$d^{(b)}_{EU} = 0.86$$
$$d^{(b)}_{CM} = 3.42 \cdot 10^{-4}$$

We find that the values of $d^{(b)}$ for all networks are approximately twice the corresponding values of $d^{(w)}$. This big numerical difference shows the entity of the information loss encountered when a weighted network is regarded as a binary one (corresponding to the maximally biased limit $z \to \infty$ as discussed in sec. 4). Our approach instead makes use of all the available information encapsulated in the weights, and ensures that the bias is minimized (corresponding to the maximum-entropy point z^*). For the four networks in our analysis, exploiting the additional knowledge of the weights has a significant 'sparsifying' effect, approximately halving the purely binary density.

6 Conclusions

In this paper we have addressed two problems of abitrariness that are systematically encountered in the analysis of weighted networks: the non-uniqueness of the generalization of binary topological properties to their weighted counterparts and that of the scale of edge weights. While in principle independent, we have shown that, when a weighted network is mapped to an ensemble of binary graphs, these two problems turn out to be intimately related. In particular, the ensemble formalism (especially when rewritten within a statistical-physics framework) provides a straightforward weighted generalization of any binary property, and at the same time allows us to find the optimal weight scale via a

Maximum Entropy criterion. It is remarkable that such a criterion cannot be invoked directly on the original system by maximizing the entropy of the weighted network, because the entropy is only defined for ensembles of graphs and not for a single instance (unless the original weighted network is trivially regarded as the only possible outcome of a deterministic ensemble with zero entropy). Therefore the transition from a single network to an ensemble of graphs is necessary in order to find the least biased scale of weights via a maximization of the entropy. Using examples of real-world socio-economic networks, we have illustrated our approach and computed the optimal scale for such networks. We have shown that this scale can be used to define the least biased generalization of any binary property to the weighted case, confirming that the problem of selecting an optimal scale and that of defining unique generalizations of binary properties are tightly interrelated within our ensemble formalism.

Acknowledgments. D.G. acknowledges support from EU FET project MULTIPLEX (317532) and the Dutch Econophysics Foundation (Stichting Econophysics, Leiden, the Netherlands) with funds from beneficiaries of Duyfken Trading Knowledge BV, Amsterdam, the Netherlands. S.E.A. was supported by The Leverhulme Trust, UK and The Royal Society, UK. G.C. acknowledges support from FET project FOC (255987) and MULTIPLEX (317532).

References

1. Caldarelli, G.: Scale-Free Networks: Complex Webs in Nature and Technology. Oxford University Press, Oxford (2007)
2. Barrat, A., Barthelemy, M., Pastor-Satorras, R., Vespignani, A.: The Architecture of Complex Weighted Networks. Proc. Nat. Acad. Sci. USA 101(11), 3747–3752 (2004)
3. Saramaki, J., Kivela, M., Onnela, J.P., Kaski, K., Kertesz, J.: Generalizations of the Clustering Coefficient to Weighted Complex Networks. Phys. Rev. E 75(2), 027105 (2007)
4. Newman, M.E.J.: Analysis of Weighted Networks. Phys. Rev. E 70(5), 056131 (2004)
5. Fagiolo, G.: Clustering in Complex Directed Networks. Phys. Rev. E 76(2), 026107 (2007)
6. Ahnert, S.E., Garlaschelli, D., Fink, T.M.A., Caldarelli, G.: Ensemble Approach to the Analysis of Weighted Networks. Phys. Rev. E 76(1), 016101 (2007)
7. Garlaschelli, D., Loffredo, M.I.: Patterns of Link Reciprocity in Directed Networks. Phys. Rev. Lett. 93(26), 268701 (2004)
8. Squartini, T., Picciolo, F., Ruzzenenti, F., Garlaschelli, D.: Reciprocity of Weighted Networks. Scientific Reports 3, 2729 (2013), http://arxiv.org/abs/1208.4208
9. Garlaschelli, D., Ahnert, S.E., Fink, T.M.A., Caldarelli, G.: Low-Temperature Behaviour of Social and Economic Networks. Entropy 15(8), 3148–3169 (2003)
10. Ahnert, S.E., Garlaschelli, D., Fink, T.M.A., Caldarelli, G.: Applying Weighted Network Measures to Microarray Distance Matrices. Journal of Physics A 41(22), 4011 (2008)
11. Park, J., Newman, M.E.J.: Statistical Mechanics of Networks. Phys. Rev. E 70(6), 066117 (2004)

12. Garlaschelli, D., Loffredo, M.I.: Generalized Bose-Fermi Statistics and Structural Correlations in Weighted Networks. Phys. Rev. Lett. 102(3), 038701 (2009)
13. Bianconi, G.: Entropy of Network Ensembles. Phys. Rev. E 79(3), 036114 (2009)
14. Stanley, W., Faust, K.: Social Network Analysis: Methods and Applications. Cambridge University Press, New York (1994)
15. Robins, G., Snijders, T., Wang, P., Handcock, M., Pattison, P.: Recent Developments in Exponential Random Graph (p^*) Models for Social Networks. Social Networks 29(2), 192–215 (2007)
16. Jaynes, E.T.: Information Theory and Statistical Mechanics. Physical Review 106(4), 620 (1957)
17. Squartini, T., Fagiolo, G., Garlaschelli, D.: Randomizing World Trade. I. A Binary Network Analysis. Phys. Rev. E 84(4), 046117 (2011)
18. Squartini, T., Fagiolo, G., Garlaschelli, D.: Randomizing World Trade. II. A Weighted Network Analysis. Phys. Rev. E 84(4), 046118 (2011)
19. RyanAir website, http://www.ryanair.com
20. Eurostat website, http://epp.eurostat.cec.eu.int
21. Newman, M.E.: The Structure of Scientific Collaboration Networks. Proc. Nat. Acad. Sci. USA 98, 404–409 (2001)

Why Do I Retweet It? An Information Propagation Model for Microblogs

Fabio Pezzoni[1], Jisun An[2], Andrea Passarella[1],
Jon Crowcroft[2], and Marco Conti[1]

[1] CNR-IIT, via G. Moruzzi, 1 - 56124 Pisa, Italy
[2] Computer Laboratory, University of Cambridge, UK
{f.pezzoni,a.passarella,m.conti}@iit.cnr.it,
{jisun.an.jon.crowcroft}@cl.cam.ac.uk

Abstract. Microblogging platforms are Web 2.0 services that represent a suitable environment for studying how information is propagated in social networks and how users can become influential. In this work we analyse the impact of the network features and of the users' behaviour on the information diffusion. Our analysis highlights a strong relation between the level of visibility of a message in the flow of information seen by a user and the probability that the user further disseminates the message. In addition, we also highlight the existence of other latent factors that impact on the dissemination probability, correlated with the properties of the user that generated the message. Considering these results we define an information propagation model that generates information cascades (i.e. flows of messages propagated from user to user) whose statistical properties match empirical observations.

1 Introduction

Online Social Networks (OSNs) have become one of the most popular services in the Web 2.0. They allow people to communicate and share content with each other, playing a fundamental role for the spread on information, ideas, and influence. In recent years, the study of the information diffusion in OSNs have attracted the attention of many researchers. A better characterisation of the phenomenon, in fact, can lead to more effective and fair use of these systems, suggest focused marketing strategies and provide insights into the underlying sociology. The properties of information diffusion (i.e. how information spreads in the social network due to communication between users) have been studied in different types of OSNs such as microblogging platforms like Twitter [1–3] and Facebook [4] and other specific Web 2.0 services, e.g. Flickr [5], blogs [6], Digg [7] and YouTube [8].

The main goal of this paper is contributing to the characterisation of the information diffusion in microblogs, analysing the role of the users' activity. For this reason we define an agent-based model to reproduce the behaviour of the users, such that the impact of the various parameters on information diffusion can be studied "in vitro". For example, one of the most important factors for the formation of information cascades is the *decaying visibility* of the content. In fact, different studies have demonstrated that the probability that a user forwards a received content decreases with time [1, 9, 10]. We believe that, for a better characterisation of the content visibility, it can not be measured only in terms of time and that the users' activity patterns should be considered too.

Focusing on Twitter, a more straightforward way for estimating the visibility of a *tweet* is considering its *position* in the tweet feed that is the result of the global users' activity. In fact, as empirically demonstrated in Sect. 2.2, the tweet's position in the feed is strongly correlated with its probability to be retweeted giving rise to *information cascades*. In addition to the position of a tweet in the user's feed, we also show that other parameters related to the user that originally generates a tweet can impact on the diffusion of information in Twitter. We collectively represent them with a unique parameter, that we call *user standing*. These properties are the base for the agent-based model we describe in Sect 3. In the model, agents simulate the users' activity in creating new messages and forwarding previously received messages. Basing on an underlying network structure, messages are dispatched to the connected agents and, based on their position in the tweet feed and the standing of the originating agent, they are probabilistically forwarded, simulating the formation of information cascades. In Sect. 5 we evaluate our model (simulating the user activity) in a network whose parameters are derived from a Twitter dataset (Sect. 4). Simulation results match empirical observations with high statistical confidence both in terms of information cascade properties and characterisation of the user influence.

2 Dataset Analysis

In this section we analyse the properties of the information diffusion as a function of Twitter users' properties using a dataset we have collected. This analysis highlights key features that determine information cascades, and it is thus the starting point for the agent-based propagation model we propose in Sect. 3.

For our analysis, we collected Twitter data from 17 October 2012 to 11 February 2013 using the Twitter REST API. Using the crawling agent described in [11] (where we also present an analysis, orthogonal to this study, of structural properties of the Twitter social network) we extract a Twitter subgraph of $2,029,143$ users. For each of them we downloaded his profile, the lists of his *followers* (people who follow the user) and *followings* (people followed by the user), and all his published tweets up to the limit of $3,200$ tweets (the maximum number of tweets that can be downloaded using the REST API). In total our dataset contains around $2,500M$ tweets that we divided in *"regular" tweets* (63.2%), *replies* (19.9%) and *retweets* (16.9%). As replies have not an active role in the propagation of information, in our analysis we consider just "regular" tweets and their retweets.

2.1 Influence in Twitter

The influence can be defined as the ability of a user to spread information in a network. In Twitter, the propagation of a message can be measured in terms of *retweet count*, that is the number of times the message has been retweeted and that is included in the metadata of each downloaded tweet. Using this information we can define the influence of a user in Twitter as the average retweet count of all tweets he created. Figure 1 displays the Complementary Cumulative Distribution Functions (CCDFs) of the retweet count and of the user influence by the solid and dotted lines respectively. These results are inline with other analysis in literature that have shown that the size of information cascades and the user influence tend to be highly skewed [1, 2, 12].

Starting from the measure of influence, we can now examine what factors are related to it using our dataset. Literature says that the structural feature that best correlates with the user influence is the number of followers [2, 10, 13] that corresponds to the in-degree of the nodes in the underlying network topology. The reason behind is that a tweet from a user with many followers reaches immediately a large audience that, possibly, will retweet it to other users. In Fig. 2 we show the log-log plot of the number of followers against the user influence. The correlation (Pearson coefficient equal to 0.532) is remarkable, however, given the same number of followers, the influence value can vary significantly. In fact, as previously mentioned, structural features of the nodes alone are not sufficient to explain the actual influence of a user in the network. Others factors should be investigated.

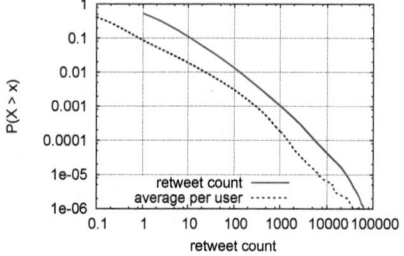

Fig. 1. CCDFs of retweet count and average retweet count per user

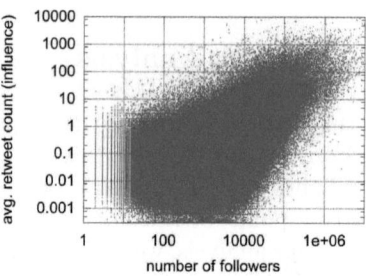

Fig. 2. Relation between # of followers and influence

2.2 Factors on Retweeting Behaviour

When a Twitter user accesses his tweet feed there are different factors that impact on his behaviour leading him to select a message to retweet. We perform our study by assuming that two main factors impact on the detailed retweeting behaviour of the users: the position of tweets in the feed, and an overall parameter describing all the properties of the creator of the tweet, that we call *user standing*.

Position in the Tweet Feed. Previous studies have inferred that visibility of the tweets is related to their probability to be retweeted [1, 9, 10]. A tweet has the maximum visibility immediately after it is received because it takes the least effort to be discovered at the top of the tweet feed. As soon as new tweets arrive in the feed, they push the old messages down in the queue reducing their visibility. We believe that the time span after receiving a tweet is a good estimator of its visibility however, it can be influenced by other factors like the temporal activity patterns of the users.

A more straightforward approach, is to analyse the actual position of the messages in the tweet feed. For this analysis we randomly selected a subset of 100,000 users from our dataset. Then for each user we have recreated his message feed joining all the published tweets of the users he follows. Successively, comparing the timestamps, we have extracted for each retweeted message its position in the tweet feed at the time of the retweet. In our analysis we have considered only the first 1,000 positions of the feed. Results in Fig. 3 show that the probability of retweeting a message in a certain position of the feed follows a power-law distribution with coefficient 1.433 estimated using the maximum-likelihood estimation (MLE).

It is worth noting that the position of the messages in a tweet feed is pretty much random, since it depends only on the time a user receives the messages and on the time he retweets. The relation between the position and the retweet probability, therefore, does not explain the variation on the user influence discussed at the end of Sect. 2.1. Visibility is, in fact, a general property of the tweets and doesn't depend on the influence or on the number of followers of the users.

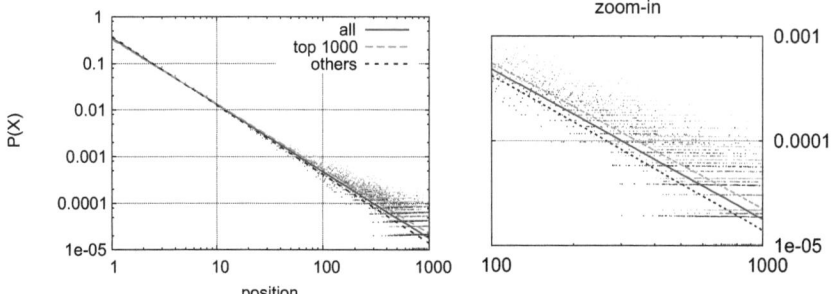

Fig. 3. Retweet probability given the position in the feed for all the tweets in the dataset ("all"), for the tweets created by the 1,000 most influential users ("top 1000") and for the tweets created by all the other users ("others")

User Standing. In order to explain mentioned variations in the user influence we have to investigate the effect of the properties of the users on the retweeting behaviour. These properties are often qualitative and, therefore, hard to quantify (e.g. credibility, expertise, enthusiasm and popularity). For this reason, we use a unique index called *user standing*, to take into consideration the joint effect of all of them.

The effect of the user standing can be observed as the variation of the retweet probability for different equally-positioned tweets. In this sense, the user standing can be considered as a sort of "favouritism in retweet" for the messages created by some users. In our case, we are interested in investigating if the tweets created by the most influential users are more likely to be retweeted than the tweets created by other users. For the analysis we ranked the users considered in the previous analysis based on their influence and then we selected the top 1,000 influential users. In Fig. 3 we plot the retweet probability of their tweets compared with the retweet probability of tweets created by all the other users. The gap between the lines appears narrow, however the fit with a power-law function has coefficient 1.389 for the influential users and 1.478 for the others. This means that, considering the same position, the most influential users have a higher probability to get their messages retweeted.

3 Activity-Based Propagation Model

The model we present in this work describes the information propagation mechanism in a microblogging social network given the topology of the network and some features of the agents that represent the users. In the model any agent interacts with the network in two different ways: creating new messages and forwarding previously received messages. The frequency with which an agent v is selected for creating and forwarding messages, is given by the parameters f_v^{cr} and f_v^{fw} respectively. Both in case of creation and forwarding, the messages are broadcast to other agents that "follow" the creator or forwarder. An agent r follows the agent v if, in the underlying network graph $G(V, E)$, a direct link between the nodes that represent agents r and v respectively exists. In this case the agent r receives all the messages created or forwarded by agent v. If an agent receives multiple copies of the same message, it keeps in memory just the first received one and discharges the others.

Assuming that an agent v is selected to perform a forwarding action at time t, the model takes the *message feed* list $F_{v,t}$ that includes all the messages received by v before time t sorted by reverse-chronological order. Then, for each message $w \in F_{v,t}$, it assigns the probability $P(w|v,t)$ to be forwarded such that $\sum_{w \in F_{v,t}} P(w|F_{v,t}) = 1$ where:

$$P(w|v,t) = \frac{\alpha_{o(w)} \varphi(\theta_{v,t}(w))}{\sum_{z \in F_{v,t}} \alpha_{o(z)} \varphi(\theta_{v,t}(z))}, \qquad w \in F_{v,t} \qquad (1)$$

$\alpha_{o(w)}$ is the *standing* of the the agent $o(w)$, who is the creator of the message w, and $\varphi(\cdot)$ is a function called *position function* that takes as a parameter the position of w in $F_{v,t}$ denoted as $\theta_{v,t}(w)$. According to (1), the probability of a message to be selected for the forward depends on: i) its position in the message feed and ii) the standing of its creator.

i) The position of the message in the feed is considered in the model since, as we demonstrated in Sect. 2.2, there is evidence that last received messages (which are on top of the message feed) are more likely to be forwarded. For this reason the position function $\varphi(\cdot)$ has to be monotonically decreasing. For example, as our analysis suggests, it can be defined as a power-law function.
ii) As discussed in Sect. 2.1, we introduced the concept of user standing that represents the joint effect of all the properties of the users that positively influence the forwarding probability of their messages. Each agent in the network v is therefore characterised, in addition to the frequencies f_v^{cr} and f_v^{fw}, also by a standing value α_v. In the next section we discuss in detail how to model the user standing.

4 Deriving the Model's Parameters

In our simulation we implement the agent-based propagation model described in previous section in order to simulate the user activity and the information diffusion of a real social network. We used the Twitter dataset described in Sect. 2 to infer both the graph structure and the agents' properties.

4.1 Social Graph

For computational reasons we selected a random subset of 100,000 users among all the active users from our dataset. We considered a user to be active if he has at least 100 followers and if he has created at least 100 tweets. These constraints allow us to avoid low-active accounts that are not relevant for the propagation of information. From this set of users, we derived the social graph that consists of 5,756,450 arcs and maintains well-known features of social networks' graphs such as high clustering coefficient and small average path length (small-world property) [14].

4.2 Position Function

As suggested in Sect. 3 we define the position function $\varphi(\cdot)$ as a power-law. In particular we use the result in Sect. 2.2 in which we have fit the retweet probability given the tweets' position with a power-law with coefficient 1.433. Considering that the position function is discrete, we define it as a ZipF Probability Mass Function with the given coefficient and limited to $N = 1,000$, which is the same number of positions we have used in our analysis.

4.3 Frequencies

For each user v we extract, from the dataset, the frequency of creating messages per day f_v^{cr} and the frequency of forwarding messages per day f_v^{fw}. Distributions of these frequencies are highly skewed since just few users have a very high activity.

4.4 User Standing

In Sect. 2.2, we defined the user standing as the joint effect of the latent factors that affect the forwarding of his messages. As previously discussed, these parameters of the model are not directly quantifiable. We could estimate them using a MLE estimator where the likelihood function is given by a sample of retweeting actions extracted from the dataset. Unfortunately, applying this method would have required to analyse the full propagation path of each and every tweet of all our users, which was not feasible due to the computational complexity and the fact that cascades can involve users not included in our dataset. Therefore, we use an approximate way to estimate the user standing, as follows.

The idea is to estimate the standing of a user as the average retweet probability of the tweets he has originated. This can be calculated as the ratio of his average retweet count (influence) to the average number of users who have received his tweets. However, the latter value is not derivable since it would require to track the full propagation trees. As approximation, we use the number of his followers instead. It is worth noting that, due to this approximation, the standing of the most influential users could be overvalued. This is because the number of followers can be significantly smaller than the number of users that received the tweets. In order to remove this bias we had to apply an exponent to the previously defined measure. As result of an extensive analysis, we set the exponent to 1/3 as this value guarantees to obtain better performance of our model. Formally, the user standing values we considered in our simulation are defined as:

$$\alpha_r = \left(\frac{\sum_{w \in W_r} \pi(w)}{|W_r| \cdot k(r)} \right)^{1/3} \qquad (2)$$

where w is a message, W_r is the set of messages created by user r, $\pi(w)$ is the number of times the message w has been forwarded and $k(r)$ is the number of followers of the node r.

5 Simulations

Using the social graph and the user parameters described in Sect. 4, we simulated a period of 30 days of user activity. We run 10 independent simulations in order to calculate the 95% confidence intervals which are shown as error bars in the figures and between square brackets in the tables and in numerical data. The simulations produced an average of $24,026,886$ [± 292] user interactions in that 77.1% ($18,515,225$ [$\pm 1,092$]) are related to the creation of new messages and the rest are forwarding messages. These proportions are consistent with those related to the dataset in Sect. 2 (excluding reply tweets). Among all created messages, 14.3% of them ($2,649,709$ [$\pm 1,128$]) have been forwarded originating cascades. In Fig. 4 we show the histogram of the depth of the cascades produced. As we can see, the trend is logarithmically decreasing with respect to the frequency. In fact, 78.7% of the forwarded messages are not propagated beyond the first level of followers. This trend is exactly the same shown in several analysis in literature [1, 2].

As discussed in Sect. 2.1, we define, for each node r in the simulations, the influence γ_r as the average retweet count of the tweets r has originated. In Fig. 5 we show the CCDFs of the number of forwards for each message as the solid line and the nodes' influence as the dashed line. Comparing these results with those in Fig. 1, we can see that the simulations replicated the presence of a small number of influential users located in the tail of the distribution[1].

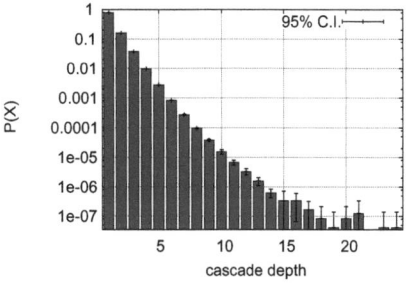
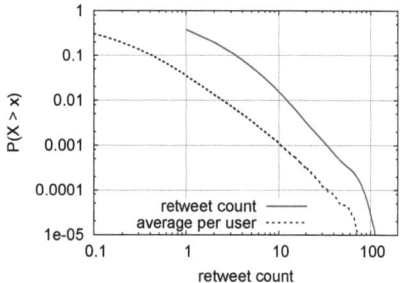

Fig. 4. Cascade depth distribution

Fig. 5. CCDFs of forwardings per message and user influence

In the column "orig" of Table 1, we summarise the results of the simulations (upper part) and the correlation of the resulting influence with other variables (lower part). In the table we refer to the the vector of the nodes' influence as γ while we use the symbol $\mathbf{k}$ for the vector of the number of followers and $\boldsymbol{\alpha}$ for the vector of the users' standing. Correlation values demonstrate that our model is able to replicate high correlation between the influence and both the number of followers and the user standing [2]. We also calculate the correlation between the simulated user influence and the influence γ^* of the selected users in the dataset described in Sect. 2. Considering that the influence from the dataset refers to the actual influence of the users in the Twitter network and that in our simulations we consider just a small subset of this network, the correlation value is remarkable and proves the ability of our model to simulate the actual user influence distribution.

5.1 Message Positioning and User Standing Impact

In order to study the impact of the message positioning in our model we run 10 simulations with the same setting described in Sect. 4, excluding the position function $\varphi(\cdot)$ from the model. Results are shown in the column "no-pos" of

---

[1] Direct comparison between the two plots is not possible, due to the large difference of the number of users in the dataset and in the simulations.

[2] Note that, while in (2) the standing is clearly a function of the influence, the values of the user standing have been computed based on the information propagation in the dataset, while influence is measured based on the simulations' results.

Table 1. Summary of results. Column "orig" refers to the results obtained using the original model; columns "no-pos" and "no-sta" refers to the results obtained without considering the position function and the user standing respectively.

	orig	no-pos	no-sta
max cascade depth	19.0 [±2.1]	121.6 [±6.0]	10.0 [±0.5]
max msg forwards	257.7 [±37.2]	10,347.4 [±304.2]	155.9 [±7.0]
max user influence	79.1 [±0.6]	1,436.0 [±187.6]	131.6 [±5.2]
	[id:41801]	[id:98020]	[id:2019]
corr(γ,k)	0.544 [±0.010]	0.100 [±0.009]	0.646 [±0.003]
corr(γ,α)	0.101 [±0.003]	0.073 [±0.004]	0.014 [±0.001]
corr(γ,γ^*)	0.595 [±0.003]	0.126 [±0.011]]	0.443 [±0.003]

Table 1. The main consequence of such change is that some messages flood the entire network and some users become extremely influential. This indicates that the decreasing visibility of the messages in social networks is fundamental for limiting the size of the information cascades.

We also studied the impact of the user standing, running 10 simulations where we have excluded it from the probability of forwarding. In this case the main change in results, shown in column "no-sta" of Table 1, is an increase in the correlation between the number of followers and a decrease in the correlation between the influence and the standing values.

In both "no-pos" and "no-sta" cases, it is noticeable the sensible decrease of the correlation between the simulated influence and the actual influence registered in the our dataset. This demonstrates the importance of considering both parameters in our model.

6 Conclusions

In this work we analysed the properties of the information diffusion in Twitter, in particular the impact of the structural features of the users and their retweeting behaviour. Using a Twitter dataset we studied the relation between the probability of a message to be retweeted and its position in the tweet feed and we concluded that this relation is described by a power-law function. We also characterised the most influential users in the network discovering that, although their ability of spreading messages is mostly given by their large number of followers, other factors have to be considered. These factors, joint effect we called user standing, have effect at the forwarding behaviour level, scaling the retweet probability given by the position of the message.

Based on these observations we proposed an agent-based information propagation model able to generate cascades whose properties match empirical observations. Agents simulate the activity of the users in a network creating and forwarding messages independently. Received messages are organised in an ordered list for reproducing the effect of the position on the forward probability.

Through simulations, we show that our model is able to reproduce information cascades statistically similar those presented in the literature and that the generated user influence is strongly correlated with the actual influence measured in the dataset. These results demonstrated that our model can thus be used to realistically study how the user activity and the forwarding mechanism influence the propagation of information.

Acknowledgements. This work was partially funded by the European Commission under the SCAMPI (FP7-FIRE 258414), RECOGNITION (FP7 FET-AWARENESS 257756), EINS (FP7-FIRE 288021) and EIT ICT Labs MONC (Business Plan 2013) projects.

References

1. Galuba, W., Aberer, K., Chakraborty, D., Despotovic, Z., Kellerer, W.: Outtweeting the twitterers-predicting information cascades in microblogs. In: Proceedings of the 3rd Conference on Online Social Networks, pp. 3. USENIX Association (2010)
2. Bakshy, E., Hofman, J.M., Mason, W.A., Watts, D.J.: Everyone's an influencer: quantifying influence on twitter. In: Proceedings of the fourth ACM International Conference on Web Search and Data Mining, pp. 65–74. ACM (2011)
3. Ye, S., Wu, S.F.: Measuring message propagation and social influence on twitter.com. In: Bolc, L., Makowski, M., Wierzbicki, A. (eds.) SocInfo 2010. LNCS, vol. 6430, pp. 216–231. Springer, Heidelberg (2010)
4. Sun, E., Rosenn, I., Marlow, C., Lento, T.M.: Gesundheit! modeling contagion through facebook news feed. In: ICWSM (2009)
5. Cha, M., Mislove, A., Adams, B., Gummadi, K.P.: Characterizing social cascades in flickr. In: Proceedings of the First Workshop on Online Social Networks, pp. 13–18. ACM (2008)
6. Leskovec, J., McGlohon, M., Faloutsos, C., Glance, N., Hurst, M.: Cascading behavior in large blog graphs. arXiv preprint arXiv:0704.2803 (2007)
7. Szabo, G., Huberman, B.A.: Predicting the popularity of online content. Communications of the ACM 53, 80–88 (2010)
8. Susarla, A., Oh, J.H., Tan, Y.: Social networks and the diffusion of user-generated content: Evidence from youtube. Information Systems Research 23, 23–41 (2012)
9. Oken Hodas, N., Lerman, K.: How visibility and divided attention constrain social contagion (2012)
10. Kwak, H., Lee, C., Park, H., Moon, S.: What is twitter, a social network or a news media? In: Proceedings of the 19th International Conference on World Wide Web, pp. 591–600. ACM (2010)
11. Arnaboldi, V., Conti, M., Passarella, A., Pezzoni, F.: Ego networks in twitter: an experimental analysis. In: The Fifth IEEE International Workshop on Network Science for Communication Networks, NetSciCom 2013 (2013)
12. Cha, M., Haddadi, H., Benevenuto, F., Gummadi, P.K.: Measuring user influence in twitter: The million follower fallacy. In: ICWSM, vol. 10, pp. 10–17 (2010)
13. Suh, B., Hong, L., Pirolli, P., Chi, E.H.: Want to be retweeted? large scale analytics on factors impacting retweet in twitter network. In: 2010 IEEE Second International Conference on Social Computing (SocialCom), pp. 177–184. IEEE (2010)
14. Newman, M.E.: The structure and function of complex networks. SIAM Review 45, 167–256 (2003)

Society as a Life Teacher – Automatic Recognition of Instincts Underneath Human Actions by Using Blog Corpus

Rafal Rzepka and Kenji Araki

Hokkaido University, Sapporo, Kita-ku, Kita 14 Nishi 9, 060-0814, Japan
{kabura,araki}@media.eng.hokudai.ac.jp
http://arakilab.media.eng.hokudai.ac.jp

Abstract. In this paper we introduce a method for generating a set of possible reasons of an action needed by an AI program for reasoning about human behavior. We achieve this goal by using web-mining and lexicons of keywords reflecting 14 instincts categories developed by psychologist William McDougall. We describe our system, the experiment and analyze its results of 78% of correct retrievals. The paper is also meant to be a message to social scientists who might be interested in testing their theories on constantly growing group of Internet users.

Keywords: causal knowledge retrieval, human instincts, text-minig.

1 Introduction

There is more than one possible reason for our behaviors and their causes are most often multidimensional even if we tend to simplify them. Although, when asked, we are able to imagine a whole range of possible reasons for a human's action or a state experienced by us or a third person. Experiences we gather from our earliest days allow us to explain things to our children and even ourselves when we have difficulties with immediate understanding. Machines do not have this capability and their attempts to reason about human acts tempt to lack necessary depth. Usually researchers working on causal knowledge retrieval concentrate on dry and easily verifiable facts as "people dry the laundry because it is sunny weather" [1], but when it comes to emotions and deeper analysis of usual or unusual acts, a computer needs a wider variety of possible causes to perform context processing. On the other hand, when, for example, a person makes a statement that *she is drinking alcohol* to a dialog system, usually there is not enough contextual data and the program needs to "imagine" why usually people drink and if a reason is confirmed (e.g. "to celebrate"), the system can more easily assume an emotional state of the person ("happy"). Retrieving affective consequences is a widely popular topic of the sentiment analysis field, but the set of instinctual causes and emotive effects is rarely a subject of knowledge acquisition. We think that our methods may be interesting not only for the AI researchers but also people from the humanities who would like to test their theories (or theories of others – like in our case) on thousands of people who share their thoughts online. In this paper we introduce our attempts to automatically categorize acts and states according to McDougall's theory of instincts.

1.1 McDougall's Categorization of Instincts

In our research on machine ethics we assume that human beings are equipped with the same instincts which build our morality. Haidt and Joseph [2] have performed a survey on common core of moral values, concerns, and issues across cultures and found three pairs: suffering/compassion, reciprocity/fairness, and hierarchy/respect. First we tried to build our system based on these pairs but we needed more sophisticated categorization and discovered works of William McDougall (1871-1938), who has been largely forgotten – until recently, with genetics and evolutionary psychology on the rise. The psychologist saw instincts as having three components. One is *perception* – human beings pay attention to stimuli relevant to our instinctual purposes, the second is *behavior* – human beings perform actions that satisfy our instinctual purposes, and the third is *emotion* – instincts have associated negative and positive emotions [3]. What is different from classic stimulus-response based behaviorism in case of McDougall's approach is purposiveness of instincts meaning that they are goal-directed. For that reason we found his theory useful for our purpose of creating a web-mining module which can "imagine" why somebody did something and what was the outcome of every possible motivation of the given act. In McDougall's opinion, all three above mentioned components work simultaneously and in concert with other instincts and we agree with him on this point – an analytical algorithm should be able to blend different instincts, not only the dominant one.

1.2 Developing a Lexicon for Blog Queries

McDougall came up with 14 types of instincts and their accompanying emotions. According to his list (in English) we created a lexicon of Japanese expressions (as we currently perform experiments limiting search span to only one culture but plan to extend the system to work with English, Chinese and Polish). We utilized phrases from Nakamura's dictionary of emotive expressions [4] and added our own query words trying to fit the explanations of instinct categories left by McDougall. The Categories are enlisted below.

- **Escape:** words associated with fear were collected, for example *scary, scared, fearful, terrifying, run away, horrifying* or *hair-raising* (21 phrases in total). The number of phrases in each category is not equal but we have already shown that it does not influence recall [5]
- **Combat:** words associated with anger, for example *get angry, furious, raging, enraged, outraged, pissed off* and *lose temper* (7 phrases in total).
- **Repulsion:** "disgust" associations (e.g. *disgusting, disgusted, disgustful, nauseating, sickening, can't believe* or *make one puke*) (18 phrases in total).

- **Parental** (protective): words associated with love and tenderness, for example *lovely, attachment, kind, friendly, nice, pleasant* or *dear* (12 phrases in total).
- **Appeal** (for help): words for matching distress and feeling of helplessness were added here, for example *weak, fragile, depressed, depressing, hopeless, powerless* or *couldn't do anything* (13 phrases in total).
- **Mating:** lust and attractiveness related words, for instance *beautiful, gorgeous woman, sexy, pretty, handsome, want to make out with* or *I'd marry* (10 phrases in total).
- **Curiosity:** words bearing meaning of feeling of mystery, of strangeness and of the unknown, e.g. *interesting, surprising, worth checking, rare, peculiar, strange* or *want to know* (8 phrases in total).
- **Submission:** words for feeling of subjection, inferiority, devotion, humility or negative self-feeling, for instance *ashamed, embarrassed, guilty, inferior, bashful, shy* or *blush* (10 phrases in total).
- **Assertion:** words for feeling of elation, superiority, masterfulness, pride and positive self-feeling, for example *happy, glad, easygoing, feeling good, good mood, satisfied* or *grin* (17 phrases in total).
- **Gregariousness:** words expressing feeling of loneliness, isolation or nostalgia – *lonely, crying, nostalgic, lonesome, tears, hurt, grieve*, etc. (16 phrases in total).
- **Food-seeking:** expressions for appetite or craving as *tasty, looking tasty, want to eat* or *wish to eat* (6 phrases in total).
- **Hoarding:** words expressing feeling of ownership and greed – *want to have, want to own, want to get, want to collect, don't want to lose*, etc. (7 phrases in total).
- **Construction:** expressions bearing meaning of feeling of creativeness, making, or productivity, for instance *would like to make, want to create, felt good to make, wanted to give birth, want to produce*, etc. (20 phrases in total).
- **Laughter:** words for amusement, carelessness, relaxation, for example *funny, laughed, feel relief, feel peaceful, peaceful* or *peace of mind* (19 phrases in total).

Above lexicon divided into 14 subsets was then used for matching process described in the next section.

2 Retrieval Process

Our system takes an input phrase consisting of a noun, a Japanese particle[1] and a verb. Then it automatically creates 9 queries by modifying the verb into conditional and continuative forms. These queries (as exact matches) are then used for retrieving sentences from 5 billion sentences blog corpus [7]. In order to avoid noisy inputs we use a hand-made facemarks database to cut strings which

---

[1] A suffix in Japanese grammar that immediately follow the modified noun and indicate if it is an object, topic, place, etc.

had no periods; we also set length limits to avoid too short and too long (or most probably wrongly divided into sentences) strings. The search is made by Apache Solr[2] and the retrieved sentences are automatically cleaned by removing ornate characters (notes, hearts, etc.) and passed to a semantic analysis tool ASA[3] for creating chunks that are more meaningful and are less prone to errors than usual N-grams. Finally, the left side chunk precedent to the query is matched against all the phrases from every subset of instincts lexicon. The hits are counted and a ranking is made. Some examples of input and output are shown in Table 1.

Table 1. Example of retrievals

Input	Top Instinct Category	Example Precedents
Kill someone	Repulsion (15)	disliked the society, felt grudge toward people, his/her state got worse...
Make a phone call	Escape (8)	mother was worried, feared about me, mom was shaking...
Help somebody	Appeal (15)	this year was in trouble, was on weak position, is in real trouble...
Tell a lie	Parental (2)	I don't like alcohol, I loved her/him
Go to restaurant	Food-seeking (8)	tasty at home, tastes nice at the sea, yummy seafood...

3 Experiment and Evaluation

In this section we introduce a preliminary test we performed to investigate how efficient are the algorithm and the used lexicon. By efficiency here we mean the number of natural instinct associations against unnatural ones. For example *sending kinds to school* is obvious in "Parental" category but does not fit "Mating" category. More than one natural categories are possible and many exceptions may happen (like "Repulsion" in case of very bad school environment), however we currently concentrate on commonsensical side of categorization. Having stated so, we are working toward a broader goal of context processing and this paper introduces an important part of deeper contextual analysis to enable machines to handle particular cases without falling in dangerous generalizations.

3.1 Experimental Setup

As an input we utilized 127 action phrases as "(to) call an ambulance" or "to kill a hero", etc. This list was an enhanced set we used in [8] for recognizing ethically

---

[2] Sorl (http://lucene.apache.org/solr/) is as fast as commercial engines and has no number of queries limitations, but obviously the scale of the data is very different. However, we have already proven that the higher precision of deeper search gives equal f-score to broader search in spite of the low recall [8].
[3] http://cl.it.okayama-u.ac.jp/study/project/asa/about_asa.html

problematic acts – this time we added more everyday life actions like "writing a book" or states like "someone in pain"[4] and removed similar entries, for example those needed for comparing *reasons for* and *reactions to* killing various kinds of animals, which in many cases have rather low hit rate in blogs.

3.2 Evaluation of the Categorization Task

First we took only the instincts with the highest score (one top instinct for an input) and the system achieved precision of 75.0% with the recall of 53.54%. After a closer look we have noticed that state describing inputs probably should not be evaluated as ones having instinctual motivations of actions[5] and we excluded them (11 phrases) together with 3 phrases written in erroneous Japanese. This time we also checked all the retrievals (553 hits), not only the top-scoring categories. Both preliminary evaluations were made by the first author and this more thorough judgement process showed that 77.78% category assignments were correct (almost 3 percentage points increase) but the recall dropped to 49.61% as we excluded state describing inputs.

3.3 Error Analysis and Possible Solutions

Rather low recall was obviously due to the restrictions we set on the retrieved data in order to avoid noisy strings for analysis. Therefore only 62,329 sentences were extracted for matching, which gave average of only 490 sentences for one input. We need to test the input against less or non-restricted data, but the low speed of semantic analysis performed by ASA using dependency parsing makes such experiments extremely time consuming.

There are obviously problematic expressions in the lexicon. For instance *liking* in "Parental" instincts category or *crying* in "Gregariousness" are too wide and can represent more than one instinct. *Worry* in "Escape" caused categorizing "make a phone call" as a fear based motivation, while this action should rather be under "Parental" instinct category. We are currently improving the lexicons by manual proofreading and automatic methods.

One Chinese character matching words should be removed as they appear in phrases with different connotations. For example there is an ideogram meaning "dislike" in a word "mood" (*go-kigen*) or "fear" character in an apology expression *kyōshuku*.

Certainly there is a need for analyzing also the right side (following chunk) of a query, because it would help disambiguate situations when such chunk can influence the meaning of the precedent (e.g. by being a negation).

---

[4] Many of phrases were taken from the logs of our online demo system http://demo.media.eng.hokudai.ac.jp/?ethical

[5] Input states were most often very broad (*being alive, laughing, enjoying*, etc.) and they were closer to being instinctual reasons for actions, not actions themselves. In future we plan to experiment with more specific state inputs.

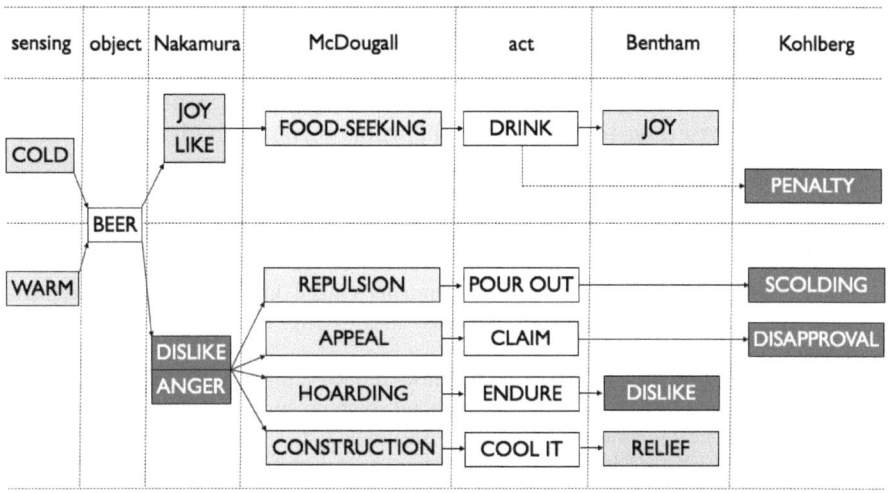

Fig. 1. Example of an act analysis using different classical theories. When an action details are input, the Web searching techniques using lexicons for each theory allow automatic retrieval of knowledge and predicting possible causes and effects. Multiple layers allow to acquire a deeper and ensure a basic understanding when one or more module fails to extract knowledge. Rules acquired in this process let the system assume what are, for example, possible outcomes when only feelings are given, or plan its own action to achieve the highest utilitarian *Pleasure* in Bentham's Felific Calculus.

4 Conclusions and Future Work

In this paper we introduced our method for automatic retrieval of possible instinctual reasons for human behaviors by using McDougall's list of instincts and web-mining and natural language processing techniques. Although our goal is to equip a machine with a capability of understanding human beings, with this research we would in addition like to suggest how interesting such approach could be also for social scientists. We have already shown usefulness of classic works of Akira Nakamura (his emotive expression dictionary [4] in sentiment analysis [6]), of Lawrence Kohlberg (his theory of moral stages development [9] in machine ethics field [10]) and currently implement Jeremy Bentham's Felific Calculus [11] into a system allowing deeper predictions of human act consequences, which we plan to use in the fields of collective behavior [12], machine self-understanding [13] and artificial empathy [14] (see Figure 1 to have a glimpse of reasoning using these theories). By this paper and above examples we wish to send social scientists a message saying that the new field of socioinformatics has plenty of opportunities for bringing back classical theories and testing them against vast number of Internet users.

As for the future work, we plan to improve our system according to the findings of above mentioned error analysis and continue developing the context recognition modules. As you can see in Figure 1, drinking a beer is usually joyful but if the social consequence retrieval algorithm finds "driving" context, the outcome may be of the opposite nature. Another task is to cooperate with social scientists on achieving more polished and wider lexicon in order to decrease the instincts categorization error rate; the numbers of query words must be also balanced according to their occurrences in the whole corpus. Finally, the evaluation experiment must be repeated with native speaker evaluators.

References

1. Inui, T., Inui, K., Matsumoto, Y.: Acquiring causal knowledge from text using the connective marker tame. ACM Trans. Asian Lang. Inf. Process. 4(4), 435–474 (2005)
2. Haidt, J., Joseph, C.: Intuitive ethics: how innately prepared intuitions generate culturally variable virtues, Dædalus, special issue on human nature: 55-66 (2004)
3. McDougall, W.: Outline of Psychology. Methuen & Co. (1923)
4. Nakamura, A.: Kanjo hyogen jiten (Dictionary of Emotive Expressions). Tokyodo Publishing (1993)
5. Ptaszynski, M., Rzepka, R., Araki, K., Momouchi, Y.: Automatically Annotating A Five-Billion-Word Corpus of Japanese Blogs for Sentiment and Affect Analysis. Computer Speech and Language (CSL). Elsevier (2013)
6. Ptaszynski, M., Dybala, P., Mazur, M., Rzepka, R., Araki, K., Momouchi, Y.: Towards Computational Fronesis: Verifying Contextual Appropriateness of Emotions. IJDET 11(2), 16–47 (2013)
7. Ptaszynski, M., Dybala, P., Rzepka, R., Araki, K., Momouchi, Y.: YACIS: A Five-Billion-Word Corpus of Japanese Blogs Fully Annotated with Syntactic and Affective Information. In: Proceedings of The AISB/IACAP World Congress 2012 in Honour of Alan Turing, 2nd Symposium on Linguistic and Cognitive Approaches To Dialog Agents (LaCATODA 2012), pp. 40–49. University of Birmingham, Birmingham (2012)
8. Rzepka, R., Araki, K.: Polarization of consequence expressions for an automatic ethical judgment based on moral stages theory. IPSJ SIG Notes 2012-NL-207(14), 1-4 (2012)
9. Kohlberg, L.: The Philosophy of Moral Development. Harper and Row (1981)
10. Komuda, R., Rzepka, R., Araki, K.: Social Factors in Kohlberg's Theory of Stages of Moral Development the Utility of (Web) Crowd Wisdom for Machine Ethics Research. In: Proceedings of the 5th International Conference on Applied Ethics, p. 14 (2010)
11. Bentham, J.: An Introduction to the Principles and Morals of Legislation. T. Payne, London (1789)
12. Rzepka, R., Araki, K.: Consciousness of Crowds - The Internet As a Knowledge Source of Humans Conscious Behavior and Machine Self-Understanding, "AI and Consciousness: Theoretical Foundations and Current Approaches", Papers from AAAI Fall Symposium, Technical Report, pp.127-128, Arlington, USA (2007)
13. Rzepka, R., Araki, K.: Artificial Self Based on Collective Mind - Using Common Sense and Emotions Web-Mining for Ethically Correct Behaviors. In: Proceedings of Toward a Science of Consciousness Conference (2009)
14. Rzepka, R., Krawczyk, M., Araki, K.: Using Empathy of the Crowd for Simulating Mirror Neurons Behavior, To Appear in the Proceedings of the 4th Workshop on Emphatic Computing, IWEC 2013, IJCAI (2013)

Diversity-Based HITS: Web Page Ranking by Referrer and Referral Diversity

Yoshiyuki Shoji and Katsumi Tanaka

Department of Social Informatics
Graduate School of Informatics, Kyoto University
Kyoto, Japan
{shoji,tanaka}@dl.kuis.kyoto-u.ac.jp

Abstract. We propose a Web ranking method that considers the diversity of linked pages and linking pages. Typical link analysis algorithms such as HITS and PageRank calculate scores by the number of linking pages. However, even if the number of links is the same, there is a big difference between documents linked by pages with similar content and those linked by pages with very different content. We propose two types of link diversity, referral diversity (diversity of pages linked by the page) and referrer diversity (diversity of pages linking to the page), and use the resulting diversity scores to expand the basic HITS algorithm. The results of repeated experiments showed that the diversity-based method is more useful than the original HITS algorithm for finding useful information on the Web.

1 Introduction

As the World Wide Web continues to rapidly expand, both Internet users and the purposes of Web documents are becoming more extensive and diverse. Users of the Web are not only computer specialists but also ordinary people, and the content on the Web has accordingly become broader, including not just informative documents but also many sub-products of communication, personal diaries, and so on. This has made Web usage increasingly more complex.

Many major Web search engines use link analysis algorithms such as HITS and PageRank to rank Web documents. For example, Google uses PageRank, which calculates popularity scores. These methods focus on the number of linking documents of each page. The popularity score is determined by recursive calculation using the simple hypothesis that pages linked by many popular pages are popular pages. The HITS algorithm uses the number of linking and linked pages to calculate the Hubs and the Authority. However, these methods focus only on the number of linking pages and are therefore unsuitable for coping with the demands of complex usages of the Web.

For example, when a novice user wants to know what a "compiler" is and inputs the query "compiler" to a search engine, the result contains many different kinds of pages, such as dictionary pages (e.g., thesaurus), encyclopedia pages (e.g., Wikipedia), introductory articles, commercial sites about specialized compilers, academic articles, and so on. All of them are popular documents and contain the term "compiler." However, the most popular documents are not necessarily useful for all users. In this case, dictionary, encyclopedia pages and introductory articles would be useful for novice and general users, but other pages would be more useful for a limited number of specialists. The problem here is that the existing link analytic methods score every document based on how many documents link to them instead of checking how it was linked. The number of linking documents expresses how popular the document is, but does not express why it is popular.

In this paper, we propose a new link analysis algorithm that considers not only the number of linking pages but also the diversity of linking pages and linked pages in order to consider the reason why the page is popular.

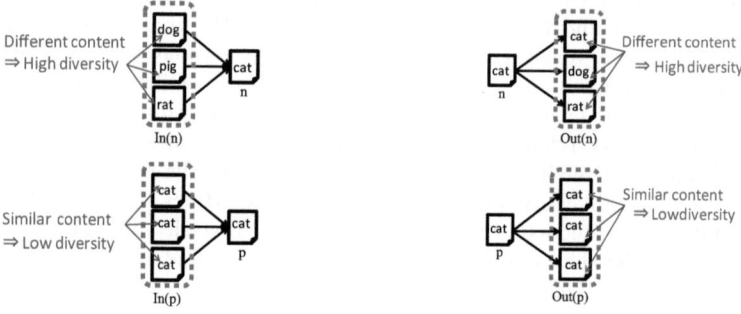

Fig. 1. Referrer diversity   **Fig. 2.** Referral diversity

Figure 1 shows an example of different ways a page is linked. There are two pages about cats, both of which are linked by three pages. Page n is the first page about cats, which is referred by documents on three different topics: a page about dogs, a page about pigs, and a page about rats. Page p is also linked by three pages, but all of them are about cats. Since each linking page has the same popularity score, PageRank gives the same popularity score to page p and page n because both of them are linked by the same number of pages. However, we suspect there is a big difference in the reason behind their popularity. Page n is linked by diverse pages. The authors of pages that link other pages may just be readers of the latter, who create the link after viewing such pages and finding them interesting. When the topic in the document reflects the interest of the author, a page linked by diverse pages must have a wide readership. An article with a diverse readership is assumed to contain information that can be interesting for many people: general information, universal information, and so on. Such information is useful not only for specialists but also for novice users. In contrast, page p has a biased, non-diverse readership, making it suitable for certain specialists or members of a specific community. The same can be said of linking sites and navigating sites(see Fig. 2). In this case, the former linking page may have been made by a user who has a broad outlook and the latter by one whose range of interest is narrow. These examples demonstrate that by considering how the linking pages and linked pages are diverse, the search algorithm can create a ranking that depends on the reason behind the popularity.

We define diversity as the dispersion of a set of pages. When each page in the set has a different topic, the diversity score is high, and when all pages are similar, the diversity score is low. We expand the HITS algorithm, which is the standard existing link analysis algorithm, with two types of diversity score: referrer diversity and referral diversity. Referrer diversity means how pages that link the page are diverse and referral diversity means how pages linked by the page are diverse. The conventional HITS algorithm calculates Hubs, which means how the page links many good authorities, and Authorities, which means how the page is linked by many good hubs. We expand the concept of Hubs and Authorities with diversity by two simple hypotheses:

– The page linking diverse Authorities is a valuable Hub
  (This Hub can be created by a well-informed generalist who has a wide range of interests.)
– The page linked by diverse Hubs is a valuable Authority
  (It must be useful not only in a specific field.)

The rest of this paper is organized as follows. Section 2 describes related work. In Section 3, we introduce the diversity-based link analysis algorithm we developed. Section 4 describes our experiments, and Section 5 evaluates our method in light of the experimental results. We conclude this paper in Section 6.

2 Related Work

In this paper, we propose a link analysis ranking algorithm that considers the diversity of linking pages and linked pages to isolate general information. We adduce three related previous studies. Section 2.1 describes research on diversity, section 2.2 describes other link analytic ranking algorithms, and section 2.3 describes methods to identify general information.

2.1 Diversity

Diversity in information science is a very active research topic. Collective intelligence is discussed particularly actively since the interactive usage of Web sites has become more common. This is called Web 2.0. Surowiecki [1] presented conditions for data and methods to realize the wisdom of crowds in his 2005 book. In his view, data need to meet three requirements: "Diversity of opinion," "Independence" and "Decentralization." When perfect data are available, they should be handled with "Aggregation," which means an algorithm to make up congregate data to knowledge. Link analytic ranking algorithms such as PageRank can be assumed as a voting model of crowds that treat the link as an acceptance. Thus, any discussion on diversity should consider the validity of the ranking.

Diversity has also been discussed in the field on Information Retrieval. One of the most active research topics in this area is the diversification of Web search result pages. These studies tackle the problem of how many diverse pages can appear in the first result page [2] [3] [4]. The diversification of Web search results can be classified by two features [5]:

- Ambiguous query terms
- Available information sources

For instance, the query "jaguar" is an ambiguous query in that it can have several meanings, such as the car manufacturer, the animal, a personal name, and so on. To diversify the search result of "jaguar," ranking algorithms should rank the pages relevant to the different meaning of "jaguar," alphabetically and without overlaps. If the aim of the query is identified, it is a problem when the ranking contains the same kind of information. Search algorithms should therefore create rankings that cover different types of content.

The research areas that discuss diversity are not limited to information sciences but include sociology, ecology, life science, economics, and so on. Stirling [6] put forth three key factors on categorical diversity: "Variety," "Balance" and "Disparity." To come up with the optimal design for the calculation of diversity, we need to place emphasis on these three factors.

2.2 Link Analysis Algorithms

Here, we present the new link analysis-based algorithm we developed. There are many past examples that expand PageRank and HITS for each aspect.

For instance, the topic sensitive PageRank [7] deals with the topic of the query and documents. The TrustRank [8] algorithm uses PageRank for spam filtering on the basis of the simple theory that spam pages link to both good and spam pages but good pages link only to good pages. Another idea is building the concept of time and space into PageRank to measure the historic impact [9].

Alternative approaches featuring Web graph analysis have been proposed. The HITS algorithm [10] uses the link structure on the Web to locate communities in the Web graph. It models the Web graph as a bipartite graph and calculates the importance of Web pages by convergence calculation. It uses two types of scores: a Hub score and an Authority score. Authorities are pages that show good information, and they obtain a higher score if the page links many highly scored Hub pages. Hubs are pages that link to good Authority sites. A typical example of a good Hub site is a linking site or a good search result page. The generalized co-HITS algorithm [11] is a HITS-based algorithm that extends the conventional HITS algorithm from Web links to general bipartite graphs such as a paper-and-author pair. SALSA [12] has similar algorithms and makes bipartite graphs based on Hubs and Authorities and has a score calculated based on random walk. Fast random walk with restart [13] is similar, too, and is used to compute the similarity between nodes.

Our algorithm is a HITS-based algorithm with weighting. The unique part is that it defines the weight of the nodes depending on the diversity.

2.3 Finding Information for Novice Users

The aim of this research is to enable novice users to find useful information on the Web. There are a few previous studies that focus on the comprehensibility or specialty of a given document. Our method finds the information for beginners by means of a link analytic approach. There has also been much research on estimating the specialty of a document by a content analysis approach. In particular, estimating the specialty of terms included in documents is a hot topic.

In the field of natural language processing, several methods that extract the special terms (i.e., technical terminology and jargon) from documents have been proposed. Nakatani et al. [14] proposed a link analytic method using the Wikipedia category structure to extract special terms. These studies used big corpora or document sets of limited specialized fields and structural information to measure the specialty of a term. Our method does not aim to find special terms but rather special Web pages without using particular datasets. The existing methods can be used to increase the accuracy of our method in a complementary style. There is a previous similar study that aims to find comprehensible Web pages by the link analytic approach. Akamatsu et al. [15] proposed a TrustRank-based method built on one simple rule: comprehensible pages are more likely to link comprehensible pages. General pages that we want to find are similar to comprehensible pages, but this method is based on PageRank, which focuses only on the number of links and not on how they link.

3 Proposed Method

In this section, we explain our link analysis method based on diversity in detail. The proposed method is composed of two parts. The first is calculating the diversity of the set of pages. For this part, we propose a method to quantify how each page in the set is different from the others. This quantification method creates a feature vector of each page with LDA and then uses the sum of the distance between the centroid and each page in the set. The second part is expanding the HITS algorithm by using the diversity of the referrer and referral documents of each document. The method calculates the diversity of document links to the document as the "referrer diversity" and of the document linked by the document as the "referral diversity." We set these two diversity scores in the HITS algorithm as the weight of the edge.

The purpose of the proposed method is to find pages that are useful for everyone: not only specialists but novice users as well. We assume a simple HITS-based hypothesis: the Hub page that links to diverse Authorities and the Authority page that is linked by diverse Hubs must feature a wide readership and widespread interest and therefore be of general interest to everyone.

3.1 Determining Diversity

To calculate diversity, each document has to be expressed as a feature vector. In our method, we take topics from the main text of the document and define the diversity as how different the topics of a page set are. The most simple way to express the document as the vector is just counting all the terms in the main text, but when the number of documents increases, the number of vector dimensions explodes. We used LDA (Latent Dirichlet Allocation) to compress the dimension. Each term in the dataset is assumed to belong to one topic of i types of topics based on the topic model, where i is the given number of dimensions. LDA classifies terms into the topics to which they belong. The frequency of topic occurrence in each document is used as the feature vector of the document. Each document can be expressed by an i-dimensional vector. The length of documents in the dataset is not constant, so the feature vector has to be normalized.

Note that to create a feature vector, we can use other information in addition to the topic of the main text, such as the degree of confirmation or denial, the stance of the author, sentiments, and so on [5]. If another kind of diversity is needed, the algorithm can deal with it by switching function just as well based on a differently constructed feature vector.

We propose a diversity function $d(P)$ to calculate the diversity of the document set P. This function takes a high value when each document in P has a different topic and a low value when all documents have a similar topic.

Every document is defined as $n = (n_1, n_2, n_3, n_4, \ldots, n_k)$, that is, documents in the dataset are explained as a k dimension feature vector based on the frequency of the topic found in their main text. The diversity, which means how the documents in the set P are diverse, is defined as

$$d(P) = \frac{1}{|P|} \sum_{p \in P} dist(p, mean(P)), \tag{1}$$

where $mean(P)$ is the arithmetic mean of the set of documents P and $dist(a, b)$ is the Euclidean distance between vectors a and b. $mean(P)$ is

$$mean(P) = \frac{1}{|P|} \sum_{p \in P} p, \tag{2}$$

and $dist(a, b)$ is

$$dist(a, b) = \sqrt{\sum_{i=1}^{k} (a_i - b_i)^2}, \tag{3}$$

where a and b are vectors expressed as $a = (a_1, a_2, a_3, \ldots, a_k), b = (b_1, b_2, b_3, \ldots, b_k)$. The definition of $d(P)$ in our method is the normalized sum of the difference between each page in P and its mean. When the dispersion of the documents in P is high, this value become high. The value drops into $[0, \frac{\sqrt{2}}{2}]$ when the norm of the feature vectors is normalized to 1. $d(P)$ gets its maximum value when all the articles in P have a different topic and gets its minimum value, 0, when all documents in P have the same content or the number of documents in P is less than two. We call $d(In(n))$ the referrer diversity of n, which means how pages linking n are diverse, where $In(n)$ is the set of pages that link to n. Likewise, we call $d(Out(n))$ the referral diversity of n, which means how pages linked by page n are diverse, where

$Out(n)$ is the set of pages linked by n. It is important to note that this calculus equation is not so novel. You can see the same equation in the K-means clustering as the value to minimize. It is used to express the cohesiveness of the cluster. When the dimension of the vector is one, this equation means the variance.

3.2 Diversity-Based HITS Algorithm

In this section, we explain the method to calculate the Hub score and the Authority score by using the diversity-based HITS algorithm considering referral diversity and referrer diversity.

First, we have to prepare the root set that is used for the result page of the given query. We also need a graph for the link analysis, so a base set was created as the sum set of the root set itself, linking document set and linked document set of the root set.

The original HITS algorithm defines Hubs and Authorities by mutual recursion, as

$$hub(p) = \sum_{q, p \to q} auth(q) \qquad (4)$$

$$auth(p) = \sum_{q, q \to p} hub(q), \qquad (5)$$

where both p and q is a page in the base set. It can be expressed as matrix calculation below:

$$h = Aa \qquad (6)$$
$$a = A^T h, \qquad (7)$$

where A is the adjacency matrix of the data set and h and a are the vectors of the Hubs and Authorities, respectively.

$$A_{ij} = \begin{cases} 1 \text{ if } i \text{ links } j \\ 0 \text{ otherwise.} \end{cases} \qquad (8)$$

In this case, A_{ij} means the link between node j and node i.

Our method modifies this HITS calculation by using diversity-based factors $d(In(p))$ and $d(Out(p))$ as

$$dhub(p) = d(Out(p)) \sum_{q, p \to q} dauth(q) \qquad (9)$$

$$dauth(p) = d(In(p)) \sum_{q, q \to p} dhub(q), \qquad (10)$$

where $In(p)$ is the set of pages that links page p, and $Out(p)$ is the set of pages linked by page p. Two scores: $dhub(p)$ and $dauth(p)$ mean diversity-based Hubs and diversity-based Authorities. We call $d(In(p))$ as "referrer diversity" of page p, and $d(Out(p))$ as "referral diversity" of page p. This formula supports the two diversity-based hypotheses above, that is, "The Hub that links diverse Authorities is a good Hub" and "The Authority that is linked by diverse Hubs is a good Authority."

We can replace the adjacency matrix A of the HITS algorithm. We propose two diversity-based adjacency matrixes. One is based on referral diversity and the other on referrer diversity. The expanded adjacency matrix taking referral diversity is as below:

$$N_{ij} = \begin{cases} d(In(j)) & \text{if } i \text{ links } j \\ 0 & \text{otherwise,} \end{cases} \quad (11)$$

where $In(j)$ is the set of pages that links page j. In this matrix, links to the document that are linked by many different documents are weighted highly and receive a higher score. In contrast, a higher score is not correlated with pages linked by many similar pages. The other diversity-based adjacency matrix, O, which takes the referral diversity, is

$$O_{ij} = \begin{cases} d(Out(i)) & \text{if } i \text{ links } j \\ 0 & \text{otherwise,} \end{cases} \quad (12)$$

where $Out(i)$ is the set of pages linked by page i. This matrix means that the weight of the link from the page linking diverse pages become high, and the weight of the link becomes low when the link is from a page linking similar pages. To consider referral diversity, A should be replaced with O. A can be replaced with N to consider referrer diversity.

$$dh = Oda \quad (13)$$
$$da = N^T dh, \quad (14)$$

where da and dh are the vectors of the diversity-based Hubs and diversity-based Authorities, respectively. In the original HITS algorithm, the coefficient of propagation is the same with the Hubs to Authorities propagation and the Authorities to Hubs propagation. The proposed method replaces A and A^T individually with diversity-based matrixes. It makes an asymmetric link weighted bipartite graph (see Fig.3). To calculate diversity-based HITS, it uses referral diversity score for back link propagation from Authority page to Hub page, and referrer diversity score for link propagation from Hub page to authority page.

The expanded HITS algorithm with a diversity-based adjacency matrix can be solved by the power method.

$$dh = ON^T dh \quad (15)$$
$$da = N^T Oda. \quad (16)$$

Vectors da and dh converge to Authorities and Hubs when they are normalized by every phase. Authority-based ranking can be used to find documents and Hub-based ranking can be used to find good linking sites or navigating sites.

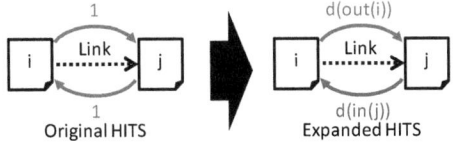

Fig. 3. Asymmetric propagation value

4 Experiment

To compare the methods explained in section 3 with the original HITS algorithm and variant methods, we conducted an experimental Web ranking evaluation. The pages for the given query were sorted by the Authority score of each method as a search result ranking. Each of these rankings was evaluated by bucket-based evaluation. As stated previously, the aim of the proposed method is to enable both novice and expert users to find information that is useful. We used one participant who played a novice and classified documents sampled by each ranking into useful and useless documents after reading the main text.

4.1 Variant Methods

To clarify the effect of referral diversity and referrer diversity particularly, we have prepared two variant methods.

One is the referral diversity-based method. It replaces A with O, and A^T is unchanged. It considers only referral diversity and not referrer diversity. This supports the hypothesis that the "The Hub that links diverse Authorities is a good Hub."

Another one is the referrer diversity-based method. It replaces A^T with N^T, and A is unchanged. It considers only the referrer diversity. This supports the hypothesis that the "The Authority that is linked by diverse Hubs is a good Authority."

We compared them to proposed method as baseline methods.

4.2 Data Set

We used The ClueWeb09-JA Dataset, which contains over 67 million pages with 400 million links between them. We prepared eight queries, shown in Table 1. These included two types of query: those about difficult topics and those about easy topics. The top 1,000 pages on BM25 sorted ranking were extracted by each query as the root set. Pages linking to a page included in the root set and linked by pages in the root set were used as the base set. The root set was then sorted by each method. The page evaluated by the participant is a sample of the root set.

We compared four methods below.

- **Both** is the proposed method based on both diversity factors. Pages are ranked by the Authority score calculated by Eqn. 9. It uses the referrer diversity to calculate Authorities and the referral diversity to calculate Hubs.
- **Referrer** is the variant method based only on referrer diversity. It uses the referrer diversity factor on the propagation from Authorities to Hubs.
- **Referral** is the variant method based only on referral diversity. It uses the referral diversity factor on the propagation from Hubs to Authorities.
- **HITS** is the baseline method. It is the original HITS algorithm.

Each method is compared using the Authority-based score because in this experiment we want to find the document but not the linking page.

To compare rankings by these methods, we evaluated sample pages of each ranking. First, we split the ranking into 5 buckets: the top 10 pages and pages ranked from 11–50, 51–200, 201–500, and 501–1,000. We took 10 sample pages from each bucket. The sampling rate of each bucket was not constant: the upper part of the ranking was sampled in high density and the bottom part was sampled coarsely. All of the top 10 pages were evaluated by one participant, but only 2 % of the bottom pages were evaluated. The total number of evaluated pages was 1,366 by 8 queries and 5 methods after removing duplicates. The participant evaluated each document in terms of whether or not it was useful for a novice user. Sample pages were sorted randomly for every query. Each page in the dataset was shown as plain text.

We used GibbsLDA++[1] as an implementation of LDA. We classified the terms in the dataset into 100 topics. The LDA sampling was iterated 2,000 times.

Table 1. Queries

Queries	Type
Postal service privatization	Easy
France trip	Easy
The Sagrada Familia	Easy
Fish called by different names in life stage	Easy
Parkinson's disease	Specialist
Game theory	Specialist
Compiler	Specialist
Machine learning	Specialist

4.3 Result

The results are shown in Table 2 and Fig. 4. The original HITS method ranked many correct pages in the middle of the ranking. All expanded methods were influenced by the original method. In the ideal case, it is hoped that many

---

[1] http://gibbslda.sourceforge.net/

correct pages appear in the top part of the ranking and that a small number of correct pages appear in the bottom. The **both** method, which uses both types of diversity, found more useful pages for novices in the top part of the ranking than the other methods. The **referrer** method seems a little bit better than the original HITS method in the top part of ranking, but it ranks many correct pages in the bottom part. The **referral** method had almost the same accuracy as the original HITS method. The ratio of correct pages in the data set was 0.22 through all queries.

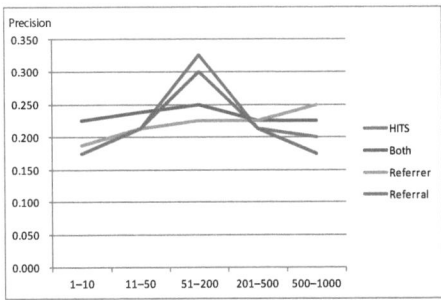

Table 2. Precision of each bucket through all queries

Bucket	HITS	Both	Referrer	Referral
1–10	0.175	0.225	0.188	0.175
11–50	0.213	0.238	0.213	0.213
51–200	0.325	0.250	0.225	0.300
201–500	0.213	0.225	0.225	0.213
500–1000	0.200	0.225	0.250	0.175

Fig. 4. Results for all queries

The eight queries used in the experiment can be separated into easy queries and specialist queries. Figure 5 shows the results for two types of query. In the easy query case, the ratio of correct pages is high: 0.35 through 4 queries. When the search task was easy, the search result contained useful pages for novice users. Each method had similar precision in the buckets of the upper part of the ranking. The **both** method found more correct pages than other methods in the middle part of the ranking. The **referral** method had findings throughout the ranking. The **referrer** method performed worse than the original HITS. In the specialist query case, the total number of correct pages in the dataset was small, with a ratio of just 0.09. It is assumed that pages on specialist topics are commonly not written in a language easy enough for novice users to understand.

On the whole, the **both** method, which uses both referral and referrer diversity, works well, especially when the query is specialist. The HITS algorithm ranked correct pages in the middle part of the ranking, and the two **referral** or **referrer** methods used the original adjacency matrix A. They were influenced strongly by original HITS result. Although the total number of correct pages was small, the **both** method ranked many correct pages in the top part of the ranking and not many of them in the bottom.

5 Discussion

The proposed method works better when the query is specific. For example, when the query is "machine learning," the proposed method finds three suitable

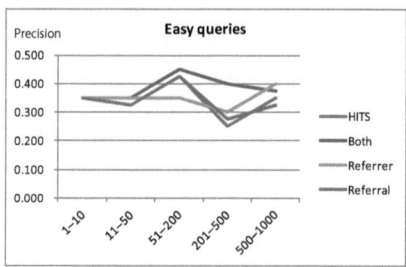 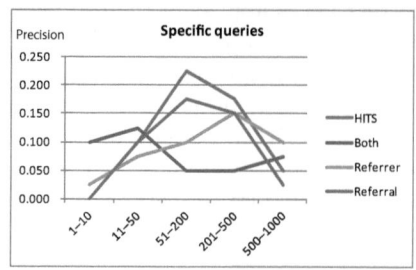

Fig. 5. Results by two types of queries

documents while basic HITS can not find any suitable documents. The detail of pages judged as correct document are online dictionary sites, or introductory works written by academic society. Dictionary sites and encyclopedia sites are frequently linked by many individual personal blogs. The author of each blog has different interests. Authors that are incidentally interested in "machine learning" will create links to those pages. Another day, they write about their interests, and link other interested pages. Then these pages have links not only to "machine learning" pages.

On the other hand, pages judged as not suitable are documents deemed too difficult, low-quality pages, spam and pages not relevant to the query. In this case, some social bookmarking sites were found in the top part of the basic HITS ranking. They do not contain useful information. These sites strongly connected with themselves by internal links. They are characterized by a high score for the both Hubs and Authority. The basic HITS algorithm is weak to such kind of link structures. Pages from social bookmarking services feature a similar design template. Our method estimated their topics as similar to each other.

The proposed method did not work well when the query was easy. The number of correct pages found is same to HITS in the top part of ranking. Our method found more correct page in the middle and bottom part of ranking. On this task, the number of pages suitable for novice users is inherently large. When the query is general, relevant pages are general too. For instance, when the query was "France trip," each method yielded results mostly containing content from major travel agency sites, all of which are about hotels, touristic hot spots or itineraries. When a judgment is determined only by the relevancy between the query and page, HITS-based algorithms may not be suitable even if it was expanded. There is a possibility that the diversity factor couses reverse effect, that is, the document linked only by documents about a trip may be relevant to the trip. Then, non-diverse tight links provide higher relevancy.

The graph of basic HITS algorithm has its peak in the middle of the ranking, where it scores good pages. In the bottom of ranking, there are many pages linked by few pages. Most of these pages are spam pages, low-quality pages, or minor pages. In the top part of ranking, a lot of individual pages of major online service sites appeared. They are linking to each other. Some of them have no content inside, i.e., private pages in social bookmarking websites, message pages in online fora with the aim to drive communication, product introduction pages of big company websites and so on. These pages are perceived as spam, or are not relevant to the query.

Even though our method works better than original HITS algorithm in this exeriment, it's accuracy is not high enough yet. The method has to be optimized by fixing and tuning parameters. For instance, the method used in this expetiment are not tuned, so it did not take into account the weight of propagation score and diversity score, the distribution of diversity score, the tuning of LDA and so on.

6 Conclusion

We proposed a diversity-based Web ranking method that expands on the HITS algorithm to include two diversity-based hypothesis: 1) that a page linking diverse Authorities is a valuable Hub, and 2) that a page linked by diverse Hubs is a valuable Authority. The objective of the diversity-based HITS algorithm is to enable not limited specialist users but general users to find suitable documents, that is, documents that are useful for novice users. We defined diversity as how the topic of each document in a set of documents is different from the topics of the other documents. We call the diversity of pages linking to the page referrer diversity and the diversity of pages linked by the page referral diversity. We expanded the HITS algorithm by replacing the adjacency matrix with two diversity-based matrixes. The proposed methods were compared with the original HITS algorithm by their authority scores in terms of finding useful pages for novice users. The method that uses both referral and referrer diversity could rank more good pages high, especially when the search query was specific.

As future work, we intend to expand the diversity-based methods further. Our method abandoned many factors to simplify the model. Of course, the method itself is built around the idea of diversity, not popularity, so it is necessary to focus on the number of linking documents, the amount of information, and the power of influence pages. Moreover, diversity can be defined not only from the topic of pages: for example, we can define it as authors' property, sentiment of documents, temporal-spatial metadata, and so on. We will tackle these issues with additional diversity-based methods.

Acknowledgments. This work was supported in part by the following projects: Grants-in-Aid for Scientific Research (Nos. 24240013) from MEXT of Japan and by JSPS Fellows (24 · 5417).

References

1. Surowiecki, J.: The wisdom of crowds. Anchor (2005)
2. Carbonell, J., Goldstein, J.: The use of mmr, diversity-based reranking for reordering documents and producing summaries. In: Proceedings of the 21st Annual International ACM SIGIR Conference on Research and Development in Information Retrieval, SIGIR 1998, pp. 335–336. ACM, New York (1998)

3. Wang, J., Zhu, J.: Portfolio theory of information retrieval. In: Proceedings of the 32nd international ACM SIGIR Conference on Research and Development in Information Retrieval, SIGIR 2009, pp. 115–122. ACM, New York (2009)
4. Capannini, G., Nardini, F.M., Perego, R., Silvestri, F.: Efficient diversification of web search results. Proc. VLDB Endow. 4(7), 451–459 (2011)
5. Minack, E., Demartini, G., Nejdl, W.: Current approaches to search result diversification. In: Proceedings of The First International Workshop on Living Web at the 8th International Semantic Web Conference (ISWC) (October 2009)
6. Stirling, A.: A general framework for analysing diversity in science, technology and society. Journal of the Royal Society Interface 4(15), 707–719 (2007)
7. Haveliwala, T.: Topic-sensitive pagerank: a context-sensitive ranking algorithm for web search. IEEE Transactions on Knowledge and Data Engineering 15(4), 784–796 (2003)
8. Gyöngyi, Z., Garcia-Molina, H., Pedersen, J.: Combating web spam with trustrank. In: Proceedings of the Thirtieth International Conference on Very Large Data Bases, VLDB 2004. VLDB Endowment, vol. 30, pp. 576–587 (2004)
9. Takahashi, Y., Ohshima, H., Yamamoto, M., Iwasaki, H., Oyama, S., Tanaka, K.: Evaluating significance of historical entities based on tempo-spatial impacts analysis using wikipedia link structure. In: Proceedings of the 22nd ACM Conference on Hypertext and Hypermedia, HT 2011, pp. 83–92. ACM, New York (2011)
10. Kleinberg, J.M.: Authoritative sources in a hyperlinked environment. J. ACM 46(5), 604–632 (1999)
11. Deng, H., Lyu, M.R., King, I.: A generalized co-hits algorithm and its application to bipartite graphs. In: Proceedings of the 15th ACM SIGKDD International Conference on Knowledge Discovery and Data Mining, KDD 2009, pp. 239–248. ACM, New York (2009)
12. Lempel, R., Moran, S.: The stochastic approach for link-structure analysis (salsa) and the tkc effect. Computer Networks 33(1-6), 387–401 (2000)
13. Tong, H.: Fast random walk with restart and its applications. In. In: ICDM 2006: Proceedings of the 6th IEEE International Conference on Data Mining, pp. 613–622. IEEE Computer Society (2006)
14. Nakatani, M., Jatowt, A., Ohshima, H., Tanaka, K.: Quality evaluation of search results by typicality and speciality of terms extracted from wikipedia. In: Zhou, X., Yokota, H., Deng, K., Liu, Q. (eds.) DASFAA 2009. LNCS, vol. 5463, pp. 570–584. Springer, Heidelberg (2009)
15. Akamatsu, K., Pattanasri, N., Jatowt, A., Tanaka, K.: Measuring comprehensibility of web pages based on link analysis. In: Proceedings of the 2011 IEEE/WIC/ACM International Conferences on Web Intelligence and Intelligent Agent Technology, WI-IAT 2011, pp. 40–46. IEEE Computer Society Press, Washington, DC (2011)

The Babel of Software Development: Linguistic Diversity in Open Source

Bogdan Vasilescu*, Alexander Serebrenik, and Mark G.J. van den Brand

Eindhoven University of Technology, The Netherlands
{b.n.vasilescu,a.serebrenik,m.g.j.v.d.brand}@tue.nl

Abstract. Open source software (OSS) development communities are typically very specialised, on the one hand, and experience high turnover, on the other. Combination of specialization and turnover can cause parts of the system implemented in a certain programming language to become unmaintainable, if knowledge of that language has disappeared together with the retiring developers.

Inspired by measures of linguistic diversity from the study of natural languages, we propose a method to quantify the risk of not having maintainers for code implemented in a certain programming language. To illustrate our approach, we studied risks associated with different languages in Emacs, and found examples of low risk due to high popularity (e.g., C, Emacs Lisp); low risk due to similarity with popular languages (e.g., C++, Java, Python); or high risk due to both low popularity and low similarity with popular languages (e.g., Lex). Our results show that methods from the social sciences can be successfully applied in the study of information systems, and open numerous avenues for future research.

1 Introduction

Open source software (OSS) development is typically characterised as a decentralised, self-directed, highly interactive, and knowledge-intensive process [14]. In OSS, programmers with different skill sets and skill levels, supporters, and users organise themselves in virtual (online) communities, and voluntarily contribute to a collaborative software project [22].

OSS communities are typically very specialised [26, 29, 37]: contributors focus on few activity types and are very territorial, touching only few parts of the system. OSS communities also co-evolve together with the associated OSS systems [22]: faced with turnover [28], these communities are sustained and reproduced over time through the progressive integration of new members [6]. However, with the abandonment of existing developers, OSS communities lose human resources with knowledge of the system or of some of its components, or, stated differently, with mastery of certain programming languages. Ensuring the heterogeneity of an OSS community in terms of the skills of its members is

---

* The author is supported by the Dutch Science Foundation through the project "Multi-Language Systems: Analysis and Visualization of Evolution—Analysis" (612.001.020).

important for a project's survival and performance [9]. To further put this issue into context, software systems are increasingly developed using multiple programming languages, as illustrated by the growing proportion of multi-language software developed in the United States from 1998 (30%) [16] to 2008 (50%) [17]. In addition, as languages become obsolete and development teams are faced with the problem of maintaining legacy code, or migrating it in order to survive, finding developers with knowledge of obsolescent technologies becomes more challenging. As the case may be, OSS communities are exposed to the *risk of not finding suitable contributors with knowledge of certain programming languages*.

Although new to software maintenance research, quantifying the risks associated with knowledge of programming languages in OSS communities around multi-language systems is related to the well-known concept of linguistic diversity from the study of natural languages [11]. Drawing inspiration from measures of linguistic diversity (Section 2), in this paper we attempt to quantify the aforementioned risk, associated with a given programming language in an OSS community (Section 3). Our model assumes that contributors are polyglot, i.e., they can "speak" more than one programming language. Moreover, analogously to dialects of a natural language being regarded as similar (mutually intelligible), our model also considers certain programming languages to be related. To quantify the strength of this relation, we mine patterns of shared knowledge of programming languages from developers participating in StackOverflow, a popular Q&A website (Section 4). Such relations need not be symmetrical: just like "Swedish is more easily understandable for a Dane, than Danish for a Swede" [20], our StackOverflow-based measure considers, e.g., that a C++ developer would be able to take over code written in C with less difficulty than the other way around.

By design, we can distinguish between two types of programming languages: those causing *high risk* within an OSS community (due to limited spread and low "similarity" with other more popular languages), and those causing *low risk* (either due to their popularity, or to their closeness to other more popular languages known to members of the community). Finally, to illustrate our risk measure, we track its evolution throughout the evolution of Emacs (Section 5).

2 Linguistic Diversity for Natural Languages

Measuring linguistic diversity for natural languages is an old research topic, dating back to Greenberg in 1956 [11]. For a given geographical area, Greenberg considers the probability that two randomly-chosen individuals *do not* speak the same language as a measure of the region's linguistic diversity. In this model (the first in a series of eight such measures proposed by Greenberg), if everyone speaks the same language, the probability that two randomly-chosen individuals speak the same language is, naturally, 1. Similarly, if everyone speaks a different language, this probability is 0. In general, for a language ℓ, the probability p_ℓ that a randomly-chosen individual speaks ℓ is the proportion of ℓ-speakers to the total population, i.e., $p_\ell = \frac{|S_\ell|}{|P|}$, where S_ℓ is the set of ℓ-speakers, P is the entire population, and $|\cdot|$ denotes cardinality. Consequently, the probability that two randomly-chosen individuals speak ℓ is p_ℓ^2, hence the probability that two randomly-chosen individuals speak the same language is $\sum_{\ell \in L} p_\ell^2$, where L is the set of all languages spoken in that region. The Greenberg linguistic diversity index A [11], corresponding to this simple model, is defined as[1]

$$A = 1 - \sum_{\ell \in L} p_\ell^2 \qquad (1)$$

A reaches its minimum of 0 when everyone speaks the same language (linguistic uniformity). Similarly, A reaches its maximum of 1 when everyone speaks a different language (maximal linguistic diversity). However, this model is overly simplistic, since (i) it does not consider *mutual intelligibility between different languages* (linguistic diversity should be lower in areas where related languages or dialects are spoken), and (ii) it does not consider *polyglotism* (a member of the population can speak more than one language). Concerns (i) and (ii) above are orthogonal. To account for (i), Greenberg proposed B [11], defined as

$$B = 1 - \sum_{\ell,m \in L} p_\ell p_m \cdot sim(\ell, m), \qquad (2)$$

where $sim(\ell, m)$ is a measure of mutual intelligibility interpreted as the similarity between languages ℓ and m (ranging between 0 when ℓ and m are completely independent, and 1 when $\ell = m$). Clearly, B reduces to A if $sim(\ell, m) = 1$ when $\ell = m$, and 0 otherwise.

To account for (ii), if polyglot, an individual is considered equally probable to speak any of the languages she commands, hence the expressions for A and B above are adjusted accordingly. Let $\mathcal{L}$ be the power set (set of all subsets) of L, excluding the empty set; $|\mathcal{L}| = 2^n - 1$, where $n = |L|$. For example, if $L = \{A, B, C\}$, then $\mathcal{L} = \{A, B, C, AB, AC, BC, ABC\}$. Let X_ℓ be the set of *exclusive ℓ speakers*, i.e., individuals that speak ℓ but do not speak any other language besides ℓ. For a subset $s \in \mathcal{L}$, by abuse of notation, we write X_s to denote the set of individuals that speak the combination of languages in s exclusively (i.e., they speak all languages in s, but no other languages besides those). By definition, $\sum_{s \in \mathcal{L}} |X_s| = |P|$. Index B becomes [11]

$$F = 1 - \sum_{s,t \in \mathcal{L}} p_s p_t \cdot \frac{\sum_{\ell \in s, m \in t} sim(\ell, m)}{|s| \cdot |t|}, \qquad (3)$$

where $p_s = \frac{|X_s|}{|P|}$, for all $s \in \mathcal{L}$. Clearly, F reduces to B in the monolingual case, since $|s| = 1$ for all $s \in \mathcal{L}$. B further reduces to A as discussed above.

Indices B and F as defined above interpret mutual intelligibility as similarity and hence assume it to be symmetric. However, it is well-known that mutual intelligibility is not necessarily symmetric: e.g., Swedes have more difficulties understanding Danish as opposed to Danes attempting to understand

---

[1] In the original paper [11] Greenberg only describes the linguistic diversity indices, but does not formalise them. The current formalisation is ours.

Swedish [20]. Therefore, one can define $sim(\ell, m) = \max(mi_\ell(m), mi_m(\ell))$ where $mi_\ell(m)$ is the measure of intelligibility of the language m for speakers of language ℓ. Similarly to $sim(\ell, m)$ we require $mi_\ell(m)$ to range between 0 and 1, such that $mi_\ell(m) = 0$ if m is unintelligible for the speakers of ℓ and $mi_\ell(m) = 1$ if m is intelligible for all speakers of ℓ. In particular, if $\ell = m$ then $mi_\ell(m) = 1$.

3 Risk of Using a Programming Language

There are many risks impacting software development, and many methods to estimate them [23]. In this paper we do not aim to cover all possible facets of risk, but rather focus on a particular scenario. For a (open source) software project using multiple programming languages, we study the readiness of the developer community to take over code implemented in a certain language, and evaluate the risk of not finding contributors that can "speak" that language.

Based on the discussion of linguistic diversity above, we require that a measure of this risk be *domain-specific*, i.e., aware of relations between programming languages. To simplify our model, we assume *perfect fluency* of developers in all the features of the languages they speak, even as the languages evolve. In addition, we assume *constant knowledge in time* (i.e., once a developer speaks a certain programming language, she never "forgets" how to speak it). For the purpose of empirically illustrating the risk measure (Section 5), we need to approximate at each point in time the developers with knowledge of a certain programming language. To this end, we furthermore assume *instant fluency*: a developer is said to "speak" a certain programming language at time τ if she has performed at least one change to a source code file in that language, prior to τ. Relaxing these assumptions is considered as future work.

3.1 Risk Measure

Let S be a multi-lingual software system, and let D be its developer community at time τ. Let L be the set of programming languages in use in S at time τ, i.e., those for which there exist source code files at time τ that need to be maintained. Similarly to the formalisation of the Greenberg indices from Section 2, let $\mathcal{L}$ be the power set of L, excluding the empty set. Let X_ℓ be the set of developers at time τ that speak ℓ *exclusively*, i.e., they speak ℓ but do not speak any other language besides ℓ. For a subset $s \in \mathcal{L}$, let X_s be the set of developers at time τ that speak the combination of languages in s exclusively (i.e., they speak all languages in s, but no other languages besides those). By definition,

$$\sum_{s \in \mathcal{L}} |X_s| = |D|. \qquad (4)$$

For a programming language $\ell \in L$, we define the risk of S at time τ of not finding developers that *can* speak ℓ as

$$risk_\ell = 1 - \sum_{s \in \mathcal{L}} p_s \cdot \max_{k \in s} mi_\ell(k), \qquad (5)$$

where $p_s = \frac{|X_s|}{|D|}$ is the probability at time τ that a developer speaks the combination of languages in s exclusively, and $mi_\ell(k)$ is an asymmetric mutual intelligibility measure as above. To illustrate the need for an asymmetric measure recall, for example, that C was originally a subset of C++ (the version of C defined by C89 is commonly referred to as the "C subset of C++" [30]), hence we perceive C to be more similar to C++ than C++ is to C. Therefore, assuming fluency of developers in all language features, and comparable complexity of the different components, we expect a C++ developer to be able to take over C code with less difficulty than the other way around.

As opposed to Greenberg [11] who is interested in an "average" case (i.e., as discussed in Section 2, if polyglot, an individual is considered equally probable to speak any of the languages she commands) when computing the linguistic diversity index F (3), we opt for the $\max(\cdot)$ function in (5) to denote that if polyglot, it is the language most intelligible to the language in question that will influence how difficult it is for a developer to take over that code.

To obtain a better understanding of how (5) can provide insights in the risk of not finding developers that can speak ℓ, we distinguish between developers D_ℓ that speak ℓ, and developers $D_{\neg \ell} = D \setminus D_\ell$ that do not speak ℓ. Similarly, let $\mathcal{L}_\ell$ be a subset of $\mathcal{L}$ such that $\forall s \in \mathcal{L}_\ell, \ell \in s$, and let $\mathcal{L}_{\neg \ell} = \mathcal{L} \setminus \mathcal{L}_\ell$. Then, we can rewrite (5) as $risk_\ell = 1 - \sum_{s \in \mathcal{L}_\ell} p_s \cdot \max_{k \in s} mi_\ell(k) - \sum_{s \in \mathcal{L}_{\neg \ell}} p_s \cdot \max_{k \in s} mi_\ell(k)$ which, given that $mi_\ell(k) = 1$ if $k = \ell$, and $\max_{k \in s} mi_\ell(k) = 1$ for all $s \in \mathcal{L}_\ell$, further simplifies to:

$$risk_\ell = \frac{|D_{\neg \ell}|}{|D|} - \sum_{s \in \mathcal{L}_{\neg \ell}} p_s \cdot \max_{k \in s} mi_\ell(k) \qquad (6)$$

Closer inspection of (6) reveals that the risk of not finding developers that can speak ℓ is *high* if very few developers speak ℓ (i.e., $\frac{|D_{\neg \ell}|}{|D|} \simeq 1$) and other languages are very distinct from ℓ rendering ℓ barely intelligible for "speakers" of those languages (i.e., $\sum_{s \in \mathcal{L}_{\neg \ell}} p_s \cdot \max_{k \in s} mi_\ell(k) \simeq 0$ because the languages in the collection are very different from ℓ, $\max_{k \in s} mi_\ell(k) \simeq 0$). By a complementary argument, two typical *low*-risk scenarios are when almost everybody can speak ℓ (i.e., $\frac{|D_{\neg \ell}|}{|D|} \simeq 0$, hence $p_s \simeq 0$ for $s \in \mathcal{L}_{\neg \ell}$), or when almost nobody can speak ℓ (i.e., $\frac{|D_{\neg \ell}|}{|D|} \simeq 1$) but popular languages make ℓ easily understandable (i.e., $\max_{k \in s} mi_\ell(k) \simeq 1$ for $s \in \mathcal{L}_{\neg \ell}$). To distinguish between these two scenarios in the empirical evaluation (Section 5), we also consider the percentage of developers that do not speak ℓ, $\frac{|D_{\neg \ell}|}{|D|}$.

4 Similarity and Mutual Intelligibility between Programming Languages

Mutual intelligibility, while being distinct from traditional notion of similarity, is still close to it. Therefore, in this section we mostly focus on measures of similarity of natural languages [7, 11] and their counterparts in programming linguistics [8] (the study of programming languages), and introduce our mutual intelligibility measure based on analysing StackOverflow[2].

4.1 Approaches to Similarity between Programming Languages

In linguistics, two complementary approaches to compute similarity between languages are commonly pursued. First, a similarity measure can be obtained "using an arbitrary but fixed basic vocabulary, e.g., the most recent version of the glottochronology list", by computing "the proportion of resemblances between each pair of languages to the total list" [11] (a similar approach has been recently pursued to study asymmetric mutual intelligibility [20]). Second, a similarity measure can be obtained using the distance between the branches languages fall into in a classification tree [7]. Using this approach, the more features two languages have in common, the more similar they are.

In programming linguistics, the approaches above are to a large extent unfeasible. First, application of approaches based on a common vocabulary would require establishing an agreed list of universal concepts present in all programming languages, akin to the Swadesh list for the natural languages [33]. The "word list" approach is being criticized in linguistics [13]; moreover, it introduces the need for identifying so-called cognates, or etymologically related words. The process of identifying cognates is complicated, since cognates do not necessarily look similar and words that look similar are not necessarily cognates [12]. Choosing the word list approach for similarity of programming languages assumes presence of universal concepts common to all (or at least most) programming languages. Even if compilation of such a list is possible at any given moment, it would rapidly become obsolete, since programming languages emerge much faster than natural languages. Moreover, the word list approach can be expected to trigger similar discussions about possible cognates, e.g., whether notions of a function in Lisp and C should be considered cognates or not.

One could also base a similarity measure on the shared concepts that underlie the design of both languages (e.g., data and types, variables and storage) and the paradigms to which they adhere (e.g., imperative or object-oriented) (cf. [7]). "We can master a new programming language most effectively if we understand the underlying concepts that it shares with other programming languages" [38, p4]. Again, the more attributes two languages would share in common, the more similar they would be considered. However, selecting the right attributes is challenging, to say the least. Most reliably, one could make use of taxonomies of programming languages [27]. However, as languages evolve, such taxonomies become inherently out of date and their categories change [15].

As an alternative to word-list and classification-tree approaches, one may consider recent studies [4, 18] that targeted the joint usage of programming languages. Karus and Gall [18] studied 22 open-source systems and observed two groups of languages for which the source code files frequently co-change, namely XML, XML Schema, WSDL (Web Service Definition Language) and Java on the one hand, and JavaScript and XSL (e.g., XSLT, XPath) on the other hand. In a

---

[2] http://stackoverflow.com

larger-scale study of 9,997 projects, Delorey et al. [4] observed that JavaScript and PHP, Java and JavaScript, C and C++, and C and Perl are commonly used both by the same authors as well as in the same projects.

As opposed to the actual usage, reflected in implementation of software systems, Doyle and Stretch [5] studied services offered by British software companies. The authors considered two programming languages to be similar if multiple companies offered these languages as part of their services. While Doyle and Stretch [5] employ the term "related by usage" to describe this relation, we prefer to call it "related by knowledge" and to reserve the term "related by usage" to such approaches as [4,18]. Indeed, companies offering multiple programming languages as part of their services do not necessarily use these languages in the same project. Instead, these companies have employees, potentially different, knowledgeable about each of these languages.

Both the "related-by-usage" and "related-by-knowledge" approaches can be seen as pertaining to pragmatics of programming languages which, together with semantics, are considered the most decisive for quantifying the similarity between programming languages [38, p5]. Therefore, we also expect that as opposed to both the word-list and classification-tree approaches, *pragmatic similarity* more accurately reflects developer expertise and ease of switching from one programming language to another. In Section 4.3 we also propose a pragmatics-pertaining mutual intelligibility measure, refining the "related-by-knowledge" insights of Doyle and Stretch [5]. Specifically, we base the mutual intelligibility measure on shared knowledge of the programming languages, as reflected in StackOverflow tags representing programming languages.

4.2 StackOverflow

StackOverflow (SO) is a free programming questions and answers (Q&A) site known to foster knowledge sharing among the developers [34,36]. When posting a question, SO users associate at least one and at most five different tags to it, and, in turn, become associated with these tags. When answering a question, SO users inherit all the tags associated with this question. Therefore, while each question can have at most five tags, a user can inherit an arbitrarily large collection of tags from all the questions she asked and answered. Tags can be related to programming languages (e.g., c#, java, php), operating systems (e.g., windows, linux), specific frameworks or technologies (e.g., hibernate, grails), specific versions of either of the above (e.g., c#-4.0, windows-7, hibernate-4.x), cross-cutting concerns (e.g., logging, algorithm), or other topics. SO tags can be collaboratively edited: while anyone can suggest an edit to question tags, only higher ranked users can review and edit tags, ensuring quality and reliability of the tags. Here we explore the public SO data from September 2011 (2,010,348 questions and 756,694 users).[3] The data is organised such that one can distinguish between *question tags* and *user tags*; one can also distinguish between user tags collected from asking questions, and user tags inherited by answering questions.

---

[3] http://www.clearbits.net/torrents/1836-sept-2011

Question tags. Frequent pairs of tags (e.g., `javascript`–`jquery`, `asp.net`–`c#`) indicate that these languages are commonly used together. However, this approach has several drawbacks. First, the number and skills of the users answering these questions is not considered (potentially leading to false positives). For example, there may be many questions tagged τ_1 and τ_2, suggesting that these languages are related, but only few people answering them. The relation between τ_1 and τ_2 might therefore not be representative of the entire (large) developer community (e.g., although few gurus with knowledge of both τ_1 and τ_2 exist, average developers may not possess the skills to easily switch between them).

User tags - answering questions. Since users inherit tags from questions they answer, frequent pairs of tags indicate that developers who possess knowledge of one language commonly also possess knowledge of the other. A frequent pair of tags (τ_1, τ_2) can emerge from multiple situations:

– many users inheriting τ_1 and τ_2 by answering questions tagged (τ_1, τ_2): either there are few questions tagged (τ_1, τ_2), but many users answering them (i.e., although τ_1 and τ_2 do not seem to be commonly associated in practice – e.g., they used to be but are by now obsolete, there is still a large pool of developers mastering both), or there are many questions tagged (τ_1, τ_2), and many users answering them (i.e., τ_1 and τ_2 are both commonly associated in practice, and supported by a large user base);
– many users inheriting τ_1 from questions tagged τ_1, and τ_2 from different questions tagged τ_2 (hence the pair τ_1–τ_2). In addition to an argument similar to the previous one, this also indicates languages that although rarely related in practice, are commonly mastered by developers. Hence, although seemingly unrelated, it seems easy for developers to switch from one language to another since they frequently master both.

User tags - asking questions. Users also inherit tags from all the questions they ask. A frequently occurring pair (τ_1, τ_2) indicates that developers are frequently faced with joint usage of τ_1 and τ_2, irrespective of the expertise available. In turn, this can suggest an emerging trend in relating τ_1 and τ_2 by usage. Note that as opposed to the variation across questions (many questions [of the same person]), frequent pairs (τ_1, τ_2) are now supported by large user pools (many questions of many persons). However, this does not indicate developer expertise.

Questions vs. users. In conclusion, both the questions-based and the users-based approaches are subject to potential false positives resulting from competing rather than interacting languages. However, we opt for user tags rather than of question tags, since the former can suggest relations between languages representative of the skills of the developer community (i.e., there are many users that share knowledge of both—fewer false positives), as well as relations between independent languages (i.e., languages which are seemingly unrelated, but knowledge of both is frequently shared by users—fewer false negatives).

4.3 StackOverflow-Based Mutual Intelligibility Measure

To quantify shared knowledge of programming languages by developers participating in SO discussions, we perform association rule mining [1] on SO tags representing programming languages. We say that a language k (with tag τ_k) is mutually intelligible or "related by knowledge" to a language ℓ (with τ_ℓ) if many of the SO users having inherited τ_k are also associated with τ_ℓ. As mutual intelligibility measure of language k with respect to language ℓ we choose *confidence*, one of the measures typically used to quantify the strength of association rules.

$$mi_\ell(k) = conf(\tau_k \Rightarrow \tau_\ell) = \frac{nBoth}{nLeft}, \qquad (7)$$

where *nLeft* is the number of users associated with τ_k, and *nBoth* is the number of users associated with both τ_k and τ_ℓ.

To ensure quality of the association rules, we perform a number of pre- and postprocessing steps. Preprocessing consists of filtering out potentially unreliable posts (either questions or answers with negative or zero score, as reflected by the number of votes), and infrequent pairs of tags (encountered for a single user). This limits the number of eligible SO contributors to slightly over 400,000 (out of 756,694 initially). Postprocessing is based on *lift*, another popular quality measure for association rules. If *lift* > 1, the tags appear more frequently together in the data than expected under the assumption of conditional independence [2]. Moreover, we require it to be unlikely that *lift* > 1 is observed only by chance and perform Fisher's exact test to determine statistical significance. Hence, we say that k is unintelligible to the speakers of ℓ (we redefine when $mi_\ell(k) = 0$) if *lift* ≤ 1, or *lift* > 1 is not statistically significant at 5% significance level. Approximately 7% of the pairs were removed by this filtering step (e.g., Python $\Rightarrow$ Visual FoxPro has lift 0.83; Curry $\Rightarrow$ C# has lift 1.78, $p = 0.31$). Finally, to reduce the amount of data processing required, we limit our scope to a subset of 160 programming languages, hence 160 corresponding SO tags. Our subset includes the most popular programming languages mentioned by TIOBE[4], Wikipedia[5], and the Transparent Language Popularity Index[6], as well as exotic languages such as M4 and RelaxNG, in use in Emacs. The complete list of languages included in our selection is part of the online appendix.[7]

4.4 Empirical Results

Table 1 displays values of the mutual intelligibility measure for a subset of the programming languages considered (also studied in [4, 5, 18]). An entry (*row, column*) represents the similarity of the language in *column* with respect to the one in *row*. For complete results we refer to the online appendix[8]. By definition

---

[4] http://www.tiobe.com/index.php/content/paperinfo/tpci/index.html
[5] http://en.wikipedia.org/wiki/List_of_programming_languages
[6] http://lang-index.sourceforge.net
[7] http://www.win.tue.nl/~bvasiles/languages/list.html
[8] http://www.win.tue.nl/~bvasiles/languages/list.html

Table 1. SO-based mutual intelligibility measure ([column] with respect to [row])

	Asm	C	C++	Cobol	CSS	Groovy	HTML	Java	JavaScript	Perl	PHP	Shell	XML
Asm	100%	55%	54%	1%	15%	1%	23%	39%	28%	12%	28%	1%	18%
C	8%	100%	48%	0%	12%	1%	17%	31%	21%	8%	21%	0%	13%
C++	5%	32%	100%	0%	10%	1%	15%	26%	18%	6%	18%	0%	11%
COBOL	12%	35%	40%	100%	24%	3%	29%	48%	38%	17%	37%	1%	28%
CSS	2%	10%	13%	0%	100%	1%	61%	21%	54%	5%	39%	0%	16%
Groovy	3%	15%	18%	1%	17%	100%	26%	63%	32%	7%	23%	0%	26%
HTML	2%	11%	14%	0%	46%	1%	100%	25%	56%	5%	40%	0%	18%
Java	2%	12%	15%	0%	10%	2%	15%	100%	19%	4%	16%	0%	12%
JavaScript	2%	9%	11%	0%	25%	1%	35%	20%	100%	4%	31%	0%	13%
Perl	5%	25%	27%	1%	18%	2%	26%	31%	30%	100%	31%	1%	19%
PHP	2%	9%	11%	0%	19%	1%	26%	17%	33%	4%	100%	0%	12%
Shell	12%	34%	38%	1%	19%	3%	32%	43%	33%	24%	35%	100%	24%
XML	3%	14%	19%	0%	20%	2%	29%	34%	35%	7%	31%	0%	100%

each language is perfectly mutually intelligible with itself (100% on the main diagonal). Next we observe that Assembly programmers are usually well-versed in other languages, including HTML (23%), Java (39%) and JavaScript (28%). Since there are only 44 posts (questions+answers) tagged assembly and java, these languages are unlikely to be related by usage, but are related by knowledge (more than 1000 developers with knowledge of both). Hence, if replacement developers are required for Java, one might consider the Assembly developers as candidates. We further observe that all the languages considered exhibit low intelligibility with such languages as COBOL, Groovy and Shell. This means that when COBOL, Groovy or Shell programmers leave, finding their replacement among programmers versed in other languages in Table 1 might be problematic. For COBOL one could argue that the low values can be explained by under-representation of legacy technologies on SO. This is, however, highly unlikely for Groovy, an object-oriented programming language first released in 2007. Low mutual intelligibility values of other languages with COBOL, Groovy and Shell contrast sharply with more easily replaceable developers in such languages as C, C++, HTML or Java. As expected, the table also shows a high degree of asymmetry. For instance, 63% of Groovy programmers know Java but only 1% of Java programmers know Groovy (not surprising since Groovy has been developed for Java, but constitutes only a minor fraction of the overall Java development).

Although we are measuring different things (similarity by usage in case of Karus and Gall [18] and Delorey et al. [4] versus mutual intelligibility or similarity by knowledge in our case), we expect that similarity by usage implies similarity by knowledge, since languages used together by the same person are likely to be known together by that person. Our results partly support the findings of Karus and Gall [18] (strong relation between XML and Java—34%, and limited evidence for C/C++ and XML—13% and 11%, respectively). Our results also support the findings of Delorey et al. [4], who observed strong relations between JavaScript and PHP ($\Rightarrow$ 31%, $\Leftarrow$ 33%), Java and JavaScript ($\Rightarrow$ 19%, $\Leftarrow$ 20%), C and C++ ($\Rightarrow$ 48%, $\Leftarrow$ 32%), and C and Perl ($\Rightarrow$ 8%, $\Leftarrow$ 25%).

5 Illustration of the Approach

To illustrate our approach in a real-world context, we performed a case study on Emacs [32], a popular text editor in development since the mid-1970s.

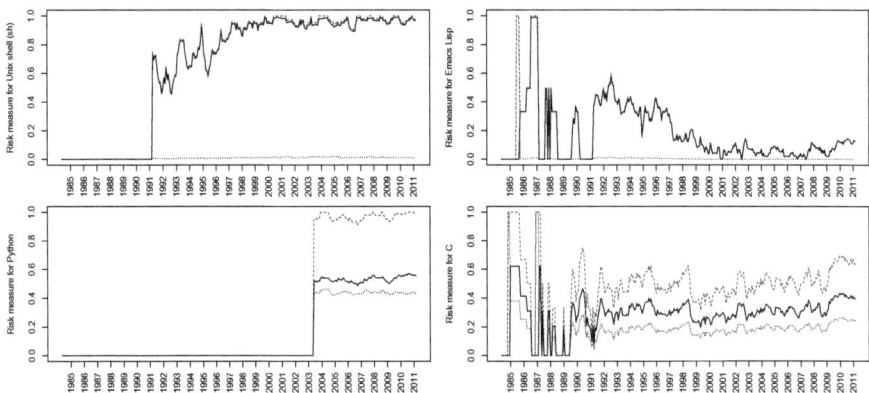

Fig. 1. The risk measure $risk_\ell$ (black solid line), the share of the community that does not speak ℓ (dashed red line), and the difference between the two (dotted blue line)

We identified 446 different (*name, email*) pairs in the *author* field for each change recorded in the Git log, corresponding to 27 years of Emacs development (1985-2012). Since there were multiple email addresses associated with the same names, and multiple names associated with the same email addresses, we performed identity merging [19, 37] (369 unique identities remained). To track the evolution of our risk measure throughout the evolution of Emacs, we extracted the programming languages used. We analysed the filename extensions of all the source files mentioned in the Git log. After filtering out files without extensions (mostly related to documentation), configuration files, make files, documentation files, and auxiliary files (e.g., used by the version control system), we uncovered the following 26 different programming languages: Assembly, Awk, Bash, Bison, C, C++, Cocoa, C shell, Emacs Lisp, Grammar, HTML, Java, Lex, Lisp, M4, Objective-C, Pascal, Perl, Prolog, Python, RelaxNG, Unix shell, SRecode, Termcap, Windows Batch, and XML. Next, we estimated the development community from which replacement developers can be sought, at one-month intervals. To filter out inactive contributors, at each point in time (e.g., February 2002), we considered that the community (per programming language) consisted of those developers who performed at least one change to a source code file (implemented in that language) in the past six months (e.g., between September 2001 and February 2002). Finally, we computed $risk_\ell$ at one month intervals.

We discuss four representative examples (Figure 1). Detailed plots for all 26 languages are available in the online appendix[9]. We start with Unix shell (top left). The risk measure and the percentage of non-speakers are very close, i.e., evolution of the risk measure can be explained predominantly by the evolution of the percentage of non-speakers. The increasing trend observed from 1991 to 2000, followed by the stabilisation from 2000 onwards reflects the diminishing proportion of Unix shell developers. Moreover, very low values of the dotted blue line indicate that Unix shell is not commonly known by developers programming

---

[9] http://www.win.tue.nl/~bvasiles/emacs/risk.html

in popular languages (e.g., Emacs Lisp and C). Emacs Lisp (top right) exhibits similarly close values of the risk measure and the percentage of non-speakers, but both values are low (below 0.2 starting from 1998). The lion's share of the development community is, hence, familiar with Emacs Lisp. In contrast, Python (bottom left), although spoken by a similarly small fraction of the Emacs community as Unix shell (dashed red line), exhibits much lower risk (black line). The high values for the difference between the two time series (dotted blue line), relatively stable in time, indicate that Python is commonly known by developers programming in popular languages. Indeed, $mi_{Python}(EmacsLisp) \simeq 0.46$, $mi_{Python}(Lisp) \simeq 0.44$, and $mi_{Python}(C) \simeq 0.23$. C (bottom right) is spoken by approximately half of the Emacs community and is also commonly known by developers programming in Lisp (0.39) and Emacs Lisp (0.38), resulting in very low risk. Emacs Lisp exhibits a similar pattern, with the difference that its low risk is mostly due to the large share of Emacs Lisp speakers within the community rather than high similarity with the other languages.

6 Conclusions

Inspired by linguistic diversity measures, we proposed a method to quantify the risk of not finding developers who can maintain code implemented in a certain programming language, and empirically illustrated it using a case study. Our method takes into account similarities between programming languages, for which we have proposed a novel measure based on shared knowledge of the developers participating in StackOverflow. By tracking the evolution of such a risk measure as projects evolve (e.g., in a dashboard-like application), risky languages can be discovered on time, and preventive action can be taken to ensure the maintainability of components implemented in those languages.

We believe the results obtained so far to be a promising start. Our new dimension to risk assessment, bordering software maintenance and the social sciences, *does* offer additional insights into the evolution of a software system, and *does* open up many avenues for future research. For example, to offer a more complete understanding of evolution, our risk model should be refined to incorporate the number of artifacts in a certain programming language (the risk seems higher if a large proportion of a system is implemented in a risky language), their role (examples may be less important than the core implementation), their stability as reflected in a version control system (files not changed for a long time seem less risky than recently changed ones, cf. [25]) or their algorithmic or linguistic complexity (files implementing more complex behaviour or using more exotic language features seem to be more risky). We also plan to include a more refined language-tag mapping such that tags corresponding to versions and dialects (e.g., `c#-2.0` and `swi-prolog`) can be accounted for. A further refinement would include distinction between tags representing different technologies (e.g., `ejb`, `hibernate` and `swing`). Apart from being all implemented in Java, such technologies share little in common and likely require different skills to maintain.

We also would like to introduce a project-level risk measure $risk_P$, being the maximum of $risk_\ell$ for all languages ℓ in the project P, indicating the highest risk of not having developers who can take over code implemented in a certain language. Similarly, for a given project P we can identify and rank developers that—should they decide to leave P—would contribute most to increase of $risk_P$. This ranking can be seen as an alternative interpretation of the "bus factor" [10] and a way to quantify the developers' contributions [3].

Beyond the boundaries of *linguistic diversity* is the general concept of diversity, and its measurement in several biological, physical, social, and management sciences [24]. Some of these techniques have recently been applied in context of software engineering as well [21, 31, 35]. A detailed comparison of these techniques with the risk measure proposed in the current paper is considered as future work.

References

1. Agrawal, R., Srikant, R.: Fast algorithms for mining association rules in large databases. In: Very Large Data Bases, pp. 487–499. Morgan Kaufmann (1994)
2. Brijs, T., Vanhoof, K., Wets, G.: Defining interestingness for association rules. Information Theories & Applications 10(4), 370–375 (2003)
3. Capiluppi, A., Serebrenik, A., Youssef, A.: Developing an h-index for OSS developers. In: Lanza, M., Di Penta, M., Xi, T. (eds.) MSR, pp. 251–254. IEEE (2012)
4. Delorey, D., Knutson, C., Giraud-Carrier, C.: Programming language trends in open source development: An evaluation using data from all production phase Sourceforge projects. In: WoPDaSD (2007)
5. Doyle, J.R., Stretch, D.D.: The classification of programming languages by usage. Man-Machine Studies 26(3), 343–360 (1987)
6. Ducheneaut, N.: Socialization in an open source software community: A sociotechnical analysis. Computer Supported Cooperative Work 14(4), 323–368 (2005)
7. Fearon, J.D.: Ethnic and cultural diversity by country. J. Econ. Growth 8(2), 195–222 (2003)
8. Gelernter, D., Jagannathan, S.: Programming linguistics. MIT Press (1990)
9. Giuri, P., Ploner, M., Rullani, F., Torrisi, S.: Skills, division of labor and performance in collective inventions: Evidence from open source software. International Journal of Industrial Organization 28(1), 54–68 (2010)
10. Goeminne, M., Mens, T.: Evidence for the Pareto principle in Open Source Software Activity. In: SQM. CEUR-WS workshop proceedings (2011)
11. Greenberg, J.: The measurement of linguistic diversity. Language 32(1), 109–115 (1956)
12. Handel, Z.: What is Sino-Tibetan? Snapshot of a field and a language family in flux. Language and Linguistics Compass 2(3), 422–441 (2008)
13. Heggarty, P.: Beyond lexicostatistics: How to get more out of "word lis" comparisons. Diachronica 27(2), 301–324 (2010)
14. Hemetsberger, A., Reinhardt, C.: Learning and knowledge-building in open-source communities a social-experiential approach. Management Learning 37(2), 187–214 (2006)
15. Jepsen, T.C.: Just what is an ontology, anyway? IT Professional 11(5), 22–27 (2009)
16. Jones, C., Jones, T.: Estimating software costs, vol. 3. McGraw-Hill (1998)
17. Jones, C.: Applied Software Measurement: Global Analysis of Productivity and Quality. McGraw-Hill (2008)

18. Karus, S., Gall, H.: A study of language usage evolution in open source software. In: MSR, pp. 13–22. ACM (2011)
19. Kouters, E., Vasilescu, B., Serebrenik, A., van den Brand, M.G.J.: Who's who in Gnome: Using LSA to merge software repository identities. In: ICSM, pp. 592–595. IEEE (2012)
20. Moberg, J., Gooskens, C., Nerbonne, J., Vaillette, N.: Conditional entropy measures intelligibility among related languages. In: Proceedings of Computational Linguistics in the Netherlands, pp. 51–66 (2007)
21. Mordal, K., Anquetil, N., Laval, J., Serebrenik, A., Vasilescu, B., Ducasse, S.: Software quality metrics aggregation in industry. Software: Evolution and Process (2012)
22. Nakakoji, K., Yamamoto, Y., Nishinaka, Y., Kishida, K., Ye, Y.: Evolution patterns of open-source software systems and communities. In: IWPSE, pp. 76–85. ACM (2002)
23. Neumann, D.E.: An enhanced neural network technique for software risk analysis. IEEE Trans. Softw. Eng 28(9), 904–912 (2002)
24. Patil, G.P., Taillie, C.: Diversity as a concept and its measurement. Journal of the American Statistical Association 77(379), 548–561 (1982)
25. Poncin, W., Serebrenik, A., van den Brand, M.G.J.: Process mining software repositories. In: CSMR, pp. 5–14. IEEE (2011)
26. Posnett, D., D'Souza, R., Devanbu, P., Filkov, V.: Dual ecological measures of focus in software development. In: ICSE, pp. 452–461. IEEE (2013)
27. Rechenberg, P.: Programming languages as thought models. Struct. Program. 11(3), 105–116 (1990)
28. Robles, G., González-Barahona, J.M.: Contributor turnover in libre software projects. In: Damiani, E., Fitzgerald, B., Scacchi, W., Scotto, M., Succi, G. (eds.) Open Source Systems, vol. 203, pp. 273–286. Springer, Heidelberg (2006)
29. Robles, G., González-Barahona, J.M., Merelo, J.J.: Beyond source code: the importance of other artifacts in software development (a case study). Journal of Systems and Software 79(9), 1233–1248 (2006)
30. Schildt, H.: C/C++ Programmer's Reference, 2nd edn. McGraw-Hill (2000)
31. Serebrenik, A., van den Brand, M.G.J.: Theil Index for Aggregation of Software Metrics Values. In: ICSM, pp. 1–9. IEEE (2010)
32. Stallman, R.M.: EMACS the extensible, customizable self-documenting display editor. SIGPLAN Not 16(6), 147–156 (1981)
33. Swadesh, M., Sherzer, J., Hymes, D.: The Origin and Diversification of Language. Adeline Transaction (1971)
34. Vasilescu, B., Filkov, V., Serebrenik, A.: StackOverflow and GitHub: associations between software development and crowdsourced knowledge. In: SocialCom, pp. 188–195. ASE/IEEE (accepted 2013)
35. Vasilescu, B., Serebrenik, A., van den Brand, M.G.J.: You can't control the unfamiliar: A study on the relations between aggregation techniques for software metrics. In: ICSM, pp. 313–322. IEEE (2011)
36. Vasilescu, B., Serebrenik, A., Devanbu, P., Filkov, V.: How social Q&A sites are changing knowledge sharing in Open Source software communities. In: CSCW. ACM (accepted 2014)
37. Vasilescu, B., Serebrenik, A., Goeminne, M., Mens, T.: On the variation and specialisation of workload–A case study of the Gnome ecosystem community. In: Empirical Software Engineering, pp. 1–54 (2013)
38. Watt, D.A., Findlay, W.: Programming language design concepts. Wiley (2004)

Using and Asking: APIs Used in the Android Market and Asked about in StackOverflow

David Kavaler*, Daryl Posnett*, Clint Gibler
Hao Chen, Premkumar Devanbu*, and Vladimir Filkov*

Univ. of California, Davis Davis, CA 95616, USA
{dmkavaler,dpposnett,cdgibler,chen}@ucdavis.edu,
{devanbu,filkov}@cs.ucdavs.edu

Abstract. Programming is knowledge intensive. While it is well understood that programmers spend lots of time looking for information, with few exceptions, there is a significant lack of data on what information they seek, and why. Modern platforms, like Android, comprise complex APIs that often perplex programmers. We ask: which elements are confusing, and why? Increasingly, when programmers need answers, they turn to StackOverflow. This provides a novel opportunity. There are a vast number of applications for Android devices, which can be readily analyzed, and many traces of interactions on StackOverflow. These provide a complementary perspective on *using* and *asking*, and allow the two phenomena to be studied together. How does the market demand for the USE of an API drive the market for *knowledge* about it? Here, we analyze data from Android applications and StackOverflow together, to find out what it is that programmers want to know and why.

1 Introduction

The cottage industry of Android application development is booming, and the number of applications, colloquially apps, is growing exponentially [5]. While writing good apps is contingent upon solid understanding of the Android OS, many app writers begin programming with little skill and learn on the fly. Naturally, the Android components include self-documentation, which aids amateur developers in their task. However, such documentation often needs supplementing with other sources. Amateur developers supplement their learning by reading online documentation and participating in knowledge exchanges in online question and answer communities, like StackOverflow. One recent field study [14] reports that StackOverflow has more examples, better coverage, and is visited by real developers more often than the online documentation. In this distributed, or crowdsourced, education system, people ask questions and others answer them, usually for no remuneration. Here we are interested in the relationship between the amount of API *use*, i.e. API popularity, and the *questions asked about* APIs, i.e. help requested for those APIs. Some APIs attract more questions and recedive more answers from the community than others. As APIs see more usage, more questions are asked about them, so popularity plays a role. But does API complexity drive a greater nimber of questions? Furthermore, does the amount of available documentation influence the number

---

* These authors contributed equally to this work.

of questions asked about an API? Such questions have previiously been difficult to approach quantitatively due to the many APIs, and because a large number of applications are required to ensure sufficient statistical variation in API use.

However, it is now possible to mine *app marketplaces*, with hundreds of thousands of applications; many of them are free, so it is feasible to download a large number for analysis. Furthermore, Android byte code is fairly high-level, so it is possible to determine exactly which APIs are used, and to what extent. This data, together with the readily-available data from StackOverflow, provides a a fresh opportunity to relate the degree to which programmers *use* APIs to the degree to which they *ask* about them. This data provides a starting point to define the popularity of an API, from the app marketplace, and the amount of help requested on StackOverflow for that API. Our contributions in this paper are as follows:

- We show that careful data gathering and cleaning will produce a data set of linked Android Market data and StackOverflow traces; this data set is of independent interest to both software engineers and documentation writers;
- We show that there is a non-trivial, sub-linear relationship between class popularity in Android apps and the requests for their documentation in StackOverflow, i.e. the number of questions on StackOverflow about a class lags its usage in the apps.
- We find a complex relationship between the size of a class, its internal documentation and the StackOverflow documentation supply and demand. The size of a class and the *total* amount of available documentation for a class both have sizable and significant impact on the number of questions, *i.e.* larger classes and more documented classes are associated with more questions. The situation is reversed for answers, although much smaller in magnitude.
- We give a case study examining several classes in detail to gain insights in the relationship between usage and documentation requests. A secondary goal of the case study is to understand the relationship between *mentions* of a class in the text of a question, and *salience* of the class to the meaning of the question.

2 Related Work

As APIs and libraries increase in complexity there has been an increased focus on leveraging, combining, and understanding multiple sources of developer documentation. Dagenais and Robillard study the decisions open source developers make while creating developer documentation for frameworks and libraries [9]. They find that developer choices to document their own code during the code creation process can not only affect the quality of documentation, but, in fact, the quality of code itself. Their results are relevant here as we are describing a relationship between internal API code, and the discussions that developers have regarding that code.

Following the public release of the StackOverflow data, researchers have studied how developers gain knowledge and expertise through online question and answer sites. Research in this area has studied the success of Q and A sites, the growth of askers and answerers, and how developers use online resources. Our focus is on the last category. Treude *et al.* evaluated the nature of questions asked on StackOverflow with respect to

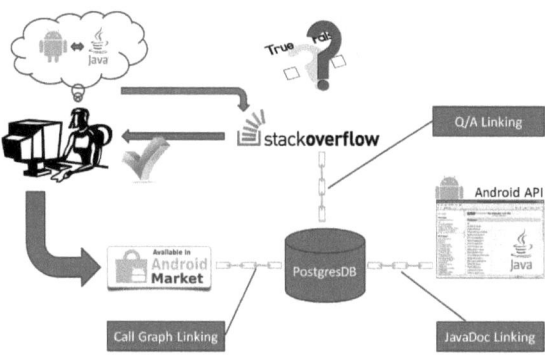

Fig. 1. Programming involves consulting documentation, internal, *viz.*, embedded in the code, and external, *e.g.*, in StackOverflow. Shown here are the data sources and their inter-connections.

programming language, question tags, and question types determined by manual coding of a sample of questions. They concluded that the community answers review, conceptual, and how-to questions more frequently than any other type, with review questions achieving a 92% accepted answer rate [21]. Posnett *et al.* studied the nature of expertise among Stack Exchange participants and found that participation in online Q/A sites does not necessarily lead to increased expertise as measured by peer scoring [16]. These two results may be viewed as mutually supportive suggesting that programmers answer questions about what they know. However, in what way does existing documentation impact the tendency to ask or answer questions about an API? There has been some previous work on this question. Parnin and Treude study availability of documentation across APIs within StackOverflow and find that it is very uneven, some APIs are well covered, while others receive very little attention [20]. Jiau and Yang address this concern and assert that more obscure API components benefit from a trickle down affect of similar API components [11].

The work most similar to ours is recent work by Parnin *et al.* who also evaluated the degree to which specific APIs are discussed within the community [15]. Our work differs from theirs in two ways. First, we use a larger pool of free apps that contain both open and closed source code, and second, we gather metrics of existing documentation and study these disparate data sources in a multiple regression context. This allows us to measure impact of some variables on a response while controlling for other covariates.

3 Theory

Writing software is a knowledge-intensive activity. Understanding the structure and purpose of large software systems well requires significant effort; this has been an active area of research [7,13]. The creation and use of software knowledge impacts developer productivity, such as the upfront time cost of writing documentation, updating documentation as the software evolves, and the time spent in informal explanations between colleagues. This knowledge barrier to entry arguably leads to the "thin spread of knowledge", where relatively few developers [8] in large teams have deep knowledge about the system under development.

A recurring challenge in software engineering is the difficulty of studying the "demand" for program understanding. Which components of the system are most frequently asked about? What influences the demand for knowledge about these components? Generally, previous studies of program understanding have employed human studies that have yielded valuable insights [12,18]. Given the cost of doing such studies however, it is difficult to get large-sample data that allows us to distinguish how the programmer demand for information varies from element to element.

The advent of on-line forums like StackOverflow, and powerful search engines, have changed the way programmers work. It has by now become a routine experience to probe a search engine with a compiler or run-time error and finding a remedy on StackOverflow. Such experiences have left numerous traces for researchers to mine and study. Thus, it is now possible to investigate programmers' quest for knowledge, and the factors that influence their various quests, using a large-sample mining methodology over StackOverflow. In this work, we focus on the relationship between Android API-focused questions on StackOverflow and the actual usage of those APIs in real world, published apps. This allows us a novel opportunity to examine how a large body of developers use an API in practice, and how this affects the frequency of mention of these APIs in StackOverflow. Figure 1 illustrates, in a simplified fashion, the process of Android app programming and documentation mining.

We expect a relationship between API usage and the frequency of questions on StackOverflow (SO), as the more developers use a given API, the more likely there are to be associated questions. We expect a relationship between the amount of documentation for a class and the number of questions about it, as classes with less documentation may be harder for developers to understand. One must control for the usage of the API, however, as highly used APIs will likely be the source of many questions, regardless of documentation. Finally, after controlling for usage, we expect fewer answerers for questions about classes with less documentation. The intuition is that fewer developers will have expertise on classes with less documentation, so questions on those classes will have a smaller set of answerers. We study the following specific questions.

Research Question 1: What kind of relationship exists between uses of classes in free apps and the mentions of those classes in SO questions and answers?

Research Question 2: Do the properties of the existing Java documentation influence the relationship between SO and usage? In other words, how does the volume of documentation relate to the number of questions in SO?

Research Question 3: How do the number of available answers depend upon the USE of a class, and the volume of available documentation?

We note several unexpected issues when attributing questions to API classes, including stack traces, code blocks, poorly formatted questions and answers (posts), incorrect user-defined tags, multi-word class name splitting, highly coupled classes, and base classes with many subclasses.

We also present a case study of the relationship between usage and demand for documentation for specific classes at different ratios of usage to demand. In this case study, we look at three clusters of API classes: those that have equal numbers of questions and uses, those with more uses than questions and vice versa. In addition, we look at random points on the major trend line. We randomly sample ten questions for each class in each group, and draw conclusions based on our findings.

4 Data and Methodology

Data. Android class invocation data was taken from a pool of 109,273 free applications. The majority of the applications in our data set were pulled from the official Android market as well as several third-party English and Chinese markets.[1] 82% of the applications were from English markets with 66% of those coming from the official Android market. 18% of the applications were taken from Chinese markets.[2] Using data from multiple markets diversifies our data set, as different groups of developers may post to different markets. Thus our data is robust against bias that may be introduced due to application market politics.

StackOverflow provides data dumps, under a Creative Commons license, sporadically on their blog [19]. Our data is based on a data dump from August 2011. We imported this data dump into our database and used each post's content and title to perform class name matching. We also used the *tags* column in our data gathering process outlined in Section 4.

The Android documentation and lines of code information were extracted by running *Javadoc* with a custom Doclet on Android source code [10]. The StackOverflow data includes questions and answers from Android releases up to version 3.2.2 (as Android version 3.2.2 was released in August 2011).

However, not all of Android version 3.x source code is available through the repository. Android 3.x was intended for use on tablets, and Google felt this code was incomplete and didn't want developers building on the source [3]. To address this issue, whenever possible, we took documentation and lines of code data from Android 2.3.5 source code; otherwise, we took data from Android 4.0.1 source code. This resulted in 90% of our documentation and lines of code data coming from Android 2.3.5. We believe this is an acceptable compromise as the number of new classes between versions is minimal in comparison to the total number of classes, reducing the influence that this problem has on data gathered from our large data set. We included deprecated classes in our data set because although the use of deprecated classes is not encouraged by API developers, users still write programs utilizing them, and thus, ask questions about them.

Linking Methodology. To gather our data, we used a process adopted from Parnin [15]. In our case, a *link* is a connection between a StackOverflow question and an Android class. The types of links are *Tag links*: A class name match occurring in the tags section

---

[1] The applications were acquired between December 2011 and October 2012.
[2] Due to changes in the crawling process these are rough estimates.

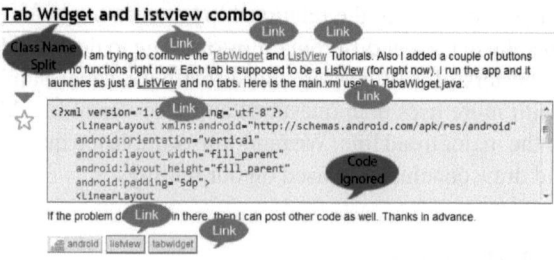

Fig. 2. An example of a question with multiple links. This also illustrates an issue with multi-word class names being split in natural language, not generating a link.

of a StackOverflow question, *Word links*: A class name match occurring directly in the natural language text of a question, and *Href markup links*: A class name match enclosed by HTML <a></a> tags. For our purposes, all link types were treated the same as shown in Figure 2 (*i.e.* we union all links and treat them without priority).

Empty string word boundaries were used to isolate class names within question bodies. Negative lookahead (*viz.*, making sure a regex pattern is *not* followed by another specified regex pattern) was used to deal with problems in our question parsing with inner class definitions. Without negative lookahead, a search for a class such as *SharedPreferences* would also result in an incorrect match for *SharedPreferences.Editor*. All parsing was done in a case insensitive manner.

A single-word class is defined by camel case, *e.g. Activity* is a single word class, whereas *ImageView*, is a two word class. For single-word classes, we include every type of link except *word links* as single-word classes are often commonly used words that turn up a large number of false positives. However, in contrast with Parnin's method, we did not consider any class name matches appearing within <code></code> tags. Novice users often post complete code blocks in with their questions; this leads to frequent confounding mentions of popular classes, just because they happen to occur in many code blocks. In addition, we attempted to remove standard Android log output from the body of questions as users will sometimes post entire log traces of errors they encounter, occasionally generating false positives. Also, all links to subclasses are attributed to their parent, thus maintaining a consistent level of aggregation.

Stack Overflow. We parsed the online Android developer javadoc based documentation [1] to generate a list of android classes. We pulled all API calls that existed as of Android version 3.2.2, API level 13, as this was the latest Android version released as of the date of our StackOverflow data dump[3].

Class names were used to generate links within StackOverflow questions as outlined above. We aggregated these links and recorded the class name that triggered each link along with the associated post *ID*. It is possible for a question to have multiple links, and each was recorded. Figure 2 clarifies this process. If a single question has multiple link

---

[3] Manual samples showed that most questions were of a more general form, so the use of classes from a single Android version is a reasonable choice.

types for the same class, only a single link for that class-question pair is recorded. We used the same method for gathering data from StackOverflow answers as for questions, with a change in generating *tag links*. In StackOverflow, there is no concept of an answer having a user-defined tag. To account for this, we linked each answer to the linked tag of the corresponding question.

This process results in data on a per-class basis that includes the number of questions, number of unique askers, number of answers, number of answer threads, and number of unique answerers.

Android. We converted Android application code (APK) to a human-readable format using APKTool [2] APKTool generates a folder from each APK file containing the Android Manifest file, resources, and a directory consisting of byte code files corresponding to original source files. We processed the extracted byte code files looking for function invocations (`invoke-virtual`, `invoke-super`, `invoke-direct`, `invoke-static`, and `invoke-interface`). We also gathered package names, class names, method names, return types, arguments, and number of calls per APK. We used this function invocation data to calculate the total number of calls for each class as well as the number of distinct APKs that call each class. This process is depicted in Figure 1. We joined the resulting APK data with our StackOverflow question links to generate the finalized data used to answer our research questions.

4.1 Model Building

For several of our research questions we wanted to understand the impact of a set of *explanatory variables* on a *response* while controlling for known covariates that might influence the response. We use *negative binomial regression*, NBR, a generalized linear model used to model non-negative integer responses.

Of concern is that many of the Android classes do not have quesztions; in other words, our data contains many rows with a response value of zero which presents a challenge for the standard NBR model. *Zero inflated negative binomial regression* and *hurdle regression* are two methods designed to address this challenge by modeling the existence of excess zeros [4]. It is common practice to fit both types of models, along with the standard NBR model, and compare model fits to ascertain which structure is the most appropriate [4]. Since these models cannot be viewed as nested models we employ both *Akaike's Information Criterion*, (AIC), and Vuong's test of non-nested model fit to determine appropriateness of method [22]. We employ *log* transformations to stabilize the variance and improve model fit when appropriate [6]. Because explanatory variables are often highly correlated, we consider the VIF, *variance inflation factor*, of the set of predictors and compare against the generally accepted recommended maximum of 5 to 10 [6]. Finally, when we consider whether explanatory variables should remain in a model we use either a Chi-Square goodness of fit test or the aforementioned Vuong's test of non-nested along with AIC to ensure that inclusion of the additional variables is justified [6].

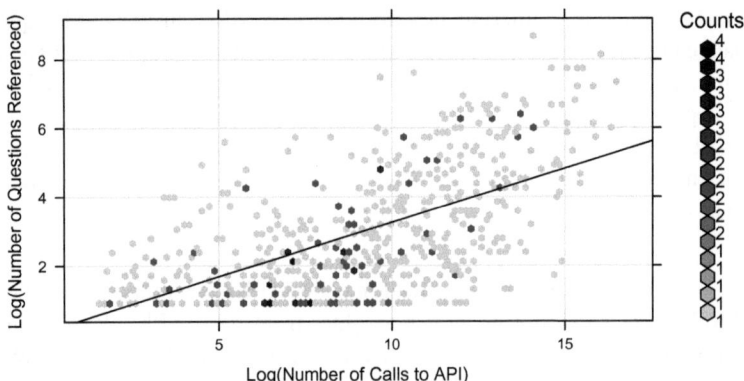

Fig. 3. A log-log scatter plot of how much each class is used vs how many questions are asked about it. The apparent linear trend with slope smaller than 1 (≈ 0.31) points to a saturation in the number of questions needed to document a class.

5 Results and Discussion

Class Popularity vs Questions Asked About It. Each class in our data has an associated total number of calls from Android applications which we use as a proxy for its popularity and a total number of StackOverflow questions that mention the class. Figure 3 is log-log scale scatterplot showing the relationship between a class's popularity (x-coordinate) and the number of mentions (y-coordinate) in questions. The color indicates how many data points share a region on the plot. The distribution of the points, is suggestive of the relationship between popularity (or usage) of a class and the demand for external documentation about it. The plot shows a strong linear trend between $log\ x$ and $log\ y$, with a slope of 0.31, which translates into a sublinear dependence of y on x. As might be expected, a class' popularity, as measured by the number of APK calls, is strongly associated with the number of questions generated about it. However, it is noteworthy that the trend becomes sub-linear toward higher uses, and we conclude that:

Result 1: *On the average, the demand for class documentation from SO grows slower than class usage in the code.*

One possible interpretation for this phenomenon is that the more a class is used, or called, by APKs the more people are using it. And thus, the more people ask about it. Since SO make available, online, all questions asked and all answers to them, it is likely that after a certain number of questions, a person asking a question will usually find that his/her question (or a very similar one) has already been asked and answered on SO. So, instead of asking it again, s/he will use the answers already provided.

Table 1. The relationship between uses of an API call and discussion of that call in both questions and answers is mediated by both class and method documentation for the API call

	Hurdle Model:		Count Model:	
	numquestions (1)	numanswers (2)	numquestions (3)	numanswers (4)
log(numquestions+0.5)		1.662***		1.021***
		(0.140)		(0.015)
log(numcalls+0.5)	0.330***	0.254***	0.333***	0.036***
	(0.029)	(0.043)	(0.020)	(0.008)
log(sourceloc+0.5)	0.176*	−0.139	0.241***	−0.044*
	(0.096)	(0.129)	(0.078)	(0.025)
log(numinnerclasses+0.5)	0.189	−0.121	0.440***	0.074**
	(0.174)	(0.249)	(0.142)	(0.032)
log(classdocloc+0.5)	0.328***	0.302**	0.141**	−0.051***
	(0.095)	(0.129)	(0.067)	(0.019)
log(avgmethdocloc+0.5)	−0.073	−0.020	0.007	0.079**
	(0.126)	(0.177)	(0.102)	(0.031)
Constant	−2.677***	−1.344**	−1.264***	−0.096
	(0.492)	(0.634)	(0.439)	(0.131)

Note: *p<0.1; **p<0.05; ***p<0.01

API Self-documentation vs StackOverflow Questions. To address this question we model a count of the number of questions that mention a particular class as a *response* against *explanatory variables* that measure the level of API documentation for that class while controlling for the usage and the size of the class. The API documentation for an Android API class was captured in two variables, the line count for the class documentation proper, as well as the mean line count of documentation for each method. To control for the usage of the class we included the number of calls to that class found in the APKs. To control for the size, we used the number of lines of code in the class. We *log* transformed all variables. In addition to stabilizing variance, the transformation induces a natural interpretation; a small increase to a short document is, on average, more likely to improve understanding than a small increase to a long document. In all cases, Vuong's test of non-nested model fit favored the transformed variant yielding a better fit to the data with p-value < 0.001. One cause of concern is that the mean of method documentation size is significantly rank (spearman) correlated with class documentation size (0.69 p=value < 0.001). The VIF, *variance inflation factor*, of these variables in a model, however, was well within bounds at < 1.35 (See Sec. 4.1).

We compared standard NBR models with their zero-inflated and hurdle counterparts. The standard NBR model slightly underestimated the zero count whereas the zero inflated models overestimated them. The hurdle models had the lowest AIC and precisely predicted the observed number of zeros. In all cases Vuong's method preferred the hurdle models with p-value < 0.001.

The count component of the model is presented in the third and fourth columns of Table 1. As expected we can see a positive relationship between the usage of a class (numcalls) and the number of questions that mention that class. On the other hand, class documentation is a positive and significant predictor for number of questions, but is slightly negative, while still significant, for number of answers referencing that class. We conclude that class documentation is important in driving the number of questions that reference a class for askers. Interestingly, the average method documentation is not significant. We also note that both the size of the class and the number of inner classes have a positive effect on the number of questions referencing the class, a result not surprising having in mind that both of those predictors are a facet of complexity. Larger, potentially more complex classes are mentioned more often in questions.

The zero portion of the hurdle model is presented in the first two columns of Table 1. Here we model the factors that influence the ability of a class to cross the "hurdle" of obtaining even a single mention in a question (first column) or answer (second column). The coefficients represent the affect of that variable on overcoming the zero hurdle. Usage also has a positive effect in the model for both number of questions and number of answers. This means that the more popular a class, the greater the chance the class has of overcoming the hurdle of acquiring a reference in either question or an answer. Interestingly, total class documentation is highly significant and strongly positive in the zero portion of the model (left column), indicating that more documentation is associated with more questions. It can be argued that when faced with impersonal pages of documentation, one may be driven to asking a question on Stack Overflow where a personalized response is minutes away.

Class size, number of inner classes, and method documentation size, on the other hand, are not significant to overcoming the hurdle. This suggests that available class documentation, and not size/complexity, drive questions for at least one person. Taken together with the count model, the average size of method documentation drives the first mention to a large extent, but subsequent mentions are also dependent on class size. Positive coefficients suggest that greater volume of documentation tends to be associated with more questions, rather than fewer.

Result 2: *Class documentation size and class size have a relationship with question mentions on Stack Overflow, while controlling for the usage in free APKs.*

API self-documentation vs StackOverflow Answers. We asked the same questions with respect to the number of answers for a given question and the count model results are presented in the third and fourth columns of Table 1. In addition to the controls we used for questions, when modeling answers we also control for the number of questions. This is necessary, despite potential multicollinearity issues, as the number of answers will have a strong dependence on the number of questions. In fact, VIF values for this model were below 3, which is within acceptable limits [6]. We see that, while still positive, the effect of usage has a much lower impact on the number of answers than it does on the number of questions. This suggests that people who answer questions are *motivated more by questions than by the usage of the classes*; SO is a "gamified" environment where people are motivated to accrue higher scores by providing more valuable answers to more questions. The dependence of answers on class documentation is also somewhat different; the coefficient for class documentation being very significant and slightly negative. Given that a class is already mentioned in an answer, greater class documentation is associated with fewer references in answers. Perhaps most interesting is that class size has no significant impact on answer references. Larger classes drive more questions, but not more answers, when controlling for the number of questions.

We turn now to the zero model presented in the first and second columns of Table 1. The coefficient for the number of questions is positive which implies that the more questions that mention a class, the more likely that at least one of the answers will also mention the class. The number of calls to a class has somewhat lower influence on the occurrence of an answer that mentions the class, than in the zero model for questions. It is a small difference, however, its lower value suggests (as above) that answerers will be somewhat less impacted by usage than simply by the existence of the question. As for questions, documentation size has a significant affect on the first answer, with a slightly smaller coefficient. Documentation size, therefore is important for answerers, when giving the first answer to questions about a given class. Perhaps this is an effect of the gamified platform underlying Stack Overflow, where the best response gets the answerer more points, thus being first to answer provides competitive advantage. The size of the class documentation, arguably, only makes the competition more interesting, *viz.* who can digest it and provide an answer first, within a short turnaround time.

Result 3: *Class documentation size has a strong positive relationship with getting the first answer which mentions it on Stack Overflow, controlling for usage in free APKs.*

5.1 Case Study: Class Usage and Their Mentions in Questions

To gain insight into our data, we sampled, and manually evaluated, classes falling in different regions of the plot in Figure 3. We identified three distinct clusters in Figure 3: one outlier group above the regression line, *top-left cluster*, one outlier group below the regression line, *bottom-right cluster*, and a group with extremal values in both, *top-right cluster*. To isolate these groups, we examined the plot of the data points and manually selected the clusters by range (nearby points). We then sampled classes corresponding to these data points and referenced the Android documenation for each class. After randomly sampling ten StackOverflow questions linked to each class name in each group, we categorized the classes in each cluster in an attempt to identify any systematic differences. We also manually marked each sampled question with a flag indicating whether or not the linked class name was the question's topic. The classes mentioned below are notable examples from this study.

Bottom Right Cluster. The bottom right region contains classes that are not explicitly created and called by users in Java code, so their function invocation counts are high, but their question references are low. This group also includes self-explanatory classes that have relatively low barriers to entry or classes or calling conventions that are analogous to standard Java or other common platforms, reducing question references. This latter cateogory includes classes such as *Editable* and *Environment*. Programmers find such classes relatively straightforward, despite frequent use largely owing to their abstract nature and simple role.

Top Left Cluster. The top left region contains classes that are more application-specific, and less frequently used. Still, they have a fair number of question references as they may have confusing APIs and relatively small amounts of documentation for their complexity. For example, the class *MediaStore* deals with audio files and has 30 static public fields - a considerable number in comparison to the average of 4.98 public fields across other classes. Similarly, the class *Parcelable* deals with serialization. Efficient serialization is often a complex topic owing to pointers and recursive data. Finally, although the class *ContactsContract* is likely application-specific, it is not necessarily uncommon. A close investigation revealed that this class is a base class with numerous subclasses that provide for interaction with a device's contact information. This base class is seldom directly called when accessing contact information, explaining its low use count in our APK data. Interestingly, there are only a total of 22 question references combined for all 61 subclasses but 131 question references for the base class. This indicates that users will sometimes ask about the base class rather than the subclass which is actually used.

Top Right Cluster. The top right cluster contains classes that are necessary to use the Android platform and thus have high invocation counts. These classes also have many question references as they are where newcomers to the platform begin their understanding. To spare repetition, we only discuss one of the sampled classes as all classes in this cluster have a similar explanation for their existence in this group. The *Activity* class is the core of Android UI; each screen or page in an application extends the Activity class. Thus, the fact that it has a very high invocation count and has a large number of question references is not surprising.

6 Threats and Caveats

In this study we considered only one topic of questions, the Android platform. It is not clear whether conclusions generalize beyond this setting. Second, our study cumulates effects (questions, answers, uses, documentation lines, lines of code) over time, and thus can only point out associations, not causes and effects. A more detailed time-series study would be required to reveal, for example, whether more questions lead to more class documentation or vice versa; and likewise, whether average documentation lines per method is a cause or effect of more answers.

The remaining threats listed below arise from the fact that StackOverflow is unstructured natural language text, and extracting information therefrom is naturally a noisy process. We anticipate that emerging work in the area of salience and named entity recognition can be fruitfully used in this context, and help address the issues below.

We primarily consider the *mentions* of a class name in answers and questions. We have taken care to ensure that the mentions are actual mentions of a class name in the natural language text, and not merely in code included in the question or the answer. However, a *mention of a class* in an SO question or answer does not necessarily mean that the message is *about* that class; but it does mean that the message is relevant somehow to the class. Manual examination suggests that while a mention indicates some relevance, more often than not, a message that mentions a class isn't necessarily mostly "about" that class. There are some new, emerging tools that promise to find the most salient mentions [17].

Code fragments included within questions and answers present a quandary. Including complete code fragments (that are pasted into Q & A) in the analysis would be tantamount to sampling mentions of classes twice, once in the code itself, and once in the code pasted into the unstructured text. On the other hand, smaller fragments may include more relevant mentions of classes than full fragments. We plan to redo the analysis with smaller fragments included. A related issue is the posting of code without <code></code> tags; this is a minor concern, however, as it is usually corrected by moderators.

There are multiple sources of possible error in class name recognition within the unstructured StackOverlow text. First, some classes have common English language class names, *e.g. Security*. The word "security" is often mentioned in posts that have nothing to do with the *Security* class, thus giving rise to false positives. Secondly, users sometimes split multi-word class names with a space, *e.g.* referring to *TabWidget* as *Tab Widget* as shown in Figure 2. This can cause problems when attempting to match a question with a class name. Fortunately, this problem was not common in our case study.

6.1 Conclusion

We have studied the relationship between the *use* of Android API classes in over 100,000 applications, and the *questions* and *answers* that mention these classes. We find that: a) questions do increase with use, although there is a non-linear saturation effect, suggesting that knowledge of the most popular classes do get internalized to some extent, or become easier to find; b) the *total* amount of available documentation for a class has no impact on the number of questions, but the size of classes does, suggesting that more complex classes increase the need for information; and c) that the situation is reversed for answers: the total amount of class documentation decreases the number of answers per question while class size has no effect, and the documentation per method increases the number of questions. To our knowledge this is one of the first studies clearly relating Android market place usage of APIs to the occurrence of these APIs in StackOverflow questions. We believe that this is a fruitful area of study, and we invite other colleagues in joining us to improve data cleaning, examine causes and effects via time-series analysis and so on. Clear findings on the causes of documentation demand, and the effects of good and bad documentation, can have a profound impact on the effective use of complex platforms like Android.

References

1. Android Documentation,
   http://developer.android.com/reference/classes.html
2. APK Tool, http://code.google.com/p/android-apktool
3. businessweek.com, http://www.businessweek.com/
   technology/content/mar2011/tc20110324_269784.html **com**

4. Cameron, A.C., Trivedi, P.K.: Regression analysis of count data, vol. 30. Cambridge University Press (1998)
5. CNN, http://edition.cnn.com/2013/03/05/business/global-apps-industry
6. Cohen, J.: Applied multiple regression/correlation analysis for the behavioral sciences. Lawrence Erlbaum (2003)
7. Corbi, T.A.: Program understanding: Challenge for the 1990s. IBM Systems Journal 28(2), 294–306 (1989)
8. Curtis, B., Krasner, H., Iscoe, N.: A field study of the software design process for large systems. Communications of the ACM 31(11), 1268–1287 (1988)
9. Dagenais, B., Robillard, M.P.: Creating and evolving developer documentation: understanding the decisions of open source contributors. In: Proceedings of 18th ACM SIGSOFT International Symposium on Foundations of Software Engineering, pp. 127–136. ACM (2010)
10. Javadoc Documentation, http://docs.oracle.com/javase/6/docs/technotes/guides/javadoc/doclet/overview.html
11. Jiau, H.C., Yang, F.-P.: Facing up to the inequality of crowdsourced api documentation. ACM SIGSOFT Software Engineering Notes 37(1), 1–9 (2012)
12. Ko, A.J., DeLine, R., Venolia, G.: Information needs in collocated software development teams. In: Proceedings of the 29th International Conference on Software Engineering, pp. 344–353. IEEE Computer Society (2007)
13. Letovsky, S.: Cognitive processes in program comprehension. Journal of Systems and Software 7(4), 325–339 (1987)
14. ninlabs blog, http://blog.ninlabs.com/2013/03/api-documentation/
15. Parnin, C., Treude, C., Grammel, L., Storey, M.-A.: Crowd documentation: Exploring the coverage and the dynamics of api discussions on stack overflow. Georgia Institute of Technology, Tech. Rep.
16. Posnett, D., Warburg, E., Devanbu, P., Filkov, V.: Mining stack exchange: Expertise is evident from earliest interactions
17. Rigby, P.C., Robillard, M.P.: Discovering essential code elements in informal documentation. In: ICSE 2013 (2013)
18. Sillito, J., Murphy, G.C., De Volder, K.: Asking and answering questions during a programming change task. IEEE Transactions on Software Engineering 34(4), 434–451 (2008)
19. StackOverflow Data, http://blog.stackoverflow.com/category/cc-wiki-dump/
20. Treude, C., Barzilay, O., Storey, M.-A.: How do programmers ask and answer questions on the web?: Nier track. In: 2011 33rd International Conference on Software Engineering (ICSE), pp. 804–807. IEEE (2011)
21. Treude, C., Storey, M.-A.: Effective communication of software development knowledge through community portals. In: Proceedings of the 19th ACM SIGSOFT Symposium and the 13th European Conference on Foundations of Software Engineering, pp. 91–101. ACM (2011)
22. Vuong, Q.H.: Likelihood ratio tests for model selection and non-nested hypotheses. Econometrica: Journal of the Econometric Society, 307–333 (1989)

Temporal, Cultural and Thematic Aspects of Web Credibility*

Radoslaw Nielek[1], Aleksander Wawer[2], Michal Jankowski-Lorek[1], and Adam Wierzbicki[1]

[1] Polish-Japanese Institue of Information Technology,
ul. Koszykowa 86., 02-008 Warsaw, Poland
{nielek,fooky,adamw}@pjwstk.edu.pl
[2] Institue of Computer Science Polish Academy of Science
ul. Jana Kazimierza 5, Warsaw, Poland
axw@ipipan.waw.pl

Abstract. Is trust to web pages related to nation-level factors? Do trust levels change in time and how? What categories (topics) of pages tend to be evaluated as not trustworthy, and what categories of pages tend to be trustworthy? What could be the reasons of such evaluations? The goal of this paper is to answer these questions using large scale data of trustworthiness of web pages, two sets of websites, Wikipedia and an international survey.

Keywords: trust, language, Wikipedia, temporal, national, credibility.

1 Introduction

In the early 90s there was a need to organize an increasing number of websites. People had problems with navigating a still expanding Internet. Search engines and Internet catalogues have flourished in order to address this issue. At that time, not only content credibility but even Internet frauds were not serious issues. Increasing number of less proficient Internet users and lowering costs of publication became a driving force in this change and have stimulated the bloom of Internet frauds.

Warning against potentially harmful websites constitutes a very useful feature but, for some time now, no longer sufficient. The broad spectrum of information, ranging from completely non-credible (e.g. theories that earth is flat) to very credible, can be found on the Internet and, therefore, users need a support while deciding whether to trust a particular information or not. It is the subject of dynamic research (e.g. Reconcile project) and also some commercial and semi-commercial projects (Hypothesis, FactLink, Web of Trust).

On the one hand, web content credibility gets more and more important for the Net. On the other hand, not much is known about temporal, cultural and

---
* Research supported by the grant "Reconcile: Robust Online Credibility Evaluation of Web Content" from Switzerland through the Swiss Contribution to the enlarged European Union.

thematic patterns of websites credibility. Since the early 50s concept of credibility has been widely studied by psychologists, media experts and economists. Most publications focus on either persuasive effect of source credibility[1-3] or media credibility[4] or importance of credibility for economic theories[5]. Most researchers agree that credibility is not a property of object, person or piece of information but it is rather a perceived quality[6].

Researchers argue whether credibility is a subjective or objective matter. Tseng[6] proposed four types of credibility: presumed, reputed, surface and experienced. The first two types are based on either stereotypes or third-party reports. The last two are derived from individuals own experiences. Some people can argue that those categories are an essentially heuristic use to assess credibility and do not define different types of assessed variable. This view seems to be strengthened by the definition of credibility as believability, given by Fogg[7].

People use heuristics to assess credibility. A physically attractive person is perceived more credible[8]. People constantly use some signals to estimate credibility of others and the same type of mechanism exists also in terms of assessing web content credibility. Study conducted on over 2500 Internet users at Stanford University revealed 18 areas that people notice while assessing web site credibility[9]. Almost 50% participants pointed at design and look, the one forth on information design and information structure. Bias of information and tone of the writing are present only in ca. 10% comments. On the other hand, asking people explicitly about features they use to assess credibility can only reveal heuristics they are aware of. People also adapt their heuristics specifically to web sites. They use position in an Internet search engine (higher position indicates more reliable information[10]) or following graphs and presence of shortened URLs for tweets[11].

The prominence-interpretation theory[12] tries to combine signals with peoples motivation and perceptibility. The theory assumes that first, user has to notice particular feature and only then he starts evaluating it. This process is repeated many times by a single user for a single web site and its efficiency depends strongly on users motivation and experience. The prominence-interpretation theory is mainly focused on conscious processing and ignores preapprehension and feelings in general. It is worth noticing the fact that some features can be difficult to notice by people (e.g. number of question marks or punctuations) but still may be highly correlated with content credibility[13].

Most studies are focused on an attempt to understand factors influencing a credibility at an individual level (either web site or person) but large-scale systems supporting credibility evaluation build in the last few years create an opportunity to take a closer look on credibility of the Internet (or at least a huge amount of web sites) and dependencies interrelation between time, subject of a web site, language and credibility. Many questions seem important from sociological point of view. Among them:

– Are the web sites becoming more credible?
– What is the most credible subject on the Internet?
– Is web sites credibility evaluation related with a trust level in societies?

This paper is devoted to an attempt to answer these questions. Such answers may also have many practical applications, namely can be used for improving automatic credibility evaluation by incorporating additional context information (e.g. subject, language etc.).

No large-scale study of trustworthiness and credibility of web sites, which focus on such dimensions like language, time or culture, exists. Therefore, the paper is intended to fill this gap. The main assumption was to analyze existing credibility ratings and real web sites instead of orchestrating a dedicated surveys or craft special content. The particular attention was placed in assuring the scale that justifies generalization and makes drawing conclusions for the whole Internet possible. In total, more than 600 thousands web sites have been analyzed.

The rest of the paper is organized as follow. In the next chapter, are described datasets used in this paper . Results obtained for these datasets are presented in chapter three. Chapter four is focused on discussions about hypothesis that may explain the results. The last chapter summarizes the paper and proposes some interesting topics for further investigation.

2 Datasets

2.1 Article Feedback Tool

For our analysis we have used dataset build upon results of Wikimedia Foundation experiment with the feature to capture reader quality assessments of articles. Article Feedback v4 (AFT) was allowing users of English Wikipedia to rate every article with 4 different dimensions (trustworthy, objective, complete, well-written). AFT is a survey for article feedback to engage Wikimedia readers in the assessment of article quality. Reader is presented with short survey below every article and he can submit his ratings about four different aspects of article by choosing on 0 to 5 stars scale.

Original AFTv4 dump contains over 11M articles ratings based on 5.6M different revisions of over 1.5M distinct articles collected between July 2011 and July 2012. For comparison we have selected two subsets of data containing first 3 and last 3 months of article ratings and aggregated them grouping by article. Then we excluded pages that werent present in both datasets.

2.2 Web of Trust

WOT is a crowdsourcing system started in 2005 by two post-graduate students from Finland. Every logged user can evaluate each visited web site on four dimensions (trustworthiness, privacy, vendor reliability, and safety for children) and may also add comment to the evaluation. WOT aggregates all evaluation and show a pair of values (ranging from 0 to 100) for each dimension level of confidence and value of evaluation. However, the WOT does not disclose how this numbers are calculated. According to the company's blog numbers presented to the users are the average of left evaluations weighted with rater's credibility.

Up to now users have evaluated 43 million web sites and every month more than 500 thousands new web sites are evaluated.

WOT neither publish the dataset nor make it open for scientists. The only way to access information about web site credibility is via an open API. A researcher can send a question about a particular web site and he will obtain the same answer as plugin users. Datasets studied in this paper have been collected by multiple sending requests to WOT API and recording results. There is no publicly available list of web sites evaluated in WOT system, so an external lists have to be used: 1M most popular web sites from the Alexa and the DMOZ catalogue.

Only a part of domains from both lists have evaluations and this fact is also strongly correlated with its popularity. For the first few thousands most popular domains only a small fraction does not have evaluation (less than 5%) which is not very surprising (the more people visit the web site it is more likely that someone will evaluate it). On the other hand only one-third of domains in the second half of the Alexa rating list have at least one evaluation. In average, 41% domains on the list are evaluated.

Quite often web sites have been evaluated at only one or two dimensions (instead of four). Correlation level between different evaluation dimensions is very high and exceed 0.95 for all but child safety dimmension. That is why in this paper only two dimensions are studied – trustworthiness and child safety.

2.3 Category Detection

All domains have been assigned to categories with help of the AlchemyAPI `www.alchemyapi.com`, a services which is based on NLP tools. For academic accounts Alchemy limits the number of requests to 30 thousands per day. The motivation to use AlchemyAPI instead of other methods (e.g. TF IDF with a manually tagged corpus) was the intention to make these results easily replicable for other scientists.

2.4 Nation Level Data

The procedure to obtain the dataset was as follows. We started with all tokens, known as correct words in each language in the Aspell library. That limited our list of languages. We then replaced accented characters with their non-accented (ascii) equivalents, because of domain names restrictions. The lists of tokens generated this way were submitted to WOT query API. Unfortunately, this seemed the only plausible procedure since WOT does not enable to browse the data it collects and one needs to query for specific domain.

3 Results

3.1 Effect of Time on Web Sites Credibility

Freshness is one of factors that influence content credibility[14]. Time can affect web site credibility in many ways. Even if the web site content has not change, new facts or discoveries may make it irrelevant or wrong (the same effect can be observed for science as well as for sport). Many web sites are regularly updated and each update may change web sites credibility (people may struggle to improve published content but may also use previously gained reputation to sell products or misinformation).

An attempt to trace credibility changes of a particular web site faces many difficulties. The most obvious is that evaluations have to be done at many points of time (we cannot ask people for evaluate an old version even if we have stored content, because the passage of time might change content credibility). To solve this problem, credibility rating from WOT has been collected every two days for almost four months. Results for first and last run are presented in Table 1 (for Alexa) and Table 2 (for DMOZ catalogue).

Table 1. Average trustworthiness (trust) and child safety (safety) for two snapshots in time ("old" – September 2012 and "new" – January 2013) for domains from the Alexa. Statistically significant differences are denoted with stars.

Category	no. of domains	Trust_old	Trust_new	Safety_old	Safety_new
Arts&entertainment	29257	79.88	79.92	73.11	73.09
Business	36218	78.14	78.18	80.60	80.59
Computer&internet	41497	76.09*	76.03*	76.40*	76.36*
Culture&politics	16098	73.17	73.20	60.98	60.96
Gaming	7491	79.90	79.91	77.46	77.44
Health	6122	77.43	77.45	77.85	77.87
Law&crime	1535	74.43*	74.24*	70.97	70.83
None	66125	75.55	75.57	75.60	75.62
Recreation	34522	76.51	76.49	72.46*	72.42*
Religion	5251	80.52	80.45	80.58	80.50
Science&technology	11149	81.84	81.86	82.18	82.13
Sports	7389	82.60*	82.70*	82.87*	88.96*
Weather	131	86.32	86.24	88.85	88.83
All	263444	77.19	77.19	75.46*	75.44*

A relatively short time span between measurements, high number of web sites and the fact that WOT returns only an aggregated credibility score for all evaluations (very old and relatively new) causes that big differences in aggregated credibility for categories should not be expected. In fact, differences are small and in most cases statistically not significant. As can be seen in Table 1 , average credibility has increased for eight categories but only for one – sport – the difference is statistically significant. Among five categories, which have lower credibility in a second run, only for Computer&Internet and Law&Crime differences are statistically significant. Similar, small changes can be also observed for the dimension child safety. The main difference is that an average child safety for all web sites has slightly decreased but this change in opposition to trustworthiness is statistically significant.

In Table 2 is presented a comparison of trustworthiness level for two runs for domains from the DMOZ catalogue. Only for two categories – culture&politics and science&technology — can be observed statistically significant differences and, in both cases, web sites are getting more trustworthy. Very small change in average trustworthiness of all domains is visible but cannot be confirmed as statistically significant.

On the other hand very interesting pattern exists for the 20% most trustworthy web sites in the Alexa. Trustworthiness has increased from 95.28 to 95.32 and this change is statistically significant on the level 0.005. The same effect can be observed for the AFT dataset where average trustworthiness has risen from 84.13 to 84.35 (statistically significant on the level 0.00003). It may be an interesting point in the discussion about the rich get richer hypothesis. For the AFT dataset such an effect does not occur.

Other interesting source making a temporal analysis of trustworthiness evolution for a big number of web sites possible, is the AFT dataset , described in details in previous chapter. In table 4 are presented results for two subsets (first three months and last three months). In opposition to results for the Alexa and the DMOZ for almost all categories average trustworthiness is decreasing and results are statistically significant. The same effect can be also observed for all articles.

Table 2. Average trustworthiness for two snapshots in time (old – September 2012 and new – January 2013) for random sample of domains from the DMOZ catalogue. Statistically significant differences are denoted with stars.

Category	no. of domains	Trust_old	Trust_new
Arts&entertainment	17656	73.93	73.96
Business	34754	71.81	71.81
Computer&internet	25660	73.08	73.02
Culture&politics	10570	74.19*	74.28*
Gaming	1923	75.68	75.79
Health	7113	72.27	72.28
Law&crime	2225	72.32	72.30
Recreation	15706	72.47	72.48
Religion	7963	73.39	73.42
Science&technology	8634	75.30*	75.46*
Sports	8045	73.45	73.46
Weather	54	77.07	77.76
All	141232	73.05	73.04

3.2 The Most Credible Websites Are about...

All web sites in both datasets have been assigned to one of twelve categories (plus None). Differences in trustworthiness between all categories presented in the tables 1 and 2 are statistically significant. Web sites about weather forecast are evaluated as the most credible for both datasets. On first sight it may look

Table 3. Average trustworthiness for two snapshots in time (old – first three months in dataset and new – last three months) for random sample of 50% articles with evaluation from the AFT dataset. Statistically significant differences are denoted with stars.

Category	no. of domains	Trust_old	Trust_new
arts&entertainment	53461	3,02*	2,99*
business	17151	2,80*	2,75*
computer&internet	12626	2,73*	2,66*
culture&politics	27691	2,83*	2,78*
gaming	3131	3,19*	3,16*
health	6725	2,74*	2,70*
law&crime	830	2,65	2,75
recreation	4739	2,99*	2,94*
religion	14992	2,86*	2,83*
science&technology	46790	2,79*	2,74*
sports	8414	3,13	3,12
ALL	196550	2,89*	2,84*

a little bit counterintuitive but it may be explained with a limited expectation. People know that weather forecasts are only a scientifically supported guess and do not validate them (although it is very easy).

Another plausible explanation with evidence in the data is the view that trustworthiness correlates with the amount of intentional deception possible in a category. People tend not to believe cultural and political communication, have little trust in legal and crime-related websites. In all these types of communication deception occurs intentionally. Perpahs, the most striking evidence of importance of the intentional trust component is the highest trust put in weather forecast websites. The content of weather forecasts is related to the reality in a limited way as the accuracy of weather models is still far from being perfect. However, the "deception" of weather forecasts is unintentional. Moreover, it is likely that no spam or other dubious activity takes place around weather forecasts, as opposed to politics and crime categories, for instance.

A comparison between trustworthiness ratings for DMOZ and Alexa datasets reveals some similarities but also many big differences. Categories like law & crime and health are among less credible in both datasets (Science & technology next to weather forecasts are very trustworthy) but for others, like sport, religion or business, rating are missing a common pattern. Business is the least credible category for DMOZ catalogue and in the middle for the Alexa. Although the category for both datasets is the same, there are huge differences concerning business web sites between the DMOZ catalogue and the Alexa. Putting web sites into the DMOZ catalogue requires just filling out the form. Being in the Alexa is reserved only for established companies with a considerable number of visitors.

The same patterns are not clearly noticeable in articles from Wikipedia. The first problem is that some categories are not represented in this dataset (or there are only a few pages in a category e.g. one page in weather forecast category).

Second, subjects of articles in the Wikipedia are not related one to one with web sites either from the DMOZ catalogue or the Alexa. On the other hand, in AFT dataset, similarly as in the Alexa, web sites about sport are among the most and law & crime among the least trustworthy.

Spearman rang correlation calculated for average trustworthiness for categories between the Alexa and the DMOZ is 0.517. Even higher is the correlation between the AFT dataset and the Alexa – 0.536. Positive correlation shows that web site category can be used at least as provisional filter which can help to identify web sites for an in-depth study (with other methods e.g. NLP).

3.3 Search for Information in Estonian

European Social Survey (ESS) is a rigorous cross-national attitude survey that tries to trace changes in social behaviors in time and spots differences between countries. The biggest advantage of ESS is that results from many countries can be compared, because there has been used the same methodology. Results are freely accessible[1]. Data from the round five (finished in 2010) has been used in this paper.

Spearman rank correlation between average trustworthiness for countries top level domains and percentage of people in population who selected values from 8 to 10 as an answer for a question Most people try to be fair is 0.264. Figure 3 presents average results of WOT data, aggregated for each country, plotted against average results of trust question of the ESS 2010. The question, a 10-point scale depicting whether people do not trust others (0) or are trustful (10).

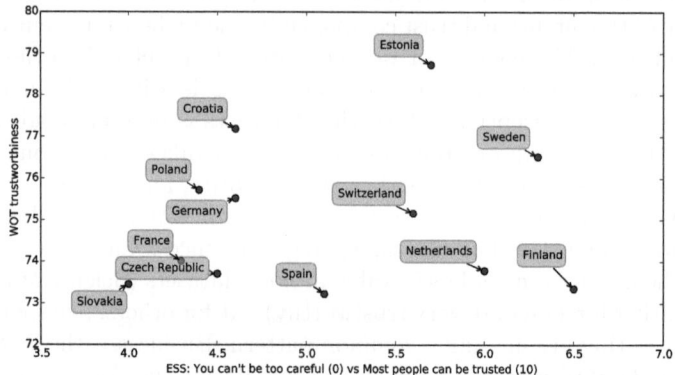

Fig. 1. Nation-level trust data: European Social Survey (ESS) and Web of Trust (WOT)

High levels of ESS trust can be found in northern Europe countries such as Finland, Sweden, but also Netherlands, Estonia and Switzerland. From this

---
[1] http://ess.nsd.uib.no/ess/round5/

group, only web contents of Estonia and to the lesser degree Sweden, are also highly trusted according to WOT. Central and eastern Europe countries such as Poland, Germany, France, Czech Republic and Slovakia are similar in their WOT and ESS trust levels.

4 Discussion

Trustworthiness of the AFT (wikipedia) data is on average decreasing. This observation has limited relevance for the Wikipedia and may be related to various phenomena related to artices life cycle. In DMOZ, differences between time points are not significant and in Alexa overall trustworthiness did not change. Although a specific subset of web sites (20% most credible) trustworthiness increase may be observed. As it is visible for the Alexa and the DMOZ catalogue but not for the AFT dataset it may be a consequence of the fact that in the WOT (in opposition to Wikipedia) users know an existing evaluation before they evaluate themselves. This so-called anchor effect may particularly strong for very credible web sites. The conclusion to draw from these observations is that it is very likely that the internet, overall, is not getting more trustworthy. This conclusion needs to be treated with certain caution: the datasets analysed represent large and important portions of the web, nevertheless still miss certain spots of communication, such as social media or microblogs falling out of the scope of this paper. Assuming that the observation is nevertheless true, we may further hypothesize that the web has reached its trust limits, alternatively no mechanisms that could successfully raise the level of trust towards web contents were introduced.

5 Conclusion

When the temporal aspect is considered, results reveal that the levels of trust are either constant in time (Alexa, DMOZ) or slightly decrease (Wikipedia). However, certain types of categories do become more trustworthy. Also, the trust to the most trustworthy websites tends to increase, which leads to the conclusion that overall variability of WOT trust levels increases.

Trust levels tend to exhibit certain patterns between topics or categories. Perhaps the most surprising finding is that the most trustworthy websites are weather forecasts. A possible explanation of this fact, as well as other patterns of trust differences between categories, is that people relate trust, at least partly, to intentional rather than factual dimension.

The paper investigated also nation-wide trust measurements of the European Social Survey to find them to be only partially reflected in the WOT data.

Reported results have strong foundations because size of the datasets is significant. The conclusions are based on comparing millions of webpages, identified with well-established data sources such as Alexa and DMOZ, and backed by many individual trust evaluations of each website.

Some of the reported results demand further investigations and pose many research questions. For instance, it is not entirely clear what are the reasons of trust decrease in the case of Wikipedia pages. In the case of other data sets, why are trust levels constant and why is trust variability increasing. Perhaps, some of the answers could be formulated by measuring the influence of linguistic and textual content of webpages on their trust levels. Another issue that remains to be addressed is how to measure the state of trust in social media and microblogs in a way comparable to WOT ratings for the web.

References

1. Sternthal, B., Dholakia, R., Leavitt, C.: The Persuasive Effect of Source Credibility: Tests of Cognitive Response. J. of Consumer Research 4(4), 252–260 (1978)
2. Hovland, C.I., Weiss, W.: The Influence of Source Credibility on Communication Effectiveness. Public Opinion Quarterly 15(4), 635–650 (1951)
3. Pornpitakpan, C.: The Persuasiveness of Source Credibility: A Critical Review of Five Decades' Evidence. J. of Applied Social Psychology 34(2), 243–281 (2004)
4. Gaziano, C., McGrath, K.: Measuring the Concept of Credibility. Journalism Quarterly 63(3), 451–462 (1986)
5. Sobel, J.: A Theory of Credibility. Rev. of Economic Studies 52(4), 557–573 (1985)
6. Tseng, S., Fogg, B.J.: Credibility and computing technology. Commun. ACM 42(5), 39–44 (1999)
7. Fogg, B.J., Tseng, H.: The elements of computer credibility. In: Proceedings of the SIGCHI Conference on Human Factors in Computing Systems1999, pp. 80–87. ACM, Pittsburgh (1999)
8. Patzer, G.L.: Source credibility as a function of communicator physical attractiveness. Journal of Business Research 11(2), 229–241 (1983)
9. Fogg, B.J., et al.: How do users evaluate the credibility of Web sites?: a study with over 2,500 participants. In: Proceedings of the 2003 Conference on Designing for User Experiences 2003, pp. 1–15. ACM, San Francisco (2003)
10. Schwarz, J., Morris, M.: Augmenting web pages and search results to support credibility assessment. In: Proceedings of the SIGCHI Conference on Human Factors in Computing Systems 2011, pp. 1245–1254. ACM, Vancouver (2011)
11. Morris, M.R., et al.: Tweeting is believing?: understanding microblog credibility perceptions. In: Proceedings of the ACM 2012 conference on Computer Supported Cooperative Work 2012, pp. 441–450. ACM, Seattle (2012)
12. Fogg, B.J.: Prominence-interpretation theory: explaining how people assess credibility online. In: CHI 2003 Extended Abstracts on Human Factors in Computing Systems 2003, pp. 722–723. ACM, Lauderdale (2003)
13. Olteanu, A., Peshterliev, S., Liu, X., Aberer, K.: Web credibility: Features exploration and credibility prediction. In: Serdyukov, P., Braslavski, P., Kuznetsov, S.O., Kamps, J., Rüger, S., Agichtein, E., Segalovich, I., Yilmaz, E. (eds.) ECIR 2013. LNCS, vol. 7814, pp. 557–568. Springer, Heidelberg (2013)
14. Dai, N., Davison, B.D.: Freshness matters: in flowers, food, and web authority. In: Proceedings of the 33rd International ACM SIGIR Conference on Research and Development in Information Retrieval 2010, pp. 114–121. ACM, Geneva (2010)

Social-Urban Neighborhood Search Based on Crowd Footprints Network

Shoko Wakamiya[1], Ryong Lee[2], and Kazutoshi Sumiya[1]

[1] Graduate School of Human Science and Environment, University of Hyogo, Japan
s.wakamiya@gmail.com, sumiya@shse.u-hyogo.ac.jp
[2] Korea Institute of Science and Technology Information (KISTI), Korea
lee.ryong@gmail.com

Abstract. Neighborhood is generally a geographically localized community often with face-to-face social interactions. However, modern cities and the widespread social networks have been drastically changing the concept of neighborhood, much beyond spatial constraint. Specifically, due to the complicated urban structures with entangled transportation network and the resulting spatio-temporally extended crowd activities, it is a non-trivial task to examine neighborhood areas from a location of interest. As a promising approach to investigate such a social-urban structure, we propose a social-urban neighborhood search which aims at identifying neighborhood areas from a specific location particularly considering social interactions between urban areas. We especially examine crowd movings through location-based social networks as an important indicator for measuring social interactions. We also introduce a data structure for aggregation of crowd movings as a simplified graph, with which we can easily analyze crowd movements in a large scale urban area. In the experiment, we will look into neighborhoods for several urban areas of our interests in terms of social interactions significantly focusing on how they are distorted from general localized vicinity.

1 Introduction

Neighborhood is generally regarded as a geographically localized community usually made from face-to-face social interactions. However, the complicated transportation infrastructures in modern cities and the widespread social networks have been drastically changing the nature of neighborhood, much beyond spatial constraint. Specifically, due to the complicated urban structures with entangled transportation network and the resulting spatio-temporally extended crowd activities, it is a non-trivial task to examine neighborhood areas from a location of interest. However, the lack of understanding neighborhood to an unfamiliar place may mislead our decision makings. For instance, location-based recommendation systems have been providing near-by search which is calculated based on physical distance between areas as shown in Fig. 1 (a). On the other hand, by considering social interactions such as crowd commuting by railways, we can find out social-urban neighborhoods, which may seemingly look distant, but practically easy to reach much quickly as depicted in Fig. 1 (b).

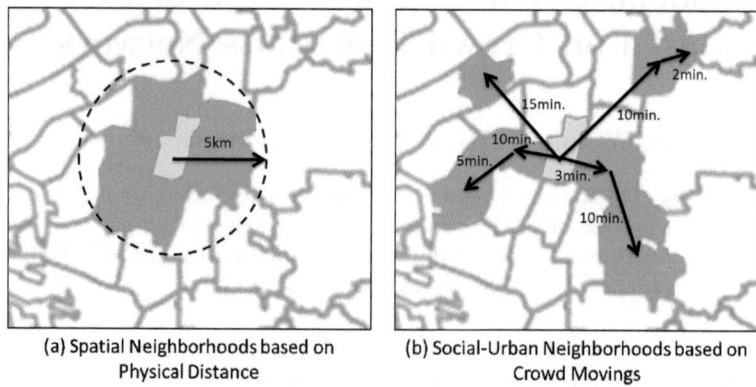

(a) Spatial Neighborhoods based on Physical Distance

(b) Social-Urban Neighborhoods based on Crowd Movings

Fig. 1. Spatial Neighborhoods vs. Social-Urban Neighborhoods

However, social-urban neighborhood search requires examining such movement time among urban areas, which is hard to be done due to complicated urban structures and dynamically updating urban situations. In order to challenge the issue, we attempt to exploit massive crowd local experiences over location-based social networks, with which we can easily collect a large number of individual footprints and furthermore observe urban utilization in terms of crowd behavior.

In this paper, as a promising approach to investigate social-urban neighborhoods, we propose a method to identify neighborhoods from any location by observing social interactions between urban areas. Especially, we examine crowd movings as an important indicator for measuring social interactions. Specifically, we also introduce a data structure, **Crowd Footprints Network (CF-Net)**, to aggregate and examine crowd movings as a simplified graph, with which we can easily analyze crowd movements between urban areas. In the experiment with crowd footprints collected from Twitter, the representative microblogging and location-based social network, we will examine neighborhoods from several areas of our interests, significantly looking into how these social-urban neighborhoods are distorted from general localized vicinity.

The remainder of this paper is as follows. Section 2 describes our research model to conduct social-urban neighborhood search by exploiting crowd moving experiences and summarizes related work. In Section 3, we introduce a data structure, **CF-Net**, for aggregating and investigating crowd moving experiences from location-based social networks represented by a graph. In Section 4, we explain the detailed procedure to search social-urban neighborhoods on a constructed **CF-Net** graph filled with crowd experiences. Section 5 shows experimental results conducted using a real dataset collected from Twitter. Finally, Section 6 concludes this paper with a brief description about future work.

2 Where Are My Neighborhoods?: Social-Urban Neighborhood Search with Crowd Footprints on Location-Based Social Networks

In this section, we describe our proposed research model for realizing crowd-sourced social-urban analytics and summarize related work.

2.1 Social-Urban Neighborhood

In our daily urban lives, the perception of urban neighborhood is quite critical, since we always need to obtain resources for living, and meet people by moving around urban areas. Conventional near-by search such as looking for nearest restaurants has been mostly conducted based on geographic distance. Thus, we often lost chances to find out better solutions from much distant places, which seemingly look so far away, but practically close with public transportations.

While such distorted urban space usually happens due to transportation network, it must be a daunting task even for individuals to be aware of neighborhoods in the today's big cities, where numerous urban elements such as buildings, roads, and daily changing social and natural phenomena will make it difficult and much more untouchable without vast amount of cost and continuous monitoring to the ever-updating urban situations. Obviously, if we are able to figure out urban neighborhoods for a region/location of an interest, it will greatly change the way of conventional near-by search. For instance, even searching for a restaurant which is arrivable in an hour, we can have more options in the final result which probably include much desirable places in the extended search range. We may consider the time tables of urban railways to estimate much distant places reachable in a time condition. However, it is generally hard to say that railways are the only available transportation for every city. Therefore, we need much generalized urban traveling information, with which we can estimate place-to-place moving time.

In order to challenge the exploration of urban structure particularly looking into the social-urban neighbor in terms of social interactions based on crowd movements, we will approach collecting crowd footprints through location-based social networks and examining relative closeness between urban areas regarding moving time, particularly with crowd moving experiences through location-based social networks.

2.2 Social Ichnology

Generally, ichnology [4] studies animals' shapes or behavioral characteristics with the remaining footprints or burrows. Likewise, by utilizing digitized crowd footprints which are now prevalent over location-based social networks, we are particularly interested in studying urban dynamics rather than monitoring and infringing individual lives. In detail, we aim at identifying urban area's social

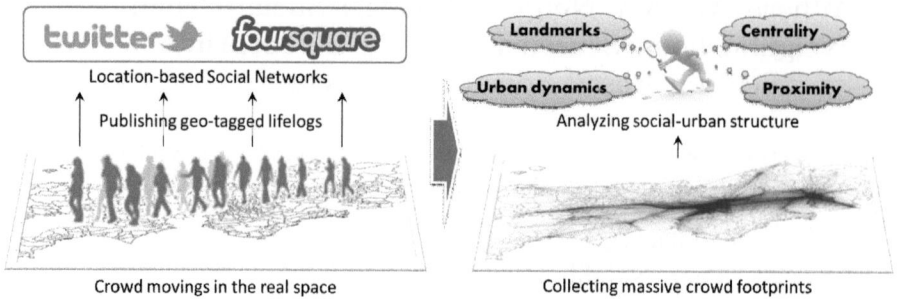

Fig. 2. Social Ichnology based on LBSN

structure, which must be strongly related to citizen lives and their urban utilization. While we focus on the footprints about crowd's whereabouts and movement behavior, we can take advantages of the social network's massive data which can reveal unprecedented scale of participatory crowd's urban utilization. Among the various types of social networks, we are strongly motivated the location-based social networks, which are recently being advanced on behalf of prevalent location-sensing mobile devices. Thus, we can easily gather crowd's digitized footprints, which can occur privacy infringements, but if appropriately anonymized by only focusing on their collective set rather than individuals, we are able to conduct various analyses regarding to crowd behavioral patterns and moreover social-urban utilization.

In this work, as depicted in Fig. 2, we present a research model for conducting social ichnology in cooperation with location-based social networks. At first, crowd's whereabouts are widely accessible from the current most location-based social networks such are Twitter[1] and Foursquare[2]. To find out movements, we need to know each user's current and last locations at minimum. Afterward, each movement can be just an activity hidden from personal identification. In order to intensively monitor a local area's crowd footprints, we can set a boundary for monitoring. Interestingly, the collective footprints can shed light on various social analytical problems such as landmark identifications, measurement of congestions, urban utilization, etc. Specifically, each moving can deliver two fundamental information regarding to moving time and distance. Furthermore, we can estimate second-order information such as moving speed, staying time, hourly traffics, etc. If we combine the footprints with other map-based local information, we can further conduct studies on urban dynamics.

2.3 Related Work

In our daily activities, we are required to conduct various location-based decision makings such as neighborhood search for the purpose of shopping, traveling, and

---

[1] Twitter: http://twitter.com/
[2] Foursquare: https://foursquare.com/

commuting. Generally, spatial neighborhoods based on distance between locations would be different from substantial neighborhoods involving in our lifestyles and urban transportations. mySociety.org [6] generates interactive maps which are able to search for geographic areas to meet users' needs in terms of travel time and house prices. According to the constructed map, spatial neighborhoods are different from practical neighborhoods. In this paper, motivated by such study, we challenge to explore non-spatial neighborhoods by borrowing crowd moving experiences, which reflects practical urban lives.

Numerous crowds are publishing their digital footprints over location-based social networks thanks to the prevalence of smartphones. By exploiting the remained footprints similarly as ichnology, we are able to observe crowd behavior in a large scale of geographic areas in terms of how to live in urban areas, how to move among urban areas, and how to utilize there. Cheng et al. [2] explores human mobility patterns by analyzing crowd footprints based on check-ins in terms of the spatial, temporal, social, and textual aspects. Wei et al. [10] proposes a method for representing popular routes from uncertain trajectories of crowds over location-based social networks. In addition, based on crowd behavior, we can conduct crowd-based urban analytics such as characteristics of urban areas and relationships among urban areas. In our previous work [5,8], we proposed methods for characterizing urban clusters by extracting crowd behavioral patterns over Twitter. As for social-urban relationships, we analyzed distortion between physical proximity and a sense of proximity which is MDS-based proximity measured based on moving distance, moving time, and moving flux between urban clusters [7,9]. Comparing with these methods, in order to efficiently deal with substantial social-urban interactions between urban clusters based on crowd footprints, we propose a social-urban neighborhood search on the **CF-Net**, which is a graph constructed by summarizing the crowd footprints. Furthermore,Cranshaw et al. [3] attempted to explore urban dynamics by clustering crowd check-ins and identifying livehoods based on social proximity as well as spatial proximity. In addition, they investigated relations between the generated livehoods and municipal boundaries by interviewing residents living in the targeted area. A livehood is defined as a consecutive geographic space having similar character in terms of facilities and users' behavior. On the other hand, in our work, we aim at exploring neighborhoods which are distributed across discontinuous geographic space by analyzing crowd movements as one of social-urban interactions.

3   Aggregation of Massive Footprints

In this section, we present a data management model for supporting various social-urban analytics with crowd footprints over location-based social networks. Characteristically, the location-based social networks are emitting numerous footprints which can be very helpful in revealing social-urban utilization as well as patterns of crowd's urban lives. While the data size of each footprint is trivial, the increasing number of populations with such mobile devices can cause scalable

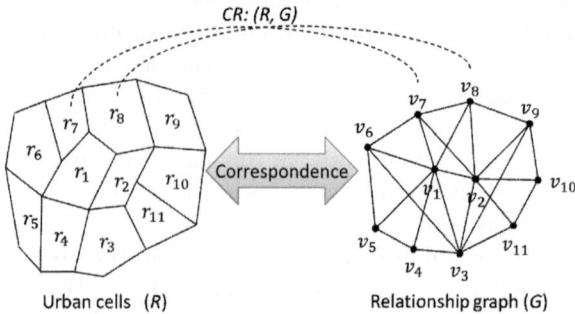

Fig. 3. Data Model for Crowd Footprints Network

problems when collecting and aggregating the data. For the reason, we need to develop a data structure enabling us to explore geographic and social features by mapping and arranging crowd footprints according to urban areas. In order to answer these requests for enabling practical urban analytics with probably ever-increasing crowd-sourced data, we present a graph-based data structure, with which we are able to conduct geographic and social analytics. We first describe the concept and basic definition of **Crowd Footprints Network (CF-Net)** which captures urban crowd's activities currently focusing on movements. We also depict the detailed procedure to construct the network. Then, we explain how geographic and social analytics can be conducted with the generated graph-based summary.

3.1 Crowd Footprints Network

With the massive amount of crowd footprints being published in the location-based social networks, we attempt to investigate local and global crowd behavioral patterns which can eventually answer to social-urban neighborhood search. First, in terms of locality, we would like to identify where are the most crowded places which are eligible to be a kind of social landmarks. For this, we will find clusters based on the locations of footprints. Second, with the found clusters, we examine social interactions between them. That is, we distinguish local movements from rather global movements. Especially, local movements are less significant long-distant movements. Thus, we will look into inter-cluster movements. To achieve the local and global movement analysis, we developed a data structure which absorbs every monitored movement into a graph in the proposed data structure.

As shown in Fig. 3, we generate two different types of data structure; on the left of the figure, it represents an example of area partitioning by cells, while the right-side figure shows its dual graph, which is a kind of dimension-reversing involution or so-called graph embedding. It can be made by putting a vertex within a cell enclosed by a cycle of edges. In our geo-social graph generation, we do not put any special constraint to set the urban cells, since we would like to cover various kinds of social-urban analytics regardless of the urban districts settings such as grid, administrative regions, clusters, etc. Fundamentally, a target real-space R is partitioned into cells. Then, for each cell r_i which is a non-overlap area each other, we put a vertex v_i and connect the vertices according to relationship $e_{ij} : (v_i, v_j)$. Finally, we define the **CF-Net** as $CF\text{-}Net : (R, G, CR)$ based on correspondence (CR) between the two data structures R and G.

The advantage of this combined structure is that we can conduct two types of query processing: 1) in the cell-based spatial data management, we can expect to apply various spatial index, aggregation and analyses, and 2) in the graph-based relationship management, we can represent various inter-cell relations such as distance, the number of inter-movements, etc. Furthermore, with the relation CR, we can traverse one structure to their other structure following the correspondence. In addition, we embed various aggregation regarding to crowd experiences into the graph. For instance, in each vertex in the graph, it holds information about the number of occurrence in itself, thereby quickly to be answerable. On the other hand, each edge will have information about the degree of movement paused by the directional path. Likewise, with the informative graph and the corresponding cell information, we can support social network analytics as well as spatial query processing.

3.2 Procedure of Social-Urban Graph Construction

We describe the details of how the proposed graph structure can be constructed with movement-based crowd experiences. As aforementioned, we rely on a given region-based and relationship-based data structures, which are complementary each other. In order to raise the adaptability of the proposed algorithm, we do not restrict the cell-based space partitioning only to a specific type such as grid or any irregular type clustering. Instead, we put a simple constraint that the whole region of interest is partitioned into non-overlapping cells on the place for convenience. Then, we can construct an initial relationship graph as a dual graph to the given cells. Consequently, the constructed graph and the correspondence to the cells are ready to import crowd moving information. Before going into the loading of movements into the graph, we detail the graph $G : (V, E)$ as follows.

i) a vertex's attributes $(v_i \in V)$:
 - occurrence: the total occurrence of data
 - inflow: the count of movement, whose destination is in v_i
 - outflow: the count of movement, whose departure is in v_i
 - neighbors
 - movement feature for intra-movements

ii) an edge's attributes $(e_{ij} \in E)$:
 - movement feature for inter-movements

Next, for each experience, specifically represented by a movement as mov_x : (t_s, t_e, p_s, p_e), we first examine whether it is intra- or inter-cell movement, by checking if the moving segments completely included in a cell. In such test, we can make it trivial, if cells is a convex, with which we can use the own property of convex, that is, any line segments whose endpoints exist in a convex area is also enclosed inside the convex. Thus, if we specify the cells as convex, we can easily check if a movement is intra- or inter-, only by looking into the endpoints' corresponding cells. Eventually, if $p_s \in r_i$, $p_e \in r_j$, and $r_i = r_j$, then we can say that the movement is the intra-movement. Generally, grid-based or voronoi-based space partitioning can follow the convex property. In this paper, we will adopt voronoi-based cells for efficient movement checking as well as its resulting reasonable partitioning.

Subsequently, we need to consider how to deal with the two types of movements; intra- and inter-movements. While each type is used differently in our work, we make a statistical summary for both cases in the same way. First, we count the number of movements by a variable N. Second, for further spatial and temporal analytics, we add two summaries by linear sum (LS) and square sum (SS) to the measurements of moving time ($t_e - t_s$), thus calculating two values, LST and SST, as follows.

$$MF : (N, LST, SST), \; LST = \Sigma(t_e - t_s), \; SST = \Sigma(t_e - t_s)^2 \qquad (1)$$

Thus, for aggregating movements, we generate a movement feature MF by means of the values. If a movement is given, we only need to increase the count N one more and add its spatial and temporal summary such as $LST' = LST + (t_e - t_s)$. Based on this feature, we can easily calculate the average moving time $Avg.MovTime$ as follows.

$$Avg.MovTime = \frac{LST}{N} \qquad (2)$$

Depending on the type of movements, we allocate this movement feature to a vertex or an edge corresponding to the relevant cells. That is, $v_{iMF_{new}} = v_{iMF_{old}} + (1, mov_x.t, mov_x.t^2)$, if mov_x is an intra-movement in the cell r_i. Likewise, $e_{ijMF_{new}} = e_{ijMF_{old}} + (1, mov_x.t, mov_x.t^2)$, if mov_x is an inter-movement relevant to the edge $e_{ij} : (v_i, v_j)$, and $v_i \neq v_j$. Here, $mov_x.t$ means a moving time calculated by $t_e - t_s$ of mov_x.

As one of most used queries in many location-based systems, route search is often considered with a network on a map. In our data structure, we partitioned a large region into a set of cells and constructed an initial network among them focusing on the centers of cells. Thus, it is somewhat bound to calculate a route search only possible on the constructive network. However, the network we constructed is from the experiences of crowds as well as distances in real space. Particularly, we developed the movement feature MF for enabling much better estimation of taken time and moving distance. In fact, we already projected every movement into the graph summarizing the crowd movements as the feature. With this feature, we can easily speculate movement information along an arbitrary path P_i as follows.

$$Avg.MovTime(P_i) = \Sigma_{k \in P_i} \frac{LST_k}{N_k} \qquad (3)$$

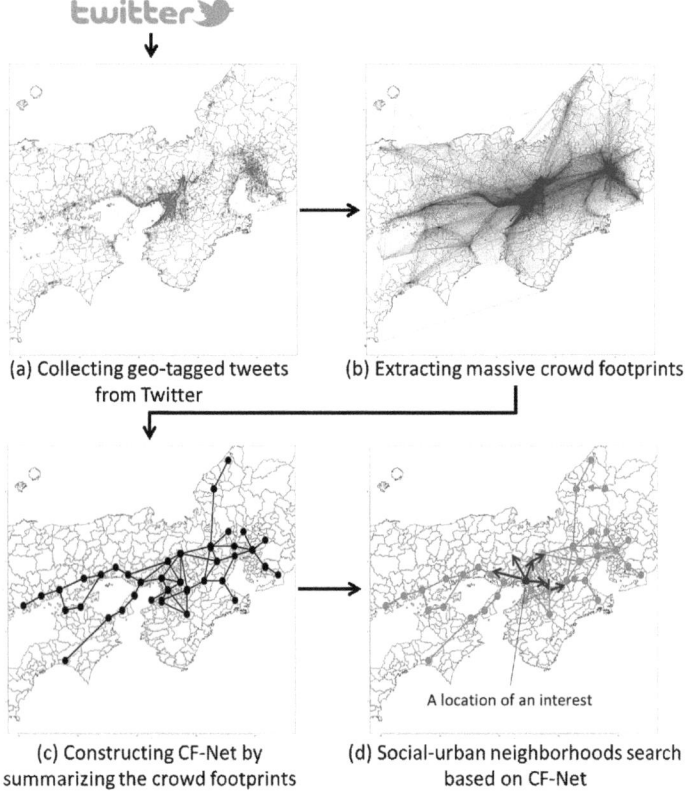

Fig. 4. Process of Social-Urban Neighborhood Search

4 Social-Urban Neighborhood Search on CF-Net

In this section, we describe the detailed procedure for finding neighborhoods with the aforementioned **CF-Net** based crowd footprints.

4.1 Following Crowd Walkways

In order to look for neighborhoods from a location of interest as depicted in Fig. 4, we characteristically depend on crowd moving experiences. In particular, with **CF-Net**, we can explore cluster-based movements which have two types of intra- and inter-movements. In this paper, we assume that a user would like to know much other areas going beyond a cluster's boundary. In other words, starting from a specified cluster, we examine the costs to reach other clusters. Specifically, for traveling from a cluster to the others, we measure the costs with the **CF-Net**, where edges already know the number of movements and can approximately compute the average moving time as described in Section 3. By means of the graph, assuming that enough crowd footprints data constructed a

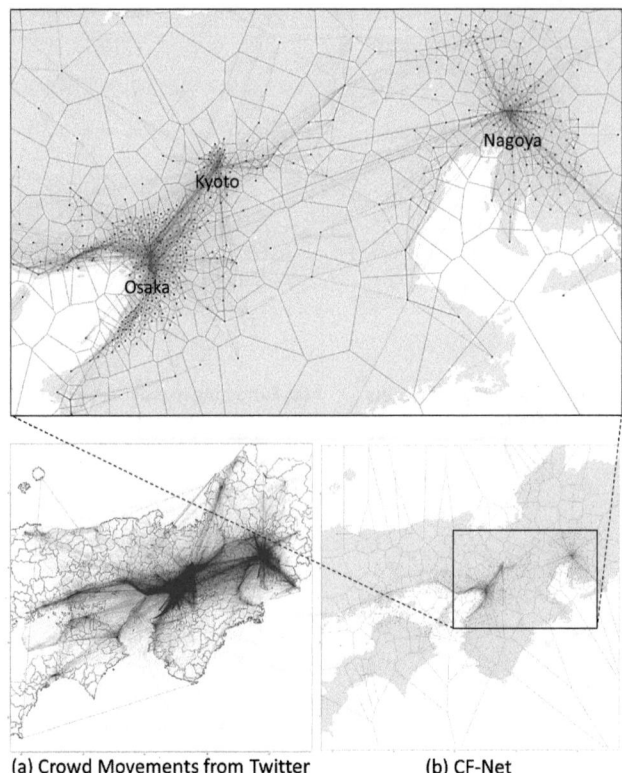

(a) Crowd Movements from Twitter (b) CF-Net

Fig. 5. Crowd Footprints Network with k-means based Voronoi Cells

connected graph, we can thus calculate the minimum traveling path and its cost to the other cluster. Although the path-following cost estimation would be an approximate approach, our exploratory investigation regarding to social-urban neighborhoods can enhance the opportunity to find out much better socially near-by places.

4.2 k-Nearest Neighborhoods and Conditional Neighborhoods

In practice, to derive appropriate number of neighborhoods, we need to consider how we specify our requests. In terms of conventional neighborhood search, we can consider two types of queries as follows.

1) **k-Nearest Neighborhoods:** In most web search and location-based systems, we usually look up the most top-k answers to satisfy a query. In the same way, we can specify k-nearest neighborhoods which answer the closest k neighborhoods to a given location. In our proposed **CF-Net**, this type of search can be easily conducted by graph-traversal. Starting from a node, we can adopt the breadth-first search measuring the costs to each other node, until k nodes are found.

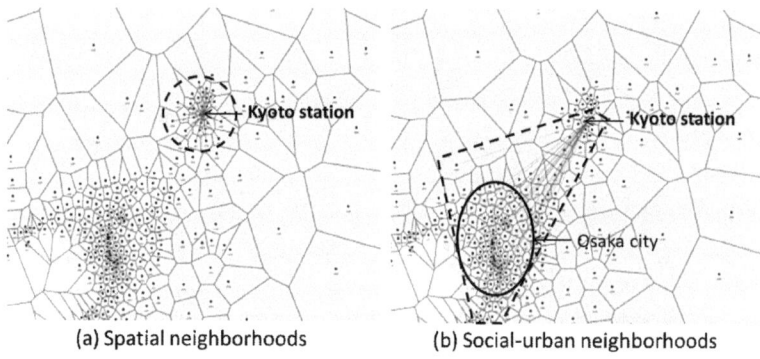

Fig. 6. Spatial Neighborhoods vs. Social-Urban Neighborhoods

2) **Conditional Neighborhoods:** The other query type that we practically need to specify is conditions which the neighborhoods should meet. In our case, a user can specify a time constraint such as 'areas reachable in an hour.' Similar to the above type of query, we can also travel the **CF-Net** graph until there is no more paths available, eventually looking for every node under the given upper-bound. In case of specification of lower and upper bounds of traveling time, we also traverse the graph, while ignoring nodes under the lower bound, and accepting nodes under upper-bound.

5 Experiment

In this section, we describe our experiment to construct a **CF-Net** with a practical dataset and to examine the proposed social-urban neighborhoods for some areas which are familiar to us for evaluating and reasoning the result. As a test dataset, we prepared a Twitter data collection, whose tweets are intentionally collected and filtered only for geo-tagged messages. The total amount of geo-tagged tweets was about 2.5 million collected for about a month between 2013/4/19-5/18, in a range centered on Osaka in Japan with longitude range: [133.0, 137.5] and latitude range [32.5, 36.5] as shown in Fig. 5 (a). Indeed, at the first glimpse on its geographic appearance as illustrated in Fig. 5 (b), the collection of the footprints vividly reflects the expected urban relationships of major cities such as Osaka, Kyoto, and Nagoya.

In our first experiment to arrange and make an order the uselessly collected movements, we first begin by preparing an initial **CF-Net** by means of k-means clustering with $k = 600$ condition. That is, we first identified 600 clusters with the method. Then, by applying the Voronoi diagram [1] with the cluster centers as the centers of urban cells, we examined each cell's region as illustrated in Fig. 5 (a). Next, with the generated **CF-Net** as depicted in Fig. 5 (b), we conducted several examinations about how spatial neighborhoods are changed. Expectedly, we were able to observe a quite different, but meaningful result as shown in Fig. 6. For this, we set our focus at a cell including Kyoto Station and examined

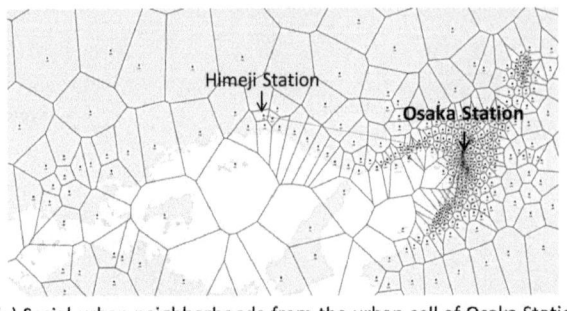

(a) Social-urban neighborhoods from the urban cell of Osaka Station

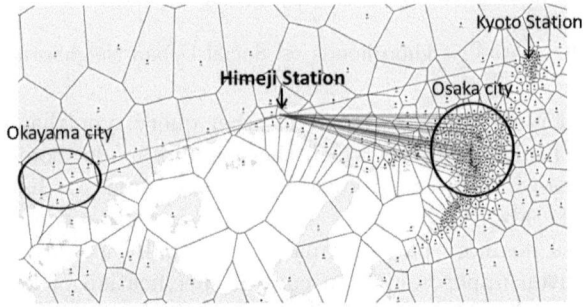

(b) Social-urban neighborhoods from the urban cell of Himeji Station

Fig. 7. Social-Urban Neighborhoods from the Cell of Himeji Station (Ekimaecho, Himeji, Hyogo) vs. from the Cell of Osaka Station (Umeda, Kita Ward, Osaka)

30-nearest neighbors from the focused cell based on the geographic distance. Consequently, the spatial neighborhoods existed in a regular shape of region where all the found cells are quite close to the focused cell. On the other hand, in the examination over the constructed **CF-Net**, we looked for neighborhoods which are reachable over the graph. Obviously, crowd movements are free from the spatial constraint and stretched out to the farther cells, which of them are mostly located around Osaka city. That is, we can say that many cells in Osaka city can be regarded as social-urban neighborhoods in terms of our **CF-Net** based survey.

Additionally, we looked into the difference of neighborhoods according to the focused cells. To make it explicit, we examined West Japan Railway Company's two stations, where a limited express stops; Osaka Station and Himeji Station, we searched for 30-nearest neighbors following the **CF-Net**. Eventually, most of the found neighboring cells of Osaka Station are located in the same city, implying that the station becomes a local center connecting to the near-by districts as shown in Fig. 7 (a). On the other hand, Himeji Station located in a rather small city has their neighborhoods mostly to other far cities such as Osaka, Kyoto, and Okayama as depicted in Fig. 7 (a). It means that the station becomes a bridge delivering many people to the other cities rather than to local districts. In fact, Himeji city is a suburban area, where lots of people are commuting to the other

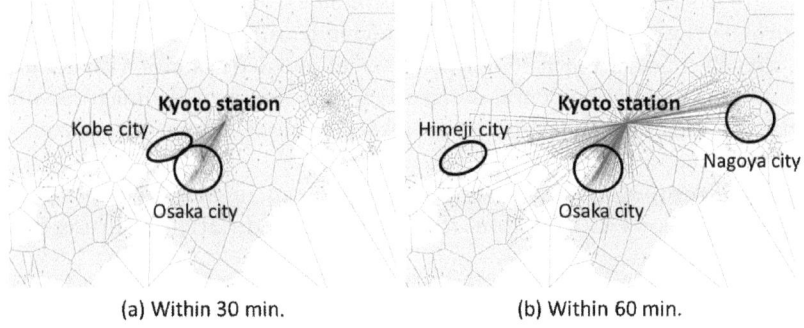

Fig. 8. Social-Urban Neighborhoods from the Cell of Kyoto Station (Shimogyo Ward, Kyoto) based on Travel Time Constraints

cities and a tourist site gathering people from other cities than the local area, thanks to one of the World Heritage Sites, Himeji castle.

Finally, we examined the time-constraint based conditional neighborhoods. For this, we focused on a cell including Kyoto Station as depicted in Fig. 8. With a 30 min. condition, we found neighborhoods mostly in Kobe and Osaka as shown in Fig. 8 (a). We found the result solely based on the **CF-Net**, comparing with the moving feature to derive the average moving time. In fact, Twitter-based moving time includes noises which cannot be controlled. While this research dealt with the noise moving data, we are somewhat successful to know the moving time information approximately measured in the **CF-Net** which can give hints about the time-constant reachability through massive crowd movements logs. Indeed, with the extended time constraint of 60 min. condition in Fig. 8 (b), we can observe much enlarged reachability from the focused cell of Kyoto Station. At least, in case of sightseeing places recommendation, we can obtain much more candidates in the enhanced areas.

6 Conclusions and Future Work

In this paper, we proposed a social-urban neighborhood search as one of applications of social ichnology by exploiting crowd footprints over location-based social networks. Especially, we introduced a data structure **CF-Net** for aggregating and analyzing crowd footprints efficiently. Then, we examined social-urban neighborhoods from urban cells of our interests by applying graph searching on the **CF-Net**. In the experiment with a real dataset from Twitter, we showed explicit differences between spatial neighborhoods and social-urban neighborhoods. This method would be useful when finding out candidate locations for moving and traveling to unfamiliar space, and even for advertising. Although we generated the **CF-Net** with k-means based voronoi cells in this paper, the proposed data structure can be applied to any type of urban area partitioning methods.

In future work, we will study urban morphology deeply by extracting meaningful urban syntax by exploiting crowd moving features on the **CF-Net**. In addition, we will consider much complicated social-urban dynamics with additional social and natural environmental data, which probably provide much reasonable contexts, bridging the gaps between social science and computer science.

Acknowledgments. This research was supported in part by the Grant-in-Aid for JSPS Fellows 24.9154 from the Ministry of Education, Culture, Sports, Science, and Technology of Japan.

References

1. Aurenhammer, F.: Voronoi diagrams — A survey of a fundamental geometric data structure. ACM Computing Surveys 23(3), 345–405 (1991)
2. Cheng, Z., Caverlee, J., Lee, K., Sui, D.Z.: Exploring millions of footprints in location sharing services. In: Proceedings of Fifth International AAAI Conference on Weblogs and Social Media, ICWSM 2011, pp. 81–88 (2011)
3. Cranshaw, J., Schwartz, R., Hong, J., Sadeh, N.: The livehoods project: Utilizing social media to understand the dynamics of a city. In: Proc. of the 6th International AAAI Conference on Weblogs and Social Media, pp. 58–65 (2012)
4. Ekdale, A., Bromley, R., Pemberton, S.: Ichnology–trace fossils in sedimentology and stratigraphy. Society of Economic Paleontologists and Mineralogists (SEPM) Short Course (1980)
5. Lee, R., Wakamiya, S., Sumiya, K.: Urban area characterization based on crowd behavioral lifelogs over twitter. Personal and Ubiquitous Computing, 1–16 (March 2012)
6. MySociety. More travel-time maps and their uses (2007), http://www.mysociety.org/2007/more-travel-maps/
7. Wakamiya, S., Lee, R., Sumiya, K.: Crowd-sourced cartography: Measuring socio-cognitive distance for urban areas based on crowd's movement. In: Proc. of the 4th International Workshop on Location-Based Social Networks, LBSN 2012, pp. 935–942 (2012)
8. Wakamiya, S., Lee, R., Sumiya, K.: Exploring reflection of urban society through cyber-physical crowd behavior on location-based social network. In: Proc. of 3rd International Workshop for Social Network and Social Web Mining, SNSM 2012, pp. 168–179 (2012)
9. Wakamiya, S., Lee, R., Sumiya, K.: Analyzing distortion of geo-social proximity using massive crowd moving logs over twitter. In: Proc. of the 4th Social Computing Symposium, SoC 2013, pp. 23–28 (2013)
10. Wei, L.-Y., Zheng, Y., Peng, W.-C.: Constructing popular routes from uncertain trajectories. In: Proc. of 18th SIGKDD Conference on Knowledge Discovery and Data Mining, KDD 2012, pp. 195–203 (2012)

A Notification-Centric Mobile Interaction Survey and Framework

Jonas Elslander and Katsumi Tanaka

Department of Social Informatics
Graduate School of Informatics, Kyoto University, Japan
{jonas,tanaka}@dl.kuis.kyoto-u.ac.jp

Abstract. In this paper we describe the results of a survey amongst smartphone owners into the use and perception of mobile notifications against 160 parameters. We conclude that not all notifications should be created or treated equally by mobile operating systems. The current generation of notifications proves not diverse enough and doesn't fit the needs and preferences of most smartphone users. Based on our findings, we offer a framework of design guidelines for more effectively engaging users with interactions initiated by the system.

Keywords: mobile, notifications, interaction, social, behaviour, perception, survey, questionnaire, framework, design, guidelines.

1 Introduction

As the smartphone adoption rate continues to grow in most countries around the world [1], it becomes increasingly easy for people anywhere to be connected to and interact with information available on the Internet. Both user and system can initiate this communication channel in order to transfer connected information. The user can actively request information – e.g. when visiting a specific website or when checking the temperature in a dedicated application. Alternatively, the system can initiate the information transfer by means of push-notifying the user, inviting him to interact in order to gain access to the complete message. In this paper, we refer to all communication started by the system as notifications. The rapid adoption of smartphones has lead to an explosion of these notifications. These are generally handled by the operating system – in more than 90% of the cases Android or iOS [2] – and are all communicated in a similar way; smartphone notifications look and behave largely the same.

By conducting the survey described in this paper, we gained greater insight into the perception of and stance towards different mobile notifications in absolute terms and by weighing them against parameters including their related timeframe, application category and social connections. We prove the hypothesis that all notifications should not be treated as equipollents by the mobile operating system. This conclusion funnels directly into further research defining an improved architecture for providing access to structured and contextually relevant pieces of mobile accessible information initiated by the mobile operating system.

To the best of our knowledge, related works largely consider all mobile notifications as equal entities, just like current operating systems do. Our work also differentiates itself by using data obtained from an extensive preliminary survey into the perception of mobile notifications, whereas most other related surveys focus on the broader mobile experience and only spend a few general questions on the topic of notifications. This approach has yielded new insights into a vast difference in perception and the preferred use of distinct mobile notifications.

The remainder of this paper is structured as follows. Section 2 explicates the conducted survey into the perception and use of mobile notifications. Survey methods, general insights, user behaviour and internal influences are highlighted and discussed. In section 3, we propose 5 design guidelines based on our findings in the previous chapter. These form a framework that can be implemented to better design systems relaying mobile notifications.

2 A Preliminary Survey into the Perception of Notifications

2.1 Survey Methods

Survey Creation and Distribution
Based on research into mobile activity, interface design and notification systems, and originating from the need for more specific academically substantiated insights into the concept of present-day mobile notifications, we constructed a questionnaire consisting of 163 questions; three defining the participating respondents and 160 questions gauging various aspects of the use of and stance towards mobile notifications on their current primary smartphone. The survey was created with smartphone users in mind and thus only accessible to participants indicating their primary mobile device as a smartphone. We distributed the questionnaire to present and former graduate students of selected universities in Japan and Belgium, who were invited through social media and direct mailing to participate.

Sample Description
We attracted 61 respondents of whom 26 are aged less than 25 years old (43%) and 33 are aged between 25 and 34 years old (54%). Accumulated, 97% of all respondents are aged less than 35 years old. Only these aforementioned two age categories were subsequently used in comparative studies. 72% of all respondents are male, leaving 28% female respondents. Analysis involving gender as a parameter were conducted with this unequal distribution in mind. We regard the total number of respondents too low for detailed gender-based crosstabulations, but big enough for some general insights. Finally, we asked participants to select the operating system of their current primary smartphone. Half (52%) indicated using iOS while 43% are Android users. In fact, 18% of all respondents are users of an older breed of the Android operating system, recognizable by 2.x a version number, while 25% use the latest 4.x Android version (with version 3.x only being available for tablets). In further analysis, both groups are combined and considered as generic Android users. Out of all respondents,

95% uses either iOS or Android as their primary mobile operating system and subsequently only these groups were used in comparative studies.

Survey Analysis
The resulting 9900 answers were automatically transcoded using an office suite and analyzed using statistical software. For the scope of this research, only descriptive analytics were used to gain insight into the frequencies of given answers, as well as crosstabulations between selected answer sets. The source files of the questionnaire and survey results dataset can be freely downloaded[1].

2.2 General Insights

In this section we aim to shine some light on more general numbers we could abstract from our dataset, such as the declared number of notifications smartphone users receive daily and their frequency of checking for notifications. We also propound the preferred level of visualization embodying mobile notification.

Notifications per day
Roughly half (45%) of smartphone users receive less than 10 notifications per day. Only 20% of smartphone users receives less than 5 notifications per day while an enumerated 82% of all participants claims to receive less than 20 notifications per day. In absolute terms, the current average number of mobile notifications per user per day is not very high. This is important to keep in mind in relation to the frequency of checking for notifications and the perception of distinct notifications, both discussed later in this paper.

Our data shows that younger smartphone users (aged below 25) tend to get notified on more occasions than older users (aged between 25 and 35). Male respondents are twice as likely (23% compared to 12%) as female respondents to belong to the category receiving less than 5 notifications per day while women (25% compared to 16%) are more likely to receive more than 20 notifications per day. One in three (32%) of iOS users receives less than 5 notifications per day versus only 4% of all Android users. Also, just under half of iOS users (45%) gets more than 10 notifications daily while two thirds (69%) of Android users do.

Frequency of Checking for Notifications
Nearly all (95%) respondents admit to activating their phone's screen in order to check for missed notifications. Sixty percent of all participants states to only do so sporadically – defined as at most a few times a day – while 35% of respondents checks their phone at least once an hour.

Youngsters check more often: half (46%) of all smartphone users aged under 25 claims to check their primary mobile device at least once an hour. That's almost twice as many as respondents aged over 25 (26%). Female users also check more often: one in three (32%) male respondents checks their smartphone at least once an hour, versus almost half (44%) of female participants. The Android mobile operating system does

---

[1] Questionnaire and survey results: http://bit.ly/mobilenotificationsurvey

not only provide it's users on average more notifications, it's users also check their phone more frequently for ones they missed out on. Half (54%) of all Android users checks his or her phone at least hourly, compared to only 16% of iOS users.

Preferred Level of Visualization

For a question probing the preferred level of visualization for generic notifications, we defined 4 distinct levels. These levels are constructed as to be gradually richer in information about the same piece of content the system wants to notify the user of. The first level only includes a reference to the application while the second level adds snippet information to that. The third and fourth levels throw a relevant picture into the mix, respectively as a thumbnail or full-screen image.

Almost all (95%) respondents prefer notifications to at least include snippet information and just over half (56%) wishes a visual representation in the form of a picture included. That is a significantly high number given the most recent version of iOS does not offer any visual representation beyond text and icons for notifications. The latest build of Andoid allows notifications to include a thumbnail image. The preference for notifications enhanced with a picture tends to be higher amongst respondents aged under 25 (62%) than amongst respondents aged over 25 (51%). Three quarters of female respondents (75%) prefer an image-based notification – compared to only half (52%) of male respondents. iOS users are far more likely (25%) to prefer full-screen notifications, while only one (4%) participant using the Android does.

2.3 User Behaviour

How smartphone users act upon receiving mobile notifications offers us an insight into the underlying behavioural traits specific to interactions initiated by the system. We analyzed why people check their phone for notifications, what the main drives are behind not taking action upon receiving notifications and how uncomfortable smartphone users feel after being unable to check for notifications within a given window of time.

Reasons to Check for Notifications

Two thirds (66%) of all participants states to check for missed notifications out of habit, while just shy of half gives being bored (49%), checking the time (43%) or the need for information (43%) as a reason. One third of respondents affirms a fear of missing out on important information (34%) or as a response to external impulses (31%) as a reason for spontaneously checking for notifications.

Reasons for Not Taking Action Upon Receiving Notifications

When given the choice, only two participants did not acknowledge the lack of time or the lack of importance as a reason for not taking immediate action upon receiving notifications. Female respondents (94% versus 59% for men) are more likely to perceive themselves as not having enough time for immediate action, while more male participants (80% versus 59% of women) view low importance as a the key reason.

Discomfort over Time

The percentage of respondents feeling uncomfortable about not being able to check for missed notifications changes in relation to the last time of checking. We proposed three levels of discomfort: "somewhat uncomfortable", "uncomfortable" and "very uncomfortable" in addition to the choice of feeling "not at all uncomfortable". Figure 1 provides a more detailed insight in the data the following conclusions are founded on.

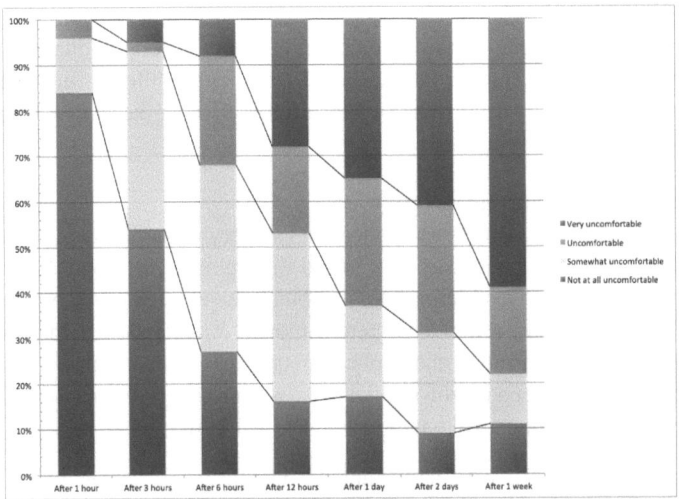

Fig. 1. Discomfort over time

The perceived comfort – defined by the respondents as feeling not at all uncomfortable – drops rapidly over the first 6 hours since the last time of checking, with losses of about 30% for every interval of 3 hours. Half of smartphone users feels uncomfortable to some degree when checking their phone less than every three hours. Feeling "uncomfortable" stabilizes between 20% and 30% from six hours onwards. The choice for "very uncomfortable" sharply rises from 8% to 28% after half a day of unavailability to check for notifications, and keeps on rising as extra (half) days are added, eventually peaking out at 59% of smartphone users feeling very uncomfortable after a week without access to notifications (but with access to other functions of their mobile phone). Half of the respondents (47%) identify themselves as feeling uncomfortable after half a day. In the first six hours of notification unavailability however, feeling "very uncomfortable" is a rare occurrence (with a maximal score of 8% within that timeframe).

Participants aged older than 25 are 7% less likely to feel "very uncomfortable" after one day of notification unavailability. After two days, the difference has grown to 12% and after a week to 21%. Men are less dependent on notifications: the number of male respondents feeling comfortable for every given timeframe is always significantly higher than the number of female respondents, except within the first hour. After two days, all female respondents feel uncomfortable to some degree, while even after a week some male respondents (14%) still feel at ease. When compared against the operating system, one in four iOS users (24%) feels uncomfortable to some extent within the first hour, while only 8% of Android users do. In light of the earlier mentioned "half (54%) of all Android users checks his or her phone at least hourly, compared to only 16% of iOS users", this is a surprising discovery. iOS users also tend to reach higher levels of discomfort – defined as the sum of feeling "uncomfortable" and feeling "very uncomfortable" – sooner than their Android using counterparts.

2.4 Internal Influences

By measuring the influence of the application category, social connection and timeframe of notifications on various behavioural aspects of mobile notifications, we conclude that the perception of notifications is influenced to great extent by parameters intrinsic to the relayed message.

Influence of the Application Category

In order to show the influence of the application category on notification preferences, we defined 7 parameters – expounded below – and distinguished 11 application categories: calls, messages, kudos & likes (defined as 'any binary expression of endorsement'), news & sports, coupons & deals, reminders, games, money & receipts, recommendations, invites, accounts & synchronization and application updates. We asked respondents to judge based on the average perception of notifications within any given application category. The parameters were propounded to the participants as follows:

— **Importance:** How relatively important are notifications related to the following application categories to you? Possible answers: very important, important, somewhat important, not important.
— **Action:** What action do you generally take upon receiving a notification related to the following application categories? Possible answers: click through immediately, click through later (not further specified), ignore.
— **Timing:** Do you prefer receiving real-time or periodical notifications given the following application categories? Possible answers: Real-time notifications, periodical (not further specified) notifications.
— **Selection:** Do you elect receiving all or only important notifications for the following application categories? Possible answers: all notifications, only important (not further specified) notifications.
— **Push/pull:** Do you prefer receiving information related to the following application categories through a push-style notification or through a pull-style widget? Possible answers: notification, widget, both.
— **Vibrations:** How often do you prefer notifications in the following application categories to be accompanied by vibrations? Possible answers: always, only when I switch vibrations for notifications on, never.
— **Sound:** How often do you elect notifications in the following application categories to be accompanied by sound of any kind? Possible answers: always, only when I switch the sound for notifications on, never.

Table 1, depicted below, provides a more detailed insight in the data our conclusions are based on.

Table 1. Influence of the application category

Application Category	Parameters						
	Importance	*Action*	*Timing*	*Selection*	*Push/pull*	*Vibrations*	*Sound*
Calls	95% more 65% very	92% imm. 8% later	100% real-time	93% all	95% not. 47% widget	75% always 2% never	31% always 6% never
Messages	82% more 49% very	77% imm. 23% later	87% real-time	87% all	88% not. 44% widget	66% always 5% never	25% always 15% never
Kudos	95% lesser 61% not	7% imm. 57% later	15% real-time	23% all	29% not. 83% widget	10% always 77% never	3% always 77% never
News	93% lesser 55% not	7% imm. 33% later	12% real-time	5% all	20% not. 88% widget	2% always 77% never	0% always 74% never
Deals	98% lesser 82% not	2% imm. 18% later	5% real-time	5% all	19% not. 88% widget	3% always 84% never	0% always 82% never
Reminders	69% more 30% very	64% imm. 26% later	73% real-time	67% all	79% not. 40% widget	49% always 16% never	16% always 34% never
Games	98% lesser 80% not	2% imm. 20% later	10% real-time	3% all	21% not. 86% widget	3% always 82% never	0% always 78% never
Money	80% lesser 60% not	7% imm. 29% later	18% real-time	22% all	32% not. 78% widget	7% always 72% never	2% always 72% never
Recommen-dations	98% lesser 70% not	0% imm. 38% later	5% real-time	3% all	22% not. 86% widget	3% always 87% never	0% always 84% never
Invites	84% lesser 34% not	8% imm. 70% later	35% real-time	36% all	49% not. 66% widget	16% always 56% never	8% always 64% never
Accounts	72% lesser 47% not	8% imm. 44% later	25% real-time	18% all	44% not. 71% widget	5% always 75% never	2% always 75% never
Updates	80% lesser 52% not	5% imm. 61% later	13% real-time	25% all	44% not. 67% widget	7% always 81% never	0% always 80% never

Only three application categories score well on all or almost all parameters for at least 2 out of 3 respondents: calls, messages and reminders. Only notifications related to applications in these categories yield widespread positive results when using the current delivery model used by operating systems such as the present versions of iOS and Android. For notifications originating from applications belonging to other categories, vast improvements can be made by making significant amendments to the current notification distribution model.

For every application category, at least half of smartphone users expressed a liking of pull-style notifications. This number rises to more than two out of three respondents when the notifications are not related to the three best received application categories. In all other categories, the preference for pull-style notifications outweighs the preference for push-style notifications with at least 50%, except for the category "invites" (where the difference amounts to 35%). It is imaginable however that users do not prefer an individual widget for every single application encompassed by these application categories.

Calls are perceived as the single most important application category, described as "very important" by two out of three respondents (65%). We take immediate action upon receiving them (92%), which we prefer in real-time (100%), in all cases (93%)

and push-styled (95%). Messages are considered the second most important application category. 9 out of 10 respondents opts to receive all related notifications (87%) pushed to them (88%) in real time (87%). Reminders are regarded as the third most important category. 7 out of 10 smartphone users considers them of greater importance – defined as the sum of perceiving them as "important" and as "very important" – (69%), takes immediate action (64%) and prefers to receive unfiltered (67%) real-time (73%) push-style notifications (79%).

Kudos & likes, News & sports, coupons & deals, games, money & receipts, recommendations, invites, accounts and synchronization & application updates are unpopular application categories in respect of their current-day notification behaviour across the board of tested parameters. Minor exceptions for a few category-parameter combinations apply, but generally they would all greatly benefit from a new notification model.

Influence of the Social Connection

In pursuance of uncovering the influence of the social connection of a notification on its perception, we defined seven parameters – described under 'Influence of the application category' – and distinguished five social categories: notifications related to a life partner, notifications related to a family member, notifications related to a friend, notifications related to an acquaintance and notifications not involving a person. We asked respondents to judge based on the average perception of notifications within any given social category. Table 2, on the next page, provides a more detailed insight in the data the following conclusions are based on.

Table 2. Influence of the social connection

Social connection	Parameters						
	Importance	Action	Timing	Selection	Push/pull	Vibrations	Sound
Partner	98% more 68% very	98% imm. 2% later	97% real-time	92% all	93% not. 51% widget	70% always 2% never	33% always 10% never
Family member	89% more 44% very	80% imm. 20% later	85% real-time	79% all	90% not. 52% widget	62% always 3% never	25% always 10% never
Friend	93% more 33% very	74% imm. 26% later	87% real-time	74% all	85% not. 55% widget	55% always 3% never	25% always 13% never
Acquaintance	75% lesser 15% not	29% imm. 66% later	35% real-time	29% all	48% not. 69% widget	22% always 30% never	5% always 36% never
No person involved	91% lesser 49% not	12% imm. 50% later	15% real-time	8% all	27% not. 87% widget	8% always 53% never	0% always 58% never

Overall only a small difference in preference of receiving notifications about a life partner, a family member or a friend is noticeable. Generally, a life partner is considered more important than both other categories. Between family members and friends, there is no significant difference for any parameter. The main difference however exists between notifications involving persons and those not involving persons, and between those involving friends and those involving acquaintances. It is thus important for the system to generally define three influence factors: one for each of the two distinct levels of social connections plus one for all non-social notifications.

For every category, at least half of the respondents expresses a liking of pull-style notifications. This number rises to more than two out of three respondents when the notifications involve acquaintances (69%) or do not involve a person at all (87%). At the same time, the acceptance of push-style notifications falls below half – from around 9 out of 10 – smartphone users when they only refer to acquaintances (47%) or do not refer to a person at all (27%). These clear distinctions suggest that behavioural styles in respect of push versus pull for mobile notifications can successfully be partly defined by the related social connection.

Notifications related to our life partners scored overall very high. Nearly all smartphone users want all related notifications (92%) pushed (93%) in real-time (97%) and take immediate action (98%) upon receiving them. As mentioned before, the respondents made no significant difference between family members and friends. Hence, notifications related to persons in either or both categories are allowed to behave in the same way. It is important however to distinguish friends from all other acquaintances. Few (25%) consider them of greater importance, with 15% view them as not important at all, compared to around 90% for the former category. This has big implications for all other parameters: two thirds of smartphone users prefer pull-style notifications for acquaintances, as they like to take action later (66%). Only one third of the respondents prefers to be presented all acquaintance related notifications (29%) and to receive them in real-time (35%). Notifications not involving a person score very low in comparison to those related to people we know.

Influence of the Time Period

Two timestamps characterize any notification: the moment at which it is delivered to the user and the period in time the notification refers to – e.g. a reminder for an event tomorrow. To discover the influence of the corresponding time period of a notification on its perception, we selected two parameters – importance and push versus pull – and distinguished 13 time periods an exemplary notification could refer to. In order to meaningfully analyze the push/pull parameter, we clustered some time periods. We asked respondents to judge based on the average perception of notifications related to the given time periods. Table 3 provides a more detailed insight in the data the following conclusions are based on. In general, smartphone users – based on two thirds or more of the respondents of this survey – only give significant importance to events that happened in the past hour or will happen in the coming hour. Just over half of the participants (55%) likes to see notifications related to events from 3 hours ago until 3 hours from now delivered to them as well. Only when something is happening right now, more than half of smartphone users will perceive it as very important (on average across all application categories and related social connections). 9 out of 10 (90% or higher) smartphone users prefer pushed notifications when the event took place in the past hour or will take place in the coming hour. For future events happening in the two subsequent hours, three out of four respondents (76%) like to receive push-style notifications. For events older than 1 hour but more recent than 3 hours, that's only half (50%) of the respondents.

Table 3. Influence of the time period

Time period	Parameters	
	Importance	*Push/pull*
Older than a day	95% lesser 59% not	31% notification 83% widget
6 – 24 hours old	88% lesser 37% not	
3 – 6 hours old	65% lesser 12% not	
1 – 3 hours old	55% more 5% very	50% notification 62% widget
11 – 60 minutes old	76% more 30% very	90% notification 35% widget
1 – 10 minutes old	83% more 42% very	
Happening right now	93% more 62% very	
1 – 10 minutes from now	83% more 37% very	93% notification 31% widget
11 – 60 minutes from now	78% more 27% very	
1 – 3 hours from now	55% more 10% very	76% notification 55% widget
3 – 6 hours from now	65% lesser 23% not	37% notification 79% widget
6 – 24 hours from now	80% lesser 40% not	
More than a day from now	85% lesser 56% not	

2.5 Survey Discussion

The used sampling method presumably skews the obtained results in favor of more advanced smartphone users as they are better represented on social media, one of our recruitment methods, and feel more inclined to provide answers on the topic of mobile connectivity in a rather lengthy questionnaire in return for no reward. During the analysis of the survey, we kept this assumption in mind. This survey did benefit from a decent sized sample given the preliminary nature of this study into a multitude of aspects of mobile notifications. A vastly bigger sample however would not only yield more refined results on the same parameters, it would also create more possibilities for crosstabulation. The influence of parameters on each other for specific subsets of users or occurrences could be analyzed. As currently presented, the isolated parameters already provide insight into the diversity of use and perception of notifications, but it's still largely unclear which share individual parameters hold in influencing the behaviour of smartphone users, which is assumed to be determined by a combination of influencers.

Due to the focused scope of this preliminary study, certain crosstabulations possible within the obtained dataset were not examined. The results of some analyses could, for example, be clustered by the amount of total notifications received on a daily basis amongst other parameters. With the possible benefits of such and similar researches in mind, we digitally included the entire dataset with this paper.

The social connections parameter could be further expanded: two respondents suggested an option for hierarchy-based social connections and the inclusion of a category for unknown persons was omitted from this questionnaire to avoid confusion with the indistinct definition of an acquaintance, but could be included in future research. Alternately, a more dynamic label-based clustering could be used instead of the rigid current approach: aspects of the social connection as well as aspects of the target of the social connection could influence the perception and use of related notifications. Users might or might not be more engaged with messages related to people with a certain profile, physical proximity, and so on.

Additionally, a new set of possible influencers could be analyzed. These could include the time of day or the relative time to certain events or other notifications. The mood or location of the user could serve as an input as well. To successfully conduct such survey, an automatic and integrated way of capturing all this data would prove far more versatile than a more free-standing questionnaire as in the current set-up.

In order to limit the length of the survey and as to focus mostly on mobile notifications as presented by current versions of popular mobile operating systems, the preferential study into the level of visualization was reduced to one question. Evidently, an in-depth research into the visual and interactive aspect of notifications on present and conceptual mobile systems could be conducted.

2.6 Survey Conclusions

We discovered several statistical differences in the usage and perception of mobile notifications, attributable to parameters we can link to the user, the system or the message.

- We learned that young, female and Android users are more likely to receive a significant greater number of notifications per day.
- Men do not perceive notifications as interruptive as women do, but in turn women label more notifications as unimportant than men do.
- Young, female and iOS users show more flexibility when it comes to challenging the interface status quo of current mobile notifications.
- Male smartphone users feel less compelled to check their notifications as frequently and feel more at ease when not doing so.
- Young and iOS users feel uncomfortable sooner when not able to check their notifications.
- After analyzing the notification model on current mobile operating systems, we conclude that push-style notifications work rather well when they are related to calls, messages, reminders. This is also the case when life partners, family members or friends are involved and when the information timestamp falls between one hour before and one hour from the moment of delivery. In the case of all other application categories, social connections and timestamps, there exists a broad margin for improvements.

- Even when pushed notifications are desirable, smartphone users express a wish for grouping them on several levels: timing (real-time or periodical), importance (rule-based filtering), social proximity and/or the target timestamp. No current mobile operating system however takes these factors into account when handling notifications.

Based on this we propose the creation of a model that can handle these parameters in order to more effectively deliver messages to the user, while addressing the desired greater variance in visualization, benefiting the communication and interaction between user and system.

3 Proposed Design Guidelines

In this section we transpose the synthesis of our findings into a framework of 5 basic interaction design guidelines. In order to ameliorate the user experience and performance of mobile notifications, these can be applied to current or future mobile operating systems.

3.1 Design for Pull

For every application category and every social connection, half or more of the survey respondents prefer accessing notifications in a pull-style manner. This demand is very high in spite of current mobile operating systems treating widgets in a very isolated way: one widget can display information related to one application. The design for pull guideline proposes *a centralized architecture in charge of relaying and displaying pull-style information* to users, resembling existing systems for push-style notifications. For some combinations of application categories, social connections and timestamps, pushed notifications are preferable on top of a pull-style information delivery system. On that account, *the simple choice of activating push notifications past a given threshold value for those parameters should be given to the user.*

3.2 Design for Human Eyes

Although our survey devoted only one question to the participant's preference regarding the visualization of notifications, related works support the claim that mobile notifications should be rendered in a visually richer and more diverse way. Our human eyes are designed to see colors and details and to read emotions and situations. Notifications should take advantage of those capabilities in order to communicate rich information while inducing as least strain as possible. Names of people, places or events can for example be replaced with *photographic images*. Computational methods for pairing notifications with images based on both contents and metadata are the subject of our future work.

3.3 Design for Time

We don't only value the present – even though significantly more – but also past and future. A system for displaying contextual information should *incorporate the dimension of time into its interface and core interaction framework*. We perceive the aforementioned *three time periods* in distinctive ways and they should thus be communicated in a respective manner. Given the right interface and interaction model, such system would allow users to explore the near past and future in a way that enriches their lives, whereas presently these notifications are often seen as an annoyance due to their non-distinctive way of being handled and communicated by the system.

3.4 Design for Social

As discovered, social connections linked to information contained in mobile notifications have a high impact on the perception of these notifications. We concluded the importance of *three social classes*, each linked to a distinctive experience. It is preferable to *cluster and display information in relation to social entities* over linking it to a more technical unit, such as the individual application, it relates to.

3.5 Design for Context

Lastly, notification systems based on our interaction framework should include some behind-the-scenes processes managing the *filtering and publication timing of notifications*. We learned that a great variance in preference exists concerning receiving contextual information in real-time and unfiltered across different application categories or social connections. The user's context defines these preferences. Filter attributes and publication settings can be based on our survey findings.

References

[1] Nielsen, How The Mobile Consumer Connects Around The Globe (February 25, 2013), http://www.nielsen.com/us/en/newswire/2013/how-the-mobile-consumer-connects-around-the-globe.html

[2] ComScore, comScore Reports April 2013 U.S. Smartphone Subscriber Market Share (June 4, 2013), http://www.comscore.com/Insights/Press_Releases/2013/6/comScore_Reports_April_2013_U.S._Smartphone_Subscriber_Market_Share

[3] Pew Research Center, The Best (and Worst) of Mobile Connectivity (November 30, 2012), http://pewinternet.org/Reports/2012/Best-Worst-Mobile.aspx

[4] Pew Research Center, Cell Phone Activities 2012 (November 25, 2012), http://pewinternet.org/Reports/2012/Cell-Activities.aspx

[5] Greenberg, S., Rounding, M.: The notification collage: posting information to public and personal displays. In: Proceedings of the SIGCHI Conference on Human Factors in Computing Systems (CHI 2001), pp. 514–521. ACM, New York (2001)
[6] Booker, J.E., Chewar, C.M., Scott McCrickard, D.: Usability testing of notification interfaces: are we focused on the best metrics? In: Proceedings of the 42nd Annual Southeast Regional Conference (ACM-SE 42), pp. 128–133. ACM, New York (2004)
[7] Cadiz, J.J., Venolia, G., Jancke, G., Gupta, A.: Designing and deploying an information awareness interface. In: Proceedings of the 2002 ACM Conference on Computer Supported Cooperative Work (CSCW 2002), pp. 314–323. ACM, New York (2002)
[8] Miller, T., Stasko, J.: Artistically conveying peripheral information with the InfoCanvas. In: De Marsico, M., Levialdi, S., Panizzi, E. (eds.) Proceedings of the Working Conference on Advanced Visual Interfaces (AVI 2002), pp. 43–50. ACM, New York (2002)
[9] Matthews, T., Dey, A.K., Mankoff, J., Carter, S., Rattenbury, T.: A toolkit for managing user attention in peripheral displays. In: Proceedings of the 17th Annual ACM Symposium on User Interface Software and Technology (UIST 2004), pp. 247–256. ACM, New York (2004)
[10] Lim, Y.-K., Stolterman, E., Jung, H., Donaldson, J.: Interaction gestalt and the design of aesthetic interactions. In: Proceedings of the 2007 Conference on Designing Pleasurable Products and Interfaces (DPPI 2007), pp. 239–254. ACM, New York (2007)
[11] Fogarty, J., Forlizzi, J., Hudson, S.E.: Aesthetic information collages: generating decorative displays that contain information. In: Proceedings of the 14th Annual ACM Symposium on User Interface Software and Technology (UIST 2001), pp. 141–150. ACM, New York (2001)
[12] Plaue, C., Miller, T., Stasko, J.: Is a picture worth a thousand words?: an evaluation of information awareness displays. In: Proceedings of Graphics Interface 2004 (GI 2004), pp. 117–126. Canadian Human-Computer Communications Society, School of Computer Science, University of Waterloo, Waterloo, Ontario, Canada (2004)
[13] Zhang, L., Tu, N., Vronay, D.: Info-lotus: a peripheral visualization for email notification. In: CHI 2005 Extended Abstracts on Human Factors in Computing Systems (CHI EA 2005), pp. 1901–1904. ACM, New York (2005)
[14] Scott McCrickard, D., Chewar, C.M., Somervell, J.P., Ndiwalana, A.: A model for notification systems evaluation—assessing user goals for multitasking activity. ACM Trans. Comput.-Hum. Interact. 10(4), 312–338 (2003)
[15] Maglio, P.P., Campbell, C.S.: Tradeoffs in displaying peripheral information. In: Proceedings of the SIGCHI Conference on Human Factors in Computing Systems (CHI 2000), pp. 241–248. ACM, New York (2000)
[16] Plaue, C., Stasko, J.: Animation in a peripheral display: distraction, appeal, and information conveyance in varying display configurations. In: Proceedings of Graphics Interface 2007 (GI 2007), pp. 135–142. ACM, New York (2007)
[17] Schmell, R.W., Umanath, N.S.: An experimental evaluation of the impact of data display format on recall performance. Commun. ACM 31(5), 562–570 (1988)

How Do Students Search during Class and Homework?
A Query Log Analysis for Academic Purposes

Rafael López-García, Makoto P. Kato, Yoko Yamakata, and Katsumi Tanaka

Department of Social Informatics, Kyoto University, Graduate School of Informatics.
Yoshida Honmachi, Sakyo-ku, 606-8501, Kyoto, Japan
{rafael.lopez,kato,yamakata,tanaka}@dl.kuis.kyoto-u.ac.jp

Abstract. Strong points, weak points and interests of students are precious data for their teachers, but it is hard to learn them quickly, especially when students do not cooperate in class. This paper explores a method for analysing queries of students that are allowed to search during class and homework. For this purpose, we first established six hypotheses on the queries and the expertise of the students. Then, we collected 143 queries from several lectures of an IT subject at Kyoto University. 36 students of this subject had previously been profiled before each lecture by means of questionnaires. When we checked our hypotheses against this collection of queries, we found that experts and novices often search the same way, although experts send more queries about different subjects. Some students also search contents that the teacher has not presented yet.

Keywords: query log, query log analysis, education, faculty development.

1 Introduction

Passivity and lack of participation of students is one of the hardest academic challenges that teachers may find in their classes. In this case, there are few symptoms that alert the teacher about whether a certain student is having a problem or what the problem is. Furthermore, ignoring these problems or deferring their solution may have undesirable consequences such as bad academic results, loss of motivation of the members of the academic community, and reluctance to collaborate with the others.

Although teachers still have some evaluation mechanisms to tackle this problem, this unpropitious environment may also offer some other data sources that teachers hardly ever manage. Use of search engines is a good example of this, since (1) their purpose is to locate the information we are interested in, (2) it is not very difficult to collect data about their use, and (3) queries and documents are generally written as a collection of meaningful words, so it is always possible to do some kind of analysis even without employing computers.

In general, query log analysis is a difficult technique, since most of the data is anonymous, it may be incomplete and it may suffer noise. However data collected in a classroom can be easily associated to a student and combined with other sources.

In addition, by using query logs we may find other data about our students, such as the topics that motivate them, their background in other areas or just if they would be able to arouse interesting questions. This information can be used to create new conceptual relationships or to increase the participation of the students in class.

The goal of this paper is to explore a technique to analyse the queries of a classroom in order to detect and profile students that are strong in a matter, that are having problems or that are just interested in other topics. For this purpose, we apply statistical analysis to several factors in the queries we collected in several lectures of an It subject at Kyoto University, in which students have been previously profiled by means of pre-evaluation questionnaires. We also take advantage of knowing the materials used by the teachers in each moment and the contents taught in those materials, as well as the temporal relationship between queries, materials and contents.

The rest of this paper is organised as follows. Section 2 briefly discusses the related work. Section 3 studies the problem and presents our hypotheses for the analysis. Section 4 describes the educational environment in which we are collecting queries and the questionnaires we use for student pre-evaluation. Section 5 discusses the result of our analysis. Section 6 concludes and shows the future work in the matter.

2 Related Work

There is a vast literature on query log analysis [1-3], although most of these studies only address how to improve IR techniques such as re-ranking, query suggestion, etc. Some papers also try to study any social group (e.g.: children [4]) or feature (e.g.: personal interests [5-6]). However, there are not so many papers that establish a relationship between search and education [7-8], and, in fact, very few analyse a query log in order to improve education by assisting teachers or helping students.

In the context of education, most of the searches are related to a certain academic task. Therefore, analysis techniques that focus on tasks [9] may be very useful. However, most of the papers in the literature rely on concepts that are too broad. For example, they try to calculate the user intent [10-11] or even subtopics of queries [12]. Since our target is not as general as these approaches, we can use narrow these concepts and use others such as the background knowledge of a student in a certain matter or the relationship between a term and the content that is being taught.

Detecting novice and expert users is also a difficult task. Important work in the matter has been done by Lazonder et al. [13], Aula et al. [14] and White et al. [15]. An interesting feature of the latter work is that the authors try to predict expertise not only after a session, but also during it.

3 Problem and Hypotheses

As we stated in the introduction of this paper, one of the essential differences between our query log and traditional ones is that it is not anonymous and we have some extra information, such as data about the students that participate in the subject (e.g.: pre-evaluation questionnaires, final scores, etc.) or the resources that are being employed in every moment. With this in mind, our analysis method is based in three assumptions regarding contents, materials and queries:

1. For any given material, we can extract the contents taught in it.
2. The materials employed in a lecture can be expressed a sequence.
3. Contents can be classified as theoretical (appear only in the slides), practical (appear only in the wiki) or both things at the same time. Queries that reference contents can be classified in the same way.

Our hypotheses about the queries sent by the students are the following:

1. Most of the queries sent by the students during the lecture will be related to the material that is being used or with any other recent one. Probability of a query to be related with a certain material will be correlated with the age of the material.
2. Queries sent by beginners will contain a lower number of teaching contents than queries sent by experts. This is based on the idea that beginners will use their searches to clarify the definition of the contents while experts will try to know more about the relationship of two or more contents.
3. If a query is strictly related to the topic that is being taught in class, queries of experts will contain a higher number of terms that are not teaching contents. This is based on the idea that beginners will just copy and paste what they do not know, while experts may add extra terms such as "definition", "example", etc.
4. In practical sessions, queries of beginners will focus more on theoretical content while queries of experts will focus more on practical content. This does not mean that all the queries sent by beginners will be about theoretical content, but beginners need to clarify theory more than experts before or during practice work.
5. If a query or sequence of queries is about content that is in our subject but not in the sequence of used materials (i.e: that content has not been taught yet or it is in the wiki), then the student is an expert.
6. If a query or sequence of queries is about content that is neither in the lecture nor in the materials, then the student is interested in topics that are not directly related to the lecture. This may happen because the student is trying to make a relationship between the two topics or just because of the lack of interest in the lecture.

4 Experimental Setup

4.1 Syllabus and Learning Environment

We collected our query log in a subject at Kyoto University called "Fundamentals and Practice of Informatics A". This subject is not about a fixed topic, but it is a collection of 15 sessions on a wide variety of disciplines whose only common point is the use of IT to solve academic problems. More concretely, the topics are:

- Document creation (Word, LaTeX).
- Web document creation (HTML, CSS, etc.).
- Cloud computing (main concepts and popular services).

- Information representation (encoding text, colours, images, mixing colours, etc.).
- Data aggregation (basic statistics and use of spreadsheets).
- Data analysis I (correlation) and II (testing hypotheses).
- Information processing I (bitmap and vector graphics) and II (natural language).
- Information retrieval.
- Database search (SQL with MySQL and phpMyAdmin).
- Data representation (XML, XPath).
- Data mining I (Introduction to the matter and basic use of R), II (association rules and clustering) and III (decision trees and introduction to Machine Learning).

However, our study does not include the sessions about "Document creation", "Web document creation" and "Information retrieval".

Note that these lectures are not completely independent, as in some cases they are continuation of the previous one (e.g.: Data analysis I & II and Data mining I, II & III).

Sessions are conducted by three different teachers, although one of them is only in charge of the two lessons in "Information processing" while another one is only responsible for the final three lessons in "Data mining".

Each lecture is 90 minutes long. It starts with a presentation by the teacher that lasts for about 35-40 minutes. After that, a practical task is presented to the students. Since the remaining time is often insufficient to complete it, students may have to finish it at home. The deadline to submit the work is often a week (in which we will continue capturing their queries). Some tasks need a certain degree of creativity (e.g. choosing a topic and creating a web page or presentation about it), while others focus almost exclusively in the technical aspect (e.g. extract the association rules from some data).

Regarding the resources, the classroom does not have a blackboard. Teachers use a wide screen in which they mostly show slides, although they may also show web pages or any other digital resource to complete their presentations. Students are provided with a printed copy of the slides and, of course, with the software they need to solve the practical work (Microsoft Office, Gimp, etc.). In addition, the subject has a website in which students can find a wiki that contains technical information to solve the practical tasks. However, teachers may demand the student to propose an example different from the one shown there (especially in lectures 13-15) and information to solve some optional tasks may be absent. Our search engine is initially present as a sidebar of the website, although students can also access a full screen version if they want.

Students of this subject are approximately 100 freshmen (first year undergraduate) who come from very different high schools and specialities, and whose major discipline is also very disparate, from literature to computing science. Their basic knowledge in IT is also quite distinct. Only 36 of them participated in the experiment, although we have to clarify that they can skip up to 5 lectures to pass the subject.

4.2 Pre-evaluation Questionnaires

In order to evaluate the previous knowledge of our students, we have created a questionnaire per lecture. Each questionnaire consists of several questions regarding (1) the concepts that are going to be taught in the subject, (2) the concepts that are needed to understand the ones that we are going to teach, (3) the procedures that are going to be taught in the subject (use of software, etc.) and (4) similar experience in previous subjects in high school or university.

Answers can be easily quantified (e.g.: "I have experience in doing this" / "I do not have experience in doing this" and "I can explain this concept" / "I cannot explain this concept"), and we set weights in order to balance the importance of every part, not letting concepts to be much more important than procedures and vice-versa. If the teacher wants to know some special details (for example, where the student learned a certain concept), we may include non-quantifiable questions, but the answers will not be considered in our analysis.

Students who get more than 50% in a questionnaire are considered "experts" in the matter, and we will try to distinguish their behaviour from the rest of the students, to whom we will call the "novice". According to this system, "experts" change in every lecture. Fig. 1 shows the distribution of expert students and novice students:

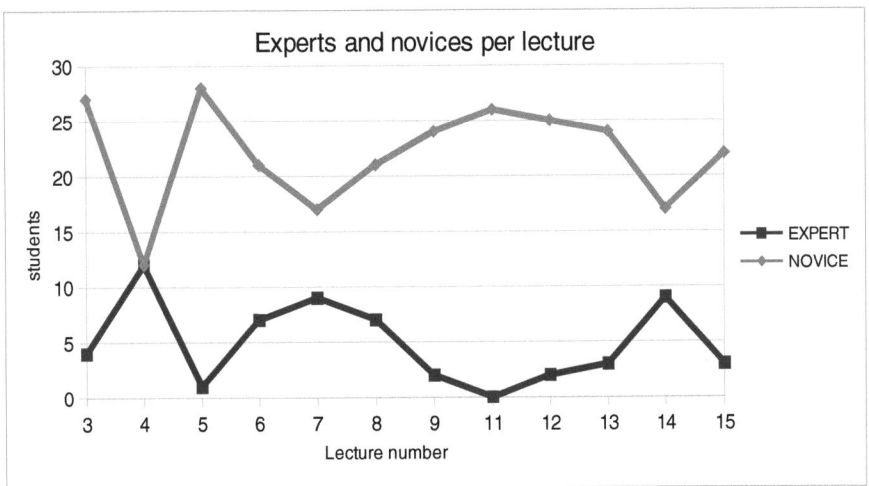

Fig. 1. Distribution of expert students and novice students per lecture

5 Analysis

We collected 181 queries in 12 lectures. However, some of them are just a visit to the previous or next page of results of a previous query, or a re-execution of the same query. If we count all these navigations as a single query, our query log is reduced to 143 queries. From these, only 30 (20.97%) were submitted by expert students.

The first impressing result we have obtained is that, in general, only 77 queries (53.84%) are related to the topic of the lecture. Expert students tend to send more queries that are not related to the lecture, and this result is statistically significant (p-value of 0.0112, 0.0198 with Yates correction). Table 1 shows this distribution:

Table 1. Are queries of students related to the lecture?

	Expert	Novice
Not related to the lecture	20 (13.99%)	46 (32.17%)
Related to the lecture	10 (6.99%)	67 (46.85%)

Contrary to what we thought when we formulated hypothesis 1, from the 47 queries sent during the presentations of the teachers, only 30 (63.83%) are related to the topic of the lecture and only 23 (48.94%) are related to the slides. In fact, it happens that in the last lecture we received a sequence of 5 queries which are related to the contents of the wiki, but not to the contents shown in the slides. There is no significant difference between novices and experts (p-value of 0.2871, 0.5421 with Yates correction).

In any case, we analysed if the queries that are related to the lecture are synchronised with the slides, obtaining the result shown in Fig. 2. Column "delay" represents the difference between the current slide and the slide that is related to the query of the student. A value of 0 means they are the same slide, while a negative value means the slide related to the query has not been shown yet.

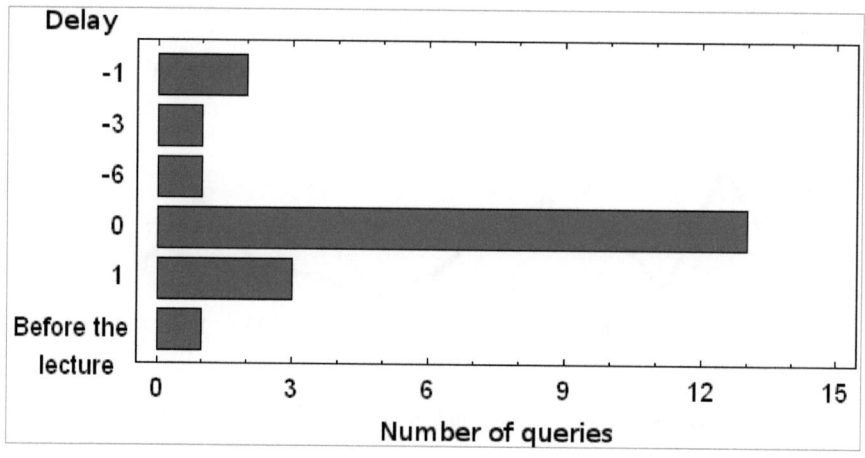

Fig. 2. Synchronization between queries and the current slide

The first result we can observe in this figure is that query sessions are normally very short, as students abandon the search if the teacher changes the slide.

The second result we find is that there are students that send queries about topics that the teacher has not taught yet. However, there is not significant difference in their expertise (p-value of 0.3679), and, in fact, there are only two queries per class of student (one expert anticipates 1 slide, another one anticipates 6 slides, one novice anticipates 1 slide and another one anticipates 3 slides). In addition, a novice anticipated a query before the lecture started. Therefore, we cannot confirm hypothesis 5.

Regarding the number of words that are teaching content ("content terms") in the queries that are related to the lecture, there is no statistical significance between experts and novices (p-value of 0.5638). However, the number of queries of experts may be too low to say something about hypothesis 2. Table 2 shows the distribution of queries according to content terms and expertise.

Table 2. Number of content terms vs. expertise of the student

	Expert	Novice
0 content terms	2 (2.60%)	6 (7.79%)
1 content term	5 (6.49%)	45 (58.44%)
2 content terms	3 (3.90%)	14 (18.18%)
3 content terms	0 (0%)	2 (2.60%)

With regard to the number of words that are not teaching content ("no content terms") in the queries that have at least one content term, there is not significant difference between experts and novices either (p-value of 0.2979). However, as in the previous case, the number of queries of experts may be too low too. Therefore, we cannot confirm or discard hypothesis 3 either. Table 3 shows the distribution of queries according to the no content terms and the expertise of the students.

Table 3. Number of no content terms vs. expertise of the students

	Expert	Novice
0 no content terms	4 (5.56%)	40 (55.56%)
1 no content term	4 (5.56%)	14 (19.44%)
2 no content terms	0 (0%)	9 (12.50%)
3 no content terms	0 (0%)	1 (1.39%)

Hypothesis 4 is related to the theoretical or practical orientation of the queries. To verify this hypothesis, we divided the queries in 4 categories: theoretical (the content appears only in the slides), practical (the content appears only in the wiki), both (the content appears both in the slides or the wiki) or none (the query is not related to the lecture). Once we excluded all the queries that are not related to the lecture, we could not find any statistical significance between these categories (p-value of 0.4991), but, once again, we have to make clear that the number of queries of experts is too low to discard the hypothesis. Table 4 shows the distribution of the queries according to their theoretical or practical orientation and the expertise of the students.

Table 4. Theoretical or practical orientation of quereis vs. expertise of the students

	Expert	Novice
Theoretical	1 (1.35%)	2 (2.70%)
Practical	3 (4.05%)	27 (36.49%)
Both	5 (6.76%)	36 (48.65%)

We have also checked the practical queries that were sent during the presentations of the teachers and they were always sent by novice students. Therefore, we can strongly reject hypothesis 5.

Queries that are not related to the lecture seem to provide an inestimable source of data about the interests of the students, supporting hypothesis 6. For example, we have located a session of one student with approximately 50 queries (including navigations between result pages and image search) containing the terms "Art Noveau", "Art Deco", "Alfons Mucha" and "William Morris". After analysing the homework of the student for that lecture, we found that it was a presentation about the aforementioned artistic movements and their most important representatives, providing us with the evidence of the interest of the student in that topic.

Another example is a student who queried "concept diagram" (5 queries related to that topic) and "Lifestyle diseases" (one query) and whose homework was a conceptual map about lifestyle diseases.

In addition, there are students that searched terms such as "Pikachu" (a character of the Japanese animation "Pokemon") or "sleep", but we could not find any evidence on their homework that confirms their interest in the matter. In the case of the first student, the task of that lecture was editing an image, so it is possible that the student considered that topic but discarded it because of the ban that the teachers established on copyrighted images.

Apart from the studies whose objective is to confirm or discard our hypothesis, we have also checked some other factors such as: (1) the type of characters used by the student (Japanese, Western or mix), (2) the use of the wiki, and (3) the number of queries that ended with a result being clicked by the student. Although in principle we detected significant difference in the two first tests, these differences were due to the fact that expert students tend to send more queries that are not related to the lecture. As soon as we analyse only the queries that are related to the lecture, the difference disappears. The third test also did not have any positive result.

6 Conclusions and Future Work

This paper has explored a statistical query log analysis for detecting the strong points, weak points and interests of students, especially of those who are often silent in class. The main differences between our methods and traditional ones are that (1) we employ not only queries, but also the materials and contents the teacher uses in class and (2) we consider the temporal relationship between the queries and the materials that are used in the lecture.

Our system makes some assumptions about the contents and the materials. For example, (1) we know what materials the teacher has been using in every moment, (2) we can express these materials as a sequence and (3) we can classify the content into theoretical, practical or both things at the same time.

With the aforementioned assumptions in mind, our paper presents a list of hypotheses about the queries of our students. These hypotheses consider the difference in behaviour we expected between expert students and novice students. For example:

- Queries of students are related to and synchronized with the slide that the teacher is currently showing in the presentation.
- Expert students try to establish more relationships between teaching contents.
- Expert students add terms that are not teaching contents to their queries in order to find what they expect with a higher probability.
- Queries of novice students will contain more theoretical contents as they will need to review them in order to solve their practical work.
- Expert students try to anticipate the contents that the teacher was going to show.
- Queries that are not related to the lecture can show the interest of the student in other matter.

However, in the experiment we performed in an IT subject of Kyoto University, we found that there is not a significant difference between the experts and the novices in most of these matters. In fact, the number of queries sent by experts is often not enough to offer any conclusion in the matter.

Regarding the temporal relationship between queries and materials, we can say that, when queries are related to the lecture, they often are related to the slide that the teacher is showing in that moment. Search sessions of the students are short. However, there are students that search for terms that are related to the lecture but the teacher has not presented yet. With this result we can set two future goals: (1) find a search interface that makes querying easier and (2) try to advance materials for students that are faster than the class.

Another significant result we have found is that expert students tend to send more queries that are not strictly related to the lecture. With this result, a teacher can try to use this kind of queries to make their authors participate more in the class, just by establishing relationships between the topic of the lecture and the topic of the queries.

Regarding the hypotheses that we could not confirm, we can conclude that, in such reduced environment, there is not statistical significance in the number of terms that are teaching contents or that are not teaching contents. However, in our future work we will add a new hypothesis about the quality of the contents: do experts use concepts that are more specific than the ones used by the novice students?

Another work is to analyse the navigations the students did and the documents they visited.

Acknowledgment. This work was supported in part by the following projects: Grants-in-Aid for Scientific Research (Nos. 24240013 and 23300311) from MEXT of Japan.

We would also like to thank Professors Masatoshi Yoshikawa, Yasuhito Asano and Masayuki Murakami for their inestimable support to our research.

References

1. White, R.W., Huang, J.: Assessing the scenic route: measuring the value of search trails in web logs. In: Proceedings of SIGIR 2010, pp. 587–594 (2010)
2. Liu, Y., Song, R., Chen, Y., Nie, J.Y., Wen, J.R.: Adaptive Query Suggestion for Difficult Queries. In: Proceedings of SIGIR 2012, pp. 15–24 (2012)
3. Cui, H., Wen, J.R., Nie, J.Y., Ma, W.Y.: Probabilistic Query Expansion Using Query Logs. In: Proceedings of the 11th International Conference on World Wide Web (2002)
4. Duarte Torres, S., Hiemstra, D., Serdyukov, P.: Query log analysis in the context of information retrieval for children. In: 33rd International ACM SIGIR Conference on Research and Development in Information Retrieval (2010)
5. Pu, H.T., Chuang, S.L., Yang, C.: Subject categorization of query terms for exploring Web users' search interests. Journal of the American Society for Information Science and Technology 53(8), 617–630 (2002)
6. Limam, L., Coquil, D., Kosch, H., Brunie, L.: Extracting User Interests from Search Query Logs: A Clustering Approach. In: Proceedings of DEXA Workshops 2010, pp. 5–9 (2010)
7. Liu, H.: Learning by Searching. In: McFerrin, K., et al. (eds.) Proceedings of Society for Information Technology & Teacher Education International Conference 2008, pp. 3843–3844. AACE, Chesapeake (2008)
8. Howard, P., Massanari, A.: Learning to search and searching to learn: Income, education, and experience online. Journal of Computer-Mediated Communication 12(3), article 5 (2007)
9. Buscher, G., White, R.W., Dumais, S.T., Huang, J.: Large-scale analysis of individual and task differences in search result page examination strategies. In: Proceedings of WSDM 2012, pp. 373–382 (2012)
10. Broder, A.: A taxonomy of web search. ACM SIGIR Forum 36(2), 3–10 (2002)
11. Rose, D.E., Levinson, D.: Understanding User Goals in Web Search. In: Proceedings of the World Wide Web Conference (WWW 2004), pp. 13–19 (2004)
12. Song, R., Zhang, M., Sakai, T., Kato, M.P., Liu, Y., Sugimoto, M., Wang, Q., Orii, N.: Overview of the NTCIR-9 Intent task. In: Proceedings of NTCIR (2011)
13. Lazonder, A.W., Biemans, H., Wopereis, I.: Difference between novice and experienced users in searching information on the World Wide Web. Journal of the American Society for Information Science 51(6), 576–581 (2000)
14. Aula, A., Jhaveri, N., Käki, M.: Information search and re-access strategies of experienced web users. In: Proceedings of WWW, pp. 583–592 (2005)
15. White, R.W., Dumais, S.T., Teevan, J.: Characterizing the Influence of Domain Expertise on Web Search Behavior. In: Proceedings of WSDM 2009, pp. 373–382 (2009)

On Constrained Adding Friends in Social Networks*

Bao-Thien Hoang and Abdessamad Imine

INRIA Nancy and Lorraine University, France
{bao-thien.hoang,abdessamad.imine}@inria.fr

Abstract. Online social networks are currently experiencing a peak and they resemble real platforms of social conversion and content delivery. Indeed, they are exploited in many ways: from conducting public opinion polls about any political issue to planning big social events for a large public. To securely perform these large-scale computations, current protocols use a simple secret sharing scheme which enables users to obfuscate their inputs. However, these protocols require a minimum number of friends, i.e. the minimum degree of the social graph should be not smaller than a given threshold. Often this condition is not satisfied by all social graphs. Yet we can reuse these graphs after some structural modifications consisting in adding new friendship relations. In this paper, we provide the first definition and theoretical analysis of the "*adding friends*" problem. We formally describe this problem that, given a graph G and parameter c, asks for the graph satisfying the threshold c that results from G with the minimum of edge-addition operations. We present algorithms for solving this problem in centralized social networks. An experimental evaluation on real-world social graphs demonstrates that our protocols are accurate and inside the theoretical bounds.

Keywords: Social networks, Graph editing, Adding friends.

1 Introduction

Online social networks (OSN) have currently become an important means for people to share personal and public information and make friendship connections with friends, family, colleagues, and even with strangers. These networks constitute live platforms exploited by huge number of users for performing large-scale computations such as conducting polls about political issues and seeking precise information on huge graph databases. To preserve data privacy during the running of these computations, recent works [8,9,10,12,27] use a simple secret sharing scheme which allows users to obfuscate their inputs. For instance, [7,9,10,12] proposed distributed polling protocols in social networks without requiring cryptographic system. Polling is the way to determine the most favorite choice amongst some options from the participants. Each participant can distribute his preference by submitting vote, and after aggregating all votes, the

---
* Funded by ANR Streams project.

majority option will be chosen as the final result. Instead of disclosing his/her vote, the user only sends a share of the vote to each friend. Thus, splitting the vote into many shares enables users to protect their choices' confidentiality.

However, secret sharing based protocols impose a threshold parameter on the number of friends, i.e. the minimum degree of the social graph should be not smaller than the given threshold. Unfortunately, not all social graphs fulfill that condition. In other words, users cannot perform any common computation (e.g. polling process) if this threshold is not achieved. To satisfy the threshold condition, we propose to enrich the social graph with new friendship relations. We assume these new relations will be accepted by all users as they are relevant to their common interest (i.e., performing any common computation in secure fashion). Indeed, these relationships are only for convenience. They do not enforce their partners to share any resource.

To achieve this graph transformation, we use a set of graph-modification operations on the initial graph in order to build other one that can be used for the purpose of deploying some secret sharing protocols. For the sake of simplicity, we consider only edge-addition modification operation. One naive approach consists in modifying a graph in such a way that each node tries to add as many edges as possible to satisfy the threshold. Nevertheless, our main concern in this work is to answer the following question: *How can a graph be minimally modified to satisfy the threshold on node degree?* The main objective is to devise efficient solutions for modifying the social graphs as the user number of social networks is blowing up exponentially. Just to demonstrate one typical example, as of now Facebook has more than one billion monthly active users and 699 million daily active users on average.[1] Accordingly, an effective way to speed up the transformation of social graphs is to minimize the number of adding friends.

Contributions. We first formally describe the adding friends problem as follows. Given an input connected graph G with N nodes and a parameter c, we want to obtain the graph G' satisfying the threshold c and resulting from G with the minimum number Φ of edge-addition operations. We propose an algorithm computing the exact value of Φ with the time complexity in the worst case $\mathcal{O}(N^4)$ (and in the best case $\mathcal{O}(cN)$). To decrease this upper bound, we prove that there exist $\frac{3}{2}$-approximation algorithms which take time $\mathcal{O}(cN^2)$. We validate our solution with a performance evaluation on real-world social graphs which shows that our protocols are accurate and inside the theoretical bounds. To our best knowledge, our work is the very first theoretical study of the adding friend problem, using only a simple edge-addition operation.

Outline. This paper is organized as follows. Section 2 describes our adding friends problem definition. Section 3 presents our protocols with their correctness properties. Section 4 illustrates our experimental results. We review related work in Section 5 and conclude the paper with future research in Section 6. Proofs for the correctness of our solutions are given in [13].

---

[1] http://newsroom.fb.com/Key-Facts

2 Problem Statement

2.1 Notations

We present the social network as the form of model of social graph. Initially, our system is described by an undirected connected graph $G = (V, E)$, where V is a set of uniquely identified nodes of size N, and $E \subseteq V \times V$ is an edge set. We denote by function $e(u, v)$ the friendship between two nodes u and v, namely, $e(u, v) = 1$ if u connects to v, and $e(u, v) = 0$ otherwise. Each node n has a set $\mathcal{F}_n$ of direct friends[2] of size d_n. Let c be an initial constant parameter in the graph G that is used to identify two kinds of nodes: weaker and normal nodes.

Definition 1. *Let $G = (V, E)$ be a graph. A node $n \in V$ is* weaker *(resp.* normal*) if $d_n < c$ (resp. $d_n \geq c$). A graph G is c-degree if all nodes are normal. Otherwise, it is called* non-c-degree *graph.*

A set of all weaker nodes is denoted by $\mathcal{W} = \{n \in V \mid d_n < c\}$. A weaker node $n \in \mathcal{W}$ has to add one or more edges to become a normal one. Thus, we define y_n as the number of these edges, i.e., $y_n = c - d_n$. A weaker node n may create links with some other weaker ones if there are no connection between n and them. We call them *available weaker nodes* (i.e., *potential friends*) of n, and denote the set of these nodes by $\mathcal{D}_n = \{v \in \mathcal{W} \mid e(n, v) = 0\}$ and its size by x_n.

2.2 Problem Definition

Given a parameter c and a non-c-degree graph $G = (V, E)$ where $c < N$. Our purpose is to transform G into a *c-degree* graph $G' = (V, E')$ that is structurally similar to G by using a graph-modification operation on G. The output G' must have the same set of nodes as G.

It should be noted that we only use edge-addition operation to modify G into G'. Moreover, once one friendship link has been established, it cannot be removed or changed later. We assume that these new friendship links will be accepted by all user nodes as they are aware these new links will enable them to achieve some common functions (e.g., polling process).

Definition 2. *Given two graphs $G = (V, E)$ and $G' = (V, E')$. The* structural difference Φ *between G and G' is the difference between sets of edges. More formally,*

$$\Phi(G', G) = |E' \setminus E| + |E \setminus E'| \tag{1}$$

As we only use edge-addition operation to modify graph, $E' \supseteq E$ and G' is a *supergraph* of G. Consequently

$$\Phi(G', G) = |E' \setminus E| + |E \setminus E'| = |E'| - |E| \tag{2}$$

Naively, to construct G' from G, one weaker node could create friendship with anyone by accumulating more and more friends until it becomes normal. However, this naive approach may be less accurate and unfair: (i) An unnecessary

---

[2] We use these terms "friend" and "neighbor" interchangeably henceforth.

large number of friendship links may be added, namely it is a 2-approximation algorithm. For instance, assume we have w unconnected weaker nodes and each one needs one new link to become normal node. In this case, the naive approach may add up to w new links whereas $\lceil \frac{w}{2} \rceil$ is sufficient if we link weaker nodes together; (ii) The process may not be fair in the sense that some user nodes could get more new links than others. We also observe that the naive approach takes time $\mathcal{O}(c|\mathcal{W}|)$ (it takes $\mathcal{O}(N^2)$ in the worst case when $|\mathcal{W}| = N$ and $c = N-1$).

Problem (ADDFRIENDS). *Given a graph $G=(V,E)$ and a constant parameter c, construct a c-degree graph $G' = (V, E')$ such that $\Phi(G', G)$ is minimized.*

We can recognize in our concern, minimizing $\Phi(G', G)$ is equivalent to minimize the degree difference between G' and G.

On the first view, our ADDFRIENDS problem seems generally related to the b-MATCHING problem [14,17,25,26], or the solution could be the reduction of some b-matching algorithms proposed in [1,22] but it turns out ADDFRIENDS problem and b-MATCHING problem are truly different. Indeed, the concept of a b-matching was defined by Edmonds [14] as follows: For each node $n \in V$ let $\delta(n)$ denote the set of edges in E which meet n. For each vector $x = (x_e : e \in E)$ and a subset E' of E, let $x(E') = \sum \{x_e : e \in E'\}$. Let $b = (b_n : n \in V)$ be a positive integer vector. A *b-matching* of G is a nonnegative integer vector $x = (x_e : e \in E)$ such that $x(\delta(n)) \leq b_n \; \forall n \in V$. From this definition, in the b-MATCHING problem, we have to use *edge-deletion* operation to modify G into a *subgraph* G' in such a way that each node n in the output must satisfy the condition $d_n \leq b_n$. Conversely, in the ADDFRIENDS problem we only allow *edge-addition* operation to modify G into a *supergraph* G' and each node n has to satisfy reverse condition for degree value $d_n \geq b_n$ (and $b_n = c \; \forall n \in V$). That difference is also applied for the generalization of the b-MATCHING problem, GENERAL FACTOR problem [19], where each vertex's degree of the output graph must belong to a list of possible predefined value called *degree list*. GENERAL FACTOR problem (where all degree lists may not contain gaps of length >1) and graph editing problem (where all degree lists are singleton) can be solved in polynomial time [3,20]. Nevertheless, to our best knowledge, there is no efficient algorithm solving this problem with precise complexity.

To solve ADDFRIENDS problem, we only examine the following two types of edge related to: (i) two weaker nodes, (ii) one weaker node and one normal node. It is noted adding edges is a *dynamic process* (not with respect to the original static graph), and during this process, a weaker node may become a normal node. We denote by κ and λ the number of adding edges of type (i) and (ii) respectively.

Lemma 1. *Φ is minimum iff κ is maximum, or λ is minimum.*

The proof is given in [13]. Lemma 1 means: to achieve the minimum number of adding edges, we have to target $\mathcal{W}$ by trying to create connections amongst nodes inside $\mathcal{W}$ (i.e., increasing κ and decreasing λ).

3 Protocol

In this section, we first present our algorithm, *AlgoCen*, then we show the correctness of this algorithm and provide a theoretical analysis of its complexity. Finally, we introduce two greedy algorithms, and describe the comparison in terms of accuracy and complexity between greedy algorithms and *AlgoCen*.

Algorithm 1. ADDING FRIENDS ALGORITHM

Input: A non-c-graph $G = (V, E)$
$\mathcal{W}$: Set of weaker nodes, $\mathcal{W} = \{u \in V \mid d_u < c\}$
c: Constant parameter
Output: A c-degree graph $G' = (V, E')$.

Algorithm *AlgoCen*
1   Get information about $\mathcal{D}_n, x_n, y_n$ for all $n \in \mathcal{W}$
2   $\mathcal{T} \leftarrow \varnothing$
3   **while** $\mathcal{W} \neq \varnothing$ **do**
4      $\mathcal{R} \leftarrow \{n \in \mathcal{W} \mid y_n \geq x_n \text{ and } x_n \geq 0\}$
5   **if** $\mathcal{R} \neq \varnothing$ **then**
6   **foreach** $n \in \mathcal{R}$ **do**
7   **if** $x_n > 0$ **then**
8               $\mathcal{D}_n \leftarrow \{u \in \mathcal{W} \mid e(n, u) = 0\}$
9   **foreach** $v \in \mathcal{D}_n$ **do**
10   ConnectUpdInfo(n, v)
11   **if** $d_v = c$ **then**
12   **if** $v \in \mathcal{R}$ **then** $\mathcal{R} \leftarrow \mathcal{R} \setminus \{v\}$
13                    $\mathcal{W} \leftarrow \mathcal{W} \setminus \{v\}$
14   **if** $d_n < c$ **then** $\mathcal{T} \leftarrow \mathcal{T} \cup \{n\}$
15            $\mathcal{R} \leftarrow \mathcal{R} \setminus \{n\}$
16            $\mathcal{W} \leftarrow \mathcal{W} \setminus \{n\}$
17   **else**
18   Compute score value for all weaker nodes
19   Find best tuple of weaker nodes (n, v) w.r.t. score value and $e(n, v) = 0$
20   ConnectUpdInfo(n, v)
21   **foreach** $n \in \mathcal{T}$ **do**
22   Create $(c - d_n)$ links between n and arbitrary $(c - d_n)$ nodes $v \in V$ s.t. $e(n, v) = 0$

3.1 Description of Protocol

The protocol, *AlgoCen*, is composed of two stages (see Algorithm 1).

Stage 1 (lines 1–20 in Algorithm 1). The system gets some information related to each weaker node n such as $\mathcal{D}_n, x_n, y_n = c - d_n$. We assign higher selection priority to weaker nodes having less number of potential friends than they require, i.e., first a set $\mathcal{R} = \{n \in \mathcal{W} \mid y_n \geq x_n \text{ and } x_n \geq 0\}$ is examined, and later we investigate the set $\mathcal{U} = \mathcal{W} \setminus \mathcal{R}$. We evaluate the following two cases:

1. $\mathcal{R} \neq \varnothing$ (lines 6–16): each node $n \in \mathcal{R}$ where $x_n > 0$ links to all nodes v in $\mathcal{D}_n$. The procedure ConnectUpdInfo(n, v) (line 10) simply creates a connection between n and v, then updates their information, e.g., $d_n, \mathcal{D}_n, x_n, y_n, d_v,$ $\mathcal{D}_v, x_v, y_v$. (For the sake of simplicity, we do not describe this procedure.)

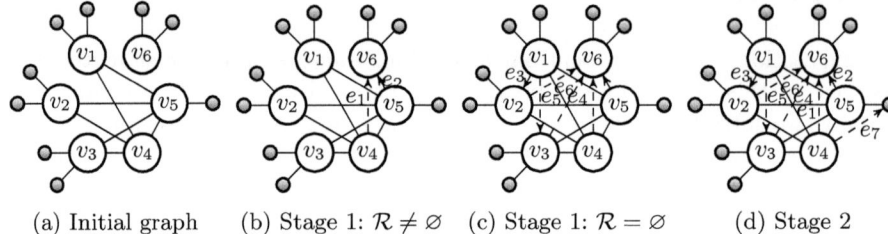

(a) Initial graph (b) Stage 1: $\mathcal{R} \neq \varnothing$ (c) Stage 1: $\mathcal{R} = \varnothing$ (d) Stage 2

Fig. 1. *AlgoCen* algorithm example

After connecting to n, if v's degree satisfies the threshold condition node v is removed out of $\mathcal{W}$ and also out of $\mathcal{R}$ if it is currently inside $\mathcal{R}$ (lines 11–13). For a node $n \in \mathcal{R}$ where $x_n = 0$, it has to request for friendship relations to normal nodes. A set $\mathcal{T}$ holds those nodes and is resolved in next stage. All nodes in $\mathcal{R}$ are finally removed out of $\mathcal{R}$ and $\mathcal{W}$ (lines 15–16).

2. $\mathcal{R} = \varnothing$: We first introduce one criterion to evaluate the selection priority of nodes. For each (weaker or normal) node n, we define a value t_n called *score* as follows:

$$t_n = \begin{cases} \dfrac{y_n}{x_n} & \text{if } x_n \neq 0 \text{ and } n \text{ is weaker} \\ 0 & \text{if } x_n = 0 \text{ and } n \text{ is weaker} \\ -1 & \text{otherwise} \end{cases} \quad (3)$$

This value expresses the level of aspiration of n to make connection to other nodes when it currently needs adding more y_n links but has a limited number x_n of weaker options. The higher score value, the higher selection priority. (If two nodes have the same score value then the one with smaller identity is considered to be higher priority.) We say a tuple of nodes (u_1, v_1) is *better* than a tuple (u_2, v_2) w.r.t. score value (assume initially $t_{u_1} \geq t_{v_1}$ and $t_{u_2} \geq t_{v_2}$) iff: (i) $t_{u_1} > t_{u_2}$, or, (ii) $t_{u_1} = t_{u_2}$ and $t_{v_1} \geq t_{v_2}$. A *best tuple* of nodes w.r.t. score value is the one that is better than any other tuples of nodes. In Algorithm 1, for this case $\mathcal{R} = \varnothing$ (lines 17–20): we first compute score values for all weaker nodes, then discover a best tuple of weaker nodes (n, v) w.r.t. score value (and they currently do not link to each other). Eventually we create a connection between them.

We repeat activities above until $\mathcal{W}$ is empty.

Stage 2 (lines 21–22 in Algorithm 1). This stage simply connects each node $n \in \mathcal{T}$ (if $\mathcal{T} \neq \varnothing$) generated from last stage to $(c - d_n)$ normal ones which currently have no connection with it.

In order to illustrate how Algorithm 1 works, we examine Example 1 below.

Example 1. Given $c = 6$ and an initial graph depicted in Fig. 1a.[3] Initial information of weaker nodes: $x_1 = 3$, $y_1 = 2$; $x_2 = 3$, $y_2 = 2$; $x_3 = 3$, $y_3 = 2$; $x_4 = 1$,

---

[3] For the sake of simplicity, in this work, we depict only subgraphs of the original connected graph. Weaker and normal nodes are drawn in white and gray color resp.

$y_4 = 2$; $x_5 = 1$, $y_5 = 1$; $x_6 = 5$, $y_6 = 4$. In Stage 1, $\mathcal{R} = \{v_4, v_5\}$. Node v_4 randomly links to v_6, and v_5 also links to v_6 by edges e_1 and e_2 (see Fig. 1b) then they are eliminated from $\mathcal{R}$. After updating information for weaker nodes: $x_4 = 0$, $y_4 = 1$, and thus, v_4 is moved to set $\mathcal{T}$. Moreover, since $y_i = 2 < x_i = 3$ for all $i \in \{1,2,3,6\}$, we have $\mathcal{R} = \varnothing$. We compute score values for each weaker node: $t_i = \frac{y_i}{x_i} = \frac{3}{2}$ for all $i \in \{1,2,3,6\}$, we can opt any tuple, e.g., (v_1, v_2), to link up together by edge e_3 (see Fig. 1c). Repeating this process until $\mathcal{W} = \varnothing$, edges e_4, e_5, e_6 are added (see Fig. 1c). In Stage 2, as $\mathcal{T} = \{v_4\}$, v_4 links to any normal node e.g., v_5, by edge e_7. The final output is illustrated in Fig. 1d. □

3.2 Correctness

This section first analyzes the accuracy of *AlgoCen* and then presents its time complexity. The proofs are presented in [13] due to space limitation.

Theorem 1. AlgoCen *produces the optimal solution.*

Proposition 1 (Time Complexity). *The time complexity of* AlgoCen *in the worst (resp. best) case is $\mathcal{O}(N^4)$ (resp. $\mathcal{O}(cN)$).*

3.3 Greedy Algorithm

This section first introduces the greedy algorithm for our ADDFRIENDS problem and then performs computational complexity analysis.

Algorithm. There are several greedy solutions for the ADDFRIENDS problem. Intuitively, we can apply following two strategies to choose weaker nodes:

1. **(CS1)** First select one weaker node with maximum degree and then choose alternative node with maximum degree of the remaining ones.
2. **(CS2)** First select one weaker node with minimum degree and then choose alternative node with minimum degree of the remaining ones.

Greedy CS1 vs. Greedy CS2
In this part, we compare the performance between CS1 and CS2, then analyze their complexities.

Definition 3 (Better/Worse Algorithm). *Given two algorithms A_1 and A_2. Algorithm A_1 is said to be better (resp. worse) than A_2 if there exist some graphs G such that after deploying A_1 and A_2 into G and receiving outputs G' and G'' respectively, we get $\Phi_{A_1}(G', G) > \Phi_{A_2}(G'', G)$ (resp. $\Phi_{A_1}(G', G) < \Phi_{A_2}(G'', G)$).*

Theorem 2. *There exist some graph structures such that CS1 is better than CS2 and vice versa.*

By Theorem 2, it is clear that using information on degree is not sufficient to solve the ADDFRIENDS problem. That is why we need to introduce another criteria to evaluate deterministically the priority amongst weaker nodes (score value presented in Section 3.1 is one of the suitable criteria).

Proposition 2 (Time Complexity). *The greedy algorithm (CS2) is done in time at most $\mathcal{O}(cN^2)$ and at least $\mathcal{O}(N^2)$.*

Greedy vs. AlgoCen

From Proposition 1 and 2, if all weaker nodes are initially fully connected together, CS2 runs slower than *AlgoCen*. Otherwise, CS2 runs faster than *AlgoCen*.

Theorem 3. *CS2 is a $\frac{3}{2}$-approximation algorithm.*

4 Experimental Evaluation

In this section we validate our solution with a performance evaluation on real-world social graphs. We start with the description of the datasets using in the experiments, then present the experimental setup, and finally show all results.

4.1 Datasets

We examined our algorithm *AlgoCen* on the following real-world social graphs which have different size and are provided by [21,24].

- **DIP (Database of Interacting Proteins):** These are experimentally determined interactions between proteins interpreted as undirected graph.
- **DBLP:** a full bibliography of publications in computer science.
- **Youtube:** a website which users can share video clips together.

Table 1 summarizes some properties of datasets. For each dataset, it shows the size of network N, number of edges $|E|$, clustering coefficient CC (i.e., that is a measure of degree to which nodes in a graph tend to cluster together), and the average number of friends $\bar{d}$ for a node in the network and $\bar{d} = 2|E|/N$.

Table 1. Datasets

Dataset	N	$|E|$	CC	$\bar{d}$
DIP	20K	41K	0.52	4.1
DBLP	511K	1.9M	0.73	7.3
Youtube	1.1M	3M	0.17	5.3

Table 2. Average execution time (in ms)

Dataset	Algorithm	Threshold c					
		31	51	71	91	121	151
DIP	AlgoCen	0.36	0.38	0.43	0.46	0.53	0.61
	CS1	0.07	0.07	0.07	0.08	0.08	0.09
	CS2	0.07	0.07	0.08	0.08	0.08	0.09
DBLP	AlgoCen	3.61	3.88	4.06	4.25	4.68	4.91
	CS1	1.78	1.85	1.89	2.09	2.25	2.46
	CS2	1.82	1.91	2.07	2.14	2.29	2.51
Youtube	AlgoCen	5.58	6.19	6.81	7.75	8.65	9.43
	CS1	3.14	3.65	3.90	4.32	4.56	4.72
	CS2	2.62	2.93	3.25	3.61	4.18	4.55

4.2 Experimental Setup

We evaluate the practical performance of our adding friend algorithms by applying them into social graphs with different value of threshold c and measure their *Percentage of number of adding edges* and *Time*. The *Percentage of number of adding edges* is computed by formula $(|E|' - |E|)/|E|$ where E' and E are respectively the set of edges of output and input graph. *Time* refers to the average time for generating one new edge between two nodes.

4.3 Results

Percentage of Number of Adding Edges. Fig. 2 shows the number of adding edges and the percentage by different methods for $c = 31, 51, 71, 91, 121, 151$. By Lemma 1 we have $\Phi = \kappa + \lambda = \frac{1}{2}\left(\sum_{n \in \mathcal{W}} y_n + \lambda\right)$. Since $0 \leq \lambda \leq \frac{1}{2}\sum_{n \in \mathcal{W}} y_n$, and $y_n = c - \overline{d}$ for all n, we have $\frac{1}{2}\sum_{n \in \mathcal{W}} (c - \overline{d}) = \frac{1}{2}|\mathcal{W}|(c - \overline{d}) \leq \Phi \leq \sum_{n \in \mathcal{W}} (c - \overline{d}) = |\mathcal{W}|(c - \overline{d})$. We observe in Fig.2a–2c value Φ is always within the theoretical bounds (two black dotted line), and *AlgoCen* always produces the lowest number of adding edges. The curve presenting *AlgoCen* nearly touches the lower bound. For two greedy strategies, we see that CS2 is better than CS1.

We illustrate the percentage of number adding edges in Fig. 2d–2f, that is $\frac{E(G')-E(G)}{E(G)}$. The lower the value is, the better the structure of the input graph is preserved. We observe that *AlgoCen* gives the smallest percentage values.

Time. Table 2 describes the average execution time computed as the total time required for processing the algorithm divided by value Φ. This also includes the time spent to load the input graph from disk. We see that *AlgoCen* is 2–6

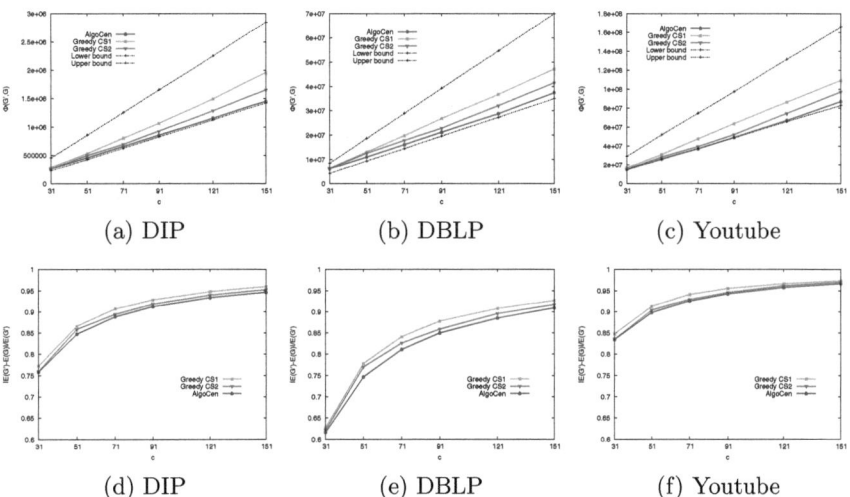

Fig. 2. Percentage of number of adding edges in the output graph

times slower than CS1 and CS2, however, it takes us only a few milliseconds for execution even for the large graph like Youtube. CS1 runs slower than CS2 for the Youtube graph, but faster than CS2 in other graphs.

5 Related Work

We present here some recent works related to graph modification and matching problem with particular attention to those do not allow edge-deletion operation.

Stable marriage problem is studied in [6,4,15,16] but they use preference lists with strict ordering of nodes which are not considered in our problem. [5] approached marriage problem where nodes are only matched to a fixed set of adjacent nodes, and so, much limited than ours.

By defining the welfare of one node as the sum of the weights of incident edges, [2] presented algorithms in social networks which try to maximize the welfare over all nodes. However this work is different from our concern. First, this work allows edge-deletion modification operation. Second, it uses an assumption about limited number of friends of one node.

Based on the k-anonymity notation [23], Liu et al. [18] studied a *graph anonymization* problem to transforms a graph to a k-anonymous one, i.e., every node shares the same degree with other $k-1$ nodes. This solution uses edge deletion operation and does not focus on threshold constraint. This also applies to the problem of identity-anonymization [11,28].

In [20], they studied some graph editing problems. In particularly, they give the algorithm for r-regular graph problem in which all nodes have the same degree r. Although the output graph from that algorithm could be r-degree graph like our conditions, the number of adding edges is not minimized.

6 Conclusion

In this paper, we formally defined the adding friends problem and proposed simple and efficient algorithms for solving it in social networks. Our algorithm produces the optimal solution. Our experimental results demonstrated that our protocols are accurate and inside the theoretical bounds. We plan to propose a protocol in distributed networks and consider a generalization of our problem: a graph consists of honest and dishonest nodes where honest weaker nodes consider their privacy when connecting to other ones, and dishonest nodes try to link to honest weaker ones to get as much their information as possible. This research direction gives us a perfect novel for adding friends problem in a secure way.

References

1. Anstee, R.P.: A polynomial algorithm for b-matchings: An alternative approach. Inf. Process. Lett. 24(3), 153–157 (1987)
2. Brandes, P., Wattenhofer, R.: On finding better friends in social networks. In: Richa, A.W., Scheideler, C. (eds.) SSS 2012. LNCS, vol. 7596, pp. 266–278. Springer, Heidelberg (2012)

3. Cornuéjols, G.: General factors of graphs. J. Comb. Theo. 45(2), 185–198 (1988)
4. Echenique, F., Oviedo, J.: A Theory of Stability in Many-to-many Matching Markets. Theoretical Economics (2006)
5. Floréen, P., Kaski, P., Polishchuk, V., Suomela, J.: Almost Stable Matchings by Truncating the Gale-Shapley Algorithm. Algorithmica 58(1), 102–118 (2010)
6. Gale, D., Shapley, L.S.: College Admissions and the Stability of Marriage. The Amer. Math. Month. 69(1), 9–15 (1962)
7. Gambs, S., Guerraoui, R., Harkous, H., Huc, F., Kermarrec, A.-M.: Scalable and secure polling in dynamic distributed networks. In: SRDS, pp. 181–190 (2012)
8. Giurgiu, A., Guerraoui, R., Huguenin, K., Kermarrec, A.-M.: Computing in Social Networks. J. Infor. and Comp.(2013)
9. Guerraoui, R., Huguenin, K., Kermarrec, A.-M., Monod, M.: Decentralized Polling with Respectable Participants. In: Abdelzaher, T., Raynal, M., Santoro, N. (eds.) OPODIS 2009. LNCS, vol. 5923, pp. 144–158. Springer, Heidelberg (2009)
10. Guerraoui, R., Huguenin, K., Kermarrec, A.-M., Monod, M., Vigfusson, Y.: Decentralized polling with respectable participants. JPDC 72(1), 13–26 (2012)
11. Hay, M., Miklau, G., Jensen, D., Towsley, D.F., Li, C.: Resisting structural re-identification in anonymized social networks. VLDB J. 19(6), 797–823 (2010)
12. Hoang, B.-T., Imine, A.: On the Polling Problem for Social Networks. In: Baldoni, R., Flocchini, P., Binoy, R. (eds.) OPODIS 2012. LNCS, vol. 7702, pp. 46–60. Springer, Heidelberg (2012)
13. Hoang, B.-T., Imine, A.: On Constrained Adding Friends in Social Networks. Report (2013), https://sites.google.com/site/hbthienosn/full_socinfo.pdf
14. Jack, E.: Maximum matching and a polyhedron with 0,1-vertices. J. Res. Nat. Bur. of Stand. Sec. B 69, 125–130 (1965)
15. Kelso Jr., A.S., Crawford, V.P.: Job matching, coalition formation, and gross substitutes. Econometric da 50(6), 1483–1504 (1982)
16. Khuller, S., Mitchell, S.G., Vazirani, V.V.: On-line algorithms for weighted bipartite matching and stable marriages. TCS 127(2), 255–267 (1994)
17. Korte, B., Vygen, J.: Combinatorial optimization. Alg. Comb. 21 (2008)
18. Liu, K., Terzi, E.: Towards identity anonymization on graphs. In: SIGMOD Conference, pp. 93–106 (2008)
19. Lovász, L.: The factorization of graphs. Act. Ma. Aca. Sc. Hung. 23, 223–246 (1972)
20. Mathieson, L., Szeider, S.: Editing graphs to satisfy degree constraints: A parameterized approach. J. Comput. Syst. Sci. 78(1), 179–191 (2012)
21. Mislove, A., Marcon, M., Gummadi, K.P., Druschel, P., Bhattacharjee, B.: Measurement and Analysis of Online Social Networks. In: IMC (2007)
22. Morales, G.D.F., Gionis, A., Sozio, M.: Social content matching in mapreduce. PVLDB 4(7), 460–469 (2011)
23. Samarati, P., Sweeney, L.: Generalizing Data to Provide Anonymity when Disclosing Information. In: PODS (1998)
24. Sommer, C.: Dblp graph, http://www.sommer.jp/graphs/
25. Tamir, A., Mitchell, J.S.B.: A maximum b-matching problem arising from median location models with applications to the roommates problem. Math. Program. 80, 171–194 (1998)
26. Tutte, W.T.: A short proof of the factor problem for finite graphs. Canad. J. Math. 6, 347–352 (1954)
27. Vu, L.-H., Aberer, K., Buchegger, S., Datta, A.: Enabling Secure Secret Sharing in Distributed Online Social Networks. In: ACSAC, pp. 419–428 (2009)
28. Zhou, B., Pei, J.: Preserving Privacy in Social Networks Against Neighborhood Attacks. In: ICDE (2008)

Social Sensing for Urban Crisis Management: The Case of Singapore Haze

Philips Kokoh Prasetyo, Ming Gao, Ee-Peng Lim, and Christie Napa Scollon

Singapore Management University
{pprasetyo,minggao,eplim,cscollon}@smu.edu.sg
http://www.larc.smu.edu.sg

Abstract. Sensing social media for trends and events has become possible as increasing number of users rely on social media to share information. In the event of a major disaster or social event, one can therefore study the event quickly by gathering and analyzing social media data. One can also design appropriate responses such as allocating resources to the affected areas, sharing event related information, and managing public anxiety. Past research on social event studies using social media often focused on one type of data analysis (e.g., hashtag clusters, diffusion of events, influential users, etc.) on a single social media data source. This paper adopts a comprehensive social event analysis framework covering content, emotion, activity, and network. We propose a set of measures for each dimension accordingly. The usefulness of these analyses are demonstrated through a haze event that severely affected Singapore and its neighbors in June 2013. The analysis, conducted on both Twitter and Foursquare data, shows that much user attention was given to the haze event. The event also saw substantial emotional and behavioral impact on the social media users. These additional insights will help both public and private sectors to prepare themselves for future haze related events.

1 Introduction

1.1 Motivation

Social sensing is an important step towards understanding how a disaster or social event affects individuals and communities. A good understanding of the event allows us to answer questions about its severity as well as the effectiveness of interventions introduced during the event. For example, during an earthquake, people may suffer from unsafe environment, inadequate supply of food and water, loss of houses, missing family members and lack of medical care. This wide range of concerns naturally become the topics of discussion among people, as well as the focuses of disaster management and rescue efforts.

In the past, social events are mostly reported by news media. One therefore can only perform the event impact assessment by analyzing news content or conducting field surveys on the event-affected people. Both approaches unfortunately require much time and resources. They also could not capture a complete view of the event due to limited reach to the larger user population.

With the popularity of social media, there is a large amount of user-generated content that can be harnessed for event analysis. Unlike traditional news media, social media content is generated directly from the user population and therefore offers direct access to both individual and community levels feedbacks. Other than its textual content, social media also records user behaviors that can be very useful in event analysis. For example, users may express their opinions online using votes and ratings, share interesting online content with their friends, etc. These non-textual behavioral data can be used to determine user activity patterns during the event.

1.2 Objective

In June 2013, Singapore experienced the worst haze in its history. The haze was a unique event that affected all people in Singapore. At the time, local media covered almost nothing but haze for several days. Office chatter and neighborly greetings were abuzz with talk of the haze. Our observations suggested that haze was on everyone's mind. But is there a way to measure the consciousness of Singaporeans during the haze to know their thoughts and feelings? Is there a way to quantify the impact of the haze on human activity?

To answer the above questions, we study social media usage by Singapore users during the haze event so as to derive some insights about peoples reactions to the haze. We adopt a social analysis framework that consists of four types of analysis on the event-relevant social media data (see Figure 1). The analysis can also conducted on data divided into different time intervals for trend analysis. The analysis techniques adopted are:

- *Content analysis*: This includes analysis on all textual social media data. The purpose is to determine the content topics, content objects (e.g., photo images, videos), representative keywords or keyphrases that help to explain the event. Content topics can be derived by clustering words or assigning them with topic labels [1].
- *Emotion analysis*: One can analyze tweet content for different emotion keywords to determine the state of user emotions as they generate the tweets. Bollen, Mao and Pepe analyzed user mood on Twitter in six dimensions using a set of words for each mood dimension [2]. In this paper, we use a similar approach using selected emotion words from Pennebaker's *Linguistic Inquiry and Word Count* (LIWC) full dictionary [3].
- *Activity analysis*: The activity behavior of users can be determined by activity words mentioned in the social media content, or by observing their actions recorded in social media data. In this paper, we use Foursquare check-in's to determine the places visited by social media users. When the activity data are geo-coded, one can even determine the locations of activities [4].
- *Network analysis*: Network analysis focuses on constructing human networks, and analyzing for each network central nodes (e.g., influential users, information gatekeepers), relationships (e.g., strong and weak ties) and communities [5]. Using network analysis, one can study how an event affects the network properties and dynamics (e.g., diffusion of information).

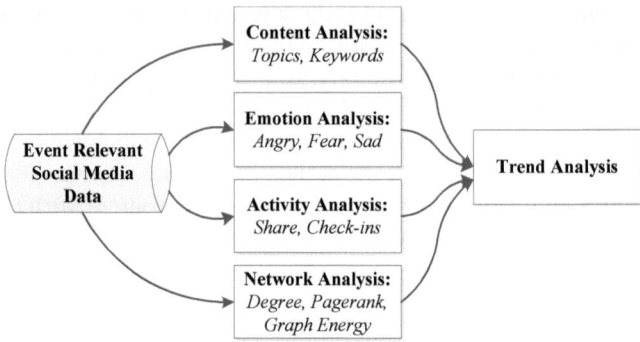

Fig. 1. Social Impact Analysis Framework

In our study, we observed that:

- The haze event attracted much attention from Singapore users only after the latter realised that the event lasted longer than expected. Substantially more people decided to tweet during the haze event.
- Users monitored the haze condition closely and depended a lot on traditional news media and government agencies for information. Nevertheless, they also demonstrated more negative emotion during the haze.
- Users reduced their outdoor activities causing fewer visits to eating places. This suggests that food and other businesses were quite badly affected by the haze event.
- The National Environment Agency (NEA) of Singapore emerged to be a central node in the network analysis during the event. Most traditional news media accounts also benefited from haze by seeing their centrality ranks improved. We however could not find any individual user gaining significant centrality rank.

1.3 Dataset Construction

The social media data used in this study are collected from Twitter and Foursquare. We collected Twitter data generated by about 130K public user accounts with Singapore stated as their user locations in June 2013. From this dataset known as SGTWITTERDATA, we selected those that contain one of the following keywords: *sghaze*, *haze*, *mustbehaze*, and *blamethehaze*. This selected subset of the data is called HAZETWITTERDATA. Using carefully selected keywords to collect event relevant tweets was also used in other works [6,7].

We also collected a month of FourSquare check-ins data in June 2013 which were generated by the same set of Singapore users. We call this data the 4SQ-DATA. Due to some crawling problem, we were not able to gather complete set of tweets on 13 June 2013. Hence, we would leave that day out of our study below. We did not include private user accounts as their profile and tweet information are not open to public.

1.4 Paper Outline

The rest of this paper is organized as follows. Section 2 describes some previous work regarding social sensing, and social sensing for crisis management. Our analysis on the content and activity dimensions of the data is presented in Section 3. More detailed analysis on emotional states is discussed in Section 4. Section 5 shows the dynamics of Singapore Twitter network during the haze crisis. Finally, we conclude our study in Section 6.

2 Related Work

Using hashtags that have been widely used to annotate tweets, Lehmann et al. studied the clusters of hashtags and their evolution over time and found four types of events: (i) those that attract attention before and during peak (measured by number of mentioned tweets); (ii) those that attract attention during and after peak; (iii) those that attract attention symmetrically around peak; and (iv) those that attract attention on a single day of the peak [8]. Crises are likely of type (ii) due to its unexpectedness and social impact.

Earle, Bowden and Guy found that sensing Twitter for earthquake events allows one to detect earthquake with human impact early among many earthquakes that have actually happened, especially in regions where the seismic sensors are not available [7]. The tweet content also provides very good contextual insights into the earthquake events.

In [6], the dissemination of rumors and news on Twitter during the 2010 earthquake in Chile was analyzed. Rumor spreading and news sharing are user behaviors prevalent in crisis events. The work analyzes about 4.7M tweets from 716K users during the event. It was found that the earthquake related content propagated very quickly on Twitter. Rumors were also found to propagate (or be retweeted) very differently from news as they are more likely to be refuted and questioned by users.

Cheong and Cheong conducted social network analysis on Twitter data related to Australia flooding events in 2010 and 2011 [9]. Two social networks were used, namely a retweet/mention network and a user-URL network. The high degree nodes in the two networks represent the *influential users* and *popular resources* respectively. These two sets of nodes lead us to find the users active and useful content in the events. Nevertheless, it also pointed out the local authorities did not manage to play influential nodes in the events.

To help emergency event-affected users to share tweets about the events, a Tweak the Tweet (TtT) syntax was proposed by Starbird and Stamberger to introduce a set of hashtags (e.g., #need, #offer, #iamok, etc.) to be used in tweets reporting the events [10]. Starbird and Palen found that very few users on the ground adopting the TtT syntax in the 2010 Haiti earthquake event but also several other users volunteering efforts to translate original tweets into ones that follow the TtT syntax [11]. These are examples of volunteerism and self-organizing behaviors that can be observed during crisis events. In the Singapore haze event studied in this paper, we unfortunately could not find tweets following the TtT syntax.

Compared with the above works, this paper adopts a comprehensive social analysis framework that covers the *content, emotion, activity* and *network* aspects. While the measures defined for each aspect are not new, there has not been any work to apply them all to a large-scale event such as the Singapore's haze event. The haze event involves haze-relevant hashtags and keywords that are of type-(ii) as most Singapore users did not expect it to happen. Unlike other disastrous events, telecommunication and transportation networks were not affected in the haze event. Users therefore can share their social media data and communicate in their usual ways.

3 Content and Activity Analyses

3.1 Overall Tweet Trend

Our first analysis tries to find out how the haze has affected the usage of social media. More than 26 million tweets are generated in June 2013 by about 130K public users in SGTWITTERDATA. We measure the number of tweets in SGTWITTERDATA each day. Figure 2 shows the daily tweet count of SGTWITTERDATA in June 2013. As shown in the figure, the daily tweets generated by Singapore users surged on 19 and 21 June when Pollutant Standards Index (PSI) hit peak numbers at 321 and 400 respectively[1]. The surge on 17 June was relatively small. This suggests that there were much more Twitter data generated by Singapore users during the haze crisis. Thus, the haze crisis indeed has affected the usage of social media.

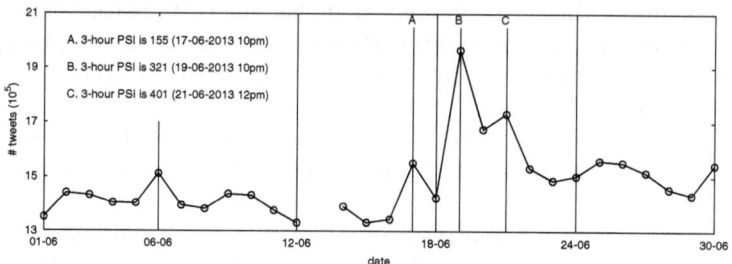

Fig. 2. Daily Tweet Count

3.2 Haze-Related Tweet Trend

Analysis on HAZETWITTERDATA was performed to learn how big was the impact of haze crisis in the overall twitter activity. Daily tweet count of HAZETWITTERDATA is shown in Figure 3(a). The figure shows that the daily tweet count of

---

[1] According to NEA, PSI reading beyond 100 is considered unhealthy.

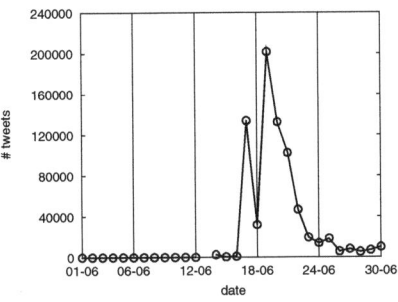 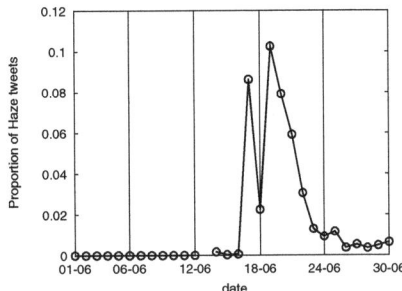

(a) Daily Haze-related Tweet Count (b) Proportion of Haze-related Tweet

Fig. 3. Daily Haze-related Tweet

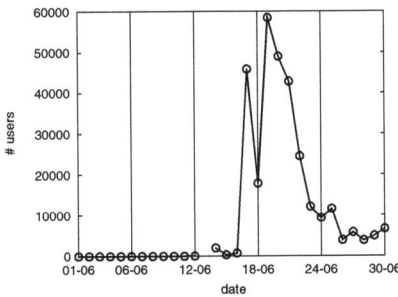 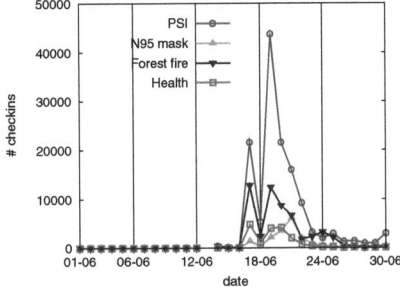

Fig. 4. Daily User Count   **Fig. 5.** Topic Trend

HAZETWITTERDATA surged on 17 June. There are almost no tweets about haze before 17 June. The number of haze related tweets reduced on June 18 possibly due to the common belief that the haze would not last long. On 19 June onwards, many more haze related tweets were generated as users realized that the haze problem was worsening and would last for a longer period. After 21 June when the haze began to subside, the number of haze-related tweets decreased substantially, but remained higher than the number before the haze.

Daily proportion of HAZETWITTERDATA is displayed in Figure 3(b). The figure shows that on 17 and 19 June, HAZETWITTERDATA accounts for about 8% and 10% of all tweets generated by Singapore users. After 19 June, the proportion of haze related tweets continued to stay substantially higher than the proportion before the haze crisis.

Daily user count in HAZETWITTERDATA can be seen in Figure 4. Very few users mentioned haze-related contents before the number surged on 17 June. The number hit the highest number of users on 19 June with more than 58,000 users. Although the number decreases substantially after the haze subsided, few thousand of users still mentioned haze-related contents until the end of the month.

3.3 Topic Analysis

Analysis on tweet content reveals few popular topics in haze-related contents. From the top frequent words that appear in HAZETWITTERDATA, we manually categorized the words into four topical categories, namely:

- PSI category: "nea", "psi"
- N95 mask category: "mask", "n95"
- Forest fire category: "forest", "fire", "Indonesia", "Malaysia", "Sumatra", "nature", "smoke", "burn"
- Health category: "asthma", "breathing", "health", "hospital", "clinic", "doctor", "sick", "respiratory"

Figure 5 shows that the overall trends of haze-related tweets under the above four topics follow that of overall haze-related tweets. The PSI reading captured the most user attention, and most tweets were about the PSI. The second largest topic was about forest fire and followed by the health topic. The N95 mask topic became more popular than the health topic on 21 June because of the high demand of N95 mask after the air quality reached its worst on 21 June.

3.4 Information Sources

URLs mentioned in tweets indicate external sources which bring in the information to social media. We examined the highly mentioned and retweeted domain names in HAZETWITTERDATA to identify the information sources people trusted during the haze crisis. There were 61,889 tweets, and 36,312 retweets that contain URL in HAZETWITTERDATA. The domain names were categorized into three categories, namely News, Government and Others.

Tables 1 shows that the top domain names mentioned and retweeted by users in June 2013 are mainstream news sites, a government agency, and several other popular social media sites. The domains marked with * are ones that appear in only one of two tables. The tables indicates that most users still referred to official news channels for information about haze.

Among the top domain names, NEA was the only government agency appeared to be an important source during the haze crisis. The NEA played an important role disseminating haze information to the public. The NEA published the 3-hour PSI reading every one hour. To observe how quickly the information got disseminated, we counted the number of retweets mentioning nea.gov.sg, and divided them into six disjoint time windows of ten minutes each across time. Figures 6 shows the retweet count of each time window on 17, 19, 20, and 21 June 2013 respectively.

The peaks in the figures indicate that most people responded (by retweeting the URL) to NEA announcements within ten minutes. This shows that people tracked NEA announcements closely. Furthermore, this also indicates that many people checked the PSI reading obsessively since most of NEA announcements were on PSI reading.

Table 1. Top Domain Names in HazeTwitterData in June 2013

(a) Top 10 Mentioned Domain Names

Rank	Domain	Category	#tweets
1	straitstimes.com	News	13,776
2	instagram.com	Others	10,460
3	nea.gov.sg	Government	7,240
4	todayonline.com	News	6,661
5	channelnewsasia.com	News	3,589
6	facebook.com	Others	2,607
7	youtube.com	Others	2,528
8	twitpic.com	Others	2,044
9	ask.fm*	Others	1,230
10	stomp.com.sg	Others	1,150

(b) Top 10 Retweeted Domain Names

Rank	Domain	Category	#retweets
1	straitstimes.com	News	11,193
2	nea.gov.sg	Government	6,421
3	todayonline.com	News	5,802
4	channelnewsasia.com	News	2,369
5	twitpic.com	Others	1,683
6	instagram.com	Others	1,543
7	youtube.com	Others	1,533
8	stomp.com.sg	Others	961
9	facebook.com	Others	922
10	yahoo.com*	News	367

3.5 Check-Ins

To study the impact of haze to user activities and businesses, we analyzed 4SQ-Data. Figure 7 shows Foursquare check-ins trend in June 2013 by Foursquare venue category. We observed four categories of check-ins:

- Food: restaurant, food, and cafe
- School: university, school, and college
- Shop: mall, shop, and department store
- Healthcare: hospital, clinic, doctor, pharmacy, drug store

The haze reduced Foursquare check-ins especially on 19-22 June 2013 (up to one day after the worst haze day). There was clear evidence that visits to shops and eating places were reduced substantially, by 20% to 50% respectively, during the three days (17, 19 and 21 June) that witnessed record breaking PSI values. Daily check-ins pattern of both shops and eating places were similar because many shops and eating places shared the same building. The reduction of check-ins to schools appeared to be less obvious. This could be due to school holidays during the haze period. At the same time, the number of check-ins to healthcare places was also reduced. However, healthcare locations typically make up a very small proportion of FourSquare check-ins in general.

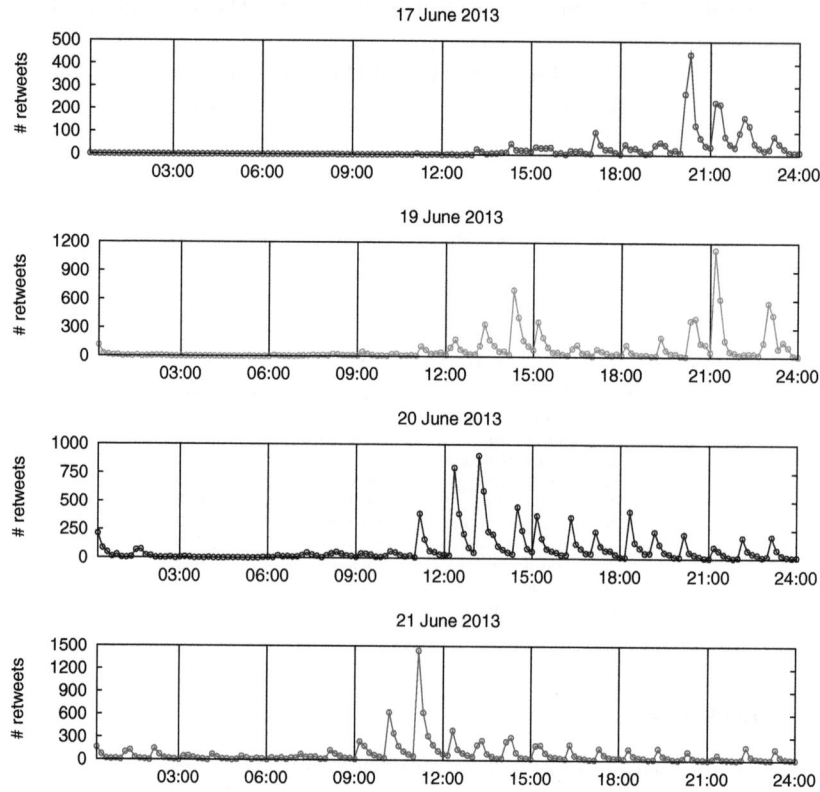

Fig. 6. Retweets of NEA Announcements

4 Emotion Analysis

Another important aspect during the haze crisis was emotion states of the people in Singapore. In this section, we present our analysis in sensing emotion state of users in Singapore during the haze crisis.

4.1 Types of Emotion

We examined the emotion state of the users by classifying the tweets into the categories below according to the emotion associated keywords from the LIWC full dictionary [3].

- Negative categories:
  - Anxiety category: "worried", "worry", "worries", "fear", "afraid", "frightened", "scared", "stress", "upset", "nervous", "anxious", "alarm", "tense", "distress", "panic", "die"
  - Anger category: "mad", "frustrate", "irritate", "annoy", "hate", "kill", "piss", "mean", "hostile", "disgust"

- Swear category: "piss", "fuck", "damn", "shit", "crap", "oh no", "OMG", "holy _", "FML"
- Low arousal negative category: "unhappy", "miserable", "sad", "depress", "hopeless", "gloomy", "tired", "sleepy", "lethargic", "fatigue", "helpless", "down", "dejected"
• Gratitude category: "thank", "thankful", "grateful", "blessed", "lucky", "fortunate", "pleased"
• Positive category: "happy", "love", "glad", "pleased", "relax", "calm", "relieve", "phew", "inspire", "proud", "joyful", "excite", "admire", "cheerful", "delight", "eager", "elate", "enthusiastic", "interest", "peaceful", "pleasant", "respect"

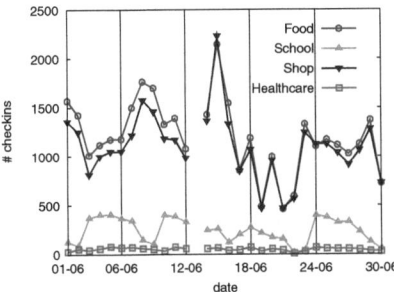

Fig. 7. Daily Foursquare Checkins

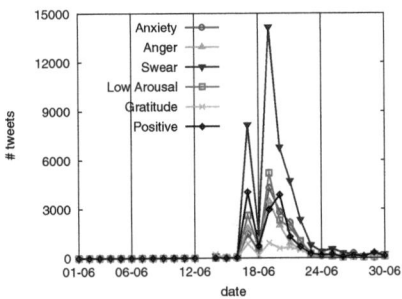

Fig. 8. Emotion Trend

4.2 Emotion Results

We observed that the negative emotions, especially the swear words, reached high peaks during the haze event as shown in Figure 8. Almost all negative emotion words, i.e., anxiety, anger, swear and low arousal words, hit the highest numbers of tweets on 19 June. There were more positive emotions expressed on 20 June before they dropped when the PSI index hit the record breaking 401 on 21 June. Grateful emotions didn't change much throughout the month.

The swear words were found about 6% and 7% of haze related tweets during the two peaks of haze event on 17 and 19 June respectively. After that, the percentage of swear words reduced to values less than 6%. On the worst haze days, i.e. 19 and 21 June, swear, low arousal, and anxiety words were the top three emotion expressed by Singapore Twitter users.

5 Network Analysis

5.1 Retweet and Reply Networks Construction

Based on HAZETWITTERDATA, we construct a *reply network* and another *retweet network* for each day. An edge (u, v) is formed in the daily reply network if user u replies at least a tweet from user v, or user v replies at least a tweet from user

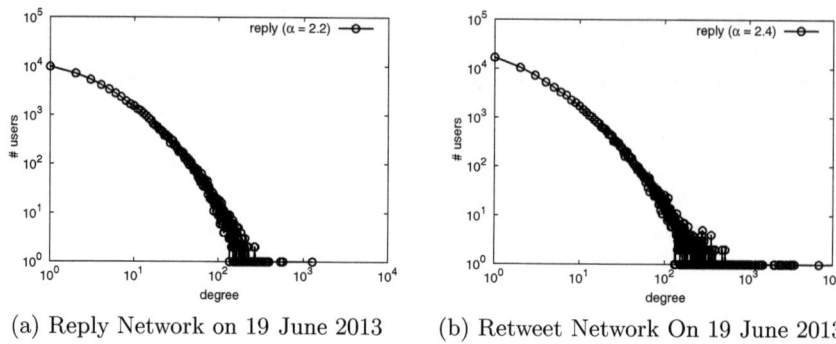

(a) Reply Network on 19 June 2013   (b) Retweet Network On 19 June 2013

Fig. 9. Degree distributions of Reply and Retweet Networks

u in the day. The edges in the daily retweet network are created in a similar manner.

By checking the degree distributions of reply and retweet networks, we can conclude that they are quite similar to scale-free networks, where the scaling parameters α are estimated by the approach presented in [12]. This is illustrated by the degree distributions of reply and retweet networks on 19 June 2013 as shown in Figure 9.

5.2 Network Robustness Analysis

Network robustness determines how well its vertices are connected to one another so as to keep the network strong and sustainable. Larger network may be more robust as it is hard to change. Since the largest CC (Connected Component) is a good representation of the whole network, the size of the largest CC is also a simple measure to evaluate the robustness of a network.

Figure 10 shows the sizes of the largest CCs on both reply and retweet networks from 1st June to 30th June (with the exception of Jun 13th when we experienced a data crawling problem). We observe that: (1) many users involved

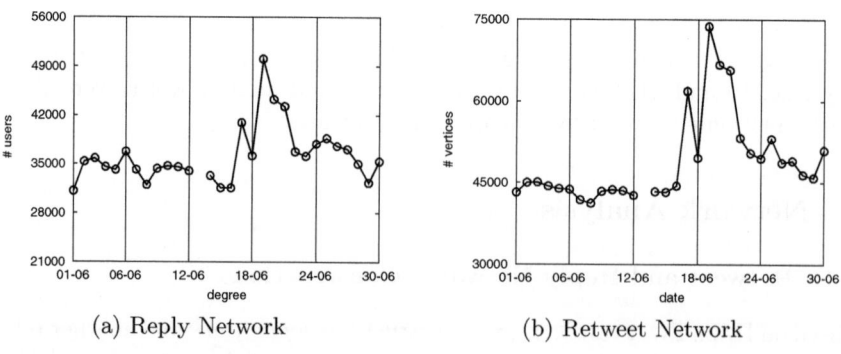

(a) Reply Network   (b) Retweet Network

Fig. 10. The Sizes of the Largest CCs on Reply and Retweet Networks

Social Sensing for Urban Crisis Management: The Case of Singapore Haze   489

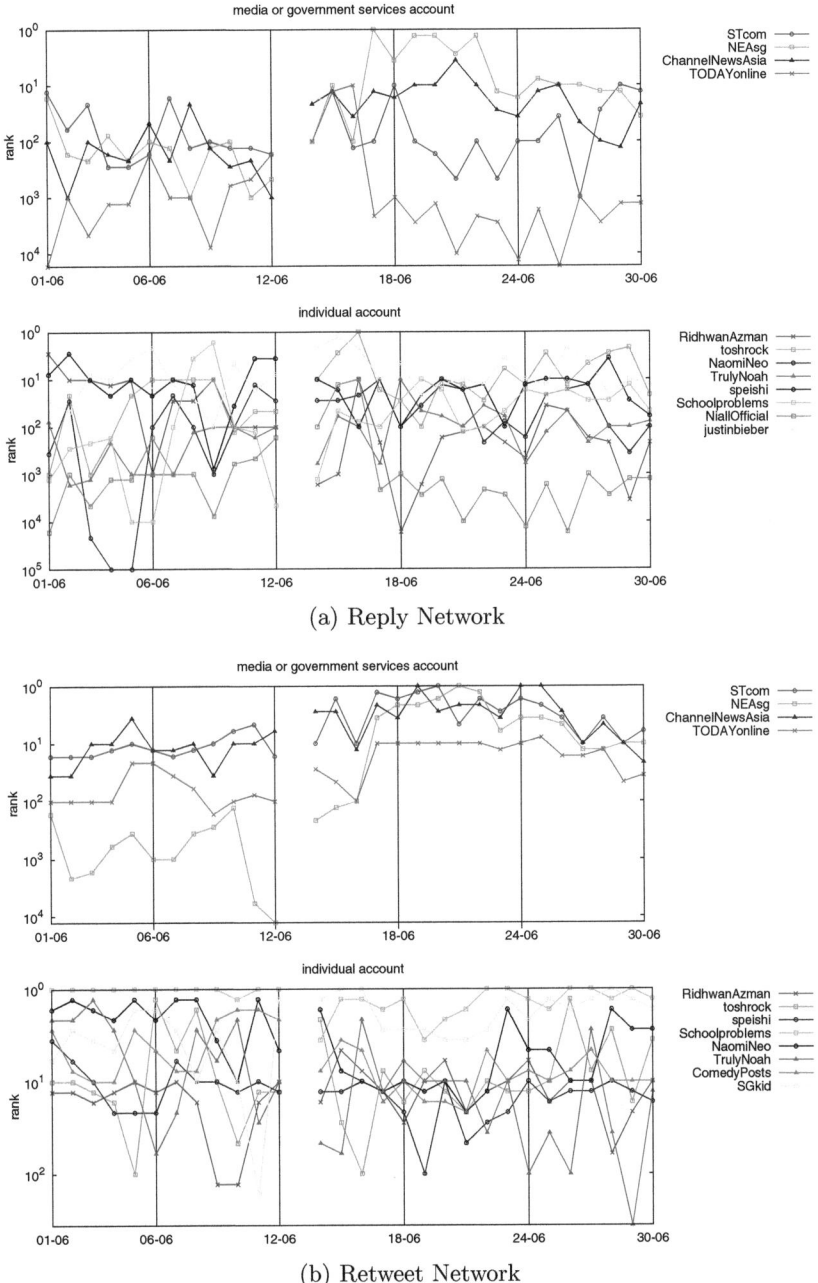

Fig. 11. Influential Users in terms of PageRank Centrality on Reply and Retweet Networks

in replying one another during the crisis, even after the crisis; (2) users were also more likely to retweet with one another during the crisis.

5.3 Centrality Analysis

Another natural question is who were the influential users and popular sources within the reply and retweet networks during the event. We therefore employ centrality measures to answer this question as the centrality of a vertex determines its relative importance within a network. Multiple centrality measures, such as degree centrality, pagerank centrality, betweenness centrality, closeness centrality, and eigenvector centrality, are widely used in network analysis [13,14]. Note that more inferential users are more likely to share, diffuse and propagate information on Twitter platform. Meanwhile, taking in account the scalability of measures, we employ the pagerank centrality, to determine the influential users and popular sources from the interaction networks.

Figure 11 shows the influential users and popular sources, measured by pagerank centrality, on the reply and retweet networks, where each curve represents the daily pagerank (in log scale) of a user from 1st June to 30th June. We find that news mediums or government services, such as *STcom*[2], *NEAsg*[3], *ChannelNewsAsia*[4], and *TODAYonline*[5], became popular. However, influential individual users are always on the top list regardless of whether there is an event. We can conclude that government services played the most important role during the crisis, followed by the new media. Except June 18, an interesting observation is that the NEA Twitter account (*NEAsg*) attracts more attention during the event, even after the event. This is due to the common belief on June 18 that the haze would not last long.

6 Conclusion

We may think of social media as a modern frivolity mainly to be used as a source of fun, but it can also supply some useful and quantifiable information. Twitter provides a window into the stream of consciousness of Singaporeans like no other technology at present can. Even large-scale surveys cannot track responses to an event as close in time to the occurrence or with such large samples. Geo-location data from FourSquare can quantify real activity to confirm or disconfirm personal observations. Our analysis of Twitter feeds found the impact of the haze on peoples lives was undeniable and intense. They drastically reduced their activity to food and shopping venues until the haze cleared. Instead of outdoor activities, people responded by turning to social media to express themselves. Their expressions were primarily ones of shock, anger, and other negative emotions. At the same time, people relied heavily on official sources of information about

---

[2] The Straits Times.
[3] NEA account.
[4] Official site of Channel NewsAsia.
[5] Singapore's most popular compact newspaper.

the haze, and they used social media to spread this information. More details about this study can be found at http://research.larc.smu.edu.sg/sghaze.

Acknowledgement. This work is supported by the National Research Foundation under its International Research Centre@Singapore Funding Initiative and administered by the IDM Programme Office, and National Research Foundation (NRF).

References

1. Blei, D.M., Ng, A.Y., Jordan, M.I.: Latent dirichlet allocation. Journal of Machine Learning Research **3** (2003) 993–1022
2. Bollen, J., Mao, H., Pepe, A.: Modeling public mood and emotion: Twitter sentiment and socio-economic phenomena. In: ICWSM. (2011)
3. Pennebaker, J.W., Francis, M.E., Booth, R.J.: Linguistic inquiry and word count: LIWC 2001. Mahway: Lawrence Erlbaum Associates (2001)
4. Noulas, A., Scellato, S., Mascolo, C., Pontil, M.: An empirical study of geographic user activity patterns in Foursquare. In: ICWSM. (2011)
5. Newman, M.E.J.: The structure and function of complex networks. SIAM Review **45** (2003) 167–256
6. Mendoza, M., Poblete, B., Castillo, C.: Twitter Under Crisis: Can we trust what we RT? In: Proceedings of the first workshop on social media analytics
7. Earle, P.S., Bowden, D.C., Guy, M.: Twitter earthquake detection: earthquake monitoring in a social world. Annals of Geophysics **54**(6) (2012)
8. Lehmann, J., Gonçalves, B., Ramasco, J.J., Cattuto, C.: Dynamical classes of collective attention in twitter. In: WWW. (2012)
9. Cheong, F., Cheong, C.: Social media data mining: A social network analysis of tweets during the 2010-2011 australian floods. In: PACIS. (2011)
10. Starbird, K., Stamberger, J.: Tweak the tweet: Leveraging microblogging proliferation with a prescriptive syntax to support citizen reporting. In: ISCRAM. (2010)
11. Starbird, K., Palen, L.: Voluntweeters: Self-organizing by digital volunteers in times of crisis. In: CHI. (2011)
12. Clauset, A., Shalizi, C.R., Newman, M.E.J.: Power-law distributions in empirical data. SIAM Review **51** (2009) 661–703
13. Freeman, L.C.: A set of measures of centrality based on betweenness. Sociometry (1977) 35–41
14. Page, L., Brin, S., Motwani, R., Winograd, T.: The PageRank citation ranking: bringing order to the Web. (1999)

Author Index

Adler, Odelia 81
Ahnert, Sebastian E. 346
Aiello, Luca Maria 40, 46
Almeida, Jussara M. 153, 309
Almoqhim, Fahad 129
An, Jisun 360
Anzengruber, Bernhard 206
Aoki, Hideto 60
Araki, Kenji 370

Burscher, Björn 333

Caldarelli, Guido 346
Cha, Meeyoung 299
Chen, Chien Chin 109
Chen, Hao 405
Chen, Pei-Yi 163
Chen, Shiuann-Shuoh 163
Chen, Wei 299
Chiarandini, Luca 40
Chuang, Yu-Wei 163
Conti, Marco 360
Cortis, Keith 284
Coscia, Michele 319
Crowcroft, Jon 360

Dalip, Daniel 153
de Melo, Pedro O.S. Vaz 309
de Rijke, Maarten 333
Devanbu, Premkumar 405
Du, Juan 75

Elslander, Jonas 443

Ferscha, Alois 206
Filkov, Vladimir 405
Fink, Thomas M.A. 346

Gao, Ming 478
Garlaschelli, Diego 346
Gavilanes, Ruth Garcia 46
Giannotti, Fosca 319
Gibler, Clint 405
Gonçalves, Marcos 153
Gottipati, Swapna 177

Hammad, Mohamed 139
Handschuh, Siegfried 284
Hemayed, Elsayed 139
Hoang, Ai Phuong 227
Hoang, Bao-Thien 467

Ikematsu, Kyohei 192
Imahori, Shinji 241
Imine, Abdessamad 467
Iwaihara, Mizuho 119

Jaimes, Alejandro 40, 46
Jankowski-Lorek, Michal 419
Jiang, Jing 177
Jung, Kyomin 299

Kato, Makoto P. 457
Kavaler, David 405
Ke, Yi-Ling 274
Kim, Hea-Jin 95
Kwon, Sejeong 299

Latif, Mohammad Ayub 16
Lee, Ryong 429
Lev-On, Azi 81
Lim, Ee-Peng 75, 227, 478
Lin, Zhuan-Yao 216
Liu, Jyi-Shane 216
Lofi, Christoph 1
López-García, Rafael 457
Loureiro, Antonio A.F. 309

Millard, David E. 129
Moraes, Felipe 153
Mori, Shinsuke 241
Morishima, Atsuyuki 60
Murata, Tsuyoshi 192
Mvungi, Basilisa 119

Naveed, Muhammad 16
Nieke, Christian 1
Nielek, Radoslaw 419
Nieminen, Jussi 206
Ning, Ke-Chih 216

Odijk, Daan 333
O'Hare, Neil 40, 46

Pappalardo, Luca 319
Passarella, Andrea 360
Pedreschi, Dino 319
Pennacchioli, Diego 319
Pezzoni, Fabio 360
Pianini, Danilo 206
Posnett, Daryl 405
Prado, Patrick 153
Prasetyo, Philips Kokoh 478

Qiu, Minghui 177

Rivera, Ismael 284
Rossetti, Giulio 319
Rowe, Matthew 30
Rzepka, Rafal 370

Salles, Juliana 309
Scerri, Simon 284
Scollon, Christie Napa 478
Serebrenik, Alexander 391
Shadbolt, Nigel 129
Shoji, Yoshiyuki 377
Sie, Yi-Jin 274
Silva, Thiago H. 309
Song, Min 95
Sugiyama, Yuichi 241

Sumiya, Kazutoshi 429
Sun, Yu-Chun 109

Tan, Biying 75
Tanaka, Katsumi 241, 377, 443, 457
Tang, Muh-Chyun 274
Tsai, Tse-Ming 268

van den Brand, Mark G.J. 391
Vasconcelos, Marisa 153
Vasilescu, Bogdan 391
Viana, Aline Carneiro 309
Vliegenthart, Rens 333

Wakamiya, Shoko 429
Wang, Jenq-Haur 255
Wang, Wen-Nan 268
Wang, Yajun 299
Wawer, Aleksander 419
Wierzbicki, Adam 419

Xie, Wei 227

Yamakata, Yoko 241, 457
Yang, Liu 177
Yang, Ping-Che 268
Ye, Ting-Wei 255
Yu, Min 163

Zaidi, Faraz 16
Zhu, Feida 75, 177, 227

MIX
Papier aus verantwortungsvollen Quellen
Paper from responsible sources
FSC® C105338

If you have any concerns about our products,
you can contact us on
ProductSafety@springernature.com

In case Publisher is established outside the EU,
the EU authorized representative is:
**Springer Nature Customer Service Center GmbH
Europaplatz 3, 69115 Heidelberg, Germany**

Printed by Libri Plureos GmbH
in Hamburg, Germany